Exosome Therapies

This book provides a comprehensive overview of the evolving field of exosome-based therapies for chronic diseases. It explores the biology of exosomes, including their discovery, composition, biogenesis, and their role in intercellular communication and targeted drug delivery. The chapters review the challenges and advancements in exosome isolation, characterization, and engineering, highlighting their potential in treating chronic diseases. The book also examines the therapeutic applications of exosomes in cancer, neurological, cardiovascular, metabolic, and inflammatory disorders, while also addressing their applications in immune modulation, wound healing, and next-generation cancer vaccines. It also presents the translational aspects of exosome-based therapies, including preclinical and clinical outcomes, integrating advanced techniques such as exosome kinetics, drug delivery systems, and clinical trials.

Key Features:

- Provides foundational knowledge of exosome biology, including their biogenesis, composition, and signaling mechanisms.
- Explores innovative methods for exosome isolation, engineering, and drug delivery applications.
- Discusses the therapeutic potential of exosomes against various chronic diseases, including cancer, cardiovascular, and metabolic disorders.
- Highlights clinical trials and translational approaches to bridge the gap between research and patient care.
- Addresses challenges in manufacturing, regulatory considerations, and future directions for exosome-based therapies.

It is useful for researchers and students of biomedical sciences, pharmacology, and biotechnology.

Exosome Therapies

From Bench to Bedside in Chronic Disease

Edited by
Madhulika Pradhan and Krishna Yadav

CRC Press
Taylor & Francis Group
Boca Raton London New York

CRC Press is an imprint of the
Taylor & Francis Group, an **informa** business

Designed cover image: Shutterstock

First edition published 2026
by CRC Press
2385 NW Executive Center Drive, Suite 320, Boca Raton FL 33431

and by CRC Press
4 Park Square, Milton Park, Abingdon, Oxon, OX14 4RN

CRC Press is an imprint of Taylor & Francis Group, LLC

ISBN: 978-1-041-03391-2 (hbk)
ISBN: 978-1-041-03392-9 (pbk)
ISBN: 978-1-003-62363-2 (ebk)

DOI: 10.1201/9781003623632

Typeset in Times LT Std
by Apex CoVantage, LLC

Contents

Foreword

The field of exosome research represents one of the most promising frontiers in modern medicine, offering unprecedented opportunities to revolutionize therapeutic approaches for chronic diseases. As we witness remarkable advances in our understanding of intercellular communication and targeted drug delivery, exosomes have emerged as nature's own sophisticated delivery vehicles, capable of transforming how we treat some of humanity's most challenging health conditions.

This comprehensive volume, *Exosome Therapies: From Bench to Bedside in Chronic Disease*, arrives at a pivotal moment when the scientific community is transitioning from foundational research to clinical implementation. The book masterfully bridges the gap between laboratory discoveries and practical therapeutic applications, providing an essential resource for researchers, clinicians, and industry professionals working to bring exosome-based treatments to patients who need them most.

Dr. Madhulika Pradhan and Dr. Krishna Yadav, along with their distinguished team of contributors, have assembled a remarkable collection of expertise that spans the entire spectrum of exosome research and development. From fundamental biology and biogenesis mechanisms to cutting-edge engineering approaches and clinical translation, this work provides an unparalleled depth of knowledge that will serve as a cornerstone reference for years to come.

The book's systematic approach to covering diverse therapeutic applications – from neurological and cardiovascular disorders to cancer, metabolic diseases, and inflammatory conditions – reflects the versatility and potential of exosome-based interventions. Each chapter contributes unique insights while maintaining a cohesive vision of advancing exosome therapeutics from experimental concepts to clinical reality.

As we stand on the threshold of a new era in precision medicine, this volume will undoubtedly accelerate progress in the field and inspire the next generation of researchers to push the boundaries of what is possible with exosome-based therapies.

Preface

The inspiration for *Exosome Therapies: From Bench to Bedside in Chronic Disease* emerged from our recognition of the transformative potential that exosome-based therapeutics hold for addressing the global burden of chronic diseases. As researchers deeply embedded in the field of pharmaceutical technology and drug delivery systems, we have witnessed firsthand the rapid evolution of exosome research and its promising trajectory toward clinical application.

Exosomes, once considered cellular waste products, have now revealed themselves as sophisticated biological messengers with extraordinary therapeutic potential. Their natural ability to facilitate intercellular communication, combined with their biocompatibility and targeting capabilities, positions them as ideal candidates for next-generation drug delivery systems. However, the journey from laboratory bench to patient bedside requires comprehensive understanding, systematic development, and careful translation of scientific discoveries into practical therapeutic solutions.

This book represents a collaborative effort among leading experts in exosome research, each contributing specialized knowledge to create a comprehensive resource that addresses both fundamental concepts and practical applications. Our objective has been to create a bridge between cutting-edge research and clinical implementation, providing readers with the knowledge necessary to advance exosome-based therapies from experimental models to tangible patient benefits.

The 14 chapters in this volume systematically explore every aspect of exosome therapeutics, beginning with foundational concepts of discovery, composition, and biogenesis and progressing through technological advances, engineering innovations, and specific therapeutic applications across multiple disease areas. Each chapter is designed to provide both theoretical understanding and practical insights that will benefit researchers, clinicians, students, and professionals working in drug development, regulatory affairs, and public health.

We have structured this work to serve multiple audiences: established researchers seeking comprehensive updates on the field, clinicians interested in emerging therapeutic options, graduate students beginning their journey in exosome research, and industry professionals developing commercial applications. The integration of insights from biology, medicine, pharmacology, and biotechnology reflects our commitment to providing a truly interdisciplinary perspective on this rapidly evolving field.

Our hope is that this book will not only serve as an authoritative reference but also inspire continued innovation and collaboration in the pursuit of effective exosome-based treatments for chronic diseases that affect millions of patients worldwide.

About the Editors

Madhulika Pradhan is Director and Professor at the School of Pharmacy, Rungta International Skills University, Bhilai, Chhattisgarh, India. She has over 15 years of teaching experience and 13 years of research experience, specializing in the design and development of novel drug delivery systems. She holds a Ph.D. in Pharmaceutical Technology from Pt. Ravishankar Shukla University, Raipur, and a Master of Pharmacy in Pharmaceutical Biotechnology.

Recognized among the World's Top 2% scientists (2025) by Stanford University and Elsevier, Dr. Pradhan is a distinguished researcher focusing on topical and transdermal drug delivery systems for psoriasis treatment. Her expertise spans preformulation studies, formulation development, and preclinical pharmacokinetics, incorporating advanced methodologies such as quality by design (QbD), design of experiment (DOE), and preclinical animal model development.

With an H-index of 28 and I-10 index of 43, Dr. Pradhan has published extensively in high-impact journals, including the *Journal of Controlled Release*, *International Journal of Biological Macromolecules*, and *Journal of Drug Delivery Science and Technology*. She has authored over 80 research articles and 20 book chapters with leading publishers such as Elsevier, Springer Nature, and CRC Press. Her innovative research has led to two granted patents (Australia and India) in nanomedicine and pharmaceutical technology.

Dr. Pradhan has received several prestigious recognitions, including the Chhattisgarh Young Scientist Award (2015), UGC-BSR Meritorious Fellowship (2012–2015), and International Travel Fellowships to present her research in the United States, Singapore, and Malaysia. She has also served as Guest Editor for reputed journals such as *Current Pharmaceutical Design* and *Drug Delivery Letters*.

Her research expertise encompasses nanoparticles, exosome delivery systems, nanovesicles, solid lipid nanoparticles (SLNs), nanostructured lipid carriers (NLCs), and hydrogels for controlled drug release. Her current research focuses on cancer-targeted delivery systems and psoriasis therapeutics, contributing to the translation of pharmaceutical nanotechnology from bench to bedside.

Krishna Yadav is Assistant Professor at the School of Pharmacy, Rungta International Skills University, Bhilai, Chhattisgarh, India, with over eight years of academic and research experience in pharmaceutics and novel drug delivery systems. He holds a Ph.D. in Pharmaceutical Technology and a Master of Pharmacy (Pharmaceutics, Gold Medalist).

Recognized among the World's Top 2% Scientists (2025) by Stanford University and Elsevier, Dr. Yadav specializes in the design and development of advanced drug delivery systems, including nanoparticles, nanovesicles, hybrid exosomes, and sustained release formulations, with a particular emphasis on cancer, dermatological, and chronic disease management.

He has authored and edited multiple books with Springer Nature and CRC Press, and serves as an Editorial Board Member of Scientific Reports (Nature Portfolio). He has also acted as Guest Editor for *Current Pharmaceutical Design* and *Drug Delivery Letters*. With more than 85 research articles, reviews, and book chapters published in leading journals (Elsevier, Wiley, Taylor & Francis, Springer), he has attained an H-index of 26 and a cumulative impact factor exceeding 186.43. He is also a recipient of international and national awards, including the prestigious 2025 LEO Foundation International Travel Grant (University of Copenhagen, Denmark).

Dr. Yadav's contributions include granted patents (Australia) and filed Indian patents in nanomedicine and pharmaceutical innovations. His research continues to advance exosome-based therapeutics, nano-enabled targeted delivery, and translational pharmaceutics from bench to bedside.

1 Introduction to Exosomes

Discovery, Composition, Biogenesis, and Historical Background

Madhulika Pradhan[1,2]
[1]Rungta College of Pharmaceutical Sciences and Research, Bhilai, Chhattisgarh, India
[2]School of Pharmacy, Rungta International Skills University, Bhilai, Chhattisgarh, India

1.1 INTRODUCTION

1.1.1 Classification: Exosomes, Microvesicles, and Apoptotic Bodies

Extracellular vesicles encompass a heterogeneous collection of membrane-encased structures shed by nearly all cell types and serve as complex tools for intercellular communication. The International Society for Extracellular Vesicles (ISEV) published a standardized naming convention that distinguishes at least three classes based on mechanisms of biogenesis and morphology [1].

Exosomes are the smallest of the three classes, and they are shed from the endosomal compartment after formation of a multivesicular body (MVB) and fusion of the MVB with the plasma membrane [2]. They maintain structural integrity and exhibit selective packaging mechanisms for their cargo, representative of their cellular origin. Microvesicles (also called ectosomes, or microparticles) emerge from the plasma membrane via direct budding, and span a larger size range of 100–1,000 nm [3, 4]. Finally, apoptotic bodies, which are the largest category, have a size range of 1–5 μm and are shed from the cell during programmed cell death as membrane-bound pieces of cellular organelles and possibly some nuclear components [5, 6].

1.1.2 Defining Features and Size Ranges

Modern characterization techniques have helped to elucidate our understanding of vesicular populations through novel techniques such as nanoparticle tracking analysis (NTA), transmission electron microscopy (TEM), and dynamic light scattering (DLS). Exosomes have a boundary that has a characteristic cup shape in morphology observed by electron microscopy, which is due to membrane dehydration from sample preparation. Exosome populations possess a characteristic size distribution, generally restricted to 30–150 nm in recognition of a peak. Also, vesicles in the exosome range can be distinguished from other extracellular vesicle populations [1, 7].

Physical characteristics can extend beyond morphology, as physical properties of density can differentiate exosomes from other populations. Exosomes have buoyant density in sucrose gradients of 1.13–1.19 g/mL, marked as low populations, and enable separation from other protein aggregates and cellular debris during isolation procedures. Exosomal membranes are also resistant to nonionic detergents, which involves the unique composition and cholesterol in exosomal membranes to resist repetition of other forms of physiological conditions of its structural integrity [8].

1.1.3 Exosomes Market Overview

The global exosomes market has emerged as a rapidly expanding segment within the healthcare and biotechnology sectors, reflecting the growing recognition of exosomes as critical players in diagnostics, therapeutics, and drug delivery. Exosomes, nanosized extracellular vesicles secreted by various cell types, are increasingly valued for their role in cell-to-cell communication, disease biomarker identification, and therapeutic potential [9]. According to market insights, the industry is projected to grow at an impressive compound annual growth rate (CAGR) of approximately 29.90% during the forecast period 2024–2032. This robust growth is primarily attributed to advancements in exosome isolation and characterization technologies, increased funding for life sciences research, and rising interest in precision medicine.

From 2019 to 2023, the market demonstrated steady expansion, with the valuation in 2024 reaching around US$4.23 billion. Moving forward, the market is expected to escalate significantly, surpassing US$33 billion by 2032. This acceleration underscores the rapid translation of exosome research from bench to bedside, with potential applications spanning oncology, neurodegenerative disorders, cardiovascular diseases, and regenerative medicine. The North American region currently accounts for the largest share of the global exosome market, driven by advanced healthcare infrastructure, extensive research initiatives, and strong investments from both public and private sectors. Other regions, including Asia Pacific and Europe, are also witnessing significant growth, supported by the rising adoption of innovative diagnostic and therapeutic approaches, as well as an increasing prevalence of chronic and lifestyle-related diseases. Middle East and Africa and South America, while relatively smaller in market size, present

DOI: 10.1201/9781003623632-1

untapped potential for future expansion due to growing research capacity and healthcare modernization [10, 11].

In terms of applications, the exosomes market is broadly categorized into diagnostic and therapeutic domains. The diagnostic segment harnesses the potential of exosomes as biomarkers for early disease detection, particularly in cancer, as they provide noninvasive liquid biopsy options with high specificity and sensitivity. Meanwhile, the therapeutic segment is witnessing rapid momentum, with exosome-based drug delivery systems and regenerative therapies gaining traction. The unique properties of exosomes, including their biocompatibility, ability to cross biological barriers, and natural origin, make them promising carriers for targeted therapeutics, including RNA-based treatments and small molecules [12–14].

The market's future trajectory is expected to be shaped by ongoing clinical trials, regulatory approvals, and strategic collaborations between academic institutions, biotech firms, and pharmaceutical companies. Moreover, technological innovations in large-scale exosome production, purification, and engineering will play a decisive role in commercial viability. Despite challenges such as standardization of protocols, scalability of production, and regulatory uncertainties, the exponential growth forecast highlights exosomes as a transformative frontier in modern medicine, with the potential to revolutionize diagnostics and therapeutics in the coming decade [15–18].

1.2 HISTORICAL DISCOVERY OF EXOSOMES

1.2.1 Early Observations in the 1980s

The scientific march toward the discovery of exosomes began with careful ultrastructural studies of reticulocyte maturation. Johnstone et al. distinguished that small vesicles were being expelled during the reticulocyte differentiation process, and they defined these structures as a form of waste disposal for cellular fragments. These pioneering studies relied upon transmission electron microscopy to visualize vesicular expulsion from multivesicular endosomes. This defining structure continues to be foundational to the criteria used to identify exosomes today [19].

Pan and Johnstone simultaneously demonstrated the externalization of the transferrin receptor via vesicular intermediates and provided some of the first evidence supporting the concept of selective protein sorting during vesicle formation. These first studies established conceptual frameworks relating endosomal maturation and vesicular release, but the biological importance of the release was largely unappreciated at the time of discovery [20].

1.2.2 Paradigm Shift from "Cellular Waste" to "Therapeutic Potential"

Exosomes underwent a conceptual shift from being just cellular debris to being functional signaling entities as evidence accumulated to demonstrate their biological activities. Raposo et al., through their study of B lymphocyte exosomes, showed that exosomes could present MHC class II molecules and, importantly, they made the case for their functional activities of exosomes, which challenged the notion of exosomes as just waste disposal by-products. Their work demonstrated that the membrane of exosomes retains functional proteins and protein complexes that can bind to their target cell receptors and activate downstream signaling events [21].

Zitvogel et al. conducted subsequent studies demonstrating that exosomes play a crucial role in activating the immune system and can stimulate T lymphocyte activation via vesicles originating from dendritic cells. This body of work helped people believe that exosomes play an active role in immune surveillance and in networks of cellular communication. The discovery of exosomal nucleic acid cargo accelerated the paradigm shift for exosomes [22].

Valadi et al. provided demonstrable evidence that exosomes were capable of carrying functional mRNA molecules and functional microRNAs, which could regulate gene expression in recipient cells. This evidence demonstrated that exosomes were capable of facilitating horizontal gene transfer between cells, further advancing the knowledge and understanding of intercellular communication systems and explaining several phenomena that would be challenging under the traditional paradigm of communication by soluble mediators only [23].

Hannafon and Ding performed research that assessed the pivotal importance of exosomes in tumor growth and metastasis. They demonstrated that tumor-derived exosomes have the capacity to transfer pro-oncogenic miRNAs and proteins to neighboring cells, modify the tumor microenvironment, increase angiogenesis, and induce drug resistance. This research examined the dark side of exosome communication disease [24].

In a substantial review, Yáñez-Mó et al. brought together the increasing body of evidence and described the wide-ranging physiological functions of extracellular vesicles, including exosomes. This research assisted in establishing the biological characteristics of exosomes, as well as their functions outside the immune system, such as in development and stem cell differentiation, and the interactions of pathogens with their host cells [25].

In recent years, research has taken a turn to investigate the therapeutic potential of exosomes. A literature review by Roszkowski organized the descriptions of mesenchymal stem cell (MSC)-derived exosomes for regenerative medicine. This study details how these exosomes are a catalyst for tissue repair, including wound healing and bone regeneration, by delivering their molecular cargo to recipient cells, which is a contemporary use of exosome biology as a clinical therapy [26].

1.2.3 Key Milestones in Exosome Research

Exosome research has experienced a sequence of significant milestones that have transformed its field from a series of descriptions to one that seeks to mechanistically understand and drive clinically relevant approaches [1, 27]. Originally seen as cellular waste disposal vesicles, exosomes emerged as key players in intercellular communication through advances in cell biology and analytical technologies. One of the earliest milestones was the establishment of standardized exosome isolation protocols, including differential ultracentrifugation and density gradients, to provide reliable methodologies for exosome purification for systematic study [28, 29]. Subsequent milestones included large-scale proteomic and transcriptomic profiling studies that characterized the cellular cargo, including proteins, RNAs, and lipids that exosomes contained, and the establishment of databases of these cargoes that could now be searched and referenced against our ongoing work [30]. As the field matured and became more rigorous, there was a further drive to examine vesicles in detail and under more physiological conditions which related to the application of new characterizing methods, including super-resolution microscopy, secretome imprinting, nanoparticle tracking analysis, and single-vesicle flow cytometry capabilities which allowed researchers to examine diversity, distribution of rates, and functional properties at the individual vesicle level versus bulk populations [31].

Recently, microfluidic isolation platforms have ushered in yet another change in the field, with high-throughput, low-cost, scalable approaches to isolate exosomes that improve on the antiquated glass and centrifuge methods, reduce processing artifacts, and legitimately optimize vesicle integrity [32–35]. With these methods collectively propelling the field away from a phenomenological description of exosomes and toward functional or mechanistic studies, current studies are primarily directed at understanding how exosomes affect a variety of biological processes, including immune modulation, angiogenesis, neurodegeneration, and tumor progression [36]. The contemporary scientific development indicates that while we continue to make strides in the complexity of exosome biology, we are also recognizing the translational potential of exosome science as researchers seek to apply exosomes as biomarkers, drug targets, drug delivery vehiclesand so on, across the healthcare continuum.

1.3 STRUCTURAL AND MOLECULAR COMPOSITION

1.3.1 Lipid Bilayer Organization

Exosomal membranes display unique lipid profiles consistent with their endosomal origin but maintain unique organizational characteristics. Lipidomic approaches have shown that exosomes have a membrane composition that is enriched in cholesterol and sphingomyelin versus phospholipids, but are certainly not deprived of them. This part of the area not only modifies the composition but also enhances membrane rigidity and accommodates stability, allowing exosomes to retain integrity when stored or circulated [37].

Another unique attribute is the asymmetry of their phospholipids, with phosphatidylserine representing a majority of the outer leaflet with a very different asymmetry compared to traditional plasma membranes. The asymmetrical distribution offers a unique recognition mechanism by phagocytic cells and may influence where in the body the exosomes would be likely to be distributed or phagocytized after systemic delivery. Another lipid modification is the concentration of lipid raft-associated molecules in exosomal membranes that further develop unique microdomains, which sort protein complexes along with areas focused on providing a zone of fusion for exosomes [38].

1.3.2 Protein Cargo: Tetraspanins, Heat Shock Proteins, Enzymes

Proteogenomic identification has identified repeated enrichment of families of proteins within exosomal preparations across many types of cells. Tetraspanins CD9, CD63, and CD81 are some of the most frequently identified exosomal markers and have a direct role in organizing the membrane, engaging in protein–protein interactions, and preparing certain proteins for the biogenesis of vesicles. As 4-transmembrane proteins, these tetraspanins facilitate the formation of tetraspanin-enriched microdomains that coalesce additional proteins (the cargo) and alter membrane curvature during vesicle biogenesis [39].

A few distinct heat shock proteins, including HSP70 and HSP90, are present in many of the exosomal studies and populations, showing a remarkable degree of conservation and suggesting a presumed role in vesicle persistence and protection of the cargo. Those molecular chaperones support the proper folding of proteins in the vesicle's confined space, and potentially serve to mediate recipient cell interactions through presentation on the surface of the vesicle. Other enzymes are included in the exosomal vesicles as well, including metabolic enzymes, proteases, and signaling kinases, and at least some remain catalytically competent after release, possibly modifying the increasingly functional extracellular matrix and recipient cell surface [40].

Cytoskeletal proteins (e.g., actin, tubulin, and regulatory proteins) also provide mechanical load but may also provide functional roles during vesicle motility and membrane deformation. The constant presence of cytoskeletal proteins indicates that the process of vesicle biogenesis is active and suggests that cargo selection may invoke active processes rather than passive effects [41].

1.3.3 Nucleic Acid Cargo: mRNA, miRNA, lncRNA, and DNA Fragments

Exosomal nucleic acid cargo is likely the most functionally significant finding in the field of vesicle biology, with exosomes being recognized as vehicles for horizontal gene transfer between cells. Whole transcriptome RNA sequencing studies confirm an incredible potential for the classification of exosomal cargo, resulting in a tremendous complexity of transcriptomic profiles reflecting parental cell expression patterns and selective enrichment of some RNA species [42–44].

microRNA species, in particular, display several interesting properties, including the preferential packaging of certain species into exosomes relative to cellular abundance. In the vesicular environment, these regulatory RNAs are stable and retain functional activity in the recipient cell involved with the modulation of cellular phenotypes and target gene expression. Long noncoding RNAs are also an important element of exosomal cargo and possibly a source of molecular scaffolds for assembling protein complexes or competitive endogenous RNA networks [45, 46].

Investigations have also recently detected double-stranded DNA fragments associated with exosomal cargo (genomic DNA, mitochondrial DNA, and possibly episomal elements). DNA fragments appear in lower abundance than RNA species, but may also facilitate mechanisms mediating horizontal gene transfer and cellular reprogramming processes. The functional significance of exosomal DNA cargo is under investigation and may further enhance our understanding of cellular communication complexity [47].

1.3.4 Lipidome and Metabolome

The lipidome and metabolome are rapidly composed aspects of exosomes with important biological properties; they not only provide a structural framework for exosome vesicle architecture, but also actively modulate intercellular communication. The development of high-resolution analytical platforms such as mass spectrometry and nuclear magnetic resonance spectroscopy has indicated that the metabolomic profile of exosomes is dynamic and an extension of the metabolic state of parent cells, but also includes some unique changes that occur during exosome biogenesis. Metabolomics may include anatomical classes, for example, amino acids, nucleotides, sugars, organic acids, and other small intermediate molecules, which can modify the metabolic activity of a recipient cell or serve to signal throughout the tissue microenvironment [48–50].

One example near the surface is the accumulation of a functional local lipidome in exosomes. In addition to their nature as fundamental constituents of membranes, exosomal lipids include a variety of signaling-active lipid molecules, for example, prostaglandins, leukotrienes, lysophospholipids, and sphingolipid derivatives (e.g., ceramides, sphingosine-1-phosphate). These metabolites retain biological activity upon delivery and modulate a recipient cell's physiology through receptor-mediated signaling (i.e., inflammation, angiogenesis, apoptosis, immunomodulation, etc.) [51, 52]. Furthermore, the presence of lipid-modifying enzymes such as phospholipases and sphingomyelinases in exosomal cargo indicates that lipid modification can also happen after exosomes are in the circulation and, therefore, can also be dynamically regulated with regard to the lipid composition and activities of the exosomes. Lipid-mediated alterations in vesicle bilayer fluidity, stability, and fusion can modify the biodistribution and uptake efficiency of lipid-modified exosomal vesicles. With both metabolites and lipids encapsulated within exosomal vesicles, fixed metabolic and lipid signatures and functional effectors can confer both directional signaling and active modifications on the original cells. Therefore, the biochemical complexity of exosomes as carriers not only summarizes the physiological state of the cell from which they originated, but it is likely that they also symbolize an active modulation of recipient cell behavior [53–56].

In short, exosomes, as abundant nanoparticles, carry lipids and metabolites and act as mediators of intercellular communication in health and disease [53].

The composition of exosomes is presented in Figure 1.1.

1.4 BIOGENESIS PATHWAYS OF EXOSOMES

1.4.1 Endosomal Sorting and MVB Formation

Exosomes originate from a tightly regulated cellular process characterized by a series of membrane activities in the endosomal compartment. The endosomal pathway begins when early endosomes are formed as a result of endocytosis, and proceeds through the maturation of the late endosomes as they begin to acidify and reorganize their membranes. During maturation of late endosomes, the limiting membrane in late endosomes starts to invaginate, ultimately resulting in the formation of intraluminal vesicles (ILVs) in multivesicular bodies (MVBs) [58].

At the MVB stage, exosomal secretion occurs versus degradation in lysosomes. The fate of the MVBs becomes regulated by numerous cellular signals, which include nutrient availability, nutrient stress/conditions, and greater cellular signaling pathways. The decision not to fuse with lysosomes in MVBs does not appear to be random but is regulated by specific Rab GTPases and SNARE proteins, which route the vesicles to be stored and secreted from the plasma membrane by decreasing or avoiding lysosomal fusion altogether. The proper routing of MVBs to the desired sites of exosome release, thereby signaling the selective release of functional vesicles in addition to utilizing lysosomal degradation processes, thus continuing to help maintain cellular homeostasis [58].

1.4.2 ESCRT-Dependent and ESCRT-Independent Mechanisms

The endosomal sorting complex required for transport (ESCRT) machinery is generally considered to be the primary regulatory mechanism that drives the generation of intraluminal vesicles (ILVs) from multivesicular bodies (MVBs), and is essential to the selective packaging and release of cargo through exosomes. This complex,

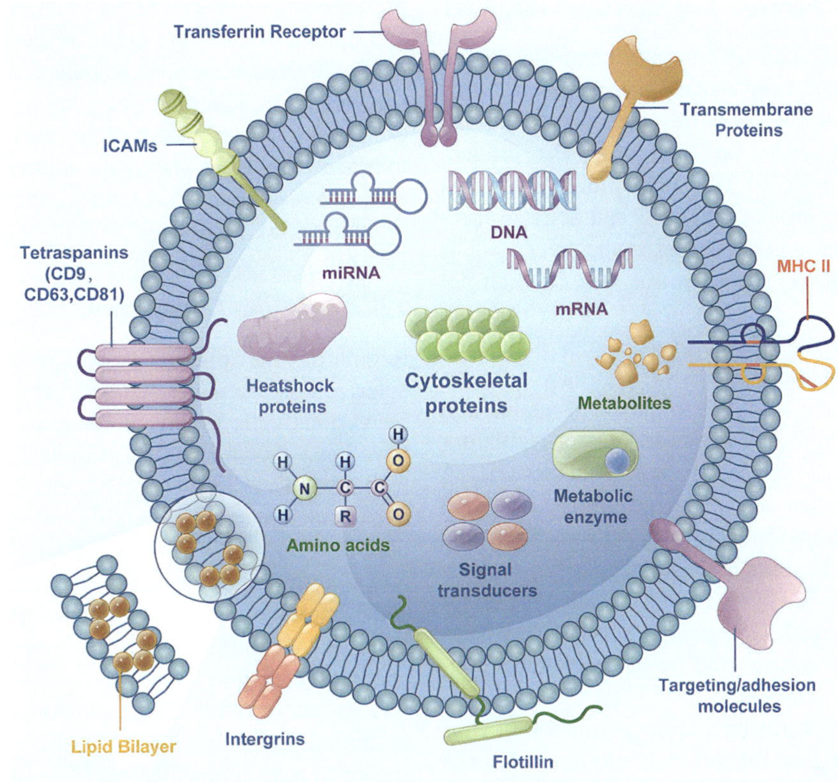

FIGURE 1.1 Exosome composition. (1) Lipid bilayers form the exosome membrane (yellow label); (2) DNA and RNA (purple label); (3) Major histocompatibility complex; MHC-I, MHC-II (orange label); (4) Transmembrane proteins (CD9, CD63, CD81), intercellular adhesion molecules (ICAM), integrin, heat shock protein, transferrin receptor, cytoskeletal protein, signal transducers (blue label); (5) metabolites such as amino acids (green label). *Source:* Adapted with permission from Ref. [57].

organized system contains four major subcomplexes, ESCRT-0, ESCRT-I, ESCRT-II, and ESCRT-III, that all behave in a linear, yet circular and coordinated fashion in order to remodel membranes and ultimately facilitate vesicle biogenesis. The ESCRT-0 and ESCRT-I complexes primarily focus on the initial recognition and clustering of "ubiquitinated" transmembrane cargo that is postured for degradation or release extracellularly, and create membrane invagination at the endosomal surface during this process. From this point, ESCRT-II acts as a functional adaptor that connects the noticed cargo to membrane deformation, recruiting ESCRT-III to the site of budding. ESCRT-III will then be used as the terminal effector complex that creates dynamic filamentous polymers to pinch or constrict budding vesicles, while scission is further regulated by an ATPase, VPS4, that disassembles or recycles ESCRT-III into subunits to be used for new rounds of vesicle formation. Although ESCRT represents the canonical way for producing ILVs, there are potential alternative biogenetic pathways to generate ILVs that do not depend on ESCRT activity [58, 59].

These pathways of ESCRT independence are mostly driven by either lipid or protein that demonstrate the biological flexibility of exosome biogenesis. For instance, there is a ceramide-dependent pathway, where neutral sphingomyelinase 2 (nSMase2) catalyzes sphingomyelin hydrolysis to create ceramide, a conical-shaped lipid which promotes negative curvature of the endosomal membrane. Upon accumulation of ceramide, the exosomal membrane stimulates the spontaneous budding process resembling vesicle release (based on biophysical properties of membranes, not a protein-induced scission). Also similarly, tetraspanins such as CD9, CD63, and CD81 create a special microdomain within the membrane to cluster cargo molecules and promote budding. The vesicle budding process opens the possibility of several complicated protein–protein and protein–lipid interactions to induce vesicle budding. ESCRT-independent form pathways are not simply alternative and redundant mechanisms. They likely represent separate evolutionarily primitive and context-specific mechanisms of biological importance playing a major role in some cell types, such as neurons, immune cells, or in some particular physiological and pathological states, such as inflammation, metabolic stress, or hypoxia. Collectively, ESCRT-dependent pathways and ESCRT-independent pathways together indicate the complexity and versatility of the biogenesis of exosomes and thus provide cells with more ways of regulating vesicle production and adapting intercellular communication to diverse biological forces [60, 61].

1.4.3 Regulatory Proteins and Signaling Pathways

The biogenesis of exosomes is carefully controlled by an array of signaling pathways and regulatory proteins that orchestrate cellular stress responses, metabolic statuses, and environmental signals that determine the production and release of vesicles. The most widely studied is the tumor suppressor protein, p53. p53 can activate transcriptional responses where the exosome secretion is modified under genotoxic stress. In this way, p53 can not only coordinate DNA damage responses in the cell, but it can also facilitate the release of exosomes that carry stress-associated molecules, effectively acting as vehicles for signaling "danger signals" to adjacent cells, and enabling the tissue to adapt to stresses. Another regulatory axis is mediated by the hypoxia-inducible factors (HIFs), transcription factors that are stabilized and activated in response to a hypoxic environment. HIF pathway activation is associated with an increase in the number and specificity of the exosomes produced by hypoxic cells, allowing adaptive pathways for intercellular signaling in ischemic tissues and tumor microenvironments that are generally under conditions of oxygen limitation [62].

Calcium signaling is involved in the regulation of exosome secretion, as higher calcium concentrations in cells may increase the rate of MVBs fusing with the plasma membrane. Upon cell activation, protein kinases and specific membrane fusion machinery are triggered, where the exocytic release of ILVs as exosomes occurs through a calcium-dependent mechanism. These regulatory processes together are used to generate an exosome secretion system that is responsive to changes in a cell's physiological state and its surrounding environment. In essence, by connecting stress responses, cellular demand, and the physiological changes to vesicle trafficking, a cell can modulate exosome production and recycling to ensure effective intercellular communication, tissue homeostasis, and adaptation to changing conditions in the microenvironment [63].

1.5 PHYSIOLOGICAL ROLES OF EXOSOMES

1.5.1 Intercellular Communication

Exosomes are regarded as advanced vehicles of communication, providing long-distance communication between cells and tissues. Exosomal communication encompasses the delivery of molecule carriers that, when released, are encased in protective packaging, providing an alternative to traditional paracrine signals through soluble mediators and allowing for a specific delivery site via surface receptor interactions. Exosomes are particularly protective for labile molecular cargo such as nucleic acid cargo, which would typically be disassembled and degraded in the extracellular space [64].

The communication specificity provided by exosomes is based on the surface protein profile of the exosomes, which is reflective of the parent cell identity, location, and activated status. The presence of cell-specific markers promotes recipient cell recognition, while select targeting and uptake by recipient cells are performed with specific uptake mechanisms, such as adhesion molecules or integrins, for tissue specificity of homing and retention. These exosome targeting mechanisms enable exosomal signals to reach recipient populations without affecting off-target recipient populations that could disrupt normal physiologic processes [65].

1.5.2 Immune Modulation

Regulating the immune system is one of the best characterized exosomal functions, and exosomes released by various cells (many of them as vesicular structures) have been shown to either stimulate or suppress immune responses. Dendritic cell (DC) exosomes are an example of exosomal antigen-presenting cells that present antigenic peptides via MHC, enabling T-lymphocyte activation in adaptive immune responses. DC antigen-presenting exosomes provide mechanisms of immune surveillance and memory that complement the benefits of direct cell-to-cell contact [66, 67].

Exosomes released by T regulatory cells (Tregs) and mesenchymal stem cells (MSCs), on the other hand, have been shown to have immunosuppressive functions; the immune suppression can organize the inflammatory response and stimulate tissue repair. These exosomes contain many anti-inflammatory exosomal signals, including regulatory proteins, anti-inflammatory cytokines, and specific miRNA (micro RNA) populations that, in total, can recondition the phenotypes of the recipient cells toward resolving inflammatory processes. Pro-inflammatory exosomal signals and regulatory exosomal signals and their appropriate balance in a given system are crucial to maintaining immune homeostasis and properly responding to tissue injury [68].

1.5.3 Tissue Regeneration and Repair

Regenerative medicine applications are increasingly relying on understanding the role of exosomes in repairs to the tissue via a triplet of complementary mechanisms. Stem cell exosomes promote angiogenesis, cellular proliferation, and matrix remodeling and therefore provide beneficial properties for therapy that do not require the directed engraftment of a cell. In addition to the absence of adverse side effects, exosomes have distinct benefits in terms of manufacturing. Exosome-mediated regenerative processes utilize a coordinated delivery of growth factors, pro-angiogenic proteins, and regulatory forms of RNAs capable of stimulating endogenous repair mechanisms [69–71].

Tissue-specific exosomal populations have time-dependent functionality reflecting their cellular origin. For example, cardiomyocyte-derived exosomes promote cardiac muscle regeneration following myocardial infarction, while neuronal-derived exosomes foster synaptic plasticity and neuroprotection during neurodegenerative processes. Once recognized, these tissue-specific functions would lay the groundwork for designing targeted therapeutic interventions that promote repair by revealing endogenous mechanisms of repair [72].

1.6 EXOSOMES IN HEALTH AND DISEASE

1.6.1 Homeostasis Maintenance

Physiological secretion of exosomes plays an important role in tissue homeostasis via multiple, integrated pathways that work to maintain equilibrium at the cellular and tissue levels. Exosomal transfer of metabolic enzymes and regulatory RNAs is analogous to metabolic regulation, where energetic production and utilization among a variety of cell populations is coordinated. A prime example of this is the important role of metabolic communication in particular tissues with high energetic demands or specialized metabolic functions [73].

Cellular quality assurance represents another critical homeostasis function. Exosomes are used for the disposal of damaged proteins and organelles from cells as a vehicle for removal. This temporary disposal, or exocytosis, occurs outside of the cells and complements their intracellular degradation processes, while preventing the toxicity of having cellular components that disrupt cellular processes. The selective character of the type of disposal implies sophisticated quality assurance processes that identify and package damaged cellular components for exosome disposal [36].

1.6.2 Alterations in Pathological Conditions

Disease states often characterize disease and are regularly associated with disruption of normal exosomal communication pathways, where changes in vesicle composition, production, and function influence disease processes. Cancer cells release increased numbers of exosomes with altered cargo profiles that can cause tumor growth, immune evasion, stimulation of angiogenesis, and preparation of a metastatic niche. Because tumor-derived exosomes carry oncogenic proteins, pro-angiogenic inflammatory factors, and oncomiRs, they influence ex vivo reprogramming of recipient cells toward a pro-tumorigenic phenotype [74, 75].

Neurodegenerative diseases involve specific marker changes in exosomal protein cargo and sequester misfolded proteins such as amyloid-β, tau, and α-synuclein in vesicles released from the diseased neural populations. Pathological exosomes can propagate disease states through misfolded protein transfer to previously unexposed cells, mimicking some of the characteristic patterns of propagation seen in neurodegenerative disease [76].

Tissue-specific inflammatory diseases often exhibit altered exosomal cytokine and immune regulatory molecule profiles that drive inflammatory signals that maintain systemic inflammation and damage tissue. Knowledge of these disease-specific alterations permits opportunities for biomarker development, diagnostic strategies, and therapeutic or intervention strategies that address pathological exosomal communication while preserving the desired communication signals and pathways [77].

1.7 SIGNIFICANCE IN TRANSLATIONAL RESEARCH

A critical step in evaluating the translational potential of exosomes is to compare their pharmacokinetic behavior with that of conventional therapeutics. Unlike small-molecule drugs, which are limited by short circulation half-lives, rapid renal clearance, and poor tissue selectivity, exosomes display unique biological stability, natural tropism for specific tissues, and efficient cellular uptake mechanisms [78]. These features directly influence their therapeutic promise by improving bioavailability, reducing off-target toxicity, and enhancing delivery to hard-to-reach sites such as the brain. To provide a consolidated overview of these distinctions, Table 1.1 summarizes the major pharmacokinetic differences between exosomes and conventional small-molecule drugs. This comparative framework offers a foundation for understanding why exosomes are increasingly viewed as a transformative platform for translational medicine and sets the context for subsequent discussion on their role in chronic disease therapy.

1.7.1 Promise of Exosomes for Chronic Disease Therapy

Exosomes are a promising therapeutic delivery vehicle for the problems of standard drug delivery systems in chronic diseases based on their favorable properties. They are naturally derived, making them biocompatible and likely lower potential for immunogenicity than synthetic nanoparticles. Their internal and external routes of trafficking and cell uptake could facilitate penetration into most tissues. Their ability to cross a variety of biological barriers while delivering a therapeutic payload creates opportunities to access normal and/or disease tissues that would otherwise be difficult to target. Exosomes can be used to deliver a cargo made up of a variety of therapeutic modalities (e.g., proteins, nucleic acids, and small molecules) simultaneously in the same delivery vehicle. This multimodal approach is valuable for complex maladies that require the coordination of multiple therapeutic targets or pathways. Finally, exosomes help address critical concerns related to conventional protein and nucleic acid (e.g., mRNA) therapeutics in terms of maintaining stability during circulation and storage, as conventional protein or nucleic acid therapies require methods of delivery and highly specialized starting formulations [88–90].

1.7.2 Bench-to-Bedside Vision

Clinical translation of exosome-based therapies will involve overcoming several substantial hurdles, including production protocols, quality control measures, and regulatory

TABLE 1.1
Comparative Pharmacokinetic Characteristics of Exosomes versus Conventional Small-Molecule Drugs

Parameter	Exosomes	Conventional Small-Molecule Drugs	References
Circulation half-life	Typically, 2–6 hours (IV); platelet-derived EVs ~5.5 hours; stability enhanced by surface modifications (PEGylation, CD47, cloaking with membranes)	Usually, minutes to hours due to rapid renal clearance (<500 Da filtered quickly); lipophilic drugs may persist longer but are metabolized	[78, 79]
First-pass uptake and biodistribution	Rapid clearance by liver and spleen macrophages via phosphatidylserine recognition; biodistribution can be tuned by engineering surface ligands	Subject to hepatic first-pass metabolism (CYP450 enzymes) or renal excretion; extensive presystemic elimination reduces oral bioavailability	[80]
Cellular entry/uptake	Endocytosis (clathrin/caveolin-mediated), macropinocytosis, phagocytosis, direct membrane fusion, and receptor–ligand interactions (e.g., integrins, tetraspanins)	Primarily passive diffusion across lipid membranes (lipophilic drugs) or carrier/ transporter-mediated uptake (hydrophilic drugs, for example, glucose transporters, OATPs).	[81, 82]
Immunogenicity and safety	Generally low immunogenicity, being endogenous; immune evasion further enhanced by CD47 ("don't eat me" signal) or autologous sourcing; risk of unwanted immune activation if from tumor cells	Variable: can trigger hypersensitivity reactions (e.g., antibiotics, NSAIDs); some require formulation with excipients that add immunogenic risks (PEG, surfactants)	[83]
Metabolism and clearance	Cleared via mononuclear phagocyte system (MPS) in liver/spleen; degradation into lipids, proteins, and nucleic acids; minimal renal excretion due to size (>30 nm)	Undergo hepatic metabolism (CYP450, conjugation) or renal filtration; metabolites may be active or toxic	[84]
Crossing biological barriers	Can cross blood–brain barrier (BBB) via transcytosis and receptor-mediated pathways; natural tropism for inflamed tissues	Limited BBB penetration (only lipophilic or actively transported drugs); often requires chemical modification or carriers	[15, 84]
Pharmacokinetic tracking tools	Novel imaging: 89Zr-labeled PET, DiR near-infrared fluorescence, nanoparticle tracking analysis (NTA), bioluminescence; ongoing development of exosome-specific tracers	Well-established methods: LC–MS/MS, PET–CT, radiolabeled tracers, plasma concentration–time profiling	[85, 86]
Therapeutic payload capacity	It can carry proteins, mRNA, miRNA, siRNA, and drugs simultaneously, providing natural protection from degradation	Restricted to chemical drug molecules; biologics require separate delivery systems	[87]

pathways. Developing cGMP protocols for the production of exosomes, through collaboration between academia, biotechnology companies, and regulatory agencies, will lead to standardized production procedures and effective quality control processes. Standardization of procedures, which include controlled biological activity, is essential for consistent product quality [16, 91].

Scalability methods present another important translation issue, with bioreactor-based and cell-free production methods in various stages of development. All manufacturing processes must be mindful of cost and the need for exosomal and therapeutic activity maintenance. Regulatory frameworks are continuously adapting to the distinct characteristics of exosome-based therapies while continuing to allow for patient safety and product efficacy [11, 92].

1.8 CONCLUSION

The remarkable journey of exosome research from initial observations of cellular debris to recognition as sophisticated communication vehicles exemplifies the dynamic nature of scientific discovery. Contemporary understanding encompasses complex molecular compositions, regulated biogenesis pathways, and diverse physiological functions that collectively establish exosomes as fundamental components of intercellular communication networks. The transition from basic biological characterization to therapeutic application development reflects both technological advancement and deeper mechanistic insight into vesicle biology.

Future research directions will undoubtedly continue expanding our understanding of exosomal complexity while advancing clinical translation efforts across multiple therapeutic domains. The integration of advanced analytical techniques, engineering approaches, and clinical validation studies promises to realize the therapeutic potential first glimpsed in those pioneering electron microscopy studies of the 1980s. As we advance through subsequent chapters exploring technological innovations and therapeutic applications, the foundational knowledge established here will provide the essential framework for understanding this exciting and rapidly evolving field.

ABBREVIATIONS

BBB	blood–brain barrier
CAGR	compound annual growth rate
cGMP	current good manufacturing practice
CT	computed tomography
DC	dendritic cell
DLS	dynamic light scattering
ESCRT	endosomal sorting complex required for transport
EVs	extracellular vehicles
HIFs	hypoxia-inducible factors
ICAM	intercellular adhesion molecules
ILVs	intraluminal vesicles
ISEV	International Society for Extracellular Vesicles
IV	intravenous
LC-MS/MS	Liquid chromatography–tandem mass spectrometry
lncRNA	long noncoding RNAs
MHC	major histocompatibility complex
miRNA	microRNA
MPS	mononuclear phagocyte system
mRNA	messenger RNA
MSC	mesenchymal stem cell
MVB	multivesicular body
nSMase2	neutral sphingomyelinase 2
NTA	nanoparticle tracking analysis
PEG	polyethylene glycol
PET	positron emission tomography
siRNA	small interfering RNA
TEM	transmission electron microscopy
Tregs	T regulatory cells

ACKNOWLEDGMENTS

The author acknowledges the support of the School of Pharmacy, Rungta International Skills University, and Rungta College of Pharmaceutical Sciences and Research, Kohka, Bhilai, Chhattisgarh, India.

AI DISCLOSURE STATEMENT

During the preparation of this work, the authors used Grammarly, QuillBot, and ChatGPT in order to enhance grammatical structure and narrative coherence. After using this tool/service, the authors reviewed and edited the content as needed and take full responsibility for the content of the publication.

CONFLICT OF INTEREST

The authors declare no conflict of interest.

REFERENCES

[1] K. Yadav, K.K. Sahu, Sucheta, S. Minz, M. Pradhan, Unlocking exosome therapeutics: The critical role of pharmacokinetics in clinical applications, *Tissue Cell* 93 (2025) 102749. https://doi.org/10.1016/j.tice.2025.102749.

[2] P. Tiwari, K. Yadav, R.P. Shukla, A.K. Bakshi, D. Panwar, S. Das, P.R. Mishra, Extracellular vesicles-powered immunotherapy: Unleashing the potential for safer and more effective cancer treatment, *Arch Biochem Biophys* 756 (2024) 110022. https://doi.org/10.1016/j.abb.2024.110022.

[3] P. Tiwari, R.P. Shukla, K. Yadav, N. Singh, D. Marwaha, S. Gautam, A.K. Bakshi, N. Rai, A. Kumar, D. Sharma, P.R. Mishra, Dacarbazine-primed carbon quantum dots coated with breast cancer cell-derived exosomes for improved breast cancer therapy, *J Control Release* 365 (2024) 43–59. https://doi.org/10.1016/j.jconrel.2023.11.005.

[4] P. Tiwari, R.P. Shukla, K. Yadav, M. Sharma, A.K. Bakshi, D. Panwar, N. Singh, N. Agarwal, M.N. Mugale, P.R. Mishra, YIGSR functionalized hybrid exosomes spatially target dasatinib to laminin receptors for precision therapy in breast cancer, *Adv Healthc Mater* 14 (2025) 2402673. https://doi.org/10.1002/adhm.202402673.

[5] G. Raposo, W. Stoorvogel, Extracellular vesicles: Exosomes, microvesicles, and friends, *J Cell Biol* 200 (2013) 373–383. https://doi.org/10.1083/jcb.201211138.

[6] R. Tenchov, J.M. Sasso, X. Wang, W.-S. Liaw, C.-A. Chen, Q.A. Zhou, Exosomes—nature's lipid nanoparticles, a rising star in drug delivery and diagnostics, *ACS Nano* 16 (2022) 17802–17846. https://doi.org/10.1021/acsnano.2c08774.

[7] H. Shen, S. Li, L. Lin, Q. Wu, Z. Dong, W. Xu, Advancements in plant-derived exosome-like vesicles: Versatile bioactive carriers for targeted drug delivery systems, *J Pharm Anal* (2025) 101300. https://doi.org/10.1016/j.jpha.2025.101300.

[8] R. Kalluri, V.S. LeBleu, The biology, function, and biomedical applications of exosomes, *Science* 367 (2020). https://doi.org/10.1126/science.aau6977.

[9] R. Szatanek, M. Baj-Krzyworzeka, J. Zimoch, M. Lekka, M. Siedlar, J. Baran, The methods of choice for extracellular vesicles (EVs) characterization, *Int J Mol Sci* 18 (2017). https://doi.org/10.3390/ijms18061153.

[10] A. Sharma, D. Khatri, D. Gugulothu, Regulatory and ethical considerations in exosome-based therapeutics, in: 2025: pp. 523–551. https://doi.org/10.1016/B978-0-443-29052-7.00020-9.

[11] Y.-S. Chen, E.-Y. Lin, T.-W. Chiou, H.-J. Harn, Exosomes in clinical trial and their production in compliance with good manufacturing practice, *Ci Ji Yi Xue Za Zhi* 32 (2020) 113–120. https://doi.org/10.4103/tcmj.tcmj_182_19.

[12] S. Bashyal, C. Thapa, S. Lee, Recent progresses in exosome-based systems for targeted drug delivery to the brain, *J Control Release* 348 (2022) 723–744. https://doi.org/10.1016/j.jconrel.2022.06.011.

[13] K. Nazimek, K. Bryniarski, W. Ptak, T. Groot Kormelink, P.W. Askenase, Orally administered exosomes suppress mouse delayed-type hypersensitivity by delivering miRNA-150 to antigen-primed macrophage APC targeted by exosome-surface anti-peptide antibody light chains, *Int J Mol Sci* 21 (2020). https://doi.org/10.3390/ijms21155540.

[14] R. Kandimalla, F. Aqil, S.S. Alhakeem, J. Jeyabalan, N. Tyagi, A. Agrawal, J. Yan, W. Spencer, S. Bondada, R.C. Gupta, Targeted oral delivery of paclitaxel using colostrum-derived exosomes, *Cancers (Basel)* 13 (2021). https://doi.org/10.3390/cancers13153700.

[15] K. Yadav, R. Vijayalakshmi, K.K. Sahu, P. Sure, K. Chahal, R. Yadav, Sucheta, A. Dubey, M. Jha, M. Pradhan, Exosome-based macromolecular neurotherapeutic drug delivery approaches in overcoming the blood-brain barrier for treating brain disorders, *Eur J Pharm Biopharm* 199 (2024) 114298. https://doi.org/10.1016/j.ejpb.2024.114298.

[16] J. Rezaie, M. Feghhi, T. Etemadi, A review on exosomes application in clinical trials: Perspective, questions, and challenges, *Cell Commun Signal* 20 (2022) 145. https://doi.org/10.1186/s12964-022-00959-4.

[17] K. Yadav, N.S. Chauhan, S. Saraf, D. Singh, M.R. Singh, Challenges and need of delivery carriers for bioactives and biological agents: An introduction, in: *Advances and Avenues in the Development of Novel Carriers for Bioactives and Biological Agents*, Elsevier, 2020: pp. 1–36. https://doi.org/10.1016/b978-0-12-819666-3.00001-8.

[18] K. Yadav, M.R. Singh, V.K. Rai, N. Srivastava, N. Prasad Yadav, Commercial aspects and market potential of novel delivery systems for bioactives and biological agents, in: *Advances and Avenues in the Development of Novel Carriers for Bioactives and Biological Agents*, Elsevier, 2020: pp. 595–620. https://doi.org/10.1016/b978-0-12-819666-3.00020-1.

[19] R.M. Johnstone, M. Adam, J.R. Hammond, L. Orr, C. Turbide, Vesicle formation during reticulocyte maturation. Association of plasma membrane activities with released vesicles (exosomes), *J Biol Chem* 262 (1987) 9412–9420.

[20] B.T. Pan, R.M. Johnstone, Fate of the transferrin receptor during maturation of sheep reticulocytes in vitro: Selective externalization of the receptor, *Cell* 33 (1983) 967–978. https://doi.org/10.1016/0092-8674(83)90040-5.

[21] G. Raposo, H.W. Nijman, W. Stoorvogel, R. Liejendekker, C. V Harding, C.J. Melief, H.J. Geuze, B lymphocytes secrete antigen-presenting vesicles, *J Exp Med* 183 (1996) 1161–1172. https://doi.org/10.1084/jem.183.3.1161.

[22] L. Zitvogel, A. Regnault, A. Lozier, J. Wolfers, C. Flament, D. Tenza, P. Ricciardi-Castagnoli, G. Raposo, S. Amigorena, Eradication of established murine tumors using a novel cell-free vaccine: Dendritic cell-derived exosomes, *Nat Med* 4 (1998) 594–600. https://doi.org/10.1038/nm0598-594.

[23] H. Valadi, K. Ekström, A. Bossios, M. Sjöstrand, J.J. Lee, J.O. Lötvall, Exosome-mediated transfer of mRNAs and microRNAs is a novel mechanism of genetic exchange between cells, *Nat Cell Biol* 9 (2007) 654–659. https://doi.org/10.1038/ncb1596.

[24] B.N. Hannafon, W.-Q. Ding, Intercellular communication by exosome-derived microRNAs in cancer, *Int J Mol Sci* 14 (2013) 14240–14269. https://doi.org/10.3390/ijms140714240.

[25] M. Yáñez-Mó, P.R.-M. Siljander, Z. Andreu, A.B. Zavec, F.E. Borràs, E.I. Buzas, K. Buzas, E. Casal, F. Cappello, J. Carvalho, E. Colás, A. Cordeiro-da Silva, S. Fais, J.M. Falcon-Perez, I.M. Ghobrial, B. Giebel, M. Gimona, M. Graner, I. Gursel, M. Gursel, N.H.H. Heegaard, A. Hendrix, P. Kierulf, K. Kokubun, M. Kosanovic, V. Kralj-Iglic, E.-M. Krämer-Albers, S. Laitinen, C. Lässer, T. Lener, E. Ligeti, A. Linē, G. Lipps, A. Llorente, J. Lötvall, M. Mančck-Keber, A. Marcilla, M. Mittelbrunn, I. Nazarenko, E.N.M. Nolte-'t Hoen, T.A. Nyman, L. O'Driscoll, M. Olivan, C. Oliveira, É. Pállinger, H.A. Del Portillo, J. Reventós, M. Rigau, E. Rohde, M. Sammar, F. Sánchez-Madrid, N. Santarém, K. Schallmoser, M.S. Ostenfeld, W. Stoorvogel, R. Stukelj, S.G. Van der Grein, M.H. Vasconcelos, M.H.M. Wauben, O. De Wever, Biological properties of extracellular vesicles and their physiological functions, *J Extracell Vesicles* 4 (2015) 27066. https://doi.org/10.3402/jev.v4.27066.

[26] S. Roszkowski, Therapeutic potential of mesenchymal stem cell-derived exosomes for regenerative medicine applications, *Clin Exp Med* 24 (2024) 46. https://doi.org/10.1007/s10238-023-01282-z.

[27] P. Tiwari, R.P. Shukla, K. Yadav, D. Panwar, N. Agarwal, A. Kumar, N. Singh, A.K. Bakshi, D. Marwaha, S. Gautam, N. Rai, P.R. Mishra, Exploring nanocarriers as innovative materials for advanced drug delivery strategies in onco-immunotherapies, *J Mol Graph Model* 128 (2024) 108702. https://doi.org/10.1016/j.jmgm.2024.108702.

[28] G.R. Willis, S. Kourembanas, S.A. Mitsialis, Toward exosome-based therapeutics: Isolation, heterogeneity, and fit-for-purpose potency, *Front Cardiovasc Med* 4 (2017). https://doi.org/10.3389/fcvm.2017.00063.

[29] Y. Zhang, J. Bi, J. Huang, Y. Tang, S. Du, P. Li, Exosome: A review of its classification, isolation techniques, storage, diagnostic and targeted therapy applications, *Int J Nanomed* 15 (2020) 6917–6934. https://doi.org/10.2147/IJN.S264498.

[30] J. V Santiago, A. Natu, C.C. Ramelow, S. Rayaprolu, H. Xiao, V. Kumar, P. Kumar, N.T. Seyfried, S. Rangaraju, Identification of state-specific proteomic and transcriptomic signatures of microglia-derived extracellular vesicles, *Mol Cell Proteom* 22 (2023). https://doi.org/10.1016/j.mcpro.2023.100678.

[31] J. Zhu, F. Wu, C. Li, J. Mao, Y. Wang, X. Zhou, H. Xie, C. Wen, Application of single extracellular vesicle analysis techniques, *Int J Nanomed* 18 (2023) 5365–5376. https://doi.org/10.2147/IJN.S421342.

[32] V. Marassi, S. Giordani, A. Placci, A. Punzo, C. Caliceti, A. Zattoni, P. Reschiglian, B. Roda, A. Roda, Emerging microfluidic tools for simultaneous exosomes and cargo biosensing in liquid biopsy: New integrated miniaturized FFF-assisted approach for colon cancer diagnosis, *Sensors (Basel)* 23 (2023). https://doi.org/10.3390/s23239432.

[33] J. Wang, P. Ma, D.H. Kim, B.-F. Liu, U. Demirci, Towards microfluidic-based exosome isolation and detection for tumor therapy, *Nano Today* 37 (2021). https://doi.org/10.1016/j.nantod.2020.101066.

[34] M.-C.N. Le, Z.H. Fan, Exosome isolation using nanostructures and microfluidic devices, *Biomed Mater* 16 (2021) 22005. https://doi.org/10.1088/1748-605X/abde70.

[35] Y. Wu, Y. Wang, Y. Lu, X. Luo, Y. Huang, T. Xie, C. Pilarsky, Y. Dang, J. Zhang, Microfluidic technology for the isolation and analysis of exosomes, *Micromachines (Basel)* 13 (2022). https://doi.org/10.3390/mi13101571.

[36] M.A. Kumar, S.K. Baba, H.Q. Sadida, S.Al. Marzooqi, J. Jerobin, F.H. Altemani, N. Algehainy, M.A. Alanazi, A.-B. Abou-Samra, R. Kumar, A.S. Al-Shabeeb Akil, M.A. Macha, R. Mir, A.A. Bhat, Extracellular vesicles as tools and targets in therapy for diseases, *Signal Transduct Target Ther* 9 (2024) 27. https://doi.org/10.1038/s41392-024-01735-1.

[37] S.S. Palakurthi, B. Shah, S. Kapre, N. Charbe, S. Immanuel, S. Pasham, M. Thalla, A. Jain, S. Palakurthi, A comprehensive review of challenges and advances in exosome-based drug delivery systems, *Nanoscale Adv* 6 (2024) 5803–5826. https://doi.org/10.1039/d4na00501e.

[38] S. Samanta, S. Rajasingh, N. Drosos, Z. Zhou, B. Dawn, J. Rajasingh, Exosomes: New molecular targets of diseases, *Acta Pharmacol Sin* 39 (2018) 501–513. https://doi.org/10.1038/aps.2017.162.

[39] Y. Fan, C. Pionneau, F. Cocozza, P.-Y. Boëlle, S. Chardonnet, S. Charrin, C. Théry, P. Zimmermann, E.

Rubinstein, Differential proteomics argues against a general role for CD9, CD81 or CD63 in the sorting of proteins into extracellular vesicles, *J Extracell Vesicles* 12 (2023) e12352. https://doi.org/10.1002/jev2.12352.

[40] R.F. Saito, C.M. Machado, A.L. Lomba, A.H. Otake, M.C. Rangel, Heat shock proteins mediate intercellular communications within the tumor microenvironment through extracellular vesicles, *Appl Biosci* 3 (2024) 45–58. https://doi.org/10.3390/applbiosci3010003.

[41] X. Hou, Y. Chen, N.D. Carrillo, V.L. Cryns, R.A. Anderson, J. Sun, S. Wang, M. Chen, Phosphoinositide signaling at the cytoskeleton in the regulation of cell dynamics, *Cell Death Dis* 16 (2025) 296. https://doi.org/10.1038/s41419-025-07616-x.

[42] L. Xu, F.N. Faruqu, Y.M. Lim, K.Y. Lim, R. Liam-Or, A.A. Walters, P. Lavender, D. Fear, C.M. Wells, J. Tzu-Wen Wang, K.T. Al-Jamal, Exosome-mediated RNAi of PAK4 prolongs survival of pancreatic cancer mouse model after loco-regional treatment, *Biomaterials* 264 (2021) 120369. https://doi.org/10.1016/j.biomaterials.2020.120369.

[43] K. Yadav, S.P. Ebenezer Gnanakani, K. Kumar Sahu, Sucheta, A. Dubey, S. Minz, W. Raza, M. Pradhan, Unleashing the potential of natural protein based nanoparticles for the delivery of therapeutic nucleic Acid: A comprehensive review, *Int J Pharm* 669 (2025) 125049. https://doi.org/10.1016/j.ijpharm.2024.125049.

[44] K. Yadav, K.K. Sahu, Sucheta, S.P.E. Gnanakani, P. Sure, R. Vijayalakshmi, V.D. Sundar, V. Sharma, R. Antil, M. Jha, S. Minz, A. Bagchi, M. Pradhan, Biomedical applications of nanomaterials in the advancement of nucleic acid therapy: Mechanistic challenges, delivery strategies, and therapeutic applications, *Int J Biol Macromol* 241 (2023) 124582. https://doi.org/10.1016/j.ijbiomac.2023.124582.

[45] S. Xie, Q. Zhang, L. Jiang, Current knowledge on exosome biogenesis, cargo-sorting mechanism and therapeutic implications, *Membranes (Basel)* 12 (2022). https://doi.org/10.3390/membranes12050498.

[46] S.-P. Li, Z.-X. Lin, X.-Y. Jiang, X.-Y. Yu, Exosomal cargo-loading and synthetic exosome-mimics as potential therapeutic tools, *Acta Pharmacol Sin* 39 (2018) 542–551. https://doi.org/10.1038/aps.2017.178.

[47] F. Zhang, A.N. Isak, S. Yang, Y. Song, L. Ren, C. Feng, G. Chen, Smartly responsive DNA–miRNA hybrids packaged in exosomes for synergistic enhancement of cancer cell apoptosis, *Nanoscale* 14 (2022) 6612–6619. https://doi.org/10.1039/D1NR08539E.

[48] J. Donoso-Quezada, S. Ayala-Mar, J. González-Valdez, The role of lipids in exosome biology and intercellular communication: Function, analytics and applications, *Traffic* 22 (2021) 204–220. https://doi.org/10.1111/tra.12803.

[49] Y. Wu, W. Chen, M. Guo, Q. Tan, E. Zhou, J. Deng, M. Li, J. Chen, Z. Yang, Y. Jin, Metabolomics of extracellular vesicles: A future promise of multiple clinical applications, *Int J Nanomed* 17 (2022) 6113–6129. https://doi.org/10.2147/IJN.S390378.

[50] K. Yadav, S.P.E. Gnanakani, K.K. Sahu, C.K. Veni Chikkula, P.S. Vaddi, S. Srilakshmi, R. Yadav, Sucheta, A. Dubey, S. Minz, M. Pradhan, Nano revolution of DNA nanostructures redefining cancer therapeutics – a comprehensive review, *Int J Biol Macromol* 274 (2024) 133244. https://doi.org/10.1016/j.ijbiomac.2024.133244.

[51] M.R. Singh, K. Yadav, N.D. Chaurasiya, D. Singh, Immune system and mechanism of immunomodulation, in: N.S. Sangwan, M.A. Farag, L.V. Modolo (Eds.), *Plants and Phytomolecules for Immunomodulation: Recent Trends and Advances*, Springer Nature Singapore, Singapore, 2022: pp. 1–31. https://doi.org/10.1007/978-981-16-8117-2_1.

[52] K. Yadav, M. Pradhan, D. Singh, M.R. Singh, Macrophage-associated disorders: Pathophysiology, treatment challenges, and possible solutions, in: S. Gupta, Y.V. Pathak (Eds.), *Macrophage Targeted Delivery Systems*, Springer, Cham, 2022: pp. 65–99. https://doi.org/10.1007/978-3-030-84164-5_4.

[53] T. Skotland, K. Sagini, K. Sandvig, A. Llorente, An emerging focus on lipids in extracellular vesicles, *Adv Drug Deliv Rev* 159 (2020) 308–321. https://doi.org/10.1016/j.addr.2020.03.002.

[54] T. Skotland, T.G. Iversen, A. Llorente, K. Sandvig, Biodistribution, pharmacokinetics and excretion studies of intravenously injected nanoparticles and extracellular vesicles: Possibilities and challenges, *Adv Drug Deliv Rev* 186 (2022) 114326. https://doi.org/10.1016/j.addr.2022.114326.

[55] K. Yadav, Ajazuddin, A. Alexander, P. Pandey, W. Raza, M.R. Singh, D. Singh, R. Tirkey, R. Jangde, P.K. Sahu, Introduction to biomaterial-inspired nanomedicines, in: *Biomaterial-Inspired Nanomedicines for Targeted Therapies*, Springer Nature, 2024. https://doi.org/10.1007/978-981-97-3925-7_1.

[56] K. Yadav, M. Pradhan, N. Singh Chauhan, Introduction to biomaterial-inspired nanomedicines, in: M. Pradhan, K. Yadav, N. Singh Chauhan (Eds.), *Biomaterial-Inspired Nanomedicines for Targeted Therapies*, Springer Nature Singapore, Singapore, 2024: pp. 1–30. https://doi.org/10.1007/978-981-97-3925-7_1.

[57] Q. Si, L. Wu, D. Pang, P. Jiang, Exosomes in brain diseases: Pathogenesis and therapeutic targets, *MedComm (Beijing)* 4 (2023) e287. https://doi.org/10.1002/mco2.287.

[58] S. Gurung, D. Perocheau, L. Touramanidou, J. Baruteau, The exosome journey: From biogenesis to uptake and intracellular signalling, *Cell Commun Signal* 19 (2021) 47. https://doi.org/10.1186/s12964-021-00730-1.

[59] Y. Yang, M. Wang, Y.-Y. Zhang, S.-Z. Zhao, S. Gu, The endosomal sorting complex required for transport repairs the membrane to delay cell death, *Front Oncol* 12 (2022) 1007446. https://doi.org/10.3389/fonc.2022.1007446.

[60] Y. Zhang, Y. Liu, H. Liu, W.H. Tang, Exosomes: Biogenesis, biologic function and clinical potential, *Cell Biosci* 9 (2019) 19. https://doi.org/10.1186/s13578-019-0282-2.

[61] E.B. Frankel, A. Audhya, ESCRT-dependent cargo sorting at multivesicular endosomes, *Semin Cell Dev Biol* 74 (2018) 4–10. https://doi.org/10.1016/j.semcdb.2017.08.020.

[62] S. Gurunathan, M.-H. Kang, J.-H. Kim, A comprehensive review on factors influences biogenesis, functions, therapeutic and clinical implications of exosomes, *Int J Nanomed* 16 (2021) 1281–1312. https://doi.org/10.2147/IJN.S291956.

[63] J. Li, Y. Zhang, P.-Y. Dong, G.-M. Yang, S. Gurunathan, A comprehensive review on the composition, biogenesis, purification, and multifunctional role of exosome as delivery vehicles for cancer therapy, *Biomed Pharmacother* 165 (2023) 115087. https://doi.org/10.1016/j.biopha.2023.115087.

[64] A. Rajput, A. Varshney, R. Bajaj, V. Pokharkar, Exosomes as new generation vehicles for drug delivery: Biomedical applications and future perspectives, *Molecules* 27 (2022). https://doi.org/10.3390/molecules27217289.

[65] N. Mukerjee, A. Bhattacharya, S. Maitra, M. Kaur, S. Ganesan, S. Mishra, A. Ashraf, M. Rizwan, K.K. Kesari, T.A. Tabish, N.D. Thorat, Exosome isolation and characterization for advanced diagnostic and therapeutic

applications, *Mater Today Bio* 31 (2025) 101613. https://doi.org/10.1016/j.mtbio.2025.101613.
[66] S. Mittal, P. Gupta, P. Chaluvally-Raghavan, S. Pradeep, Emerging role of extracellular vesicles in immune regulation and cancer progression, *Cancers (Basel)* 12 (2020). https://doi.org/10.3390/cancers12123563.
[67] E. Segura, S. Amigorena, C. Théry, Mature dendritic cells secrete exosomes with strong ability to induce antigen-specific effector immune responses, *Blood Cells Mol Dis* 35 (2005) 89–93. https://doi.org/10.1016/j.bcmd.2005.05.003.
[68] B.B. Shenoda, S.K. Ajit, Modulation of immune responses by exosomes derived from antigen-presenting cells, *Clin Med Insights Pathol* 9 (2016) 1–8. https://doi.org/10.4137/CPath.S39925.
[69] A. Sarasati, M.H. Syahruddin, A. Nuryanti, I.D. Ana, A. Barlian, C.H. Wijaya, D. Ratnadewi, T.D.K. Wungu, H. Takemori, Plant-derived exosome-like nanoparticles for biomedical applications and regenerative therapy, *Biomedicines* 11 (2023) 1053. https://doi.org/10.3390/biomedicines11041053.
[70] M.D. Hade, C.N. Suire, Z. Suo, Mesenchymal stem cell-derived exosomes: Applications in regenerative medicine, *Cells* 10 (2021). https://doi.org/10.3390/cells10081959.
[71] X. Xu, Y. Liang, X. Li, K. Ouyang, M. Wang, T. Cao, W. Li, J. Liu, J. Xiong, B. Li, J. Xia, D. Wang, L. Duan, Exosome-mediated delivery of kartogenin for chondrogenesis of synovial fluid-derived mesenchymal stem cells and cartilage regeneration, *Biomaterials* 269 (2021) 120539. https://doi.org/10.1016/j.biomaterials.2020.120539.
[72] E. Lazar, T. Benedek, S. Korodi, N. Rat, J. Lo, I. Benedek, Stem cell-derived exosomes – an emerging tool for myocardial regeneration, *World J Stem Cells* 10 (2018) 106–115. https://doi.org/10.4252/wjsc.v10.i8.106.
[73] M. Quan, S. Kuang, Exosomal secretion of adipose tissue during various physiological states, *Pharm Res* 37 (2020) 221. https://doi.org/10.1007/s11095-020-02941-6.
[74] M. Zhang, S. Hu, L. Liu, P. Dang, Y. Liu, Z. Sun, B. Qiao, C. Wang, Engineered exosomes from different sources for cancer-targeted therapy, *Signal Transduct Target Ther* 8 (2023) 124. https://doi.org/10.1038/s41392-023-01382-y.
[75] A. Javdani-Mallak, S.J. Mowla, M. Alibolandi, Tumor-derived exosomes and their application in cancer treatment, *J Transl Med* 23 (2025) 751. https://doi.org/10.1186/s12967-025-06814-7.
[76] N. Perets, O. Betzer, R. Shapira, S. Brenstein, A. Angel, T. Sadan, U. Ashery, R. Popovtzer, D. Offen, Golden exosomes selectively target brain pathologies in neurodegenerative and neurodevelopmental disorders, *Nano Lett* 19 (2019) 3422–3431. https://doi.org/10.1021/acs.nanolett.8b04148.
[77] D. Yuan, Y. Zhao, W.A. Banks, K.M. Bullock, M. Haney, E. Batrakova, A.V. Kabanov, Macrophage exosomes as natural nanocarriers for protein delivery to inflamed brain, *Biomaterials* 142 (2017) 1–12. https://doi.org/10.1016/j.biomaterials.2017.07.011.
[78] I. Kimiz-Gebologlu, S.S. Oncel, Exosomes: Large-scale production, isolation, drug loading efficiency, and biodistribution and uptake, *J Control Release* 347 (2022) 533–543. https://doi.org/10.1016/j.jconrel.2022.05.027.
[79] A. Martínez-Santillán, J. González-Valdez, Novel technologies for exosome and exosome-like nanovesicle procurement and enhancement, *Biomedicines* 11 (2023). https://doi.org/10.3390/biomedicines11051487.
[80] H. Choi, Y. Choi, H.Y. Yim, A. Mirzaaghasi, J.-K. Yoo, C. Choi, Biodistribution of exosomes and engineering strategies for targeted delivery of therapeutic exosomes, *Tissue Eng Regen Med* 18 (2021) 499–511. https://doi.org/10.1007/s13770-021-00361-0.
[81] T. Tian, Y.-L. Zhu, Y.-Y. Zhou, G.-F. Liang, Y.-Y. Wang, F.-H. Hu, Z.-D. Xiao, Exosome uptake through clathrin-mediated endocytosis and macropinocytosis and mediating miR-21 delivery*, *J Biol Chem* 289 (2014) 22258–22267. https://doi.org/10.1074/jbc.M114.588046.
[82] K.J. Svensson, H.C. Christianson, A. Wittrup, E. Bourseau-Guilmain, E. Lindqvist, L.M. Svensson, M. Mörgelin, M. Belting, Exosome uptake depends on ERK1/2-heat shock protein 27 signaling and lipid raft-mediated endocytosis negatively regulated by caveolin-1, *J Biol Chem* 288(2013) 17713–17724. https://doi.org/10.1074/jbc.M112.445403.
[83] Z. Gu, Z. Yin, P. Song, Y. Wu, Y. He, M. Zhu, Z. Wu, S. Zhao, H. Huang, H. Wang, C. Tong, Z. Qi, Safety and biodistribution of exosomes derived from human induced pluripotent stem cells, *Front Bioeng Biotechnol* 10 (2022). https://doi.org/10.3389/fbioe.2022.949724.
[84] W. Wang, N. Zhu, T. Yan, Y.-N. Shi, J. Chen, C.-J. Zhang, X.-J. Xie, D.-F. Liao, L. Qin, The crosstalk: Exosomes and lipid metabolism, *Cell Commun Signal* 18 (2020) 119. https://doi.org/10.1186/s12964-020-00581-2.
[85] M. Morishita, Y. Takahashi, M. Nishikawa, Y. Takakura, Pharmacokinetics of exosomes – an important factor for elucidating the biological roles of exosomes and for the development of exosome-based therapeutics, *J Pharm Sci* 106 (2017) 2265–2269. https://doi.org/10.1016/j.xphs.2017.02.030.
[86] Y. Takahashi, M. Nishikawa, Y. Takakura, Analysis and control of in vivo kinetics of exosomes for the development of exosome-based DDS, *Yakugaku Zasshi* 136 (2016) 49–53. https://doi.org/10.1248/yakushi.15-00227-2.
[87] H.B. Koh, H.J. Kim, S.-W. Kang, T.-H. Yoo, Exosome-based drug delivery: Translation from bench to clinic, *Pharmaceutics* 15 (2023). https://doi.org/10.3390/pharmaceutics15082042.
[88] R. Kar, R. Dhar, S. Mukherjee, S. Nag, S. Gorai, N. Mukerjee, D. Mukherjee, R. Vatsa, M. Chandrakanth Jadhav, A. Ghosh, A. Devi, A. Krishnan, N.D. Thorat, Exosome-based smart drug delivery tool for cancer theranostics, *ACS Biomater Sci Eng* 9 (2023) 577–594. https://doi.org/10.1021/acsbiomaterials.2c01329.
[89] M.-C. Didiot, L.M. Hall, A.H. Coles, R.A. Haraszti, B.M. Godinho, K. Chase, E. Sapp, S. Ly, J.F. Alterman, M.R. Hassler, D. Echeverria, L. Raj, D. V Morrissey, M. DiFiglia, N. Aronin, A. Khvorova, Exosome-mediated delivery of hydrophobically modified siRNA for huntingtin mRNA silencing, *Mol Ther* 24 (2016) 1836–1847. https://doi.org/10.1038/mt.2016.126.
[90] D. Yang, W. Zhang, H. Zhang, F. Zhang, L. Chen, L. Ma, L.M. Larcher, S. Chen, N. Liu, Q. Zhao, P.H.L. Tran, C. Chen, R.N. Veedu, T. Wang, Progress, opportunity, and perspective on exosome isolation – efforts for efficient exosome-based theranostics, *Theranostics* 10 (2020) 3684–3707. https://doi.org/10.7150/thno.41580.
[91] F. Tan, X. Li, Z. Wang, J. Li, K. Shahzad, J. Zheng, Clinical applications of stem cell-derived exosomes, *Signal Transduct Target Ther* 9 (2024) 17. https://doi.org/10.1038/s41392-023-01704-0.
[92] M.N. Huda, M. Nafiujjaman, I.G. Deaguero, J. Okonkwo, M.L. Hill, T. Kim, M. Nurunnabi, Potential use of exosomes as diagnostic biomarkers and in targeted drug delivery: Progress in clinical and preclinical applications, *ACS Biomater Sci Eng* 7 (2021) 2106–2149. https://doi.org/10.1021/acsbiomaterials.1c00217.

2 Exosome Therapies
Advances in Exosome Isolation, Characterization, and Engineering

Vinay Sagar Verma[1], *Amrita Thakur*[2], *Aakansha Pandey*[1], *and Hemant R. Badwaik*[3]
[1]Kamla Institute of Pharmaceutical Sciences, Shri Shankarcharya Professional University, Bhilai, Chhattisgarh, India
[2]School of Pharmacy, Vishwakarma University, Pune, Maharashtra, India
[3]Shri Shankaracharya Institute of Pharmaceutical Sciences and Research, Shri Shankarcharya Professional University, Bhilai, Chhattisgarh, India

2.1 INTRODUCTION

Exosomes are a type of extracellular vesicle (EV) that are derived from the endosomal system. They typically range in size from 30 nm to 150 nm and are secreted by nearly all cell types. Exosomes play a vital role in intercellular communication by mediating the transfer of proteins, lipids, and nucleic acids. This process influences various physiological and pathological processes in the body (1, 2). Exosome biogenesis starts with the inward budding of early endosomes, resulting in the formation of multivesicular bodies (MVBs), which then fuse with the plasma membrane to release exosomes into the extracellular space (3).

Exosomes are recognized as important facilitators of cell-to-cell communication. They play a vital role in both local and systemic signaling by transporting bioactive molecules to recipient cells. The molecular content of exosomes varies based on their cell of origin and the specific physiological or pathological context. As a result, exosomes are crucial players in numerous biological processes, including immune modulation, tissue regeneration, and neurodegeneration (4). Exosome-mediated communication plays a significant role in various disease mechanisms, including cancer progression, neurodegenerative disorders such as Alzheimer's and Parkinson's diseases, and cardiovascular diseases. This makes exosomes promising candidates for biomarker discovery and targeted therapies (5).

The ability of exosomes to carry disease-specific molecular signatures, including proteins, lipids, and nucleic acids, has made them valuable biomarkers for liquid biopsy-based diagnostics (6). Their presence in biofluids such as blood, urine, cerebrospinal fluid, and saliva allows for noninvasive detection of pathological conditions, facilitating early diagnosis and prognosis assessment (7).

Beyond diagnostics, exosomes are being explored as natural nanocarriers for drug delivery due to their biocompatibility, low immunogenicity, and intrinsic targeting properties. Engineered exosomes can be loaded with therapeutic molecules, including small RNAs, proteins, and chemotherapeutic agents, offering a promising approach for precision medicine (8). Furthermore, in regenerative medicine, stem cell–derived exosomes have shown significant potential in promoting tissue repair and modulating immune responses, particularly in wound healing, cardiovascular repair, and neuroprotection (9).

Despite the increasing interest in exosome research, significant challenges persist in their isolation, characterization, and bioengineering. Traditional isolation methods, such as ultracentrifugation, often suffer from low purity and yield, highlighting the need for more efficient and scalable techniques (10). Characterizing exosomes requires advanced analytical techniques to accurately determine their size, morphology, and molecular composition, ensuring that research and clinical applications are reproducible (11). Advancements in exosome engineering are crucial for enhancing their therapeutic potential by improving targeting efficiency and cargo-loading capabilities (12).

The chapter concludes with a discussion on the challenges, future directions, and ethical considerations in exosome-based technologies. By providing a comprehensive analysis of recent advancements, this chapter aims to contribute to the standardization and translational potential of exosome research in biomedical sciences.

2.2 EXOSOME ISOLATION TECHNIQUES

Efficiently isolating exosomes with high purity and yield is crucial for their subsequent characterization and therapeutic applications. Traditional isolation techniques, such as differential ultracentrifugation, have been commonly used but often suffer from low scalability and contamination with nonvesicular components. In response to these limitations, newer methods like size-exclusion chromatography (SEC),

DOI: 10.1201/9781003623632-2

precipitation-based techniques, and microfluidic technologies have been developed to improve exosome recovery. Additionally, emerging methods such as acoustic, electrophoretic, and innovative filtration techniques are gaining popularity due to their potential for high throughput and automation.

2.2.1 Ultracentrifugation-Based Methods

Ultracentrifugation is the most commonly used technique for isolating exosomes due to its effectiveness in separating particles based on their size and density (Table 2.1). There are two main types of ultracentrifugation:

A. *Differential Ultracentrifugation:* This method involves a series of centrifugation steps at both low and high speeds to eliminate larger debris before collecting the exosomes.
B. *Density-Gradient Ultracentrifugation:* This technique utilizes sucrose or iodixanol gradients to separate exosomes according to their buoyant density, resulting in improved purity compared to differential centrifugation.

Although widely used, ultracentrifugation has limitations such as vesicle aggregation, low recovery efficiency, and being time-intensive (13).

2.2.2 Size-Exclusion Chromatography (SEC)

SEC allows for the separation of exosomes from contaminating proteins and other small vesicles by passing the sample through a porous column. It provides high purity and preserves the bioactivity of exosomes (Table 2.2). Unlike ultracentrifugation, SEC does not subject exosomes to mechanical stress, making it an attractive alternative for functional studies. However, it has limitations in terms of processing large sample volumes.

2.2.3 Precipitation-Based Approaches

Exosome precipitation relies on hydrophilic polymers such as polyethylene glycol (PEG) to aggregate exosomes, allowing for their recovery via low-speed centrifugation (25). Commercial kits such as ExoQuick™ and Total Exosome Isolation™ use this principle for rapid isolation. While these methods are scalable and require minimal instrumentation, they often co-isolate protein contaminants, affecting downstream analyses.

2.2.3.1 Polyethylene Glycol (PEG) Precipitation

Polyethylene glycol (PEG) precipitation is one of the most widely used and straightforward techniques for exosome isolation. This method utilizes hydrophilic polymers, typically polyethylene glycol (PEG), to decrease the solubility of exosomes in solution, which leads to their aggregation. Once the exosomes are aggregated, they can be collected through low-speed centrifugation. This approach offers several practical advantages, such as scalability, cost-effectiveness, and reduced dependence on specialised equipment compared to ultracentrifugation or chromatography-based methods. Importantly, PEG precipitation has served as the basis for many commercially available exosome isolation kits.

Despite its strengths, PEG precipitation has significant drawbacks. One major limitation is the frequent co-isolation of protein contaminants and other nonexosomal components. These impurities can interfere with downstream analyses, leading to inaccurate interpretations of exosomal content and function. Additionally, they may

TABLE 2.1
A Summarized Description of Various Types of Ultracentrifugation Methods along with Their Limitations

Technique	Description	Advancements	Limitations	References
Differential ultra-centrifugation	Separates particles by size/sedimentation using stepwise centrifugal forces	Foundational, widely used; can process large volumes	Co-sedimentation, low purity, low recovery, time-intensive	(13, 14)
Density-gradient ultra-centrifugation (sucrose)	Uses a sucrose gradient; exosomes settle at a buoyant density	Improved purity; more homogeneous population	Time-consuming, specialised equipment, and possible osmotic effects	(13, 15)
Density-gradient ultracentrifugation (iodixanol)	Employs iodixanol gradient; less viscous and faster than sucrose	Better resolution, reduced osmotic effects, faster separation	Costly; requires expertise and equipment	(13, 16)
General limitations	Shared drawbacks of UC-based methods	Encourages the development of alternative methods	Vesicle aggregation is low recovery and time-intensive	(13, 17)

TABLE 2.2
Overview of Size-Exclusion Chromatography (SEC) Techniques for Exosome Isolation and Purification

Technique	Description	Advancements	Limitations	References
SEC (general)	Separates exosomes by size via porous columns; larger vesicles elute earlier	High-purity, gentle method, preserves bioactivity	Limited for large volumes, low concentration handling	(18)
SEC (high-resolution)	Separation based on hydrodynamic size; avoids shear stress	High-resolution fractions with structural/functional integrity	Requires column optimization; overlapping with large biomolecules; slow for large volumes	(19)
High-purity SEC	Uses optimized pore sizes, column materials, and buffers	Yields high-purity exosomes for drug delivery/biomarker studies	Still limited for very large volumes; fractionation/dilution issues	(20)
SEC versus ultracentrifugation	Comparative studies of purity, yield, and exosome integrity	SEC preserves exosome integrity and reduces contamination	May need extra concentration steps; method choice depends on the application	(21)
Scalable SEC	Large-scale SEC with optimized dimensions/flow rates	Moves SEC toward clinical/industrial use; higher throughput	High purity at scale remains challenging; costly columns	(22)
Integrated SEC workflows	SEC combined with ultrafiltration, affinity, and so on	Achieves highly pure/concentrated exosomes via hybrid workflows	More complex, risk of loss during multi-step processing	(23)
SEC versus other methods	Compares SEC with precipitation, UC, and so on	Highlights SEC's niche advantages in purity/functionality	No universal method; optimal choice varies by sample and application	(24)

artificially alter the perceived bioactivity of the isolated exosomes (25, 26).

2.2.3.2 Commercial Precipitation Kits: ExoQuick™ and Total Exosome Isolation™

Commercial precipitation kits, such as ExoQuick™ and Total Exosome Isolation™, have improved upon the basic principle of PEG precipitation. These kits are specifically designed to provide a user-friendly and rapid method for isolating exosomes. By offering simplified protocols and ready-to-use reagents, they make the process accessible even to researchers with limited expertise in exosome biology or those without specialised laboratory infrastructure. The main advantages of these kits include speed, reproducibility, and convenience, allowing for streamlined exosome recovery from various sample types (27, 28).

Nevertheless, these kits share the same limitations as general PEG precipitation. They are prone to co-precipitation of abundant proteins and other nonvesicular molecules, which reduces the purity of exosome preparations. Additionally, since the exact chemical compositions of the proprietary reagents are undisclosed, it becomes difficult for researchers to optimize or troubleshoot the protocols. Another practical concern is the cost per sample, which can be higher compared to traditional precipitation (28, 29).

2.2.3.3 Optimization of Precipitation Methods

In an effort to overcome purity-related challenges, researchers have focused on refining precipitation protocols. Modifications such as adjusting polymer concentrations, varying incubation times, incorporating additional washing steps, or combining precipitation with complementary purification techniques (e.g., size-exclusion chromatography, SEC) have been explored. These attempts aim to reduce protein contamination while maintaining yield. Although certain modifications have achieved better exosome purity, these often come at the expense of lower yields or increased protocol complexity. Furthermore, the outcomes can be highly sample-type dependent, making universal optimization difficult (30).

2.2.3.4 Comparative Studies on Precipitation

Comparative studies evaluating precipitation methods against other isolation strategies – such as ultracentrifugation, SEC, and affinity-based capture – provide critical insights. These investigations generally highlight precipitation as a technique that offers higher exosome yield but lower purity compared to more rigorous approaches like density gradient ultracentrifugation. While precipitation is advantageous for applications requiring large vesicle quantities, the consistent issue of protein contamination raises concerns for studies where exosome purity and integrity are essential (31).

2.2.3.5 Novel Polymers for Exosome Precipitation

Recent research has explored the use of alternative polymers, beyond conventional PEG, to improve exosome selectivity and reduce contamination. These novel polymers or polymer combinations are being tested for their ability to interact more specifically with exosome membranes while minimizing nonspecific binding of soluble proteins. This represents an important advancement in addressing precipitation's limitations. However, thorough evaluation across diverse biological samples and applications is still required. Additional considerations include potential cytotoxicity and interference with downstream biochemical or functional assays (13).

2.2.3.6 Precipitation Combined with Filtration

An emerging approach involves combining precipitation with filtration methods. Following exosome precipitation, filters with defined pore sizes are employed to remove larger contaminants or aggregates, enhancing the overall purity of the exosome preparation. This integration seeks to balance high precipitation yields with effective filtration size discrimination. While promising, this method is not without challenges. The choice of pore size is critical, as smaller exosomes may be lost during filtration, and soluble protein contamination often remains unresolved. Moreover, adding filtration increases protocol complexity (32).

2.2.3.7 Standardization of Precipitation Protocols

Given the variability in results obtained across different laboratories and sample types, efforts have been made to standardize precipitation-based exosome isolation protocols. Research has focused on defining optimal PEG concentrations, incubation conditions, and centrifugation parameters to enhance reproducibility. Such standardization is particularly crucial for comparative studies and for potential translation into clinical applications. Nonetheless, achieving universal standards remains challenging due to the inherent heterogeneity of biological fluids and the incomplete understanding of how polymers interact with exosomes at a molecular level (24).

2.2.4 Microfluidic and Immunoaffinity-Based Isolation

Recent advances in microfluidics have enabled high-throughput and automated exosome isolation (Table 2.3).

TABLE 2.3
Overview of Microfluidic and Immunoaffinity Techniques for Exosome Isolation

Research Area/Technique	Description	Advancements Reported	Limitations	References
Microfluidic-based isolation	Uses microchips (hydrodynamic trapping, lateral displacement, acoustic separation) for label-free exosome capture	High-throughput, automated, rapid processing; preserves exosome integrity	Limited scalability, costly fabrication, and clogging risk	(33, 34)
Immunoaffinity-based isolation	Antibodies against markers (CD63, CD81, CD9) immobilized on beads or surfaces for selective capture	High specificity, purer subsets; adaptable to microfluidics	Costly antibodies, limited scalability, and elution may affect vesicles	(34)
Advanced microfluidics	New principles like inertial focusing, viscoelastic flow, and dielectrophoresis	Improved purity, efficiency, and potential on-chip analysis	Needs expertise; scalability is still limited	(35)
Immunoaffinity with novel biomarkers	Uses additional markers (integrins, tetraspanins) for subtype capture	Enables study of specific subpopulations	Antibody availability and heterogeneity issues	(36)
Integrated microfluidic–immunoaffinity	Combines microfluidics with antibody capture in channels	Rapid, automated, specific isolation; streamlined workflows	Complex fabrication, antibody stability issues, and scaling are difficult	(37)
Liquid biopsy microfluidics	Microfluidics applied to blood/urine for diagnostics	Translational potential, point-of-care integration	Handling viscous samples, cost, and reproducibility	(38)
High-throughput immunoaffinity	Scaled systems (flow-based, high-density beads, multichannels)	Improves scalability while maintaining specificity	Needs optimization, antibody cost remains high	(39)
Multi-marker capture	Targets multiple markers simultaneously/sequentially	Higher specificity and ability to isolate diverse subtypes	Complex, costly, and requires careful antibody selection	(38)
Hybrid microfluidics	Combines multiple principles (e.g., inertial + acoustic)	Enhanced purity and resolution	Complex design and fabrication	(39)
Immunoaffinity for functional studies	Ensures vesicles remain biologically active post-isolation	Preserves exosome function for assays/therapeutics	Optimization of binding/elution is needed	(40)
Point-of-care microfluidics	Portable chips for direct exosome capture and detection	Early diagnosis has potential, is rapid, and is low-cost	Sensitivity, reliability, and regulatory hurdles	(41)
Isolation from complex fluids	Adapted methods for whole blood/viscous samples	Integrates pre-processing and enhances recovery	Fouling, nonspecific binding, and sample variability	(42)
Subpopulation isolation	Sequential/multiplex antibody-based separation of subsets	Enables study of specific functional roles	Rare subsets hard to isolate; cross-reactivity	(10)
Acoustic microfluidics	Uses acoustic forces for gentle, label-free separation	Preserves vesicle integrity, contact-free	Purity versus yield trade-off; scaling challenges	(43)

Microfluidic chips leverage techniques such as hydrodynamic trapping, deterministic lateral displacement, and acoustic manipulation to selectively capture exosomes from biofluids (32). Immunoaffinity methods, on the other hand, use antibodies targeting exosomal surface markers (CD63, CD81, CD9) to selectively enrich exosomes. While these approaches offer high specificity, their scalability remains a challenge for large-scale applications.

2.2.5 Emerging Techniques

Innovative methods such as acoustic trapping, electrophoretic isolation, and tangential flow filtration have shown promise in exosome research (also summarized in Table 2.4):

- Acoustic trapping uses sound waves to concentrate exosomes without altering their integrity (44).
- Electrophoretic separation exploits the charge properties of exosomes to isolate them in an electric field (45).
- Tangential flow filtration (TFF) is a scalable method that enhances exosome yield while maintaining purity.

2.2.6 Comparative Analysis of Exosome Isolation Methods

A comparison of exosome isolation techniques based on yield, purity, processing time, scalability, and reproducibility is summarized in Table 2.5.

2.3 CHARACTERIZATION STRATEGIES FOR EXOSOMES

Exosomes are small extracellular vesicles (EVs), typically ranging from 30 nm to 150 nm in diameter, secreted by various cell types. These vesicles are involved in intercellular communication and carry a cargo of proteins, lipids, and nucleic acids, reflecting the physiological or pathological state of the parent cell. Due to their potential as diagnostic biomarkers and therapeutic agents, comprehensive characterization of exosomes is essential. The main characterization strategies (Figure 2.1) can be grouped into morphological and structural analysis, size and concentration measurements, surface marker profiling, and functional analysis (58).

2.3.1 Morphological and Structural Characterization

2.3.1.1 Electron Microscopy (TEM, SEM, Cryo-EM)

Electron microscopy remains the gold standard for assessing the morphology of exosomes.

2.3.1.1.1 Transmission Electron Microscopy (TEM)

TEM offers high-resolution images, allowing visualization of the internal and external structures of exosomes. Exosomes appear as cup-shaped or spherical vesicles. The preparation often involves negative staining or ultrathin sectioning.

2.3.1.1.2 Scanning Electron Microscopy (SEM)

SEM provides detailed surface morphology and topography of exosomes. Although less commonly used due to sample

TABLE 2.4
Comparative Analysis of Emerging Techniques for Exosome Isolation and Characterization

Research Area/Technique	Description	Advancements Reported	Limitations	References
Acoustic trapping	Uses sound waves to trap and concentrate exosomes while preserving integrity	Label-free, gentle, maintains bioactivity, and has potential for microfluidic integration	Scalability issues: acoustic parameters must be optimized	(44)
Electrophoretic isolation	Separates exosomes by surface charge in an electric field	Label-free, size-independent, can isolate subpopulations	Scalability, heterogeneous charge, buffer-dependent, and purity concerns	(45)
Tangential flow filtration (TFF)	Parallel flow across a porous membrane retains exosomes while smaller molecules pass	Scalable, high yield, reduced fouling, suitable for large volumes	Membrane pore size critical; proteins may co-pass; may need extra purification	(46)
Acoustic trapping in microfluidics	Integrates acoustic trapping into chips for small-volume processing	Miniaturized, fast, reagent-saving, potential point-of-care	Requires precise control, risk of acoustic streaming, and limited scalability	(47)
Electrophoresis for subfractionation	Advanced electrophoresis separates exosomes by subtle charge differences	Resolves heterogeneity, isolates subtypes, and specialised buffers are developed	Sensitive to buffer pH/ionic strength; recovery optimization needed	(48)
Scalable TFF systems	Large-scale TFF for clinical/industrial biofluids	Enables therapeutic/biomarker-scale production; automated, optimized systems	Membrane fouling: pore size selection is critical; may need post-processing	(49)

TABLE 2.5
Comparison of RNA Isolation Methods: Efficiency, Cost, and Limitations

Method	Purity	Yield	Processing Time	Scalability	Cost	RNA Recovery	Key Limitations	References
Differential ultracentrifugation	Low	Medium	High (~6 hours)	Low	Low	Medium	Labor-intensive, low reproducibility; low-throughput, vesicle aggregation	(50)
Density-Gradient Ultracentrifugation	High	Low	High (~8 hours)	Low	Low	High	Time-consuming, limited scalability	(51, 52)
Size-exclusion chromatography (SEC)	High	Medium	Medium (~1 hour)	Low-medium	Medium	High	Low throughput for large sample volumes	(18, 53)
PEG precipitation	Low	High	Short (~1 hour)	High	Low	Low	Co-isolation of proteins, batch variability, and protein contamination	(25)
Immunoaffinity isolation	Very high	Low	Medium (~2–3 hours)	Low	High	High	Expensive, surface marker bias	(32, 54)
Microfluidic-based isolation	High	Medium	Short (~1 hour)	Medium	High	High	Costly, limited throughput	(55)
Acoustic isolation	High	Medium	Short (~30 minutes)	High	High	High	Requires specialised equipment	(44, 55)
Electrophoretic isolation	High	Medium	Short (~30 minutes)	Medium	Medium	Medium	High-voltage requirements, potential vesicle damage	(45)
Tangential flow filtration (TFF)	High	High	Short (~1 hour)	Very high	Medium	High	Cost of specialised membranes	(46)
Dielectrophoretic isolation	Very high	Medium	Short (~30 minutes)	Medium	High	High	Requires precise voltage control	(56)
Nanowire-based isolation	Very high	Medium	Short (~45 minutes)	Medium	High	High	Fabrication cost, potential vesicle modification	(57)

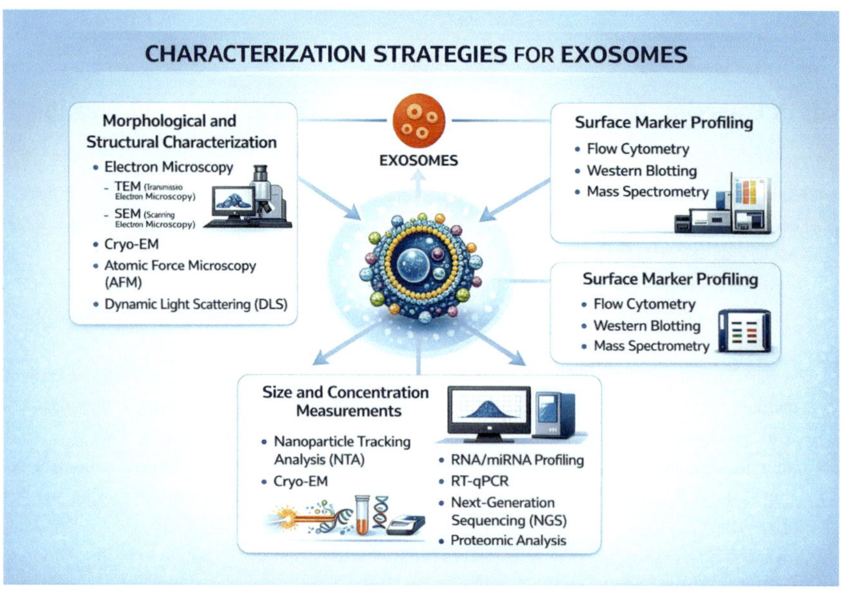

FIGURE 2.1 Various characterization strategies for exosomes.

dehydration and coating requirements, SEM gives a broader 3D perspective compared to TEM.

2.3.1.1.3 Cryo-Electron Microscopy (Cryo-EM)

Cryo-EM preserves the native structure of exosomes without the need for staining or fixation, offering near-native state visualization at nanometer resolution. It is particularly useful for understanding membrane architecture and vesicle fusion processes (59).

2.3.1.2 Atomic Force Microscopy (AFM)

AFM is a label-free technique that provides high-resolution topographic images and mechanical properties (e.g., stiffness, adhesion) of exosomes. AFM operates in a liquid environment, allowing observation of vesicles in conditions close to physiological. It can distinguish between different types of EVs based on elasticity and height measurements (60).

2.3.2 Size and Concentration Measurements

2.3.2.1 Nanoparticle Tracking Analysis (NTA)

NTA is a dynamic light scattering–based method that tracks the Brownian motion of individual exosomes in a laser-illuminated sample. It calculates the hydrodynamic diameter using the Stokes–Einstein equation and provides real-time size distribution and particle concentration.

- *Advantages:* Quantitative, real-time, high-throughput.
- *Limitations:* Cannot distinguish vesicles from similarly sized contaminants (61).

2.3.2.2 Dynamic Light Scattering (DLS)

DLS measures fluctuations in light scattering caused by particle motion in a solution to determine size distribution. It is quick and easy but less accurate for polydisperse samples due to a bias toward larger particles.

- *Advantages:* Fast, nondestructive.
- *Limitations:* Limited accuracy for heterogeneous populations like exosomes (62).

2.3.3 Surface Marker Profiling

Exosomes express specific surface markers such as tetraspanins (CD9, CD63, CD81), heat shock proteins, and integrins, which help identify their cellular origin.

2.3.3.1 Flow Cytometry

Traditionally used for cell analysis, flow cytometry has been adapted to analyze exosomes by either of the following:

- *Bead-based assays:* Exosomes are captured on antibody-coated beads and analyzed for surface markers.
- *High-resolution flow cytometry:* Newer instruments can directly detect single exosomes using fluorescence-based labeling (63).

2.3.3.2 Western Blotting

Western blotting is used to detect the presence of exosomal markers (e.g., TSG101, Alix, CD63). It is a semi-quantitative method to verify exosome enrichment and purity (64).

2.3.3.3 Mass Spectrometry (MS)

MS allows in-depth proteomic profiling of exosomes. It can identify and quantify thousands of proteins, including low-abundance biomarkers, offering insights into the biological functions of exosomes.

- *Techniques used:* LC-MS/MS, MALDI-TOF.
- *Applications:* Disease biomarker discovery, exosomal cargo profiling (65).

2.3.4 Functional Analysis

2.3.4.1 RNA/miRNA Profiling

Exosomes are rich in regulatory RNAs, including microRNAs (miRNAs), mRNAs, and long noncoding RNAs. Profiling these RNAs can help understand the signaling roles of exosomes and their utility in diagnostics.

- *Reverse transcription quantitative PCR (RT-qPCR)*
  Allows sensitive and quantitative analysis of selected RNA targets. Requires careful normalization and controls.
- *Next-generation sequencing (NGS):*
  Offers an unbiased and comprehensive view of the RNA content, including discovery of novel miRNAs and isoforms (66).

2.3.4.2 Proteomic Analysis

Proteomic profiling of exosomes involves the identification and quantification of their protein content to understand their role in cell communication, disease progression, and therapeutic potential.

- *Approaches*
  Label-free quantification, isobaric tagging (iTRAQ, TMT), and targeted proteomics (MRM/SRM).
- *Key insights*
  Proteomics helps in identifying disease-specific biomarkers and understanding mechanisms of disease, such as cancer metastasis or neurodegeneration (67).

2.4 EXOSOME ENGINEERING: ADVANCES AND APPLICATIONS

Exosome engineering has emerged as a promising field in nanomedicine, offering innovative solutions for drug delivery and therapeutic applications. Exosomes, naturally occurring nanosized vesicles, are engineered to enhance their capabilities in targeting specific tissues, improving drug delivery efficiency, and extending circulation time. These engineered exosomes (eEXOs) are being explored for their potential in treating a variety of diseases, including cancer, cardiovascular, neurological, and immune disorders.

2.4.1 Surface Modification Strategies

Bioengineered methods can easily modify the surface of exosomes. This strategy to modify exosomes is also known as the surface modification strategy. This strategy plays a very important role in the field of nanomedicine, including extended circulation, specific activity targeting, and therapeutic efficacy. Surface modifications of exosomes could be achieved using chemical modification, genetic engineering, or hybrid membrane engineering (68). In genetic engineering, as an appropriate technique for imparting exosomes with new properties, the targeting or ligand molecules are fused with the membrane proteins or lipids and subsequently overexpressed in the donor cells. The plasmid construction and protein overexpression in the donor cells are required for this technique (69). Unlike genetic engineering, chemical modification techniques can induce a large number of molecules using non-covalent or covalent interactions, which do not disrupt the exosome membrane. However, the complexity associated with membrane surfaces and issues related to additional steps of purification are key challenges that need to be addressed (70).

2.4.2 Genetic Engineering

An accessible method for introducing exosomes with new properties is genetic engineering. With gene modification techniques, it is possible to promote site-specific insertion, deletion, or modification at specific sites in the genome, which indirectly improves exosomal function (71). This is achieved by plasmid DNA or mRNA transfection of parental cells expressing genes of targeted fractions (such as peptides, receptors, and antibodies) that are fused to exosomal membrane components (such as tetraspanins and lysosome-associated membrane protein) (72). There is another approach to genetic engineering, and this approach involves the viral transduction-based strategy. This strategy is considered a delivery system, due to stable and definite transfection properties (73). Viral vectors like retrovirus, lentivirus, adenovirus, and adeno-associated virus have been extensively used as viral vectors for gene delivery. After virus infection, infected cells overexpress specific genes or regulate transcription, which could be loaded into exosomes. Following encapsulation, exosomes transport biologically active viral components to distant noninfectious cells (74). Viruses can enter the exosome biogenesis pathway, and viral RNA genome, microRNAs, and proteins are incorporated into exosomes (74). Therefore, by manipulating this process, exosomes can be modified to target the delivery of the drugs or genes of interest (75). Viral transduction is effective for delivering genetic material into a wide range of cells that are typically resistant to chemical transfection methods. However, this approach is often labor-intensive, time-consuming, and lacks a fully understood mechanism. Moreover, the potential risks of viral pathogenicity and teratogenicity in exosome-based applications warrant further investigation. Therefore, gene delivery vectors intended for clinical use must prioritize tissue specificity, non-immunogenicity, and nontoxicity as critical attributes (73).

2.4.3 Chemical and Physical Modifications

Chemical modification of exosomes involves functionalising their surface by attaching therapeutic molecules or other moieties through chemical reactions. This can enhance targeting, imaging, and other applications, such as increasing immune responses against tumor cells. Modified exosomes exhibit better passive/active targeting, cellular uptake, and immune evasion than natural exosomes (76).

Physically modifying exosomes involves altering their structure and properties using physical methods, such as magnetic manipulation, ultrasound, or mechanical stress, to enhance their functionality or therapeutic efficacy. These modifications can be utilized for targeted delivery, loading cargo, or tracking biodistribution (77, 78). Chemical and physical modifications of exosomes involve lipid conjugation and click chemistry as strategies for their modification.

2.4.3.1 Lipid Conjugation

Lipid conjugation is a crucial strategy in exosome engineering that allows for the functionalization of exosomal surfaces while maintaining their structural integrity. This technique takes advantage of the amphiphilic properties of lipids to integrate various functional molecules into the exosomal membrane. As a result, it enhances the therapeutic potential and targeting capabilities of exosomes (79, 80). One important method uses polyethylene glycol (PEG)-conjugated lipids, like DSPE-PEG, to attach targeting ligands to exosomes. Li et al. successfully incorporated aminoethyl anisamide (AA), a ligand for sigma receptors, into exosome membranes using DSPE-PEG. These engineered exosomes demonstrated a high targeting efficacy for sigma receptors, which are overexpressed in lung cancer cells. As a result, they efficiently accumulated in cancer cells following systemic administration, leading to prolonged survival in a mouse model of lung metastasis (80, 81). The incorporation of cell-penetrating peptides (CPPs) through lipid conjugation has shown potential for enhancing intracellular delivery. One study demonstrated that modifying exosomes with Tat peptide-conjugated PEG-lipids significantly improved their uptake by human umbilical vein endothelial cells (HUVECs). This highlights the effectiveness of lipid conjugation in facilitating targeted delivery (82). Lipid conjugation has been used to enhance the retention and therapeutic effectiveness of exosomes in specific tissues. For instance, Vandergriff et al. attached a cardiac-targeting peptide (CHP) to the surface of exosomes using a DOPE-NHS linker. This modification resulted in increased retention of the exosomes in the heart and a reduction in fibrosis after ischemia/reperfusion injury (80, 81).

2.4.3.2 Click Chemistry

Click chemistry, particularly the strain-promoted azide-alkyne cycloaddition (SPAAC), has emerged as a robust and biocompatible method for exosome surface modification. This copper-free reaction facilitates the covalent attachment of various functional molecules to exosomes under mild conditions, preserving their biological activity (83). A key application of click chemistry in exosome engineering is the metabolic incorporation of azide groups into the glycans on exosomal surfaces. This is followed by the conjugation of these glycans with molecules modified by dibenzocyclooctyne (DBCO) (84). One study employed this method to fluorescently label exosomes for real-time tracking both in vitro and in vivo (85). The labeled exosomes showed efficient uptake by target cells and successfully escaped from lysosomal confinement, highlighting their potential as delivery carriers for therapeutic agents (85, 86). Click chemistry has been used to improve the targeting capabilities of exosomes. Luo et al. modified exosomes by attaching azide-functionalized c(RGDyK) peptides, which have a high affinity for specific integrins that are overexpressed in tumor cells. The modified exosomes demonstrated enhanced targeting and therapeutic efficacy in a mouse model of ischemic brain injury (87). In a separate study, Jia et al. attached a neuropilin-1-targeted peptide (RGE) to the phosphatidylethanolamine on the exosomal membrane. This was accomplished using a cycloaddition reaction involving sulfonyl azide. This modification provided the exosomes with glioma-targeting properties, thereby enhancing their potential for targeted drug delivery (88). These examples illustrate the flexibility and effectiveness of click chemistry in exosome engineering, providing a valuable method for the creation of targeted and functionalized exosome-based therapeutics.

2.4.4 Cargo Loading Strategies

Cargo-loading exosomes are small extracellular vesicles secreted by various cell types and have become important tools for delivering therapeutic agents, including proteins, nucleic acids, and small molecules, to target cells (89). These vesicles are essential for intercellular communication, enabling the transfer of biological molecules and providing a novel approach for therapeutic applications (90). Recent studies indicate that the type of cargo, such as specific microRNAs (miRNAs) or proteins, can significantly affect the biological activity of exosomes and how they interact with recipient cells (91). The selective loading of miR-21 has been shown to improve the antitumor effects of exosomes in cancer therapy, indicating a potential for customized exosomal therapies that utilize specific biomolecular signatures (91). As research continues to advance, it is essential to understand the intricacies of cargo selection and loading processes for the development of more effective exosome-based therapies. Additionally, engineering exosomes to enhance their cargo-loading capacity is an emerging area of research. Innovative methodologies are being developed to improve both the efficiency and targeting specificity of exosomes. For instance, incorporating glycan-based targeting strategies has shown promise in facilitating the selective uptake of exosomes by specific cell types, thereby enhancing the therapeutic potential of the cargo delivered (92). Recent advancements in hybrid exosomes – combining natural exosomes with synthetic components – have improved loading efficiency and stability, paving the way for more effective therapeutic applications (93). These strategies not only enhance the functional payload of exosomes but also establish a foundation for personalized medicine, allowing treatments to be customized based on individual biomolecular profiles of patients, thereby improving clinical outcomes (89). As these technologies develop, they have the potential to transform exosome-based therapies, highlighting the need for ongoing research in this dynamic field. As we move toward clinical applications, combining exosome engineering with imaging techniques could greatly improve the precision of targeted therapies. This, in turn, may lead to better patient outcomes and a more personalized approach to treatment (94). As exosome research advances, the potential for developing combination therapies that utilize multiple types of cargo within a single exosome is becoming increasingly promising. This approach not only aims to enhance therapeutic efficacy by delivering synergistic agents but also addresses the challenge of achieving a more comprehensive treatment response for complex diseases such as cancer. For example, the simultaneous delivery of miRNAs and chemotherapeutic agents has shown encouraging results in preclinical models, where their combined effects led to increased apoptosis in tumor cells compared to treatments that used only a single agent (92). By utilizing exosomes to deliver various therapeutic agents, researchers can develop comprehensive treatment plans that are not only more effective but also customized to meet individual patients' needs, paving the way for advancements in personalized medicine (95). The ongoing research into exosome engineering and cargo loading strategies is vital for advancing targeted therapies. As these innovative methods develop, they are expected to lead to breakthroughs in personalized medicine and improved treatment outcomes for patients. Researchers are increasingly recognizing the potential of exosomes in addressing neurological disorders. Recent studies have shown that exosomes can effectively deliver therapeutic agents, such as siRNA, to target cells in the brain. This capability allows for the control of gene expression and opens up new possibilities for treating conditions like Alzheimer's disease and Parkinson's disease (92). By optimizing the selection of cargo and loading techniques, researchers are enhancing the therapeutic efficacy of exosomes while creating innovative treatment strategies that could greatly improve the quality of life for patients suffering from debilitating conditions. Furthermore, exosomes' ability to cross the blood–brain barrier allows for the delivery of diagnostic agents alongside therapeutic payloads, enabling a dual-function approach that could enhance patient management (96). This dual capability highlights exosomes' versatility and potential to revolutionize personalized medicine by providing tailored diagnostic and therapeutic solutions that adapt to the changing landscape of neurological disorders. Recent studies indicate that exosomes can deliver neuroprotective agents and gene-silencing therapies, such as small interfering RNAs (siRNAs), directly to affected neurons, thereby modulating the pathological processes associated with diseases like Alzheimer's and Parkinson's. Furthermore, the ability of exosomes to cross the blood–brain barrier increases their potential as vehicles for both therapeutic and diagnostic applications, allowing for simultaneous monitoring of disease progression through exosomal biomarkers (97). This dual capability represents a promising strategy for personalized medicine and highlights the transformative potential of exosomes in tackling complex neurological conditions. It paves the way for more effective and tailored treatment regimens that could significantly improve patient outcomes.

2.4.4.1 Passive and Active Loading Techniques

Exosomes are nanosized extracellular vesicles (30–150 nm) with an inherent ability to carry biomolecules such as proteins, lipids, and nucleic acids, making them attractive carriers for therapeutic delivery. However, natural exosomal cargo is limited by the donor cell's intrinsic biology. Thus, engineering techniques that allow efficient encapsulation of desired therapeutic agents are crucial. Broadly, these techniques are categorized into passive and active loading strategies.

2.4.4.1.1 Passive Loading Techniques

Passive loading is based on the simple diffusion of therapeutic agents into exosomes, driven by concentration gradients. This method typically involves incubating exosomes with a high concentration of a drug or nucleic acid, allowing spontaneous diffusion through the lipid bilayer (8). Hydrophobic small molecules such as curcumin and paclitaxel have been passively loaded into exosomes with moderate success (98, 99). This approach is advantageous for its simplicity, cost-effectiveness, and preservation of exosome structure. However, it suffers from low encapsulation efficiency, limited control over dosage, and poor reproducibility, especially for large or hydrophilic biomolecules (100).

2.4.4.1.1 Active Loading Techniques

To overcome the limitations of passive loading, *active loading* techniques apply physical or chemical stimuli to transiently disrupt exosomal membranes, thereby facilitating the incorporation of external cargos. Common active loading methods include *electroporation*, *sonication*, and *transfection*.

(a) **Electroporation:** Electroporation is one of the most widely used methods for loading nucleic acids, especially small interfering RNA (siRNA), into exosomes. It involves the application of a short electrical pulse to create temporary nanopores in the exosomal membrane, through which charged molecules can enter (101). Electroporation offers higher loading efficiency for hydrophilic molecules compared to passive methods, although it may lead to exosome aggregation and degradation of nucleic acids if not optimized (102). Proper buffer composition and pulse duration are critical for maintaining cargo stability and vesicle integrity (103).

(b) **Sonication:** Sonication utilizes ultrasound energy to transiently deform the exosomal membrane, enhancing drug diffusion into the vesicles. The mechanical shear forces and acoustic cavitation caused by ultrasound waves increase membrane permeability (104). Sonication is particularly effective for loading small molecules and proteins into exosomes. However, this method may alter the exosomal membrane structure and protein composition, potentially affecting biological activity and targeting efficiency. The sonication time, frequency, and amplitude must be carefully optimized to minimize exosome damage (105).

(c) **Transfection:** Transfection-based loading is primarily used to introduce genetic materials into exosomes by transfecting the parent (donor) cells. Common transfection agents include cationic lipids or polymers that facilitate the intracellular packaging of plasmid DNA, siRNA, or miRNA into exosomes during their biogenesis (106). Alternatively, exosome-mimetic vesicles can be generated from genetically engineered cells using extrusion or freeze–thaw cycles (107). While this method enables the incorporation of highly specific cargo, the use of transfection reagents may induce cytotoxicity or affect exosomal composition, warranting careful selection and validation (108).

2.5 APPLICATION IN DRUG DELIVERY

The unique biological origin and physicochemical properties of exosomes make them highly attractive as drug delivery vehicles. Their intrinsic ability to mediate intercellular communication, combined with their biocompatibility, low immunogenicity, and capacity to cross biological barriers, positions them as promising platforms for targeted therapy. Exosome-based drug delivery has been explored for a variety of cargo types, including small molecules, nucleic acids, proteins, and CRISPR-associated systems. Exosomes possess surface proteins such as integrins, tetraspanins (e.g., CD63, CD81), and adhesion molecules that naturally contribute to their homing ability toward specific tissues or cell types (109). This property can be leveraged for delivering therapeutic agents selectively to diseased tissues, such as tumors, minimizing systemic toxicity. For instance, exosomes derived from brain endothelial cells can traverse the blood–brain barrier (BBB), facilitating drug delivery to the central nervous system (110). Hydrophobic small-molecule drugs have shown excellent encapsulation within exosomes, particularly when loaded via passive diffusion or sonication. Exosome-encapsulated paclitaxel exhibited enhanced cytotoxic effects against multidrug-resistant cancer cells compared to free paclitaxel (111). Similarly, curcumin-loaded exosomes demonstrated improved bioavailability and anti-inflammatory activity in vitro and in vivo (112). The delivery of RNA-based therapeutics, including siRNA, miRNA, and mRNA, has seen significant advancement with exosome platforms. Electroporation has enabled the successful loading of siRNA into exosomes for targeted gene silencing. Alvarez-Erviti et al. (2011) demonstrated efficient delivery of BACE1 siRNA to neurons in mice using targeted exosomes, resulting in gene knockdown in the brain – a breakthrough for neurodegenerative disease therapy (110). Furthermore, engineered exosomes have been utilized to deliver miRNAs for cancer therapy, such as miR-21 inhibitors for glioblastoma (113). Exosomal delivery systems have also shown potential in CRISPR-Cas9 gene editing applications, where exosomes are used to deliver Cas9 mRNA and guide RNAs for precise genomic modulation (114). Exosomes can transport functional proteins and peptides either naturally or through active loading. This is particularly useful in immunotherapy, where exosomes derived from dendritic cells can present antigens to T cells and stimulate immune responses (115). Recombinant proteins have also been successfully loaded into exosomes for delivery to specific tissues, such as catalase for Parkinson's disease treatment (104). A major advantage of exosome-based delivery is its ability to bypass biological barriers that limit conventional drug carriers. Exosomes can cross the BBB, penetrate deep into tumor microenvironments, and evade phagocytosis due to their small size and endogenous origin (8). Their long circulation time and resistance to enzymatic degradation further support their utility in systemic delivery applications. Despite promising preclinical results, clinical translation of exosome-based drug delivery remains challenging. Standardization of isolation, characterization, and loading methods is crucial to ensure reproducibility and safety (108). Additionally, large-scale manufacturing, quality control, and regulatory guidelines need to be addressed before widespread clinical implementation. Nevertheless, several early-phase clinical trials are ongoing. For example, exosomes loaded with curcumin or KRAS-targeting siRNAs are under investigation for cancer therapy, highlighting growing interest in this platform (12).

2.5.1 Targeted Therapy

Targeted therapy aims to deliver therapeutics precisely to the affected tissues or cells, minimizing off-target effects and systemic toxicity. Exosomes, as natural nanocarriers, have shown promising potential in targeted delivery due to their biocompatibility, ability to cross physiological barriers, and inherent cell-targeting capabilities. Their engineering for enhanced targeting is enabling new paradigms in precision medicine, especially in diseases such as cancer, neurological disorders, and cardiovascular diseases. The application of engineered exosomes in oncology has gained substantial attention due to their ability to target tumor microenvironments, deliver therapeutic cargo, and evade immune detection. Exosomes derived from mesenchymal stem cells (MSCs), immune cells, or tumor cells themselves have been used to deliver chemotherapeutic agents, RNA therapeutics, and immune-modulating molecules to malignant tissues (7). For instance, paclitaxel-loaded exosomes have shown enhanced antitumor activity and reduced toxicity compared to free drug formulations in multidrug-resistant cancers (111). Exosomes decorated with targeting ligands such as folate or antibodies can recognize specific cancer cell receptors (e.g., EGFR, HER2) for selective uptake (116). Furthermore, the delivery of siRNA or miRNA targeting oncogenes via exosomes has led to effective tumor growth suppression in models of breast, lung, and pancreatic cancer (12). Immunoengineered exosomes have also shown potential in cancer immunotherapy. Dendritic cell–derived exosomes presenting tumor-associated antigens can activate cytotoxic T cells, providing a cell-free alternative to traditional cancer vaccines (115). One of the most challenging aspects of treating neurological diseases is the delivery of therapeutics across the blood–brain barrier (BBB). Exosomes naturally cross the BBB due to their nanosize and membrane composition, making them ideal candidates for central nervous system (CNS) drug delivery. Alvarez-Erviti et al. demonstrated that systemic administration of exosomes engineered with neuron-targeting peptides could deliver siRNA to the mouse brain, leading to gene knockdown of BACE1, a therapeutic target in Alzheimer's disease (110). Similarly, catalase-loaded exosomes administered intranasally have been shown to reduce neuroinflammation in Parkinson's disease models, illustrating their ability to improve antioxidant delivery to the brain (104). Exosomes containing neuroprotective miRNAs and anti-inflammatory agents have shown potential in models of traumatic brain injury, stroke, and epilepsy (117). These findings highlight the potential of exosome-based therapies to deliver functional materials for neuroprotection, neural repair, and gene silencing. In the context of cardiovascular diseases, exosome-mediated therapy focuses on promoting tissue repair, modulating inflammation, and delivering cardioprotective agents. Exosomes derived from cardiac cells have shown the ability to promote angiogenesis, reduce fibrosis, and enhance myocardial regeneration after ischemic injury (118). Research is investigating exosomes enriched with specific miRNAs (e.g., miR-21, miR-126, miR-210) for their potential roles in reducing apoptosis, promoting endothelial repair, and enhancing neovascularization following myocardial infarction (119). Exosomes derived from mesenchymal stem cells (MSCs), when administered intravenously, have demonstrated the ability to improve cardiac function and reduce the size of infarcts in preclinical models. The targeted delivery of these exosomes within the cardiovascular system can be enhanced through surface modifications. For instance, by conjugating exosomes with cardiac homing peptides or integrins, we can direct them specifically to ischemic myocardium. This targeting improves therapeutic localization and overall efficacy (120).

2.5.2 Regenerative Medicine and Vaccine Development

The multifaceted roles of exosomes in intercellular communication, immunomodulation, and tissue repair have positioned them as transformative agents in regenerative medicine and vaccinology. Their ability to carry nucleic acids, proteins, and lipids in a protected, biocompatible vesicular structure enables applications beyond traditional drug delivery. Exosomes offer innovative platforms for tissue regeneration and the development of next-generation vaccines due to their inherent biological compatibility, stability in circulation, and potential for surface engineering. Exosomes derived from stem cells, particularly mesenchymal stem cells (MSCs), have demonstrated regenerative capabilities across a wide array of tissues, including bone, cartilage, skin, cardiac, and neural tissues. These exosomes carry bioactive molecules – such as growth factors, cytokines, and microRNAs – that orchestrate tissue repair by promoting angiogenesis, modulating inflammation, and enhancing cell proliferation and differentiation (121). In cardiac regeneration, MSC-derived exosomes have been shown to reduce apoptosis, promote neovascularization, and improve cardiac function following myocardial infarction (122). Similarly, in cutaneous wound healing, exosomes enhance fibroblast proliferation, collagen synthesis, and epithelial regeneration (123). Neural regeneration is another area of intense research, with exosomes promoting axonal growth and neurogenesis after spinal cord injury and stroke (124). Moreover, bone marrow–derived exosomes have been used to treat osteoarthritis by inhibiting cartilage degradation and reducing inflammatory responses in joint tissues (125). The regenerative effects are mainly due to miRNAs like miR-21, miR-146a, and miR-29a, which regulate essential signaling pathways such as PI3K/AKT, Wnt/β-catenin, and NF-κB (126). Despite promising preclinical outcomes, challenges in scalability, standardization, and cargo-loading efficiency remain barriers to clinical translation. Nevertheless, tissue-engineered scaffolds integrated with exosomes are emerging as bioactive materials to enhance localized repair and regeneration (127). Exosomes have recently gained traction as novel vaccine delivery vehicles due to their natural antigen-presenting capabilities, ability to stimulate immune responses, and structural resemblance to viruses. Their lipid bilayer facilitates the display of pathogen-associated molecular patterns (PAMPs), making them effective in initiating both innate and adaptive immunity (128). Dendritic cell–derived exosomes (DEXs) are among the most studied in vaccine development. These exosomes present major histocompatibility complex (MHC) molecules and co-stimulatory signals necessary to activate T cells. Several clinical studies have demonstrated the safety and immunogenic potential of DEXs in cancer immunotherapy (129). For example, tumor antigen-loaded DEXs have been employed to prime cytotoxic T cells in melanoma and non-small cell lung cancer (130). The COVID-19 pandemic accelerated research into exosome-based vaccines. Engineered exosomes displaying SARS-CoV-2 spike proteins or carrying mRNA encoding viral antigens have been explored as stable, targeted vaccine platforms capable of inducing robust neutralizing antibody and T-cell responses (131). Furthermore, exosome-based vaccines have shown efficacy in preclinical models against infectious diseases like hepatitis B, tuberculosis, and HIV (132). Their advantages include reduced reactogenicity, enhanced cellular uptake, and the potential for oral or intranasal administration. Challenges include ensuring efficient antigen loading, avoiding immunosuppressive cargo from source cells, and scaling production under good manufacturing practices (GMP). However, continued advances in exosome isolation and engineering techniques are likely to overcome these limitations.

2.6 CONCLUSION

Recent technological advances have significantly enhanced the landscape of exosome research, particularly in the domains of isolation, characterization, and engineering. Improved isolation methods – ranging from ultracentrifugation and size-exclusion chromatography to novel microfluidic and immunoaffinity-based approaches – have enabled higher purity and yield of exosomes, critical for downstream applications. Characterization techniques have evolved with the integration of high-resolution tools such as nanoparticle tracking analysis, dynamic light scattering, and single-particle imaging, facilitating a better understanding of exosomal content and heterogeneity. Engineering strategies, including surface modification and cargo loading via electroporation, sonication, and genetic engineering, have transformed exosomes into efficient delivery vehicles for therapeutic agents. Future research should focus on standardising protocols, scaling up production, and enhancing targeting efficiency. The promising potential of exosomes in diagnostics, regenerative medicine, and targeted drug delivery paves the way for their clinical translation and commercialization, although regulatory challenges and manufacturing consistency remain critical hurdles to address.

ACKNOWLEDGEMENT

All the authors want to credit their affiliated institutes and Perpetual Association of Professional Innovations (PAPI)®, Durg, Chhattisgarh.

CONFLICT OF INTEREST

None.

AI DISCLOSURE STATEMENT

This chapter was edited with Grammarly to improve clarity and grammar. The content remains based on original research and analysis.

REFERENCES

[1] Doyle, L., & Wang, M. (2019). Overview of extracellular vesicles, their origin, composition, purpose, and methods for exosome isolation and analysis. *Cells*, ***8***(7), 727. https://doi.org/10.3390/cells8070727

[2] Kalluri, R., & LeBleu, V. S. (2020). The biology, function, and biomedical applications of exosomes. *Science (New York, N.Y.)*, ***367***(6478). https://doi.org/10.1126/SCIENCE.AAU6977

[3] Van Niel, G., D'Angelo, G., & Raposo, G. (2018). Shedding light on the cell biology of extracellular vesicles. *Nature Reviews. Molecular Cell Biology*, ***19***(4), 213–228. https://doi.org/10.1038/NRM.2017.125

[4] Théry, C., Witwer, K. W., Aikawa, E., Alcaraz, M. J., Anderson, J. D., Andriantsitohaina, R., . . . Zuba-Surma, E. K. (2018). Minimal information for studies of extracellular vesicles 2018 (MISEV2018): A position statement of the International Society for Extracellular Vesicles and update of the MISEV2014 guidelines. *Journal of Extracellular Vesicles*, ***7***(1), 1535750. https://doi.org/10.1080/20013078.2018.1535750

[5] Sen, S., Xavier, J., Kumar, N., Ahmad, M. Z., & Ranjan, O. P. (2023). Exosomes as natural nanocarrier-based drug delivery system: Recent insights and future perspectives. *3 Biotech*, ***13***(3), 101. https://doi.org/10.1007/s13205-023-03521-2

[6] Jeppesen, D. K., Fenix, A. M., Franklin, J. L., Higginbotham, J. N., Zhang, Q., Zimmerman, L. J., . . . Coffey, R. J. (2019). Reassessment of exosome composition. *Cell*, ***177***(2), 428–445.e18. https://doi.org/10.1016/j.cell.2019.02.029

[7] Whiteside, T. L. (2018). Exosome and mesenchymal stem cell cross-talk in the tumor microenvironment. *Seminars in Immunology*, ***35***, 69–79. https://doi.org/10.1016/j.smim.2017.12.003

[8] Luan, X., Sansanaphongpricha, K., Myers, I., Chen, H., Yuan, H., & Sun, D. (2017). Engineering exosomes as refined biological nanoplatforms for drug delivery. *Acta Pharmacologica Sinica*, ***38***(6), 754–763. https://doi.org/10.1038/APS.2017.12;KWRD

[9] Witwer, K. W., & Wolfram, J. (2021). Extracellular vesicles versus synthetic nanoparticles for drug delivery. *Nature Reviews Materials*, ***6***(2), 103–106. https://doi.org/10.1038/s41578-020-00277-6

[10] Li, P., Kaslan, M., Lee, S. H., Yao, J., & Gao, Z. (2017). Progress in exosome isolation techniques. *Theranostics*, ***7***(3), 789–804. https://doi.org/10.7150/thno.18133

[11] Thind, A., & Wilson, C. (2016). Exosomal miRNAs as cancer biomarkers and therapeutic targets. *Journal of Extracellular Vesicles*, ***5***(1), 31292. https://doi.org/10.3402/JEV.V5.31292

[12] Kamerkar, S., Lebleu, V. S., Sugimoto, H., Yang, S., Ruivo, C. F., Melo, S. A., . . . Kalluri, R. (2017). Exosomes facilitate therapeutic targeting of oncogenic KRAS in pancreatic cancer. *Nature*, ***546***(7659), 498–503. https://doi.org/10.1038/NATURE22341

[13] Livshts, M. A., Khomyakova, E., Evtushenko, E. G., Lazarev, V. N., Kulemin, N. A., Semina, S. E., . . . Govorun, V. M. (2015). Isolation of exosomes by differential centrifugation: Theoretical analysis of a commonly used protocol. *Scientific Reports*, ***5***(1), 1–14. https://doi.org/10.1038/SREP17319;TECHMETA

[14] Théry, C., Amigorena, S., Raposo, G., & Clayton, A. (2006). Isolation and characterization of exosomes from cell culture supernatants and biological fluids. *Current Protocols in Cell Biology*, ***30***(1). https://doi.org/10.1002/0471143030.CB0322S30

[15] Raposo, G., & Stoorvogel, W. (2013). Extracellular vesicles: Exosomes, microvesicles, and friends. *Journal of Cell Biology*, ***200***(4), 373–383. https://doi.org/10.1083/JCB.201211138

[16] Zabeo, D., Cvjetkovic, A., Lässer, C., Schorb, M., Lötvall, J., & Höög, J. L. (2017). Exosomes purified from a single cell type have diverse morphology. *Journal of Extracellular Vesicles*, ***6***(1), 1329476. https://doi.org/10.1080/20013078.2017.1329476

[17] Gardiner, C., Vizio, D. Di, Sahoo, S., Théry, C., Witwer, K. W., Wauben, M., & Hill, A. F. (2016). Techniques used for the isolation and characterization of extracellular vesicles: Results of a worldwide survey. *Journal of Extracellular Vesicles*, ***5***(1). https://doi.org/10.3402/JEV.V5.32945

[18] Böing, A. N., van der Pol, E., Grootemaat, A. E., Coumans, F. A. W., Sturk, A., & Nieuwland, R. (2014). Single-step isolation of extracellular vesicles by size-exclusion chromatography. *Journal of Extracellular Vesicles*, ***3***(1). https://doi.org/10.3402/JEV.V3.23430

[19] Hirschberg, Y., Schildermans, K., Dam, A. Van, Sterck, K., Boonen, K., Nelissen, I., . . . Mertens, I. (2021). Characterizing extracellular vesicles from cerebrospinal fluid by a novel size exclusion chromatography method. *Alzheimer's & Dementia*, ***17***(S5). https://doi.org/10.1002/ALZ.051264

[20] Yang, Y., Wang, Y., Wei, S., Zhou, C., Yu, J., Wang, G., . . . Zhao, L. (2021). Extracellular vesicles isolated by size-exclusion chromatography present suitability for RNomics analysis in plasma. *Journal of Translational Medicine*, ***19***(1), 1–12. https://doi.org/10.1186/S12967-021-02775-9/FIGURES/6

[21] Webber, J., & Clayton, A. (2013). How pure are your vesicles? *Journal of Extracellular Vesicles*, ***2***(1). https://doi.org/10.3402/JEV.V2I0.19861

[22] Bonner, S. E., van de Wakker, S. I., Phillips, W., Willms, E., Sluijter, J. P. G., Hill, A. F., . . . Vader, P. (2024). Scalable purification of extracellular vesicles with high yield and purity using multimodal flowthrough chromatography. *Journal of Extracellular Biology*, ***3***(2). https://doi.org/10.1002/JEX2.138

[23] Van Deun, J., Mestdagh, P., Sormunen, R., Cocquyt, V., Vermaelen, K., Vandesompele, J., . . . Hendrix, A. (2014). The impact of disparate isolation methods for extracellular vesicles on downstream RNA profiling. *Journal of Extracellular Vesicles*, ***3***(1). https://doi.org/10.3402/JEV.V3.24858
[24] Lobb, R. J., Becker, M., Wen, S. W., Wong, C. S. F., Wiegmans, A. P., Leimgruber, A., & Möller, A. (2015). Optimized exosome isolation protocol for cell culture supernatant and human plasma. *Journal of Extracellular Vesicles*, ***4***(1). https://doi.org/10.3402/JEV.V4.27031
[25] Rider, M. A., Hurwitz, S. N., & Meckes, D. G. (2016). ExtraPEG: A polyethylene glycol-based method for enrichment of extracellular vesicles. *Scientific Reports*, ***6***(1), 1–14. https://doi.org/10.1038/SREP23978;TECHMETA
[26] García-Romero, N., Madurga, R., Rackov, G., Palacín-Aliana, I., Núñez-Torres, R., Asensi-Puig, A., . . . Ayuso-Sacido, A. (2019). Polyethylene glycol improves current methods for circulating extracellular vesicle-derived DNA isolation. *Journal of Translational Medicine*, ***17***(1), 1–11. https://doi.org/10.1186/S12967-019-1825-3/TABLES/1
[27] Ding, M., Wang, C., Lu, X., Zhang, C., Zhou, Z., Chen, X., . . . Zhang, C. (2018). Comparison of commercial exosome isolation kits for circulating exosomal microRNA profiling. *Analytical and Bioanalytical Chemistry*, ***410***(16), 3805–3814. https://doi.org/10.1007/S00216-018-1052-4/METRICS
[28] Holcar, M., Ferdin, J., Sitar, S., Tušek-Žnidarič, M., Dolžan, V., Plemenitaš, A., . . . Lenassi, M. (2020). Enrichment of plasma extracellular vesicles for reliable quantification of their size and concentration for biomarker discovery. *Scientific Reports*, ***10***(1), 1–13. https://doi.org/10.1038/S41598-020-78422-Y;SUBJMETA
[29] Deregibus, M. C., Figliolini, F., D'Antico, S., Manzini, P. M., Pasquino, C., De Lena, M., . . . Camussi, G. (2016). Charge-based precipitation of extracellular vesicles. *International Journal of Molecular Medicine*, ***38***(5), 1359–1366. https://doi.org/10.3892/IJMM.2016.2759/HTML
[30] Taylor, D. D., & Shah, S. (2015). Methods of isolating extracellular vesicles impact down-stream analyses of their cargoes. *Methods*, ***87***, 3–10. https://doi.org/10.1016/J.YMETH.2015.02.019
[31] Pons Royo, M. del C., & Jungbauer, A. (2025). Polyethylene glycol precipitation: Fundamentals and recent advances. *Preparative Biochemistry & Biotechnology*, ***55***(8), 935–954. https://doi.org/10.1080/10826068.2025.2470220
[32] Lawrence, S. R., & Shah, K. M. (2024). Prospects and current challenges of extracellular vesicle-based biomarkers in cancer. *Biology*, ***13***(9), 694. https://doi.org/10.3390/BIOLOGY13090694
[33] Menck, K., Bleckmann, A., Schulz, M., Ries, L., & Binder, C. (2017). Isolation and characterization of microvesicles from peripheral blood. *Journal of Visualized Experiments: JoVE*, ***119***, 55057. https://doi.org/10.3791/55057
[34] Yang, F., Liao, X., Tian, Y., & Li, G. (2017). Exosome separation using microfluidic systems: Size-based, immunoaffinity-based and dynamic methodologies. *Biotechnology Journal*, ***12***(4), 1600699. https://doi.org/10.1002/BIOT.201600699
[35] The Human Protein Atlas. (n.d.). Retrieved September 13, 2025, from https://www.proteinatlas.org/
[36] Kang, H., Kim, J., & Park, J. (2017). Methods to isolate extracellular vesicles for diagnosis. *Micro and Nano Systems Letters*, ***5***(1), 1–11. https://doi.org/10.1186/S40486-017-0049-7/FIGURES/7
[37] Minami, S., Chikazu, D., Ochiya, T., & Yoshioka, Y. (2023). Extracellular vesicle-based liquid biopsies in cancer: Future biomarkers for oral cancer. *Translational Oncology*, ***38***, 101786. https://doi.org/10.1016/J.TRANON.2023.101786
[38] Panagopoulou, M. S., Wark, A. W., Birch, D. J. S., & Gregory, C. D. (2020). Phenotypic analysis of extracellular vesicles: A review on the applications of fluorescence. *Journal of Extracellular Vesicles*, ***9***(1). https://doi.org/10.1080/20013078.2019.1710020
[39] Nam, Y. H., Lee, S., Lee, S. K., Kim, J. H., & Park, J. H. (2024). Microfluidic chip with integrated separation, mixing, and concentration operations for rapid and sensitive bacterial detection utilizing synthetic inorganic antibodies. *Sensors and Actuators B: Chemical*, ***404***, 135202. https://doi.org/10.1016/J.SNB.2023.135202
[40] Gudbergsson, J. M., Johnsen, K. B., Skov, M. N., & Duroux, M. (2016). Systematic review of factors influencing extracellular vesicle yield from cell cultures. *Cytotechnology*, ***68***(4), 579–592. https://doi.org/10.1007/S10616-015-9913-6/METRICS
[41] Yang, S. M., Lv, S., Zhang, W., & Cui, Y. (2022). Microfluidic point-of-care (POC) devices in early diagnosis: A review of opportunities and challenges. *Sensors*, ***22***(4), 1620. https://doi.org/10.3390/S22041620
[42] Mukerjee, N., Bhattacharya, A., Maitra, S., Kaur, M., Ganesan, S., Mishra, S., . . . Thorat, N. D. (2025). Exosome isolation and characterization for advanced diagnostic and therapeutic applications. *Materials Today Bio*, ***31***, 101613. https://doi.org/10.1016/J.MTBIO.2025.101613
[43] Wu, M., Chen, C., Wang, Z., Bachman, H., Ouyang, Y., Huang, P. H., . . . Huang, T. J. (2019). Separating extracellular vesicles and lipoproteins via acoustofluidics. *Lab on a Chip*, ***19***(7), 1174–1182. https://doi.org/10.1039/C8LC01134F
[44] Evander, M., & Laurell, T. (2012). Acoustic trapping. *Encyclopedia of Nanotechnology*, 41–45. https://doi.org/10.1007/978-90-481-9751-4_424
[45] Králová, B. (1999). Electrophoretic methods for the isolation and characterization of enzymes. *Analytica Chimica Acta*, ***383***(1–2), 109–117. https://doi.org/10.1016/S0003-2670(98)00492-9
[46] Busatto, S., Vilanilam, G., Ticer, T., Lin, W. L., Dickson, D. W., Shapiro, S., . . . Wolfram, J. (2018). Tangential flow filtration for highly efficient concentration of extracellular vesicles from large volumes of fluid. *Cells*, ***7***(12), 273. https://doi.org/10.3390/CELLS7120273
[47] Tsan, V. hung, Fan, D., Caneva, S., Smith, C. S., & Verbiest, G. J. (2023). Low-cost acoustic force trap in a microfluidic channel. *HardwareX*, ***14***, e00428–e00428. https://doi.org/10.1016/J.OHX.2023.E00428
[48] Ouahabi, O. El, Salim, H., Pero-Gascon, R., & Benavente, F. (2021). A simple method for the analysis of extracellular vesicles enriched for exosomes from human serum by capillary electrophoresis with ultraviolet diode array detection. *Journal of Chromatography A*, ***1635***, 461752. https://doi.org/10.1016/J.CHROMA.2020.461752
[49] Scalable Tangential Flow Filtration (TFF) for Downstream Processing of Viral Vectors for Gene Therapy and Vaccine Production. (n.d.). Retrieved from https://www.sigmaaldrich.com/IN/en/technical-documents/technical-article/pharmaceutical-and-biopharmaceutical-manufacturing/viral-vector-downstream-processing/scalable-tangential-flow-filtration-downstream-processing
[50] Langevin, S. M., Kuhnell, D., Orr-Asman, M. A., Biesiada, J., Zhang, X., Medvedovic, M., & Thomas, H. E. (2019).

Balancing yield, purity and practicality: A modified differential ultracentrifugation protocol for efficient isolation of small extracellular vesicles from human serum. *RNA Biology*, ***16***(1), 5–12. https://doi.org/10.1080/15476286.2018.1564465

[51] Li, K., Wong, D. K., Hong, K. Y., & Raffai, R. L. (2018). Cushioned–density gradient ultracentrifugation (C-DGUC): A refined and high performance method for the isolation, characterization, and use of exosomes. *Methods in Molecular Biology*, ***1740***, 69–83. https://doi.org/10.1007/978-1-4939-7652-2_7

[52] Cvjetkovic, A., Lötvall, J., & Lässer, C. (2014). The influence of rotor type and centrifugation time on the yield and purity of extracellular vesicles. *Journal of Extracellular Vesicles*, ***3***(1). https://doi.org/10.3402/JEV.V3.23111

[53] Uliyanchenko, E. (2014). Size-exclusion chromatography – from high-performance to ultra-performance. *Analytical and Bioanalytical Chemistry*, ***406***(25), 6087–6094. https://doi.org/10.1007/S00216-014-8041-Z/METRICS

[54] Yarmush, M. L., Weiss, A. M., Antonsen, K. P., Odde, D. J., & Yarmush, D. M. (1992). Immunoaffinity purification: Basic principles and operational considerations. *Biotechnology Advances*, ***10***(3), 413–446. https://doi.org/10.1016/0734-9750(92)90303-Q

[55] Wu, Y., Wang, Y., Lu, Y., Luo, X., Huang, Y., Xie, T., . . . Zhang, J. (2022). Microfluidic technology for the isolation and analysis of exosomes. *Micromachines*, ***13***(10), 1571. https://doi.org/10.3390/MI13101571

[56] Boettcher, M., Jaeger, M. S., Riegger, L., Ducrée, J., Zengerle, R., & Duschl, C. (2006). Lab-on-chip-based cell separation by combining dielectrophoresis and centrifugation. *Biophysical Reviews and Letters*, ***01***(04), 443–451. https://doi.org/10.1142/S1793048006000306

[57] Fang, X., Wang, Y., Wang, S., & Liu, B. (2022). Nanomaterials assisted exosomes isolation and analysis towards liquid biopsy. *Materials Today Bio*, ***16***, 100371. https://doi.org/10.1016/J.MTBIO.2022.100371

[58] Gurunathan, S., Kang, M. H., Jeyaraj, M., Qasim, M., & Kim, J. H. (2019). Review of the isolation, characterization, biological function, and multifarious therapeutic approaches of exosomes. *Cells*, ***8***(4), 307. https://doi.org/10.3390/CELLS8040307

[59] Donoso-Quezada, J., Guajardo-Flores, D., & González-Valdez, J. (2020). Enhanced exosome-mediated delivery of black bean phytochemicals (*Phaseolus vulgaris* L.) for cancer treatment applications. *Biomedicine & Pharmacotherapy*, ***131***, 110771. https://doi.org/10.1016/J.BIOPHA.2020.110771

[60] Main, K. H. S., Provan, J. I., Haynes, P. J., Wells, G., Hartley, J. A., & Pyne, A. L. B. (2021). Atomic force microscopy – a tool for structural and translational DNA research. *APL Bioengineering*, ***5***(3). https://doi.org/10.1063/5.0054294

[61] Grobelny, J., DelRio, F. W., Pradeep, N., Kim, D. I., Hackley, V. A., & Cook, R. F. (2011). Size measurement of nanoparticles using atomic force microscopy. *Methods in Molecular Biology*, ***697***, 71–82. https://doi.org/10.1007/978-1-60327-198-1_7

[62] Zanetti-Ramos, B. G., Fritzen-Garcia, M. B., de Oliveira, C. S., Pasa, A. A., Soldi, V., Borsali, R., & Creczynski-Pasa, T. B. (2009). Dynamic light scattering and atomic force microscopy techniques for size determination of polyurethane nanoparticles. *Materials Science and Engineering: C*, ***29***(2), 638–640. https://doi.org/10.1016/J.MSEC.2008.10.040

[63] Surface Marker Analysis by Flow Cytometry – IN. (n.d.). Retrieved from https://www.thermofisher.com/in/en/home/references/protocols/neurobiology/neurobiology-protocols/surface-marker-analysis-by-flow-cytometry.html

[64] Mahmood, T., & Yang, P. C. (2012). Western blot: Technique, theory, and trouble shooting. *North American Journal of Medical Sciences*, ***4***(9), 429. https://doi.org/10.4103/1947-2714.100998

[65] Urban, P. L. (2016). Quantitative mass spectrometry: An overview. *Philosophical Transactions of the Royal Society A: Mathematical, Physical and Engineering Sciences*, ***374***(2079). https://doi.org/10.1098/RSTA.2015.0382

[66] Wang, Z., Gerstein, M., & Snyder, M. (2009). RNA-Seq: A revolutionary tool for transcriptomics. *Nature Reviews Genetics*, ***10***(1), 57–63. https://doi.org/10.1038/NRG2484;KWRD

[67] Schmidt, A., Forne, I., & Imhof, A. (2014). Bioinformatic analysis of proteomics data. *BMC Systems Biology*, ***8***(2), S3–S3. https://doi.org/10.1186/1752-0509-8-S2-S3/FIGURES/1

[68] Weng, Z., Zhang, B., Wu, C., Yu, F., Han, B., Li, B., & Li, L. (2021). Therapeutic roles of mesenchymal stem cell-derived extracellular vesicles in cancer. *Journal of Hematology and Oncology*, ***14***(1), 1–22. https://doi.org/10.1186/S13045-021-01141-Y/TABLES/5

[69] Liang, Y., Duan, L., Lu, J., & Xia, J. (2021). Engineering exosomes for targeted drug delivery. *Theranostics*, ***11***(7), 3183. https://doi.org/10.7150/THNO.52570

[70] Richardson, J. J., & Ejima, H. (2019). Surface engineering of extracellular vesicles through chemical and biological strategies. *Chemistry of Materials*, ***31***(7), 2191–2201. https://doi.org/10.1021/ACS.CHEMMATER.9B00050

[71] Damasceno, P. K. F., de Santana, T. A., Santos, G. C., Orge, I. D., Silva, D. N., Albuquerque, J. F., . . . Soares, M. B. P. (2020). Genetic engineering as a strategy to improve the therapeutic efficacy of mesenchymal stem/stromal cells in regenerative medicine. *Frontiers in Cell and Developmental Biology*, ***8***, 539056. https://doi.org/10.3389/FCELL.2020.00737/FULL

[72] Jia, X., Tang, J., Yao, C., & Yang, D. (2021). Recent progress of extracellular vesicle engineering. *ACS Biomaterials Science & Engineering*, ***7***(9), 4430–4438. https://doi.org/10.1021/ACSBIOMATERIALS.1C00868

[73] Chen, H., Wang, L., Zeng, X., Schwarz, H., Nanda, H. S., Peng, X., & Zhou, Y. (2021). Exosomes, a new star for targeted delivery. *Frontiers in Cell and Developmental Biology*, ***9***, 751079. https://doi.org/10.3389/FCELL.2021.751079/FULL

[74] Sancho-Albero, M., Medel-Martínez, A., & Martín-Duque, P. (2020). Use of exosomes as vectors to carry advanced therapies. *RSC Advances*, ***10***(40), 23975–23987. https://doi.org/10.1039/D0RA02414G

[75] Gilligan, K. E., & Dwyer, R. M. (2017). Engineering exosomes for cancer therapy. *International Journal of Molecular Sciences*, ***18***(6), 1122. https://doi.org/10.3390/IJMS18061122

[76] Akbari, A., Nazari-Khanamiri, F., Ahmadi, M., Shoaran, M., & Rezaie, J. (2022). Engineered exosomes for tumor-targeted drug delivery: A focus on genetic and chemical functionalization. *Pharmaceutics*, ***15***(1), 66. https://doi.org/10.3390/PHARMACEUTICS15010066

[77] Fu, S., Wang, Y., Xia, X., & Zheng, J. C. (2020). Exosome engineering: Current progress in cargo loading and targeted delivery. *NanoImpact*, ***20***, 100261. https://doi.org/10.1016/J.IMPACT.2020.100261

[78] Wu, D., Zhao, X., Xie, J., Yuan, R., Li, Y., Yang, Q., . . . Zhu, N. (2024). Physical modulation of mesenchymal stem

cell exosomes: A new perspective for regenerative medicine. *Cell Proliferation*, ***57***(8), e13630–e13630. https://doi.org/10.1111/CPR.13630
[79] Kundu, S., Guo, J., Islam, M. S., Rohokale, R., Jaiswal, M., & Guo, Z. (2025). A new strategy to functionalize exosomes via enzymatic engineering of surface glycans and its application to profile exosomal glycans and endocytosis. *Advanced Science*, ***12***(18), 2415942. https://doi.org/10.1002/ADVS.202415942
[80] Li, L., Wang, F., Zhu, D., Hu, S., Cheng, K., & Li, Z. (2025). Engineering exosomes and exosome-like nanovesicles for improving tissue targeting and retention. *Fundamental Research*, ***5***(2), 851–867. https://doi.org/10.1016/J.FMRE.2024.03.025
[81] Jung, I., Shin, S., Baek, M. C., & Yea, K. (2024). Modification of immune cell-derived exosomes for enhanced cancer immunotherapy: Current advances and therapeutic applications. *Experimental and Molecular Medicine*, ***56***(1), 19–31. https://doi.org/10.1038/S12276-023-01132-8;TECHMETA
[82] Huang, T., Sato, Y., Kuramochi, A., Ohba, Y., Sano, M., Miyagishi, M., . . . Teramura, Y. (2023). Surface modulation of extracellular vesicles with cell-penetrating peptide-conjugated lipids for improvement of intracellular delivery to endothelial cells. *Regenerative Therapy*, ***22***, 90–98. https://doi.org/10.1016/J.RETH.2022.12.007
[83] N'diaye, E. R., Orefice, N. S., Ghezzi, C., & Boumendjel, A. (2022). Chemically modified extracellular vesicles and applications in radiolabeling and drug delivery. *Pharmaceutics*, ***14***(3), 653. https://doi.org/10.3390/PHARMACEUTICS14030653
[84] Mukerjee, N., & Sonar, S. (2024). Signature of click chemistry in exosome modification for cancer therapeutic. *Clinical and Translational Discovery*, ***4***(4), e335–e335. https://doi.org/10.1002/CTD2.335
[85] Song, S., Shim, M. K., Lim, S., Moon, Y., Yang, S., Kim, J., . . . Kim, K. (2020). In situ one-step fluorescence labeling strategy of exosomes via bioorthogonal click chemistry for real-time exosome tracking in vitro and in vivo. *Bioconjugate Chemistry*, ***31***(5), 1562–1574. https://doi.org/10.1021/ACS.BIOCONJCHEM.0C00216
[86] González, M. I., González-Arjona, M., Santos-Coquillat, A., Vaquero, J., Vázquez-Ogando, E., de Molina, A., . . . Salinas, B. (2021). Covalently labeled fluorescent exosomes for in vitro and in vivo applications. *Biomedicines*, ***9***(1), 81. https://doi.org/10.3390/BIOMEDICINES9010081
[87] Luo, H., Huang, Z., & Shi, Y. (2023). Functionalized exosome-loaded ginsenoside Rg1 for the treatment of ischemic stroke. *Sheng wu gong cheng xue bao = Chinese Journal of Biotechnology*, ***39***(1), 275–285. https://doi.org/10.13345/j.cjb.220253
[88] Jia, G., Han, Y., An, Y., Ding, Y., He, C., Wang, X., & Tang, Q. (2018). NRP-1 targeted and cargo-loaded exosomes facilitate simultaneous imaging and therapy of glioma in vitro and in vivo. *Biomaterials*, ***178***, 302–316. https://doi.org/10.1016/J.BIOMATERIALS.2018.06.029
[89] Kim, H., Jang, H., Cho, H., Choi, J., Hwang, K. Y., Choi, Y., . . . Yang, Y. (2021). Recent advances in exosome-based drug delivery for cancer therapy. *Cancers*, ***13***(17), 4435. https://doi.org/10.3390/CANCERS13174435
[90] Wallen, M., Aqil, F., Spencer, W., & Gupta, R. C. (2023). Exosomes as an emerging plasmid delivery vehicle for gene therapy. *Pharmaceutics*, ***15***(7), 1832. https://doi.org/10.3390/PHARMACEUTICS15071832
[91] Alli, A. A. (2022). *Mechanisms of Extracellular Vesicle Biogenesis, Cargo Loading, and Release*. IntechOpen. https://doi.org/10.5772/intechopen.100458
[92] Bahadorani, M., Nasiri, M., Dellinger, K., Aravamudhan, S., & Zadegan, R. (2024). Engineering exosomes for therapeutic applications: Decoding biogenesis, content modification, and cargo loading strategies. *International Journal of Nanomedicine*, ***19***, 7137–7164. https://doi.org/10.2147/IJN.S464249
[93] Zeng, H., Guo, S., Ren, X., Wu, Z., Liu, S., & Yao, X. (2023). Current strategies for exosome cargo loading and targeting delivery. *Cells*, ***12***(10), 1416. https://doi.org/10.3390/CELLS12101416
[94] Chen, X., Wang, K., Liang, F., Jiang, R., Sun, J., Chen, X., . . . Sun, J. (2022). Exosomes drug delivery systems and their application in tumor treatment. *Progress in Chemistry*, ***34***(4), 773–786. https://doi.org/10.7536/PC210901
[95] Chakraborty, A., Mazumdar, S., Rizvi, S., & Rashid, G. (2024). Advancement of engineered exosomes to deliver therapeutic protein cargos in brain disorders. *Exosomes Based Drug Delivery Strategies for Brain Disorders*, 257–272. https://doi.org/10.1007/978-981-99-8373-5_10
[96] Muteeb, G., Zia, Q., & Alshoaibi, A. (2023). Engineered exosomes as nano-vectors against neurodegenerative disorders. *Applications of Stem Cells and Derived Exosomes in Neurodegenerative Disorders*, 291–327. https://doi.org/10.1007/978-981-99-3848-3_12
[97] Munson, P., & Shukla, A. (2015). Exosomes: Potential in cancer diagnosis and therapy. *Medicines (Basel, Switzerland)*, ***2***(4), 310–327. https://doi.org/10.3390/MEDICINES2040310
[98] Kar, R., Dhar, R., Mukherjee, S., Nag, S., Gorai, S., Mukerjee, N., . . . Thorat, N. D. (2023). Exosome-based smart drug delivery tool for cancer theranostics. *ACS Biomaterials Science & Engineering*, ***9***(2), 577–594. https://doi.org/10.1021/ACSBIOMATERIALS.2C01329
[99] Oskouie, M. N., Aghili Moghaddam, N. S., Butler, A. E., Zamani, P., & Sahebkar, A. (2019). Therapeutic use of curcumin-encapsulated and curcumin-primed exosomes. *Journal of Cellular Physiology*, ***234***(6), 8182–8191. https://doi.org/10.1002/JCP.27615
[100] Ha, D., Yang, N., & Nadithe, V. (2016). Exosomes as therapeutic drug carriers and delivery vehicles across biological membranes: Current perspectives and future challenges. *Acta Pharmaceutica Sinica B*, ***6***(4), 287–296. https://doi.org/10.1016/j.apsb.2016.02.001
[101] Kooijmans, S. A. A., Stremersch, S., Braeckmans, K., De Smedt, S. C., Hendrix, A., Wood, M. J. A., . . . Vader, P. (2013). Electroporation-induced siRNA precipitation obscures the efficiency of siRNA loading into extracellular vesicles. *Journal of Controlled Release: Official Journal of the Controlled Release Society*, ***172***(1), 229–238. https://doi.org/10.1016/J.JCONREL.2013.08.014
[102] Lamichhane, T. N., Raiker, R. S., & Jay, S. M. (2015). Exogenous DNA loading into extracellular vesicles via electroporation is size-dependent and enables limited gene delivery. *Molecular Pharmaceutics*, ***12***(10), 3650–3657. https://doi.org/10.1021/ACS.MOLPHARMACEUT.5B00364
[103] Johnsen, K. B., Gudbergsson, J. M., Skov, M. N., Pilgaard, L., Moos, T., & Duroux, M. (2014). A comprehensive overview of exosomes as drug delivery vehicles – endogenous nanocarriers for targeted cancer therapy. *Biochimica et Biophysica Acta – Reviews on Cancer*, ***1846***(1), 75–87. https://doi.org/10.1016/j.bbcan.2014.04.005

[104] Haney, M. J., Klyachko, N. L., Zhao, Y., Gupta, R., Plotnikova, E. G., He, Z., . . . Batrakova, E. V. (2015). Exosomes as drug delivery vehicles for Parkinson's disease therapy. *Journal of Controlled Release: Official Journal of the Controlled Release Society*, ***207***, 18–30. https://doi.org/10.1016/J.JCONREL.2015.03.033

[105] Colja, S., Jovčevska, I., Šamec, N., Romih, R., & Zottel, A. (2023). Sonication is a suitable method for loading nanobody into glioblastoma small extracellular vesicles. *Heliyon*, ***9***(5). https://doi.org/10.1016/j.heliyon.2023.e15674

[106] El Andaloussi, S., Mäger, I., Breakefield, X. O., & Wood, M. J. A. (2013). Extracellular vesicles: Biology and emerging therapeutic opportunities. *Nature Reviews. Drug Discovery*, ***12***(5), 347–357. https://doi.org/10.1038/NRD3978

[107] Jang, S. C., Kim, O. Y., Yoon, C. M., Choi, D. S., Roh, T. Y., Park, J., . . . Gho, Y. S. (2013). Bioinspired exosome-mimetic nanovesicles for targeted delivery of chemotherapeutics to malignant tumors. *ACS Nano*, ***7***(9), 7698–7710. https://doi.org/10.1021/NN402232G

[108] Lener, T., Gimona, M., Aigner, L., Börger, V., Buzas, E., Camussi, G., . . . Giebel, B. (2015). Applying extracellular vesicles based therapeutics in clinical trials – an ISEV position paper. *Journal of Extracellular Vesicles*, ***4***(1). https://doi.org/10.3402/JEV.V4.30087

[109] Mathieu, M., Martin-Jaular, L., Lavieu, G., & Théry, C. (2019). Specificities of secretion and uptake of exosomes and other extracellular vesicles for cell-to-cell communication. *Nature Cell Biology*, ***21***(1), 9–17. https://doi.org/10.1038/S41556-018-0250-9

[110] Alvarez-Erviti, L., Seow, Y., Yin, H., Betts, C., Lakhal, S., & Wood, M. J. A. (2011). Delivery of siRNA to the mouse brain by systemic injection of targeted exosomes. *Nature Biotechnology*, ***29***(4), 341–345. https://doi.org/10.1038/NBT.1807

[111] Kim, M. S., Haney, M. J., Zhao, Y., Mahajan, V., Deygen, I., Klyachko, N. L., . . . Batrakova, E. V. (2016). Development of exosome-encapsulated paclitaxel to overcome MDR in cancer cells. *Nanomedicine: Nanotechnology, Biology, and Medicine*, ***12***(3), 655–664. https://doi.org/10.1016/J.NANO.2015.10.012

[112] Sun, D., Zhuang, X., Xiang, X., Liu, Y., Zhang, S., Liu, C., . . . Zhang, H. G. (2010). A novel nanoparticle drug delivery system: The anti-inflammatory activity of curcumin is enhanced when encapsulated in exosomes. *Molecular Therapy*, ***18***(9), 1606–1614. https://doi.org/10.1038/mt.2010.105

[113] Ohno, S. I., Takanashi, M., Sudo, K., Ueda, S., Ishikawa, A., Matsuyama, N., . . . Kuroda, M. (2013). Systemically injected exosomes targeted to EGFR deliver antitumor microrna to breast cancer cells. *Molecular Therapy*, ***21***(1), 185–191. https://doi.org/10.1038/mt.2012.180

[114] Dalmizrak, A., & Dalmizrak, O. (2022). Mesenchymal stem cell-derived exosomes as new tools for delivery of miRNAs in the treatment of cancer. *Frontiers in Bioengineering and Biotechnology*, ***10***. https://doi.org/10.3389/FBIOE.2022.956563

[115] Viaud, S., Théry, C., Ploix, S., Tursz, T., Lapierre, V., Lantz, O., . . . Chaput, N. (2010). Dendritic cell-derived exosomes for cancer immunotherapy: What's next? *Cancer Research*, ***70***(4), 1281–1285. https://doi.org/10.1158/0008-5472.CAN-09-3276

[116] Tian, Y., Li, S., Song, J., Ji, T., Zhu, M., Anderson, G. J., . . . Nie, G. (2014). A doxorubicin delivery platform using engineered natural membrane vesicle exosomes for targeted tumor therapy. *Biomaterials*, ***35***(7), 2383–2390. https://doi.org/10.1016/j.biomaterials.2013.11.083

[117] Xin, H., Li, Y., Cui, Y., Yang, J. J., Zhang, Z. G., & Chopp, M. (2013). Systemic administration of exosomes released from mesenchymal stromal cells promote functional recovery and neurovascular plasticity after stroke in rats. *Journal of Cerebral Blood Flow and Metabolism: Official Journal of the International Society of Cerebral Blood Flow and Metabolism*, ***33***(11), 1711–1715. https://doi.org/10.1038/JCBFM.2013.152

[118] Saludas, L., Oliveira, C. C., Roncal, C., Ruiz-Villalba, A., Prósper, F., Garbayo, E., & Blanco-Prieto, M. J. (2021). Extracellular vesicle-based therapeutics for heart repair. *Nanomaterials (Basel, Switzerland)*, ***11***(3), 1–30. https://doi.org/10.3390/NANO11030570

[119] Gong, M., Yu, B., Wang, J., Wang, Y., Liu, M., Paul, C., . . . Xu, M. (2017). Mesenchymal stem cells release exosomes that transfer miRNAs to endothelial cells and promote angiogenesis. *Oncotarget*, ***8***(28), 45200–45212. https://doi.org/10.18632/ONCOTARGET.16778

[120] Mentkowski, K. I., & Lang, J. K. (2019). Exosomes engineered to express a cardiomyocyte binding peptide demonstrate improved cardiac retention in vivo. *Scientific Reports*, ***9***(1). https://doi.org/10.1038/S41598-019-46407-1

[121] Phinney, D. G., & Pittenger, M. F. (2017). Concise review: MSC-derived exosomes for cell-free therapy. *Stem Cells (Dayton, Ohio)*, ***35***(4), 851–858. https://doi.org/10.1002/STEM.2575

[122] Lai, R. C., Arslan, F., Lee, M. M., Sze, N. S. K., Choo, A., Chen, T. S., . . . Lim, S. K. (2010). Exosome secreted by MSC reduces myocardial ischemia/reperfusion injury. *Stem Cell Research*, ***4***(3), 214–222. https://doi.org/10.1016/j.scr.2009.12.003

[123] Yang, G., Waheed, S., Wang, C., Shekh, M., Li, Z., & Wu, J. (2023). Exosomes and their bioengineering strategies in the cutaneous wound healing and related complications: Current knowledge and future perspectives. *International Journal of Biological Sciences*, ***19***(5), 1430–1454. https://doi.org/10.7150/IJBS.80430

[124] Sun, G., Li, G., Li, D., Huang, W., Zhang, R., Zhang, H., . . . Wang, B. (2018). hucMSC derived exosomes promote functional recovery in spinal cord injury mice via attenuating inflammation. *Materials Science and Engineering: C*, ***89***, 194–204. https://doi.org/10.1016/J.MSEC.2018.04.006

[125] Cosenza, S., Ruiz, M., Toupet, K., Jorgensen, C., & Noël, D. (2017). Mesenchymal stem cells derived exosomes and microparticles protect cartilage and bone from degradation in osteoarthritis. *Scientific Reports*, ***7***(1). https://doi.org/10.1038/S41598-017-15376-8

[126] Ghafouri-Fard, S., Abak, A., Tavakkoli Avval, S., Rahmani, S., Shoorei, H., Taheri, M., & Samadian, M. (2021). Contribution of miRNAs and lncRNAs in osteogenesis and related disorders. *Biomedicine and Pharmacotherapy*, ***142***. https://doi.org/10.1016/j.biopha.2021.111942

[127] Shafiei, M., Ansari, M. N. M., Razak, S. I. A., & Khan, M. U. A. (2021). A Comprehensive review on the applications of exosomes and liposomes in regenerative medicine and tissue engineering. *Polymers*, ***13***(15). https://doi.org/10.3390/POLYM13152529

[128] Tan, D., Li, G., Fu, W., & Lei, C. (2024). Exosomes: The next frontier in vaccine development and delivery. *Frontiers in Immunology*, ***15***, 1435426. https://doi.org/10.3389/FIMMU.2024.1435426

[129] Luo, S., Chen, J., Xu, F., Chen, H., Li, Y., & Li, W. (2023). Dendritic cell-derived exosomes in cancer immunotherapy. *Pharmaceutics*, ***15***(8), 2070. https://doi.org/10.3390/PHARMACEUTICS15082070

[130] Besse, B., Charrier, M., Lapierre, V., Dansin, E., Lantz, O., Planchard, D., . . . Chaput, N. (2016). Dendritic cell-derived exosomes as maintenance immunotherapy after first line chemotherapy in NSCLC. *OncoImmunology*, ***5***(4). https://doi.org/10.1080/2162402X.2015.1071008

[131] Kim, J., & Thapa, N. (2023). Exosome-based COVID-19 vaccine. *Methods in Molecular Biology (Clifton, N.J.)*, ***2668***, 301–311. https://doi.org/10.1007/978-1-0716-3203-1_20

[132] Santos, P., & Almeida, F. (2021). Exosome-based vaccines: History, current state, and clinical trials. *Frontiers in Immunology*, ***12***. https://doi.org/10.3389/FIMMU.2021.711565

3 Essential Concepts of Exosome Kinetics

Overview and Biological Relevance

Krishna Yadav[1,2]
[1]Rungta College of Pharmaceutical Sciences and Research, Bhilai, Chhattisgarh, India
[2]School of Pharmacy, Rungta International Skills University, Bhilai, Chhattisgarh, India

3.1 INTRODUCTION

Exosome pharmacokinetics is the study of the absorption, distribution, metabolism, and excretion processes governing the behavior of exosomes *in vivo* when administered as therapeutic agents. Therapeutically administered exosomes differ from those of conventional small-molecule drugs and even larger biologics; exosomes are naturally produced nanocarriers that exhibit unique pharmacokinetic properties inspired by their biological origin and structural nature [1].

The pharmacokinetic behavior for exosomes can primarily be attributed to their nanoscale size (30–150 nm), lipid bilayer membrane structure, and range of surface protein molecules determining their interaction with biological systems [2]. Studies indicate exogenous exosomes have rapid elimination from systemic circulation, with half-lives typically ranging from 2 minutes to 30 minutes in most preclinical models. This unique excretion pattern indicates that exosomes can be readily eliminated from systemic circulation, which is completely different from typical nanoparticle behavior and is a significant barrier to their use for therapeutics [3].

At present, pharmacokinetic studies for exosomes use practices such as labeling fluorescence dyes (e.g., DiR, PKH26), radionuclides (89Zr, 99mTc), and protein-based markers, most often to identify and track biodistribution and clearance kinetics [4–6]. These labeled approaches may alter the behavior of exosomes, and, therefore, there is an unmet need for label-free approaches, such as LC-MS/MS-based quantification using known proteins in exosomes as biomarkers [7].

Exosome pharmacokinetics differ from conventional pharmacokinetics in many important dimensions. The distribution, absorption, and elimination of small-molecule drugs are typically recognized to be first-order kinetics with predictable trends based on their physicochemical properties (lipophilicity, molecular weight, protein binding, etc.) [8]. Exosomes, however, have multi- and biphasic kinetics that are much more variable due to their biological recognition or use of cellular uptake mechanisms and host cell immunological response.

Additionally, exosome elimination is more complicated than that of standard monoclonal antibodies or other protein therapeutics that are eliminated only through receptor-mediated clearance or incidental proteolytic degradation [9]. In contrast, exosomes are eliminated through several pathways involving endocytosis by mononuclear phagocyte system (MPS) cells, uptake by target tissues caused by cells internalizing the exosomes, or degradation in endolysosomes. These multimodal clearance mechanisms lead to complicated pharmacokinetics that are not well approximated by standard compartmental pharmacokinetics.

The volume of distribution of an exosome is also dramatically different than small-molecule drugs. Small-molecule drug distribution could be viewed as equivalent dispersion either through the total body water or compartments of specific related tissue, based predominantly on their lipophilicity [10]. In contrast, an exosome's distribution would often only be allowed through compartments of its vascular space and organs related to the MPS, and/or penetration into extravascular spaces and tissues could be very limited under physiological conditions [2, 11].

Understanding exosome kinetics is critical for maximizing the therapeutic efficacy and safety profile of these therapies. The pharmacokinetic properties of exosomes dictate how well they can reach target tissues, deliver their therapeutic payloads, and avoid off-target effects. Besides, exosome binding and subsequent exosome accumulation in the liver and spleen restrict the availability of exosomes for tumor-targeting applications [5, 12].

In addition, exosome kinetics can affect how useful exosomes can be used as biomarkers for diagnostics. For a circulating exosome to be a useful diagnostic, it should provide sufficient stability and a detectable level in biological fluids to capture pathology related to its tissue of origin [13–15]. Research has shown that exosome-transported miRNAs, such as miR-155, will have their own half-life and tissue distribution characteristics that limit their functional effects in target cells [16].

The kinetic properties of exosomes also have relevance for personalized approaches. In patients, their unique

DOI: 10.1201/9781003623632-3

metabolism, immune status, and organs may affect exosome pharmacokinetics, making it vital to manage patient-specific dosages and monitoring protocols. This chapter presents a focused, recent overview of pertinent studies on the *in vivo* kinetics and therapeutics of exosomes, analyzing how administration routes, biodistribution patterns, circulation stability, cellular uptake mechanisms, and clearance pathways influence therapeutic efficacy and safety.

3.2 ABSORPTION AND ADMINISTRATION ROUTES

3.2.1 Intravenous versus Intranasal versus Oral versus Local Delivery

The route of administration plays an important role in the pharmacokinetic behavior of exosomes, as each uptake route has its own advantages and disadvantages. Intravenous administration is the most straightforward method with obvious systemic bioavailability; however, it also leads to rapid elimination by the MPS organs [17]. Administration of exosomes via intravenous injection resulted in a peak concentration of exosomes within 10 minutes, following which exosomes were rapidly disseminated to the liver and spleen, with elimination half-lives of 2–30 minutes [10].

Intranasal administration is an increasingly used alternative approach, especially for central nervous system applications [12]. This route of administration uses both the olfactory and trigeminal nerve pathways and allows for direct nose-to-brain transport, resulting in bypass of the blood–brain barrier. In a few research studies, intranasally administered exosomes resulted in selective uptake in the brain with minimal systemic uptake. For instance, mesenchymal stem cell–derived exosomes administered intranasally preferentially disseminated to the olfactory bulb, striatum, and regions of the midbrain, within hours of administration [18].

The oral delivery of exosomes has logistical issues relating to gastrointestinal stability and absorption. Although plant-derived exosome-like nanoparticles have been reported to exhibit some oral bioavailability, mammalian exosomes can be digested in the acidic conditions of the stomach [19, 20]. Nevertheless, exosomes with specialized formulations and targeting strategies may be able to enhance oral absorption; for example, milk-derived exosomes have been well-documented to transit the whole gastrointestinal tract to gain access to the systemic circulation [21].

Local delivery approaches, such as intratumoral, intraocular, or topical administration, will have the advantage of achieving a high local exposure while mitigating systemic exposure. The local routes of delivery are ideally suited to localized applications where an extended effect at the local site, or local systemic toxicity from the alleviation of a systemic indication, is a concern [19, 20].

3.2.2 Barriers to Systemic and Tissue Absorption

Exosomes face a number of biological obstacles to absorption and distribution after administration [22]. In the case of neurotherapeutics, the blood–brain barrier is a major obstacle, preventing systemically administered exosomes from penetrating into the brain parenchyma. However, some populations of exosomes may be able to cross the blood–brain barrier by transcytosis [23, 24].

The reticuloendothelial system, currently termed the MPS, is the primary barrier to exosome distribution and delivery into the tissues [11]. Kupffer cells in the liver, splenic macrophages, and alveolar macrophages have the ability to rapidly identify and internalize circulating exosomes by various mechanisms, including (a) opsonization by complement proteins; (b) immunoglobulin binding; and (c) simply recognizing surface markers or ligands [25–27].

Renal filtration may also play a role in the elimination of exosomes, especially with smaller populations of vesicles or when exosomes partially degrade and release fragments [28]. Consequently, evidence exists of urinary elimination of exosomes following administration. In several studies utilizing radiolabeled exosomes, the urinary excretion of radioactivity was substantial following administration, and reported the urinary excretion of a fraction of the administered dose [5]. This suggests that some components of the exosome can cross the glomerular filtration barrier.

3.2.3 Influence of Surface Markers and Payloads on Absorption Dynamics

Exosome surface composition can significantly change their pharmacokinetics and tissue targeting. The characteristics of exosome interactions with different cell types and tissues are primarily determined by specific membrane proteins such as integrins, tetraspanins, and cell adhesion–related factors. For instance, exosomes with integrin $\alpha v\beta 5$ binding properties would preferentially target the liver because of binding with liver macrophages (Kupffer cells). In contrast, $\alpha 6\beta 4$ and $\alpha 6\beta 1$ integrin-expressing exosomes have a better lung-tracking profile [29].

Surface modifications, such as PEGylation and stealth approaches, can significantly alter exosome absorption and distribution. PEGylated exosomes exhibit longer circulation times (20–60 minutes), as well as a sixfold extension over unmodified vesicles' half-lives in plasma [30]. However, the degree of surface modification must be optimally set in relation to the ability to achieve longer times in circulation while still promoting richness in targeting tissues and specific interactions [31].

Surface cargo composition may also modify the pharmacokinetic profile of the exosome. Drug-loaded exosomes may exhibit variation and change in stability and elimination in comparison to native vesicles [32]. Depending on payload, exosomes that carry therapeutic proteins or nucleic acids may degrade at an accelerated rate, which may result in a greater cellular uptake than native exosomes [33].

3.3 BIODISTRIBUTION AND TARGETING

3.3.1 Organotropism and Tissue-Specific Accumulation

Exosomal biodistribution is highly predictable and illustrates the effects of passive and active targeting. Following intravenous administration, exosomes travel to far more nontarget tissues, although there are enough biodistribution patterns that typically exist, and are illustrated through exosome accumulation in organs with many mononuclear phagocytes [34]. The liver virtually always accumulates 40–70% of the dose of exosomes, and various organs typically have the following order of percentage: spleen (10–20%), lungs (5–15%), and the kidneys (2–10%) [35]. The pattern of biodistribution of exosomes is typically consistent across all sources of exosomes, *in vitro* culture conditions, and animal models, thus supporting physiological considerations as a molecular mechanism of tissue targeting as the biodistribution of exosomes.

The liver primarily accumulates exosomes through the uptake by Kupffer cells, which are resident macrophages in the liver sinusoids. These macrophages possess numerous types of scavenger receptors that recognize exosomes and internalize them through phagocytosis and endocytosis [36]. The spleen also utilizes the uptake by red pulp macrophages and marginal zone macrophages when targeting exosomes; marginal zone macrophages can also act as a secondary filter on catalogued exosomes that are in circulation [37].

Some experiments show concerning evidence of intra-brain intravasation of exosomes via systemic exposure, including high levels of exosomes in the choroid plexus and circumventricular organ-covered regions of the brain, where the blood–brain barrier is inherently less impermeable to exosome trafficking [38].

Exosome accumulation is tissue-specific, but this can be part of the engineering phase of exosome projects. Studies have demonstrated success in redirecting biodistribution to tissues or cell types of interest through modifying the exosome surface with targeting ligands or fusing with tissue or cell-type-specific peptides, or minimizing off-target accumulation [39].

3.3.2 Role of the Mononuclear Phagocyte System (MPS)

The MPS is the primary way that exosomes are cleared from the body. The MPS consists of any tissue-resident macrophage in the liver (Kupffer cells), spleen (splenic macrophages), lungs (alveolar macrophages), and many other tissues that collectively coordinate in order to limit the entry of foreign particles into the body and maintain immune homeostasis [11].

MPS can recognize exosomes using multiple mechanisms, including but not limited to opsonization with complement factors/immunoglobulin; the direct binding of molecules by pathogen-associated molecular patterns (PAMP); and scavenger receptors [40]. MPS clearance can be altered, depending on the type and membrane composition of the exosome, by exosome size and surface charge.

Studies utilizing MPS depletion or blockade approaches show the important role of the MPS on exosome pharmacokinetics; macrophage depletion using clodronate liposomes or MPS blockade using a saturating dose of inert particles prolongs exosome time in the circulation and alters the biodistribution of exosomes [40, 41]. There have been attempts to develop methods to escape or modulate MPS recognition for therapeutic purposes.

Macrophage polarization status or phenotype is also important in determining exosome clearance kinetics; in general, M1-polarized macrophages have higher phagocytic activity than M2/polarized cells, and states of disease, including inflammation or malignancy, often alter MPS function and exosome pharmacokinetics [40].

3.3.3 Strategies to Improve Targeted Delivery

Several engineering strategies have been developed to increase therapeutic exosome targeting specificities and efficiencies. The most common engineering strategy is surface modifications, that is, the addition of targeting ligands, antibodies, and peptides to exosome membranes using chemical modifications and/or genetic modifications [42]. Furthermore, there are also modifications like stealth modifications, such as PEGylation, which encompasses modifying the structure to avoid MPS recognition but make exosomes target tissue-specific. Studies have determined that optimal PEGylation molecular weight and density could enhance circulation time while simultaneously maintaining targeting tissue specificity [43]. However, on the other end of the spectrum, substantial PEGylation may inhibit cellular uptake and thus limit therapeutic potential. To achieve cell-specific targeting, researchers can either utilize the natural exosome tropism properties or use genetic manipulation to induce exosome-producing cells to express targeting moieties on the exosome surfaces. Exosomes extracted from a specific type of cell may have acquired naturally occurring target properties so that the exosome migrates specifically toward homologous cells or tissues [42]. Advanced targeting strategies also include the use of stimulus-responsive exosomes that are designed to unload or release their loaded materials for a prescriptive time period and/or microenvironment condition, such as acidic pH, temperature, or possibly enzymatic activity. These strategies may create a more selective therapy with fewer off-target effects.

3.4 CIRCULATION KINETICS AND HALF-LIFE

3.4.1 Parameters Influencing Circulation Time

The circulation kinetics of exosomes may be influenced by a number of co-related factors, which together determine plasma half-life and systemic bioavailability. Size is one such important consideration, with smaller exosomes (30–50 nm) having longer circulation time compared to larger (>100 nm) vesicles, since smaller exosomes would face lesser recognition by phagocytic cells, and have reduced clearance due to physical mechanisms [16].

Surface charge is another important consideration for circulation kinetics, as neutral or slightly negative surfaces are likely to circulate better than highly charged vesicles. Generally, positively charged exosomes can undergo rapid clearance because of uptake routes by negatively charged plasma proteins and cell surfaces, while excessive negative charge may activate complement and other dendritic processes [2].

Formation of the protein corona on exosome surfaces is another key driver of circulation kinetics. Once in blood, exosomes rapidly acquire a protein corona that includes albumin, immunoglobulins, complement factors, and other plasma proteins [44]. The composition and density of the protein corona that surrounds exosomes upon entering the blood will affect cellular recognition and clearance.

Membrane rigidity and flexibility also affect circulation time, with the implication that more flexible vesicles could circulate longer simply due to less mechanical stress and their ability to traverse narrow capillaries. Some studies suggest that properties related to membrane cholesterol and lipid compositions can alter both rigidity and flexibility components [45].

3.4.2 Serum Stability and Immune Recognition

Stability of exosomes in serum is a fundamental requirement for therapeutic use. Exosomes have some stability advantages over synthetic nanoparticles primarily due to their biological membrane composition and structure, but are still vulnerable to multiple mechanisms of degradation, such as enzymatic digestion, complement-mediated lysis, and mechanical disruption [46].

Immunological recognition of exosomes occurs through both innate and adaptive immune pathways. Immune cells, through pattern recognition receptors, can recognize exosome-associated molecular patterns that may trigger an inflammatory response, and exosomes can also experience accelerated clearance [47]. Exosomes with surface self-markers (e.g., CD47, “don’t eat me” signals) could provide some protection regarding immunological recognition.

The majority of immune-mediated exosomal clearance occurs through complement activation, as different components of the complement system can be activated through binding to exosome surface components, inducing opsonization and enhanced phagocytosis by cells of the MPS [2]. Approaches that involve inhibiting complement activation, such as incorporating complement regulatory proteins, have shown the potential of extending circulation time.

Immunogenicity varies with exosomal source physiologically and clinically; allogeneic exosomes may provoke a stronger immune response compared to autologous vesicles, but in the majority of human clinical study settings demonstrated acceptable safety profiles with the low frequency of serious, undesirable effects associated with exosome-based therapeutics [48].

3.4.3 PEGylation and Surface Modifications for Prolonged Circulation

PEGylation is the most studied method in the literature to extend the circulation time for exosomes. PEGylation is a process where polyethylene glycol (PEG) chains are covalently attached to the surface of the exosome to create a hydrophilic layer, which reduces protein adsorption and recognition by cells. PEGylation has been shown to extend the circulation half-life of exosomes to hours, versus the minutes associated with no PEGylation [30].

There are an ideal number of factors to consider to work toward defining the PEGylation strategy. This includes the PEG molecular weight, PEG density, and attachment chemistry. Ideal PEG chains are 2–5 kDa linear molecules; higher molecular weight chains reduce cellular uptake for potential therapeutic efficacy [49]. Factors related to PEG density are essential to provide “stealth” characteristics, while still maintaining functionality of the exosomes.

Zwitterionic polymers are another surface modification that can provide stealth properties similar to PEG with added benefits in terms of biocompatibility [50]. Phosphorylcholine and carboxybetaine surface modifications have been helpful in extending the circulation time of nanoparticles. One other example of a biomimetic method of extending circulation time is with coated exosomes with cellular membranes or material derived from membranes [51, 52]. When exosomes are coated with membranes of breast cancer cells, they receive extended circulation, which is supported by natural stealth properties through CD61,-83 expression, and other self-recognition markers [51, 53].

3.5 CELLULAR UPTAKE AND INTERNALIZATION KINETICS

3.5.1 Receptor-Mediated Endocytosis and Macropinocytosis

The internalization of exosomes by target cells can occur through many different mechanisms that together dictate the rate and efficiency of cellular uptake (Figure 3.1). Receptor-mediated endocytosis is the most specific process, in which a cellular receptor recognizes surface ligands on exosomes, leading to membrane invagination and ultimately vesicle formation [22]. This variety of endocytosis often happens in minutes to hours to complete and has the ability to become saturated when presented with high concentrations of exosomes.

Clathrin-mediated endocytosis provides a substantial cellular pathway for exosome uptake, in particular, small vesicles (<100 nm in diameter), where exosomes bind to specific receptors, which bring in clathrin and adaptor proteins, and ultimately form clathrin-coated vesicles [22, 49].

Although the kinetics of endocytosis are fairly rapid, clathrin-coated vesicles can be formed in about 1–2 minutes upon receptor binding.

Caveolin-mediated endocytosis provides another mechanism for internalizing exosomes, especially in cell types that express large amounts of caveolin proteins. While this pathway probably favors larger exosomes or exosomes with unique membrane compositions, it generally proceeds at much slower kinetics than clathrin-mediated uptake, typically 10–30 minutes for vesicle formation [54]. Macropinocytosis is a nonspecific mechanism for exosome uptake that forms large ruffles of membrane that then engulf extracellular fluid and other associated particles. This mechanism can internalize multiple exosomes at once, which is particularly important in cells with high phagocytic behavior, such as macrophages and dendritic cells, and its kinetics vary but generally happens in 30 minutes with exosome exposure [22].

3.5.2 Kinetics of Intracellular Trafficking and Endosomal Escape

After the process of internalization, exosomes move through many phases of intracellular trafficking, which ultimately provides their therapeutic potential and cargo transfer. Most internalized exosomes are found in early endosomes, a common sorting compartment for recycling portions of internalized materials back to the cell surface and progressing to late endosomes and lysosomes [23].

The amount of time it takes to go from early to late endosomes is around 15–30 minutes, and it undergoes changes in endosomal pH, membrane composition, and the endosomal partner proteins [56]. The changes associated with acidification may lead to conformational changes in exosomal membranes that may promote the release of any cargo or facilitate membrane-mediated fusion.

Endosomal escape is crucial for delivering therapeutics to the cytoplasm, and a very small amount of internalized exosomes escapes from endosomal compartments; the majority will simply be degraded by lysosomes. Although the endosomal escape is not well-defined, it appears to occur most frequently at 2–6 hours post-internalization based on some research [23].

There are several putative mechanisms for exosomal escape that include pH-specific membrane-mediated fusion, osmotic pressure, membrane disruption, and interaction with endosomal proteins [23, 56]. There is likely a huge variability in effectiveness across both cell types and populations of exosomes in these processes.

3.5.3 Temporal Aspects of Functional Payload Delivery

The release kinetics of functional payloads from internalized exosomes can be quite complex, exhibiting spatiotemporal dynamics that depend on the type of cargo, intracellular trafficking pathway, and target cell [57]. Protein-based cargoes may be released relatively quickly after escaping the endosome, leading to functional effects that can be detected within hours of exosome treatment. Nucleic acid cargoes (mRNA and miRNA) show different kinetics depending on the stability of the cargo and whether additional processing is required before becoming functional. In the case of miRNA delivery, functional effects can be detected within

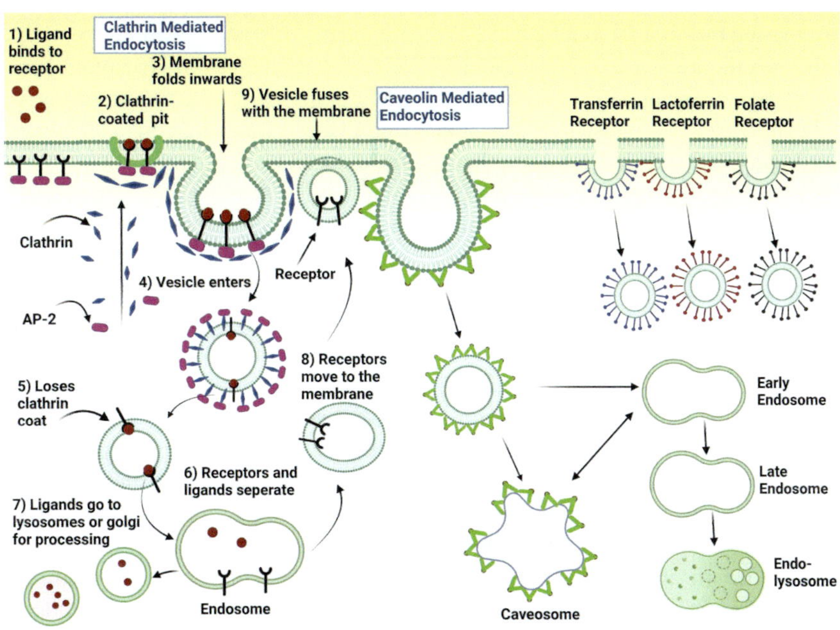

FIGURE 3.1 Receptor-mediated endocytosis of exosomes and its mechanism of cellular internalization. *Source:* Adapted with permission from Ref. [55].

6–24 hours, which is the time required for incorporation into RNA-induced silencing complexes and subsequent regulation of target mRNA. In the case of mRNA delivery, effects can be more rapid since successful cytoplasmic delivery means the mRNA is free for protein translation within 2–4 hours, so it could have relatively rapid functional effects [23, 58].

The duration of the effect of therapeutic benefit is wide-ranging (e.g., hours to days for proteins) compared to more prolonged effects (e.g., days or even weeks for genetic payloads), reflecting the stability of the genetic payload and the turnover of cellular components that the payload acts upon [49].

The kinetic aspects associated with the temporal delivery of payloads will be influenced by other variables associated with dosing (i.e., dose and frequency). Multiple dosing is key in some cases to build up therapeutic levels that last long enough to produce an effect, while for many others, dosing frequency may be only one or a few times due to some sustained intracellular effect(s) [59].

3.6 METABOLISM AND BIOTRANSFORMATION

3.6.1 Intracellular Degradation Pathways

The metabolism of exosomes within receiving cells consists of complex degradation pathways that will, ultimately, shape the fate of both the vesicle structure and the enclosed contents. The majority of exosomes taken up via endocytosis typically undergo lysosomal degradation, where acidic pH and any number of hydrolytic enzymes (also collectively referred to as hydrolases) will degrade lipid membranes, proteins, and nucleic acids [56].

Lysosomal degradation normally begins 2–4 hours after endocytosis of the exosome and may last for as long as 12–24 hours, depending on the specific exosome composition and the metabolic activity of the accepting cell. Lipid components will primarily be degraded by phospholipases and sphingomyelinases, while proteins will undergo proteolytic cleavage (by cathepsins and other lysosomal proteases) [60].

Autophagy is an alternative degradation pathway which may be particularly pertinent to larger exosome populations or those which provide an excess clearance challenge to the endosomal–lysosomal system [61]. Autophagy follows a process similar to the endosomal–lysosomal pathway, where exosomes are sequestered within autophagosomes, leading to degradation once an autophagosome fuses with a lysosome.

Some of the captured components of exosomes may escape degradation and proceed through recycling, leading to either the return of recycling exosomal components back to the surface of the accepting cell or through utilization/recycling into different cellular metabolite pathways. Lipid components can be utilized for membrane synthesis, while certain proteins can be processed for antigen presentation, or other cellular functions [60].

3.6.2 Exosomal RNA/Protein Processing in Target Cells

The processing of exosomal RNA and protein cargoes in target cells occurs through distinct pathways that affect the potential functional outcomes. mRNA molecules delivered by exosomes may be translated if they successfully escape from the endosomal compartments and enter the cytoplasm of the target cell [62]. However, this process is usually inefficient, given that most exosomal mRNA is degraded shortly after entering the endolysosomes prior to translation in the cytoplasm.

miRNA processing presents a more efficient pathway for functional exosomal RNA. Delivered miRNAs can become incorporated into RNA-induced silencing complexes (RISC) after escaping the endosomal compartment for posttranscriptional regulation of target mRNA. The kinetics of miRNA silencing usually occur between 6hours and 48 hours and can have prolonged regulatory activity [63].

The processing of protein differs widely depending on the nature of the cargo and its localization in the target cell. In general, exosomal enzymes and functional proteins may retain their activity after exosomal delivery, whereas the other cargo may be degraded or modified, affecting protein function [60]. Overall, delivered proteins have local stability that usually lasts hours to days.

Some exosomal proteins could be presented with antigen development and on MHC molecules in antigen-presenting cells, inducing immune activation [60]. This mechanism is an important basis for exosome vaccine application.

3.6.3 Impact of Host Factors on Metabolic Stability

The processing and metabolic stability of exosomes in recipient cells will greatly depend upon host factors. Age-associated changes in lysosomal function may affect exosome degradation kinetics, resulting in undegraded material, and disease states may also have significant consequences on the metabolic stability of exosomes [64]. Cancer cells may have downregulated autophagy and lysosomal activity, which could influence the cell's hoarding of therapeutic exosomes [65]. Chronic inflammation can adeptly differentiate in the degradative pathway and potentially elevate exosomal clearance. In addition to age and disease, host variation in metabolic enzymatic activity (e.g., variation in lysosomal enzymes) may also impact the degradation of exosomes and disposition of cargo release; therefore, the response of the host to therapeutic exosomes is physiologically variable and has elements of both the genetic basis for hosts and the environment of the host [60]. Environmental factors such as oxygen tension, pH, temperature, and oxidative stress may also impact exosome stability and processing. Alterations in pathological conditions that modify pH or enhance reactive oxygen species (ROS) may enhance the degradation of exosomes or alter cargo activity.

3.7 CLEARANCE MECHANISMS AND EXCRETION

3.7.1 Hepatic and Renal Elimination Pathways

Hepatic clearance is the core elimination route for exogenously administered exosomes, comprising 40–70% of the total administered dose in most studies [11]. Clearance occurs mostly by uptake by Kupffer cells, the resident macrophages of hepatic sinusoids. Kupffer cells express several scavenger receptors, such as SR-A, MARCO, and CD36, to detect and engulf circulating exosomes by phagocytosis [66].

After hepatic uptake, exosomes are intracellularly processed in Kupffer cells, where they are degraded in phagolysosomes. The degradation product is either processed further for antigen presentation or eliminated by biliary excretion. Studies of exosomes radiolabeled with radioisotopes have shown considerable biliary elimination, with radioactivity found in feces within 6–24 hours of administration [11, 66].

Renal elimination is an important mechanism for the clearance of exosomes, using mechanisms such as glomerular filtration for smaller exosome populations, or small degradation products, particularly for those smaller than the typical molecular weight cutoff of 60–70 kDa [5]. Several studies observed robust urinary excretion of exosome-associated markers. This suggests that either entire tiny vesicles were filtered or that renal clearance resulted from the elimination of processed components from larger exosome-associated vesicles. Although renal clearance kinetics typically confirm exosome persistence in the kidney, total urinary recovery peaks 2–6 hours post-administration and typically returns to baseline activity within a 24–48 hour time frame [67]. The extent of renal elimination varied among the different exosome preparations, with those containing smaller and less protein-associated vesicles showing greater urinary recovery.

3.7.2 Role of Spleen and Lung in Exosome Clearance

Splenic clearance is another pathway for removing circulating exosomes and typically comprises ~10–20% of the dose. The macrophage populations in the red pulp (RP) and marginal zones (MZ) have evolved to specifically capture and clear circulating particles [68].

Red pulp macrophages act as mechanical filters and are able to clear exosomes through phagocytosis when blood flows through the splenic sinusoids. This mechanism is particularly more efficient for larger exosomes, or those that have been shown to change deformability [69, 70]. Marginal zone macrophages capture exosomes from circulation and facilitate clearance of exosomes, but potentially serve an additional role for another level of immune recognition and processing.

Pulmonary clearance occurs mainly through alveolar macrophages that patrol the alveolar surface and the pulmonary capillaries. While accounting for a small fraction of the total clearance (5–15% on average), pulmonary clearance is not trivial for some populations of exosomes and some other pathophysiological states [11].

Kinetically, splenic and pulmonary clearance are typically parallel to hepatic clearance, with peak uptake occurring minutes to hours post-administration; however, the retention time in organs may be shorter than in the liver, with some indications of redistribution back to circulation [5].

3.7.3 Clearance Modulation for Sustained Therapeutic Effects

Several strategies to minimize exosomal clearance and prolong exosome therapeutic time have been devised. One way to accomplish this is by temporarily decreasing macrophage uptake capacity through the use of a blockade of MPS (reticuloendothelial system). This can be reached by IV administering a saturating dose of inert particles either as a bolus or by continuous infusion for an extended period of time [11]. Reports indicate threefold to tenfold increases in circulation half-lives in preclinical studies, depending on the form of the inert particle and its delivery method.

In addition to completing MPS blockade, another strategy, which is again, predominantly surface modification, involves reducing clearance mechanisms by reducing identification [2]. At the present time, we are able to provide surface modification for the exosomes, including fusing PEG, where uptake by cells will be diminished as there is decreased opsonization of protein, thus limiting cellular identification, resulting in prolonged circulation time. Further surface modification to promote clearance, or reduce the clearance mechanisms, is shown by using CD47 expression signaling on exosomes that the exosome is a "don't eat me" signal to macrophages to inhibit phagocytic activity [2].

However, less common and still experimental treatment strategies involving dosing can also impact clearance kinetics. High-dose administration can only occur for therapeutic dosing when the dose can saturate clearance pathways, and, therefore, results in nonlinear pharmacokinetics and extended elimination half-lives [71]. Importantly, this change in clearance kinetics can contribute to nonlinear pharmacokinetics, but it cannot ignore consideration for potential toxicity and immune activation.

Another option to decrease clearance and enhance prolonged therapeutic activity is utilizing combination therapies with either clearance inhibitors or immune-modulating agents. Complement inhibitors are examples that have effects to minimize opsonization, therefore clearance of exosomes, while anti-inflammatory agents can attenuate macrophage activation, and therefore eliminate phagocytic activity of the injected host macrophages [72].

3.8 QUANTITATIVE AND ANALYTICAL APPROACHES

3.8.1 Imaging and Tracing Techniques

Quantitative study of exosome pharmacokinetics requires imaging and analytical techniques that are sophisticated enough to follow these nanoscale vesicles in complicated biological systems. Fluorescent labeling is one of the most common methods to label exosomes, many of which utilize lipophilic dyes, such as DiR (1,1′-dioctadecyl-3,3,3′,3′-tetramethylindotricarbocyanine iodide), that provide near-infrared fluorescence for *in vivo* imaging [4]. DiR labeling has a number of benefits, such as allowing imaging with deep tissue penetration, very little background interference from autofluorescence, and relatively high photostability. Labeling with DiR requires incubating exosomes with the dye, which allows it to incorporate into the lipid bilayer membrane; however, there may be questions about its redistribution or transfer between vesicles, and potentially alternative exosome behavior in a prior manner [4].

Fluorescent protein-based genetic labeling strategies (GFP, RFP, as well as luciferase) have advantages in terms of stability and specificity, but require an upfront genetic manipulation to produce cells for exosome generation [73]. There is no physical labeling followed in the genetic method, but we can label exosomes intrinsically with a unique signal traceable and maintain their respective native exosome attributes.

Radionuclide labeling can provide exceptional sensitivity and quantitative accuracy, representing a rich field of study for *in vivo* and *in vitro* pharmacokinetics of exosomes. Some commonly used isotopes include 89Zr ($t1/2$ = 78.4 hours), 99mTc ($t1/2$ = 6 hours), and 111In ($t1/2$ = 68 hours); these, plus others, can provide different advantages in terms of imaging duration or spatial resolution [5, 73, 74]. Radiolabeling can be achieved by surface conjugation or by intraluminal incorporation into the exosomes; an intraluminal method would generally provide a more stable radiolabel.

3.8.2 Pharmacokinetic Modeling and Simulation

Pharmacokinetic modeling of exosomes creates special issues because pharmacokinetic modeling of exosomes must account for relatively complex multiphasic elimination kinetics and variable tissue uptake kinetics [75]. Traditional compartmental models will potentially be unable to adequately capture exosome disposition, necessitating some more advanced models such as physiologically based pharmacokinetic (PBPK) models [10].

PBPK models utilize anatomical and physiological parameters to calculate exosome elimination from and disposition within various tissues, while modeling specific organ clearance processes, blood flow rates, and tissue binding behavior. Developing PBPK models with appropriate parameters to accurately model exosomes will require a considerable amount of parameter estimates using experimental data [10].

Noncompartmental analysis, which is not bound to a specific compartmental structure of processes, is a common model-independent approach to characterizing exosomal pharmacokinetics. While noncompartmental analyses will not provide as many mechanistic predictions as compartmental or PBPK models, noncompartmental analyses do provide robust estimates of key parameters such as area under the curve (AUC), clearance, and mean retention time, for comparative assessments [7].

Lastly, population pharmacokinetic modeling may provide compelling value in generating knowledge about the interindividual variability of exosome kinetics, particularly when concomitant covariates include differences associated with age, weight, state of health, and disease, and genetic and epigenetic traits, for the purposes of developing individualized dosing strategies [10].

3.8.3 Bioanalytical Method Validation for Exosome PK

One of the main challenges to clinically translating cell-derived exosome testing has centered around the absence of validated bioanalytical methods for quantifying exosomes. The majority of methods are based on labeling with dyes, antibodies, or other ways to discern particles; in contrast, label-free methods allow for the avoidance of potential artifacts and maintenance of native features of exosomes [76].

Existing label-free methods that utilize LC-MS/MS detection of specific exosomal proteins as probabilistic surrogate markers showed significant promise as label-free-based methods of quantification. The protein is usually digested, followed by multiple reaction monitoring (MRM) of either the full protein or some unique characteristic peptides. When developing LC-MS/MS methods, it is expected that once again, specificity, linearity, accuracy, precision, and stability will be validated based on all regulatory guidelines [51].

Nanoparticle tracking analysis (NTA) can serve as a useful complement to exosome quantification. NTA technology relies on observing the Brownian motion of particles, and it can provide both concentrations of particles and size distributions, but does not provide specificity in more complex biological samples [77].

Flow cytometry methods based on providing labels on exosomes using specific antibodies may provide potentially quantitative and phenotypic information on circulating exosomes, although the detection limits on a conventional flow cytometer may require more specialized, high-sensitivity flow cytometers [78].

Lastly, novel and emerging methods based on surface plasmon resonance (SPR) and electrochemical sensing may provide further means for real-time monitoring of exosomes and potentially facilitate the development of methods for the monitoring of therapeutic drugs in the clinical field [77].

3.9 BIOLOGICAL IMPLICATIONS OF EXOSOME KINETICS

3.9.1 Kinetics in Physiological versus Pathological States

The pharmacokinetics of exosomes are mostly distinct from that of an organism in a healthy physiological status. In the case of cancer patients, the exosome pharmacokinetics are difficult to elucidate, as the tumors may disturb exosome pharmacokinetics for a number of reasons; for instance, tumors may convert macrophages into different phenotypes, affect the permeability of the vasculature, and affect complement activation [11]. Research in tumor-bearing animal models has suggested that the MPS clearance may be greater, which may be due to greater recruitment of activated macrophages [79].

Inflammation has a synergetic and more complicated influence on exosome pharmacokinetics due to the enhanced immune recognition mechanisms and altered vascular permeability. It is observed that an acute immune response extends the clearance of exosomes via activation of the complement cascade and action of macrophages, leading to phagocytosis. Inflammation, when chronic, will have effects whereby the MPS is exhausted and will allow for a longer period of circulation of exosomes [79].

The most conclusive evidence regarding altered pharmacokinetics for exosomes in disease states is in the area of diseases of the liver, particularly patient-related Kupffer cell dysfunction and changes in blood flow to the liver, which compromise the pharmacokinetics of exosomes. In disease states such as cirrhosis or hepatitis, liver disease may decrease the clearance of exosomes and result in changes to the biodistribution of exosomes [11]. In the case of kidney disease, the influence of these biological degradation products on clearance from exosomes may add to the polymerized exosome metabolites.

Age-related changes in immune function and organ physiology represent additional factors that may have an unpredictable effect on exosome kinetics. Changes have been suggested to both functional MPS activity for elderly patients, as well as changes in binding of plasma proteins (up to 33% of pharmaceutical molecules are protein-binding), and changes to tissue distribution [3]. The above factors create variability and should provide some rational dosing considerations for a population of older patients.

3.9.2 Relevance in Tumor Microenvironment, Inflammation, and Regeneration

In the tumor microenvironment, exosome-based therapeutics present unique challenges and opportunities. Heterogeneity is inherent in tumor microenvironments and can affect exosome kinetics and therapeutic utility. Tumors often have a higher vascular permeability due to the enhanced permeation and retention (EPR) effect, allowing for exosome accumulation; however, tumor-associated macrophages and tumor-associated immune status can also influence exosome kinetics and therapeutic utility [65, 79].

Exosome targeting to tumors can be considered in terms of passive targeting that relies on EPR effects that allow preferential accumulation in leaky tumor vasculature, or active targeting, which involves tumor-specific ligands or natural exosome tropism to cancer cells. As for kinetics, exosome tumor targeting would show peak accumulation typically between 6 hours and 24 hours after administration and will gradually disperse until clearance [80].

Under inflammatory conditions, exosome kinetics may be altered because of increased vascular permeability combined with increased immune cell recruitment to the tumor site. Inflammatory conditions can promote extravasation and tissue penetration, but can also lead to relatively quick clearance of exosomes, predicting immune activation to the exosomes [65, 79].

For regenerative medicine-based exosome therapy, they often utilize the accumulation of exosomes in areas of tissue injury or inflammation. The kinetics associated with delivery to these injury sites ultimately depend on the degree of vascular damage and the inflammatory state of the tissue [81]. Optimal therapeutic contribution will often rely on sustained local concentrations over a period of several days to weeks.

3.9.3 Immunogenicity and Safety Considerations

The immunogenicity of exosomes is a safety concern and may impact their biodistribution. Generally, exosomes are less immunogenic than synthetic nanoparticles, but exosomes may still induce immune responses, particularly if allogeneic or administered multiple times [82].

Clinical trials indicate that, in general, exosome-based treatment has a positive safety profile, with a low incidence of serious adverse events (0.7%) and adverse events (4.4%). However, the longer-term immunological effects of administering repeated doses of exosomes are relatively unknown [48].

If the immune system develops anti-exosome antibodies after the initial dose and after the subsequent doses, pharmacokinetic behavior may be altered due to increased clearance or altered biodistribution. This type of situation is particularly relevant for chronic therapeutic scenarios of repeated dosing [82].

There are methods to reduce immunogenicity, including autologous sources of exosomes, an immune suppression protocol, or engineering methods to remove immunogenic epitopes [48]. There are certainly trade-offs to these approaches, especially in terms of efficacy and practical considerations as part of testing shortcuts. Table 3.1 summarizes the key attributes of exosomes, defining their pharmacokinetic fate.

TABLE 3.1
Summary of Key Aspects Related to the Pharmacokinetics of Exosomes

Pharmacokinetic Phase	Mechanism of Exosome–Cell Interaction	Factors Influencing Biodistribution	*In vivo* and *In vitro* Models	Tracking and Quantification Methods	Interindividual Variability Factors	Engineering Strategies to Improve PK	Clinical Translation Potential	References
Absorption	Endocytosis (clathrin/caveolin-mediated, macropinocytosis), direct fusion, receptor–ligand binding	Vesicle size, surface charge, lipid bilayer stability, membrane proteins (integrins, tetraspanins), GI and mucosal barriers	Caco-2, IEC-6, organoids, oral/IV/intranasal rodent models	Fluorescent membrane dyes (DiR, PKH67), isotopic labeling, LC–MS/MS, NTA	Age, gastric pH, microbiota composition, disease state	PEGylation, ligand-functionalization, lipid composition tuning	Suitable for sustained oral or transmucosal delivery; improved bioavailability in preclinical cancer models	[83, 84]
Distribution	Organotropism via integrin–ligand interactions; MPS uptake; EPR effect in tumors	Cellular origin, glycosylation, lipid composition, admin route, disease microenvironment	Mouse/rat biodistribution, tumor-bearing xenografts, PDX, healthy controls	IVIS, MRI, PET/SPECT, LC–MS/MS tissue assays	Tumor burden, vascular permeability, and genetic polymorphisms	Targeting ligands (peptides, antibodies), stealth coatings, hybrid EV–NP systems	Needed for achieving precise targeted delivery in human therapy	[8, 34]
Metabolism	Lysosomal degradation, endosomal sorting, lipid/protein recycling	Membrane composition, size, uptake mechanism, and lysosomal enzyme activity	*In vitro* cell lines (macrophages, tumor cells), murine PK studies	LC–MS/MS of cargo metabolites, proteomics, RNA-seq	Lysosomal storage disorders, metabolic rate, and inflammation	Use biodegradable cargo, optimize membrane composition	Regulatory requirements for metabolic safety profiling	[56, 85]
Clearance	MPS phagocytosis (Kupffer/splenic macrophages), renal filtration, biliary excretion	Surface “self” markers (CD47), phosphatidylserine, protein corona, and immune opsonins	Macrophage-depleted versus wild-type mice; knockout models	Bioluminescence, radiolabeling (99mTc, 111In), gamma counting	Age, immune status, liver/kidney function	PEGylation, CD47 expression, and CL inhibitors	Essential to meet regulatory safety thresholds for organ-retained vesicles	[86, 87]

3.10 PHARMACOKINETIC CHALLENGES IN CLINICAL TRANSLATION

3.10.1 Interindividual Variability and Biodistribution

Individual variation in the pharmacokinetics of exosomes poses a major challenge to both basic science and clinical research. Sources of variability include genetic variations in drug-metabolizing enzymes, variations in organ functioning and capacity, immune status, and concomitant medications. Studies involving humans experience coefficient of variation values of 30–50% for significant pharmacokinetic parameters [48].

Age constitutes a major variable factor influencing pharmacokinetics, older patients can have both lower clearance capacity and altered biodistribution. The ability of the MPS to clear exosomes decreased by about 15% between 30 and 80 years, which has implications for dosing in older populations [11].

The mode of delivery may also be influenced by gender-based differences in body composition, hormones, and metabolic enzyme expression [88]. However, it has yet to be determined whether there are any clinically significant gender-based differences in the pharmacokinetics of exosomes.

An individual's disease state can also significantly influence the pharmacokinetics of exosomes and consequently influence interindividual variability in a patient group. Patients suffering from cancers may exhibit greatly altered clearance kinetics due to the high tumor burden that cancer requires, as well as concurrent chemotherapy or altered immune status [89].

3.10.2 Scaling from Animal Models to Humans

Exosome pharmacokinetics from animal models to humans is complex, and this complexity arises from differences in anatomy, physiology, and immune function between species [88]. Consequently, allometric scaling (a tool for translating data from one species to another, particularly for small-molecule drugs), while commonplace for small molecules in a drug development context, may not be as suitable for exosomes due to the manner of clearance.

Moreover, differences in the function of the MPS between species will also have an influence on the clearance kinetics, with rodents often clearing substances much faster than larger animal models and humans, which can lead to overestimation of clearance rates as one scales from mouse work [11].

Further, differences in immune systems between species will likely lead to challenges in translation, due to extensive variation in specific pattern recognition receptors, complement systems, and macrophage phenotypes, in addition to clear differences in recognizing and clearing exosomes [48].

PBPK models may provide some improved translation ability when studying exosomes with physiologic parameters based on other species (i.e., respective physiological parameters within species); however, the development of PBPK models is time-consuming, and thus validating any predictions with humans is also required [10].

3.10.3 Regulatory Expectations for Exosome-Based Therapeutics

Both regulatory authorities (like the FDA) and their expectations for exosome-based therapeutics affect the pharmacokinetic study requirements. Because exosome products are biologics, they are regulated by the Center for Biologics Evaluation and Research (CBER) as biologics, and the pharmacokinetic and biodistribution studies are developed in direct relation to the overall development program [90, 91].

Based on regulatory requirements, companies may be required to demonstrate that product quality is consistent, understand relevant pharmacokinetic characteristics in an appropriate animal model, and establish any dose–response relationships in animal models [92–94]. The heterogeneous makeup of exosome preparations makes it difficult to meet some of the regulatory requirements associated with pharmacokinetic studies.

Manufacturing product consistency is a regulatory issue that has a direct impact on pharmacokinetic studies. Variability in exosome content, size distribution, and surface properties can vary from batch to batch and change pharmacokinetic behavior and possibly the therapeutic effect [92].

In addition, regulatory agencies will require long-term monitoring for safety, which may elongate studies to determine pharmacokinetics to observe potential accumulation, delayed adverse events, and immunological effects related to longer-term or repeated administration, requiring follow-up as long as multiple months (or longer) since treatment, contingent on the therapeutic indication [95].

3.11 CONCLUSION AND FUTURE DIRECTIONS

There have been major strides in understanding exosome pharmacokinetics, but there are still important gaps in knowledge that can limit clinical translation. We need a better mechanistic understanding of exosome–cellular interactions, uptake efficiency, and endosomal escape, in addition to label-free analytical measurement and standards. Similarly, interindividual variability, long-term biodistribution, and any potential chronic toxicity must be systematically explored in population pharmacokinetic studies, ideally with annual follow-up assessments. Emerging engineering technologies, such as detection technologies, stimuli-responsive polymers, targeted ligands, combinations of targeted exosome–nanoparticle systems, and genetically modified exosome producer cells, offer great potential to improve biodistribution, circulation time, target specificity, as well as pharmacological and therapeutic potential for exosomes. Combining exosomal therapeutic delivery with personalized medicine strategies (such as companion diagnostics) and individual formulations can advance dosing optimization and therapeutic benefit. Exosome pharmacokinetics could be developed with systems pharmacology, AI, or even biomarker-based surveillance to help with therapeutic prediction, formulation, and adaptive dosing. With continued emphasis on mechanism discovery, analytical harmonization, and translational strategies, next-generation exosome therapeutics can hypothetically be developed with predicted pharmacokinetic profiles, thus allowing clinicians greater access to exosomes and their clinical applications.

ABBREVIATIONS

89Zr	zirconium-89
99mTc	technetium-99m
111In	indium-111
Autophagosomes	autophagic vesicles
Caco-2	human epithelial colorectal adenocarcinoma cell line
CD36	cluster of differentiation 36
CD47	cluster of differentiation 47
CD61	cluster of differentiation 61
CD83	cluster of differentiation 83
CL	clearance
DiR	near-infrared fluorescent dye
DNA	deoxyribonucleic acid
EPR effect	enhanced permeation and retention effect
EV	extracellular vesicle
GFP	green fluorescent protein
IEC-6	rat small-intestine epithelial cell line
In	indium (radioisotope)/inhibitor (context dependent)
IVIS	*in vivo* imaging system
MARCO	macrophage receptor with collagenous structure
MRI	magnetic resonance imaging
NP	nanoparticle
NTA	nanoparticle tracking analysis
PBPK	physiologically based pharmacokinetic
PDX	patient-derived xenograft
PKH26	fluorescent dye (red)
PKH67	fluorescent dye (green)
RFP	red fluorescent protein
RISC	RNA-induced silencing complex
ROS	reactive oxygen species
SPR	surface plasmon resonance
SPECT	single photon emission computed tomography
SR-A	scavenger receptor class A
***t*1/2**	half-life (time for half of the substance to be eliminated)

ACKNOWLEDGMENT

I would like to extend my sincere gratitude to my current institution for generously providing the resources and support that facilitated the drafting of this chapter.

AI DISCLOSURE STATEMENT

During the preparation of this work, the authors used Grammarly, QuillBot, and ChatGPT in order to enhance grammatical structure and narrative coherence. After using this tool/service, the authors reviewed and edited the content as needed and take full responsibility for the content of the publication.

CONFLICT OF INTEREST

None.

REFERENCES

[1] K. Yadav, K.K. Sahu, Sucheta, S. Minz, M. Pradhan, Unlocking exosome therapeutics: The critical role of pharmacokinetics in clinical applications, *Tissue Cell* 93 (2025). https://doi.org/10.1016/j.tice.2025.102749.

[2] N. Parada, A. Romero-Trujillo, N. Georges, F. Alcayaga-Miranda, Camouflage strategies for therapeutic exosomes evasion from phagocytosis, *J Adv Res* 31 (2021) 61–74. https://doi.org/10.1016/j.jare.2021.01.001.

[3] D.H. Kim, V.K. Kothandan, H.W. Kim, K.S. Kim, J.Y. Kim, H.J. Cho, Y.K. Lee, D.E. Lee, S.R. Hwang, Noninvasive assessment of exosome pharmacokinetics *in vivo*: A review, *Pharmaceutics* 11 (2019). https://doi.org/10.3390/pharmaceutics11120649.

[4] J. Santelices, M. Ou, W.W. Hui, G.H.B. Maegawa, M.J. Edelmann, Fluorescent labeling of small extracellular vesicles (EVs) isolated from conditioned media, *Bio Protoc* 12 (2022) e4447. https://doi.org/10.21769/BioProtoc.4447.

[5] H. Choi, M.-Y. Kim, D.-H. Kim, H. Yun, B.-K. Oh, S.-B. Kim, I.-H. Song, H.-S. Park, S.-E. Kim, C. Park, C. Choi, Quantitative biodistribution and pharmacokinetics study of GMP-grade exosomes labeled with ^{89}Zr radioisotope in mice and rats, *Pharmaceutics* 14 (2022). https://doi.org/10.3390/pharmaceutics14061118.

[6] K. Yadav, S.P. Ebenezer Gnanakani, K. Kumar Sahu, Sucheta, A. Dubey, S. Minz, W. Raza, M. Pradhan, Unleashing the potential of natural protein based nanoparticles for the delivery of therapeutic nucleic acid: A comprehensive review, *Int J Pharm* 669 (2025) 125049. https://doi.org/10.1016/j.ijpharm.2024.125049.

[7] B. Li, F. Yu, A label-free liquid chromatography–tandem mass spectrometry method for the quantitative analysis of exosome pharmacokinetics *in vivo*, *Pharmaceutics* 17 (2025). https://doi.org/10.3390/pharmaceutics17060699.

[8] O.P.B. Wiklander, J.Z. Nordin, A. O'Loughlin, Y. Gustafsson, G. Corso, I. Mäger, P. Vader, Y. Lee, H. Sork, Y. Seow, N. Heldring, L. Alvarez-Erviti, C.I. Edvard Smith, K. Le Blanc, P. Macchiarini, P. Jungebluth, M.J.A. Wood, S. El Andaloussi, Extracellular vesicle *in vivo* biodistribution is determined by cell source, route of administration and targeting, *J Extracell Vesicles* 4 (2015) 1–13. https://doi.org/10.3402/jev.v4.26316.

[9] S.G. Antimisiaris, S. Mourtas, A. Marazioti, Exosomes and exosome-inspired vesicles for targeted drug delivery, *Pharmaceutics* 10 (2018). https://doi.org/10.3390/pharmaceutics10040218.
[10] P. Kumar, D. Mehta, J.J. Bissler, Physiologically based pharmacokinetic modeling of extracellular vesicles, *Biology (Basel)* 12 (2023). https://doi.org/10.3390/biology12091178.
[11] I.V. Zelepukin, K.G. Shevchenko, S.M. Deyev, Rediscovery of mononuclear phagocyte system blockade for nanoparticle drug delivery, *Nat Commun* 15 (2024) 4366. https://doi.org/10.1038/s41467-024-48838-5.
[12] A.M. Tolomeo, G. Zuccolotto, R. Malvicini, G. De Lazzari, A. Penna, C. Franco, F. Caicci, F. Magarotto, S. Quarta, M. Pozzobon, A. Rosato, M. Muraca, F. Collino, Biodistribution of intratracheal, intranasal, and intravenous injections of human mesenchymal stromal cell-derived extracellular vesicles in a mouse model for drug delivery studies, *Pharmaceutics* 15 (2023). https://doi.org/10.3390/pharmaceutics15020548.
[13] A.M. Beshbishy, S. Alghamdi, T.G.E. Onyiche, M. Zahoor, N. Rivero-Perez, A. Zaragoza-Bastida, M.A. Ghorab, A.K. Meshaal, M.A. El-Esawi, H.F. Hetta, G. El-Saber Batiha, Biogenesis, biologic function and clinical potential of exosomes in different diseases, *Appl Sci (Switzerland)* 10 (2020). https://doi.org/10.3390/app10134428.
[14] K. Yadav, Ajazuddin, A. Alexander, P. Pandey, W. Raza, M.R. Singh, D. Singh, R. Tirkey, R. Jangde, P.K. Sahu, Introduction to biomaterial-inspired nanomedicines, in: *Biomaterial-Inspired Nanomedicines for Targeted Therapies*, Springer Nature, 2024. https://doi.org/10.1007/978-981-97-3925-7_1.
[15] M. Pradhan, K. Yadav, N. S. Chauhan (Eds.), *Biomaterial-Inspired Nanomedicines for Targeted Therapies* (1st ed.). Springer, Singapore, 2024. https://doi.org/10.1007/978-981-97-3925-7.
[16] S. Bala, T. Csak, F. Momen-Heravi, D. Lippai, K. Kodys, D. Catalano, A. Satishchandran, V. Ambros, G. Szabo, Biodistribution and function of extracellular miRNA-155 in mice, *Sci Rep* 5 (2015). https://doi.org/10.1038/srep10721.
[17] T. Smyth, M. Kullberg, N. Malik, P. Smith-Jones, M.W. Graner, T.J. Anchordoquy, Biodistribution and delivery efficiency of unmodified tumor-derived exosomes, *Journal of Controlled Release* 199 (2015) 145–155. https://doi.org/10.1016/j.jconrel.2014.12.013.
[18] T. Ikeda, M. Kawabori, Y. Zheng, S. Yamaguchi, S. Gotoh, Y. Nakahara, E. Yoshie, M. Fujimura, Intranasal administration of mesenchymal stem cell-derived exosome alleviates hypoxic-ischemic brain injury, *Pharmaceutics* 16 (2024). https://doi.org/10.3390/pharmaceutics16040446.
[19] S.S. Palakurthi, B. Shah, S. Kapre, N. Charbe, S. Immanuel, S. Pasham, M. Thalla, A. Jain, S. Palakurthi, A comprehensive review of challenges and advances in exosome-based drug delivery systems, *Nanoscale Adv* 6 (2024) 5803–5826. https://doi.org/10.1039/d4na00501e.
[20] Q. Miao, S. Li, W. Lyu, J. Zhang, Y. Han, Exosomes in oral diseases: Mechanisms and therapeutic applications, *Drug Des Devel Ther* 19 (2025) 457–469. https://doi.org/10.2147/DDDT.S505355.
[21] A.M. Timofeeva, A.P. Paramonik, S.S. Sedykh, G.A. Nevinsky, Milk exosomes: Next-generation agents for delivery of anticancer drugs and therapeutic nucleic acids, *Int J Mol Sci* 24 (2023). https://doi.org/10.3390/ijms241210194.
[22] W.A. Banks, P. Sharma, K.M. Bullock, K.M. Hansen, N. Ludwig, T.L. Whiteside, Transport of extracellular vesicles across the blood-brain barrier: Brain pharmacokinetics and effects of inflammation, *Int J Mol Sci* 21 (2020) 1–21. https://doi.org/10.3390/ijms21124407.
[23] S. V. Krylova, D. Feng, The machinery of exosomes: Biogenesis, release, and uptake, *Int J Mol Sci* 24 (2023). https://doi.org/10.3390/ijms24021337.
[24] D. Yuan, Y. Zhao, W.A. Banks, K.M. Bullock, M. Haney, E. Batrakova, A. V. Kabanov, Macrophage exosomes as natural nanocarriers for protein delivery to inflamed brain, *Biomaterials* 142 (2017) 1–12. https://doi.org/10.1016/j.biomaterials.2017.07.011.
[25] M.W.A. Hussain, S. Jahangir, B. Ghosh, F. Yesmin, A. Anis, S.N. Satil, F. Anwar, M.H. Rashid, Exosomes for regulation of immune responses and immunotherapy, *J Nanotheranostics* 3 (2022) 55–85. https://doi.org/10.3390/jnt3010005.
[26] M.R. Singh, K. Yadav, N.D. Chaurasiya, D. Singh, Immune system and mechanism of immunomodulation, in: N.S. Sangwan, M.A. Farag, L.V. Modolo (Eds.), *Plants and Phytomolecules for Immunomodulation: Recent Trends and Advances*, Springer Nature Singapore, Singapore, 2022: pp. 1–31. https://doi.org/10.1007/978-981-16-8117-2_1.
[27] K. Yadav, M. Pradhan, D. Singh, M.R. Singh, Macrophage-associated disorders: pathophysiology, treatment challenges, and possible solutions, in: S. Gupta, Y.V. Pathak (Eds.), *Macrophage Targeted Delivery Systems*. Springer Nature Switzerland AG, Cham, 2022. https://doi.org/10.1007/978-3-030-84164-5_4.
[28] S. Alvarez, C. Suazo, A. Boltansky, M. Ursu, D. Carvajal, G. Innocenti, A. Vukusich, M. Hurtado, S. Villanueva, J.E. Carreño, A. Rogelio, C.E. Irarrazabal, Urinary exosomes as a source of kidney dysfunction biomarker in renal transplantation, *Transplant Proc* 45 (2013) 3719–3723. https://doi.org/10.1016/j.transproceed.2013.08.079.
[29] E.S. Grigoryeva, O.E. Savelieva, N.O. Popova, N. V. Cherdyntseva, V.M. Perelmuter, Do tumor exosome integrins alone determine organotropic metastasis? *Mol Biol Rep* 47 (2020) 8145–8157. https://doi.org/10.1007/s11033-020-05826-4.
[30] S.A.A. Kooijmans, L.A.L. Fliervoet, R. Van Der Meel, M.H.A.M. Fens, H.F.G. Heijnen, P.M.P. Van Bergen En Henegouwen, P. Vader, R.M. Schiffelers, PEGylated and targeted extracellular vesicles display enhanced cell specificity and circulation time, *J Control Release* 224 (2016) 77–85. https://doi.org/10.1016/j.jconrel.2016.01.009.
[31] S.E. Emam, N.E. Elsadek, A.S. Abu Lila, H. Takata, Y. Kawaguchi, T. Shimizu, H. Ando, Y. Ishima, T. Ishida, Anti-PEG IgM production and accelerated blood clearance phenomenon after the administration of PEGylated exosomes in mice, *J Control Release* 334 (2021) 327–334. https://doi.org/10.1016/j.jconrel.2021.05.001.
[32] S. Salunkhe, Dheeraj, M. Basak, D. Chitkara, A. Mittal, Surface functionalization of exosomes for target-specific delivery and *in vivo* imaging & tracking: Strategies and significance, *J Control Release* 326 (2020) 599–614. https://doi.org/10.1016/j.jconrel.2020.07.042.
[33] F.N.A. Jahromi, R. Dowran, R. Jafari, Recent advances in the roles of exosomal microRNAs (exomiRs) in hematologic neoplasms: Pathogenesis, diagnosis, and treatment, *Cell Commun Signal* 21 (2023). https://doi.org/10.1186/s12964-023-01102-7.
[34] H. Choi, Y. Choi, H.Y. Yim, A. Mirzaaghasi, J.-K. Yoo, C. Choi, Biodistribution of exosomes and engineering strategies for targeted delivery of therapeutic exosomes, *Tissue Eng Regen Med* 18 (2021) 499–511. https://doi.org/10.1007/s13770-021-00361-0.

[35] Y.W. Yi, J.H. Lee, S.-Y. Kim, C.-G. Pack, D.H. Ha, S.R. Park, J. Youn, B.S. Cho, Advances in analysis of biodistribution of exosomes by molecular imaging, *Int J Mol Sci* 21 (2020). https://doi.org/10.3390/ijms21020665.

[36] A.T. Nguyen-Lefebvre, A. Horuzsko, Kupffer cell metabolism and function, *J Enzymol Metab* 1 (2015) 101.

[37] Y. Nakayama, Y. Masuda, T. Mukae, T. Mikami, R. Shimizu, N. Kondo, H. Kitagawa, N. Itoh, M. Konishi, A secretory protein neudesin regulates splenic red pulp macrophages in erythrophagocytosis and iron recycling, *Commun Biol* 7 (2024). https://doi.org/10.1038/s42003-024-05802-9.

[38] Q. Si, L. Wu, D. Pang, P. Jiang, Exosomes in brain diseases: Pathogenesis and therapeutic targets, *MedComm (Beijing)* 4 (2023). https://doi.org/10.1002/mco2.287.

[39] J.J.J. Liu, D. Liu, S.K.Y. To, A.S.T. Wong, Exosomes in cancer nanomedicine: Biotechnological advancements and innovations, *Mol Cancer* 24 (2025) 166. https://doi.org/10.1186/s12943-025-02372-0.

[40] J. Lu, X. Gao, S. Wang, Y. He, X. Ma, T. Zhang, X. Liu, Advanced strategies to evade the mononuclear phagocyte system clearance of nanomaterials, *Exploration* 3 (2023). https://doi.org/10.1002/EXP.20220045.

[41] J.E. Bader, R.T. Enos, K.T. Velázquez, M.S. Carson, M. Nagarkatti, P.S. Nagarkatti, I. Chatzistamou, J.M. Davis, J.A. Carson, C.M. Robinson, E.A. Murphy, Macrophage depletion using clodronate liposomes decreases tumorigenesis and alters gut microbiota in the AOM/DSS mouse model of colon cancer, *Am J Physiol Gastrointest Liver Physiol* 314 (2018) G22–G31. https://doi.org/10.1152/ajpgi.00229.2017.

[42] J. He, W. Ren, W. Wang, W. Han, L. Jiang, D. Zhang, M. Guo, Exosomal targeting and its potential clinical application, *Drug Deliv Transl Res* 12 (2022) 2385–2402. https://doi.org/10.1007/s13346-021-01087-1.

[43] J.S. Suk, Q. Xu, N. Kim, J. Hanes, L.M. Ensign, PEGylation as a strategy for improving nanoparticle-based drug and gene delivery, *Adv Drug Deliv Rev* 99 (2016) 28–51. https://doi.org/10.1016/j.addr.2015.09.012.

[44] M. Heidarzadeh, A. Zarebkohan, R. Rahbarghazi, E. Sokullu, Protein corona and exosomes: New challenges and prospects, *Cell Commun Signal* 21 (2023). https://doi.org/10.1186/s12964-023-01089-1.

[45] S.M. Kong, D.F. Costa, A. Jagielska, K.J. Van Vliet, P.T. Hammond, Stiffness of targeted layer-by-layer nanoparticles impacts elimination half-life, tumor accumulation, and tumor penetration, *Proc Natl Acad Sci USA* 118 (2021). https://doi.org/10.1073/pnas.2104826118.

[46] Y. Jin, K. Chen, Z. Wang, Y. Wang, J. Liu, L. Lin, Y. Shao, L. Gao, H. Yin, C. Cui, Z. Tan, L. Liu, C. Zhao, G. Zhang, R. Jia, L. Du, Y. Chen, R. Liu, J. Xu, X. Hu, Y. Wang, DNA in serum extracellular vesicles is stable under different storage conditions, *BMC Cancer* 16 (2016). https://doi.org/10.1186/s12885-016-2783-2.

[47] Y. Li, Y. Wu, E.A. Federzoni, X. Wang, A. Dharmawan, X. Hu, H. Wang, R.J. Hawley, S. Stevens, M. Sykes, Y.G. Yang, CD47 cross-dressing by extracellular vesicles expressing CD47 inhibits phagocytosis without transmitting cell death signals, *Elife* 11 (2022). https://doi.org/10.7554/ELIFE.73677.

[48] M. Van Delen, J. Derdelinckx, K. Wouters, I. Nelissen, N. Cools, A systematic review and meta-analysis of clinical trials assessing safety and efficacy of human extracellular vesicle-based therapy, *J Extracell Vesicles* 13 (2024) e12458. https://doi.org/10.1002/jev2.12458.

[49] L. Digiacomo, S. Renzi, A. Pirrottina, H. Amenitsch, V. De Lorenzi, D. Pozzi, F. Cardarelli, G. Caracciolo, PEGylation-dependent cell uptake of lipid nanoparticles revealed by spatiotemporal correlation spectroscopy, *ACS Pharmacol Transl Sci* 7 (2024) 3004–3010. https://doi.org/10.1021/acsptsci.4c00419.

[50] X. Han, S. Li, X. Li, Q. Zhan, Y. Zhan, J. Zhao, X. Hou, X. Yuan, The effect of zwitterionic surface content on blood circulation time of nanocapsule, *J Biomater Appl* 35 (2020) 371–384. https://doi.org/10.1177/0885328220935381.

[51] P. Tiwari, R.P. Shukla, K. Yadav, N. Singh, D. Marwaha, S. Gautam, A.K. Bakshi, N. Rai, A. Kumar, D. Sharma, P.R. Mishra, Dacarbazine-primed carbon quantum dots coated with breast cancer cell-derived exosomes for improved breast cancer therapy, *J Control Release* 365 (2023) 43–59. https://doi.org/10.1016/j.jconrel.2023.11.005.

[52] K. Yadav, K.K. Sahu, Sucheta, S.P.E. Gnanakani, P. Sure, R. Vijayalakshmi, V.D. Sundar, V. Sharma, R. Antil, M. Jha, S. Minz, A. Bagchi, M. Pradhan, Biomedical applications of nanomaterials in the advancement of nucleic acid therapy: Mechanistic challenges, delivery strategies, and therapeutic applications, *Int J Biol Macromol* 241 (2023) 124582. https://doi.org/10.1016/j.ijbiomac.2023.124582.

[53] P. Tiwari, R.P. Shukla, K. Yadav, D. Panwar, N. Agarwal, A. Kumar, N. Singh, A.K. Bakshi, D. Marwaha, S. Gautam, N. Rai, P.R. Mishra, Exploring nanocarriers as innovative materials for advanced drug delivery strategies in onco-immunotherapies, *J Mol Graph Model* 128 (2024) 108702. https://doi.org/10.1016/j.jmgm.2024.108702.

[54] A.L. Kiss, E. Botos, Endocytosis via caveolae: Alternative pathway with distinct cellular compartments to avoid lysosomal degradation? *J Cell Mol Med* 13 (2009) 1228–1237. https://doi.org/10.1111/j.1582-4934.2009.00754.x.

[55] P. Tiwari, K. Yadav, R.P. Shukla, S. Gautam, D. Marwaha, M. Sharma, P.R. Mishra, Surface modification strategies in translocating nano-vesicles across different barriers and the role of bio-vesicles in improving anticancer therapy, *J Control Release* 363 (2023) 290–348. https://doi.org/10.1016/j.jconrel.2023.09.016.

[56] B.S. Joshi, M.A. de Beer, B.N.G. Giepmans, I.S. Zuhorn, Endocytosis of extracellular vesicles and release of their cargo from endosomes, *ACS Nano* 14 (2020) 4444–4455. https://doi.org/10.1021/acsnano.9b10033.

[57] D. Wang, Y. Bai, G. Cheng, S. Shen, G. Xiao, D. Ma, G. Zhao, W. Chen, T. Li, L. Zhang, X. Ge, Exosome-drug conjugates delivery: A promising strategy for ameliorating the pharmacokinetic profile of artesunate, *Front Bioeng Biotechnol* 12 (2024) 1437787. https://doi.org/10.3389/fbioe.2024.1437787.

[58] K. Yadav, S.P.E. Gnanakani, K.K. Sahu, C.K. Veni Chikkula, P.S. Vaddi, S. Srilakshmi, R. Yadav, Sucheta, A. Dubey, S. Minz, S. Minz, M. Pradhan, Nano revolution of DNA nanostructures redefining cancer therapeutics – a comprehensive review, *Int J Biol Macromol* 274 (2024). https://doi.org/10.1016/j.ijbiomac.2024.133244.

[59] A.S. Strik, Y.M.C. Wang, L.E. Ruff, W. Yashar, B.T. Messmer, D.R. Mould, Individualized dosing of therapeutic monoclonal antibodies – a changing treatment paradigm? *AAPS J* 20 (2018). https://doi.org/10.1208/s12248-018-0257-y.

[60] S. Vanacova, R. Stef, The exosome and RNA quality control in the nucleus, *EMBO Rep* 8 (2007) 651–657. https://doi.org/10.1038/sj.embor.7401005.

[61] D. Glick, S. Barth, K.F. Macleod, Autophagy: Cellular and molecular mechanisms, *J Pathol* 221 (2010) 3–12. https://doi.org/10.1002/path.2697.

[62] Z. Iqbal, K. Rehman, A. Mahmood, M. Shabbir, Y. Liang, L. Duan, H. Zeng, Exosome for mRNA delivery: Strategies

and therapeutic applications, *J Nanobiotechnol* 22 (2024) 395. https://doi.org/10.1186/s12951-024-02634-x.
[63] Q. Huang, J. Yang, J. Zheng, C. Hsueh, Y. Guo, L. Zhou, Characterization of selective exosomal microRNA expression profile derived from laryngeal squamous cell carcinoma detected by next generation sequencing, *Oncol Rep* 40 (2018) 2584–2594. https://doi.org/10.3892/or.2018.6672.
[64] A.M. Gleason, E.G. Woo, C. McKinney, E. Sidransky, The role of exosomes in lysosomal storage disorders, *Biomolecules* 11 (2021). https://doi.org/10.3390/biom11040576.
[65] C. Hu, M. Chen, R. Jiang, Y. Guo, M. Wu, X. Zhang, Exosome-related tumor microenvironment, *J Cancer* 9 (2018) 3084–3092. https://doi.org/10.7150/jca.26422.
[66] K. Sato, C. Hall, S. Glaser, H. Francis, F. Meng, G. Alpini, Pathogenesis of Kupffer cells in cholestatic liver injury, *Am J Pathol* 186 (2016) 2238–2247. https://doi.org/10.1016/j.ajpath.2016.06.003.
[67] M. Yu, J. Liu, X. Ning, J. Zheng, High-contrast noninvasive imaging of kidney clearance kinetics enabled by renal clearable nanofluorophores, *Angew Chem Int Ed* 54 (2015) 15434–15438. https://doi.org/10.1002/anie.201507868.
[68] H. An, Y. Huang, Z. Zhao, K. Li, J. Meng, X. Huang, X. Tian, H. Zhou, J. Wu, Q. Dai, J.-R. Zhang, Splenic red pulp macrophages eliminate the liver-resistant *Streptococcus pneumoniae* from the blood circulation of mice, *Sci Adv* 11 (2025) eadq6399. https://doi.org/10.1126/sciadv.adq6399.
[69] I. Freitas, E. Boncompagni, R. Vaccarone, C. Fenoglio, S. Barni, G.F. Baronzio, Iron accumulation in mammary tumor suggests a tug of war between tumor and host for the microelement, *Anticancer Res* 27 (2007) 3059–3065.
[70] S.G. Elci, Y. Jiang, B. Yan, S.T. Kim, K. Saha, D.F. Moyano, G. Yesilbag Tonga, L.C. Jackson, V.M. Rotello, R.W. Vachet, Surface charge controls the suborgan biodistributions of gold nanoparticles, *ACS Nano* 10 (2016) 5536–5542. https://doi.org/10.1021/acsnano.6b02086.
[71] P.M. Glassman, V.R. Muzykantov, Pharmacokinetic and pharmacodynamic properties of drug delivery systems, *J Pharmacol Exp Ther* 370 (2019) 570–580. https://doi.org/10.1124/jpet.119.257113.
[72] Q. Hu, C.J. Lyon, J.K. Fletcher, W. Tang, M. Wan, T.Y. Hu, Extracellular vesicle activities regulating macrophage- and tissue-mediated injury and repair responses, *Acta Pharm Sin B* 11 (2021) 1493–1512. https://doi.org/10.1016/j.apsb.2020.12.014.
[73] C. Bao, H. Xiang, Q. Chen, Y. Zhao, Q. Gao, F. Huang, L. Mao, A review of labeling approaches used in small extracellular vesicles tracing and imaging, *Int J Nanomed* 18 (2023) 4567–4588. https://doi.org/10.2147/IJN.S416131.
[74] A.A. Khan, F. Man, F.N. Faruqu, J. Kim, F. Al-Salemee, A. Carrascal-Miniño, A. Volpe, R. Liam-Or, P. Simpson, G.O. Fruhwirth, K.T. Al-Jamal, R.T.M. De Rosales, PET imaging of small extracellular vesicles via [^{89}Zr]Zr(oxinate)$_4$ direct radiolabeling, *Bioconjug Chem* 33 (2022) 473–485. https://doi.org/10.1021/acs.bioconjchem.1c00597.
[75] M. Morishita, Y. Takahashi, M. Nishikawa, Y. Takakura, Pharmacokinetics of exosomes – an important factor for elucidating the biological roles of exosomes and for the development of exosome-based therapeutics, *J Pharm Sci* 106 (2017) 2265–2269. https://doi.org/10.1016/j.xphs.2017.02.030.
[76] P. Tiwari, R.P. Shukla, K. Yadav, M. Sharma, A.K. Bakshi, D. Panwar, N. Singh, N. Agarwal, M.N. Mugale, P.R. Mishra, YIGSR functionalized hybrid exosomes spatially target dasatinib to laminin receptors for precision therapy in breast cancer, *Adv Healthc Mater* n/a (2025) 2402673. https://doi.org/10.1002/adhm.202402673.
[77] A. Al Mannai, T. Al-Ansari, K.M. Saoud, Quantification of serum exosome biomarkers using 3D nanoporous gold and spectrophotometry, *Sensors* 22 (2022). https://doi.org/10.3390/s22176347.
[78] Y. Chen, D. Gao, Q. Zhu, B. Chu, J. Peng, J. Wang, L. Liu, Y. Jiang, Rapid exosome isolation and in situ multiplexed detection of exosomal surface proteins and microRNAs on microfluidic platform, *Analyst* 148 (2023) 2387–2394. https://doi.org/10.1039/D3AN00335C.
[79] S. Huang, F. Yan, Y. Qiu, T. Liu, W. Zhang, Y. Yang, R. Zhong, Y. Yang, X. Peng, Exosomes in inflammation and cancer: From bench to bedside applications, *Mol Biomed* 6 (2025) 41. https://doi.org/10.1186/s43556-025-00280-9.
[80] C. Lyu, H. Sun, Z. Sun, Y. Liu, Q. Wang, Roles of exosomes in immunotherapy for solid cancers, *Cell Death Dis* 15 (2024). https://doi.org/10.1038/s41419-024-06494-z.
[81] S. Gotoh, M. Kawabori, M. Fujimura, Intranasal administration of stem cell-derived exosomes for central nervous system diseases, *Neural Regen Res* 19 (2024) 1249–1255. https://doi.org/10.4103/1673-5374.385875.
[82] Y. Xia, J. Zhang, G. Liu, J. Wolfram, Immunogenicity of extracellular vesicles, *Adv Mater* 36 (2024) e2403199. https://doi.org/10.1002/adma.202403199.
[83] A.K. Agrawal, F. Aqil, J. Jeyabalan, W.A. Spencer, J. Beck, B.W. Gachuki, S.S. Alhakeem, K. Oben, R. Munagala, S. Bondada, R.C. Gupta, Milk-derived exosomes for oral delivery of paclitaxel, *Nanomedicine* 13 (2017) 1627–1636. https://doi.org/10.1016/j.nano.2017.03.001.
[84] M. Samuel, P. Fonseka, R. Sanwlani, L. Gangoda, S.H. Chee, S. Keerthikumar, A. Spurling, S. V. Chitti, D. Zanker, C.S. Ang, I. Atukorala, T. Kang, S. Shahi, A.L. Marzan, C. Nedeva, C. Vennin, M.C. Lucas, L. Cheng, D. Herrmann, M. Pathan, D. Chisanga, S.C. Warren, K. Zhao, N. Abraham, S. Anand, S. Boukouris, C.G. Adda, L. Jiang, T.M. Shekhar, N. Baschuk, C.J. Hawkins, A.J. Johnston, J.M. Orian, N.J. Hoogenraad, I.K. Poon, A.F. Hill, M. Jois, P. Timpson, B.S. Parker, S. Mathivanan, Oral administration of bovine milk-derived extracellular vesicles induces senescence in the primary tumor but accelerates cancer metastasis, *Nat Commun* 12 (2021). https://doi.org/10.1038/s41467-021-24273-8.
[85] S. Gurung, D. Perocheau, L. Touramanidou, J. Baruteau, The exosome journey: From biogenesis to uptake and intracellular signalling, *Cell Commun Signal* 19 (2021). https://doi.org/10.1186/s12964-021-00730-1.
[86] F.N. Faruqu, J.T.W. Wang, L. Xu, L. McNickle, E.M.Y. Chong, A. Walters, M. Gurney, A. Clayton, L.A. Smyth, R. Hider, J. Sosabowski, K.T. Al-Jamal, Membrane radiolabelling of exosomes for comparative biodistribution analysis in immunocompetent and immunodeficient mice – a novel and universal approach, *Theranostics* 9 (2019) 1666–1682. https://doi.org/10.7150/thno.27891.
[87] M. Morishita, Y. Takahashi, M. Nishikawa, Y. Takakura, Pharmacokinetics of exosomes – an important factor for elucidating the biological roles of exosomes and for the development of exosome-based therapeutics, *J Pharm Sci* 106 (2017) 2265–2269. https://doi.org/10.1016/j.xphs.2017.02.030.
[88] K. Yadav, Advancing pediatric drug safety: The potential of physiologically based pharmacokinetic modeling, *Drug Metab Bioanal Lett* 18 (2025) 122–165. https://doi.org/10.2174/0118723128367217250602073115.
[89] K. Yadav, R. Vijayalakshmi, K. Kumar Sahu, P. Sure, K. Chahal, R. Yadav, Sucheta, A. Dubey, M. Jha, M. Pradhan, Exosome-based macromolecular neurotherapeutic drug delivery approaches in overcoming the blood-brain barrier

for treating brain disorders, *Eur J Pharm Biopharm* 199 (2024) 114298. https://doi.org/10.1016/j.ejpb.2024.114298.
[90] J. Saint-Pol, M. Culot, Minimum information for studies of extracellular vesicles (MISEV) as toolbox for rigorous, reproducible and homogeneous studies on extracellular vesicles, *Toxicol In vitro* 106 (2025) 106049. https://doi.org/10.1016/j.tiv.2025.106049.
[91] CBER Product Jurisdiction | FDA, 2025. https://www.fda.gov/about-fda/center-biologics-evaluation-and-research-cber/cber-product-jurisdiction (accessed August 13, 2025).
[92] C.-K. Wang, T.-H. Tsai, C.-H. Lee, Regulation of exosomes as biologic medicines: Regulatory challenges faced in exosome development and manufacturing processes, *Clin Transl Sci* 17 (2024) e13904. https://doi.org/10.1111/cts.13904.
[93] K. Yadav, N.S. Chauhan, S. Saraf, D. Singh, M.R. Singh, Challenges and need of delivery carriers for bioactives and biological agents: An introduction, in: *Advances and Avenues in the Development of Novel Carriers for Bioactives and Biological Agents*, Elsevier, 2020: pp. 1–36. https://doi.org/10.1016/b978-0-12-819666-3.00001-8.
[94] K. Yadav, M.R. Singh, V.K. Rai, N. Srivastava, N. Prasad Yadav, Commercial aspects and market potential of novel delivery systems for bioactives and biological agents, in: *Advances and Avenues in the Development of Novel Carriers for Bioactives and Biological Agents*, Elsevier, 2020: pp. 595–620. https://doi.org/10.1016/b978-0-12-819666-3.00020-1.
[95] P. Rana, B. Hollingshead, R. Mangipudy, Rethinking the necessity of long-term toxicity studies for biotherapeutics using weight of evidence assessment, *Regul Toxicol Pharmacol* 153 (2024) 105710. https://doi.org/10.1016/j.yrtph.2024.105710.

4 Exosome Therapies Innovations in Cell Therapy and Targeted Drug Delivery

Biba Vikas[1], Advik Vikas[2], Vishnupriya. K. Sweety[3], Lekshmi K. Edison[4], Amartya Raj Suraj[3], and Sukumaran Anil[5]
[1]NextGen Precision Health, University of Missouri, Columbia, MO, USA
[2]Oral Health Institute, Hamad Medical Corporation, Qatar University, Doha, Qatar
[3]Pushpagiri Institute of Medical Sciences, Thiruvalla, Kerala, India
[4]Department of Comparative Medicine, College of Veterinary Medicine, University of Florida, Gainesville, FL, USA
[5]Department of Dentistry, Hamad Medical Corporation, Oral Health Institute, Qatar University, Doha, Qatar

4.1 INTRODUCTION

The landscape of modern medicine is continuously evolving, marked by paradigm shifts that introduce novel therapeutic modalities with unprecedented potential. Following the era of small molecules and biologics and the subsequent rise of cell-based therapies, the scientific community now stands at the precipice of a third wave: cell-free therapeutics, headlined by exosomes [1]. Exosomes, a specialized subset of extracellular vehicles (EVs) with diameters ranging from 30 nm to 150 nm, have emerged as transformative tools in therapeutic science. Initially regarded as cellular debris in reticulocyte maturation, exosomes are now recognized as critical mediators of intercellular communication, exerting profound effects across various physiological and pathological processes. Derived from the endosomal multivesicular body (MVB) pathway and secreted by nearly all cell types, these vesicles encapsulate a curated set of bioactive molecules, including lipids, proteins, mRNAs, microRNAs, and long noncoding RNAs [2]. Their unique biological features, such as intrinsic biocompatibility, ability to traverse biological barriers like the blood–brain barrier (BBB), low immunogenicity, and preferential homing to diseased tissues, position exosomes as naturally evolved nanocarriers for targeted therapeutic delivery [3, 4]. These advantages make exosomes superior to synthetic nanoparticles and viral vectors, which are often limited by immunogenicity, toxicity, and poor tissue penetration.

When positioned against conventional synthetic drug delivery systems, such as liposomes and polymeric nanoparticles, the unique advantages of exosomes become strikingly clear. While synthetic carriers offer high reproducibility and can be loaded with substantial amounts of therapeutic cargo, they are fundamentally foreign materials that elicit immune responses and face toxicity and rapid clearance challenges [5]. In contrast, exosomes, being of endogenous origin, exhibit remarkable biocompatibility, low immunogenicity, and an innate stability in circulation [6]. The complex process of exosome biogenesis is illustrated in Figure 4.1, showing the formation from multivesicular bodies and subsequent secretion. Perhaps most critically, they possess a natural and unparalleled ability to traverse formidable biological barriers, most notably the blood–brain barrier (BBB), which has long stymied the development of effective treatments for neurological diseases. The key differences between endogenous and exogenous engineering approaches are compared in Table 4.1. This inherent biological sophistication makes them superior candidates for sensitive therapeutics to hard-to-reach tissues. However, the promise of natural exosomes is tempered by significant practical limitations. Unmodified exosomes often suffer from low production yields, nonspecific biodistribution, leading to rapid accumulation in clearance organs such as the liver and spleen, and an insufficient concentration of therapeutic cargo to elicit a potent clinical effect. This reality gives rise to the central thesis of this chapter: sophisticated bioengineering is not merely an incremental enhancement but an absolute prerequisite for unlocking the full clinical potential of exosome-based therapies [7]. These natural nanovesicles can be transformed into precision-guided medicines through deliberate modification, enhancing their stability, equipping them with targeting moieties, and loading them with potent therapeutic payloads. Despite these inherent strengths, unmodified (native) exosomes face critical limitations in clinical application. Heterogeneity in size, cargo composition, and origin-related variability hamper reproducibility [8]. Furthermore, passive targeting mechanisms and inefficient loading of large or sensitive therapeutics (e.g., siRNA, CRISPR-Cas systems) reduce their functional efficacy and scalability [9].

The therapeutic versatility of engineered exosomes is particularly relevant to chronic diseases, which involve complex molecular dysregulation and necessitate precision-targeted interventions [15]. Engineered exosomes, with their

DOI: 10.1201/9781003623632-4

customizable biological interface and tunable pharmacokinetics, represent a paradigm shift in personalized medicine, capable of bridging the gap between molecular diagnostics and site-specific therapeutic intervention [6, 16]. The primary therapeutic applications of engineered exosomes across different disease areas are outlined in Table 4.2.

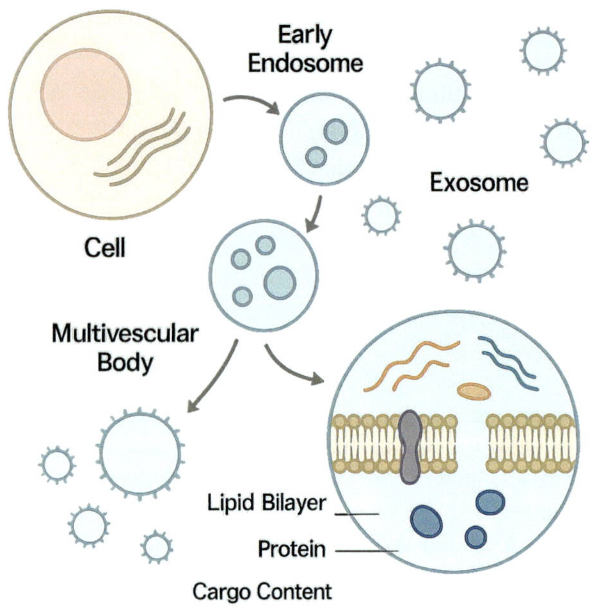

FIGURE 4.1 Schematic of exosome biogenesis and structure, showing formation from multivesicular bodies, cargo packaging, and secretion. Credit: Biorender.com.

Engineered exosomes have been applied in the following disorders or diseases:

- *Cancer* – delivering siRNA, chemotherapeutics, or immune modulators to tumors
- *Neurodegenerative diseases* – crossing the BBB to deliver neuroprotective agents
- *Cardiovascular and metabolic diseases* – modulating inflammation and fibrosis
- *Rare and genetic disorders* – delivering corrected genes or enzymes

To address these constraints, the field has embraced exosome engineering, a multidisciplinary approach that, unlike synthetic drug carriers, which are often rapidly cleared by the mononuclear phagocyte system or trigger immune reactions, exosomes are naturally evolved delivery vehicles with high specificity, stability, and intracellular trafficking capabilities. Their membranes, enriched in sphingomyelin, ceramide, cholesterol, and tetraspanins (CD9, CD63, and CD81), facilitate membrane fusion and cell-specific uptake. The transition of engineered exosomes from bench to bedside is accelerating, with several candidates advancing through preclinical and early-phase clinical trials. Current research includes applications in regenerative medicine, vaccine platforms, and gene editing. However, this progress is tempered by challenges in large-scale production, batch-to-batch consistency, quality control, and regulatory classification.

TABLE 4.1
Comparison of Endogenous versus Exogenous Exosome Engineering Approaches

Feature	Endogenous Engineering	Exogenous Engineering	References
Method	Modify parent cells	Modify isolated exosomes	[10, 11]
Examples	Genetic transfection, CRISPR, metabolic labeling	Electroporation, sonication, and click chemistry	[12, 13]
Pros	Biologically integrated, stable modifications	High flexibility, immediate customization	[14]
Cons	Biologically integrated, stable modifications	High flexibility, immediate customization	[14]

TABLE 4.2
Key Therapeutic Applications of Engineered Exosomes

Disease Area	Engineered Function	Example Cargo/Feature	Benefit	References
Oncology	Tumor-targeting ligand	siRNA, doxorubicin	Tumor regression, immune activation	[17, 18]
Neurological disorders	BBB-penetrating peptide	Curcumin, anti-Aβ agents	Cognitive preservation, plaque reduction	[19, 20]
Cardiovascular disease	CXCR4 homing ligand	miRNA-21	Improved cardiac repair	[21, 22]
Metabolic syndrome	Anti-inflammatory exosomes	IL-10	Reduced insulin resistance and fibrosis	[23]

4.2 MECHANISMS OF EXOSOME FORMATION AND THEIR BIOLOGICAL FUNCTIONS

Exosomes are generated through a highly orchestrated process involving the endosomal sorting complex required for transport (ESCRT) machinery. This process begins with the inward budding of the early endosomal membrane, resulting in the formation of multivesicular bodies (MVBs) that contain intraluminal vesicles (ILVs). These ILVs, once released into the extracellular space following MVB fusion with the plasma membrane, are referred to as exosomes [24]. The initial step involves invading the plasma membrane to form early endosomes, which mature into late endosomes. Within these late endosomes, inward budding of the limiting membrane generates ILVs, which become the cargo-rich exosomes upon MVB fusion with the plasma membrane [25]. This inward budding and ILV formation is primarily mediated by the endosomal sorting complexes required for transport (ESCRT) machinery, which comprises four sequential protein complexes (ESCRT-0 to ESCRT-III) along with associated ATPases, such as VPS4, and accessory proteins like ALIX and TSG101 [26]. ESCRT-independent pathways have also been described, often involving tetraspanins (e.g., CD9, CD63, CD81) and lipid raft-associated components such as ceramide, suggesting redundancy and complexity in exosome biogenesis [27].

The molecular architecture of exosomes is highly specialized, reflecting both their endosomal origin and the selective cargo sorting mechanisms that govern their composition. Lipid components of exosomes are enriched in cholesterol, sphingomyelin, ceramide, and phosphatidylserine, contributing to their membrane rigidity and fusogenic properties [28]. The proteomic profile is dominated by tetraspanins (CD9, CD63, CD81), heat shock proteins (HSP70, HSP90), ESCRT-associated proteins (ALIX, TSG101), integrins, and annexins. These proteins serve as canonical exosomal markers and play functional roles in membrane trafficking, cell adhesion, and immune modulation [29]. Exosomes are also rich in nucleic acid cargo, including messenger RNAs, microRNAs (miRNAs), long noncoding RNAs (lncRNAs), and even DNA fragments, which are selectively sorted through RNA-binding proteins such as hnRNPA2B1 and YBX1. This cargo can be horizontally transferred to recipient cells, exerting epigenetic, transcriptional, or posttranscriptional regulatory effects [30].

Functionally, exosomes are essential mediators of intercellular communication, enabling the transfer of bioactive molecules over short and long distances without the need for direct cell–cell contact. They play crucial roles in immune regulation, including antigen presentation, modulation of cytokine signaling, and either promoting or suppressing immune responses, depending on their cellular origin [31]. For example, exosomes derived from mesenchymal stem cells (MSCs) have demonstrated profound immunosuppressive and pro-regenerative properties in models of inflammation and autoimmune disease. In tissue repair, exosomes facilitate angiogenesis, matrix remodeling, and cellular proliferation, often by activating signaling pathways such as PI3K/Akt, MAPK/ERK, and Wnt/β-catenin in recipient cells [23]. Moreover, their ability to cross biological barriers and exert effects in distal tissues underscores their relevance in systemic regulation and pathophysiological processes, including those associated with chronic diseases such as cancer, neurodegeneration, and metabolic syndrome. As intrinsic conveyors of molecular information and modulators of cellular behavior, exosomes represent a fundamental component of the secretome and a blueprint for developing biomimetic delivery systems. A comprehensive understanding of their biogenesis and native function is thus indispensable for harnessing their full therapeutic potential.

4.3 INNOVATIVE APPROACHES IN EXOSOME ENGINEERING

The advancement of exosome-based therapeutics has been propelled by the advent of sophisticated engineering strategies that enable the customization of exosomes for enhanced delivery efficacy, specificity, and stability. These strategies broadly encompass modification of the exosome-producing donor cells through genetic engineering, direct manipulation of isolated exosomes via surface or cargo modification, and the creation of hybrid systems that combine the functional benefits of exosomes and synthetic nanocarriers [11]. Each of these approaches presents unique advantages and challenges in the context of targeted drug delivery and cell therapy.

4.3.1 Genetic Engineering of Donor Cells

One of the most biologically congruent approaches to exosome customization involves the genetic engineering of donor cells, allowing for the incorporation of desired biomolecules into the exosome lumen or membrane during biogenesis. This method ensures that engineered components are naturally sorted into exosomes via endogenous pathways, thus preserving vesicle integrity and biological function. Transfection of parental cells with expression vectors encoding therapeutic RNAs (e.g., siRNA, miRNA, lncRNA) or proteins (e.g., cytokines, tumor suppressors) can lead to the selective enrichment of these payloads in secreted exosomes. For example, overexpression of anti-oncogenic miR-122 in donor hepatocytes resulted in exosomes that suppressed hepatocellular carcinoma progression via targeted gene silencing [32]. Similarly, fusion constructs involving exosomal membrane proteins such as Lamp2b, CD63, or CD9 tagged with targeting ligands (e.g., neuron-specific RVG peptide) have enabled cell-specific delivery, particularly in neurodegenerative models [14]. CRISPR/Cas-based genome editing has been utilized to program donor cells to produce exosomes with a precise molecular composition. This includes not only the delivery of CRISPR-Cas9/sgRNA complexes via exosomes for gene editing in recipient cells but also the modulation of exosome-producing machinery to alter sorting pathways and enhance therapeutic cargo incorporation [12]. Such genome-level reprogramming holds potential for disease-modifying interventions in monogenic and complex chronic diseases.

4.3.2 Direct Exosome Surface Modification

Direct chemical or biochemical modification of exosome membranes after isolation enables surface functionalization without altering the donor cells. This is particularly advantageous for clinical translation, as it provides modular control over targeting moieties and pharmacokinetics [11]. Click chemistry, particularly copper-free strain-promoted azide–alkyne cycloaddition (SPAAC), provides a bio-orthogonal and efficient approach to conjugating targeting ligands, fluorescent probes, or imaging agents to exosome surfaces without compromising vesicle integrity [13]. Similarly, PEGylation – the covalent attachment of polyethylene glycol chains – has been utilized to enhance exosome circulation time by reducing opsonization and subsequent clearance by the mononuclear phagocyte system [33]. For active targeting, ligands such as RGD peptides, folic acid, or monoclonal antibodies can be conjugated to the exosomal membrane, facilitating selective binding and uptake by target cells [34]. These modifications not only enhance biodistribution but also mitigate off-target effects, a crucial consideration in the treatment of systemic chronic diseases and malignancies.

4.3.3 Cargo-Loading Techniques

In addition to modifying exosomal surfaces, various strategies have been developed to load exogenous therapeutic agents into exosomes post-isolation. These can be broadly categorized into passive and active loading techniques, with hybrid approaches combining the advantages of both [35]. Passive loading relies on the natural diffusion of small hydrophobic molecules or nucleic acids into exosomes during incubation under optimized physicochemical conditions. While minimally disruptive, this method often yields low encapsulation efficiency and is unsuitable for large or charged biomolecules [36]. Active loading techniques, including electroporation, sonication, and extrusion, facilitate membrane permeabilization to enhance cargo entry. Electroporation transiently disrupts the exosomal membrane using electric pulses, enabling loading of siRNA, miRNA, or small drugs; however, it may induce vesicle aggregation and cargo degradation. Sonication applies mechanical shear forces, while extrusion forces exosomes through nanoporous membranes; both methods promote high loading efficiencies but risk altering membrane integrity and functional properties [37]. Hybrid exosome systems represent a cutting-edge approach in which exosomes are fused with synthetic liposomes or polymeric nanocarriers to generate vesicles with enhanced mechanical stability, increased cargo capacity, and stimuli-responsive release properties. For instance, fusogenic liposomes containing anticancer agents can integrate with exosomes to produce hybrid nanovesicles capable of deep tumor penetration and controlled payload delivery under acidic or enzymatic conditions [38]. Together, these strategies form the technological backbone of modern exosome bioengineering, enabling the rational design of highly specialized exosome formulations tailored to the therapeutic demands of chronic, refractory, and systemic diseases.

4.4 ENGINEERING EXOSOMES FOR ENHANCED THERAPEUTIC FUNCTIONS

The engineering of exosomes extends beyond the simple incorporation of therapeutic cargo or surface moieties; it encompasses a strategic reprogramming of their biophysical and biochemical attributes to optimize performance in targeted delivery, systemic circulation, controlled release, and immune modulation [10, 39]. These enhancements are crucial for overcoming the inherent limitations of native exosomes and aligning them with the stringent requirements of clinical translation in the treatment of chronic diseases.

4.4.1 Enhanced Targeting and Cellular Uptake

Native exosomes exhibit some degree of cell tropism; however, their unmodified surfaces often result in suboptimal targeting efficiency, especially in heterogeneous or poorly vascularized tissues. Engineering strategies have addressed this by decorating exosome membranes with targeting ligands such as peptides, aptamers, antibodies, and small molecules that bind with high specificity to surface receptors overexpressed on diseased cells [40]. For instance, the conjugation of RGD peptides to exosome surfaces enhances binding to $\alpha v\beta 3$ integrins, commonly upregulated in tumor neovasculature, thereby improving homing and internalization [34]. Similarly, displaying neuron-specific RVG peptides via fusion with exosomal membrane proteins, such as Lamp2b, has enabled the precise delivery of therapeutic RNA to neuronal cells, thereby bypassing the blood–brain barrier. Cellular uptake of engineered exosomes is also modulated by surface charge, lipid composition, and membrane fluidity, parameters that can be tuned through biochemical modification or hybridization with synthetic vesicles [41]. These design principles enhance receptor-mediated endocytosis and macropinocytosis, thus ensuring robust intracellular delivery of the therapeutic payload.

4.4.2 Prolonged Circulation Time and Stability

The in vivo application of exosomes requires enhanced pharmacokinetics, specifically prolonged systemic circulation and stability under physiological conditions. Native exosomes are rapidly cleared by the mononuclear phagocyte system (MPS), primarily through uptake by liver and spleen macrophages. To mitigate this, PEGylation of exosomal surfaces has been extensively employed, forming a steric barrier that reduces protein opsonization and recognition by immune cells [42]. Moreover, incorporation of sialic acid residues or CD47 proteins, which act as "don't eat me" signals, into exosomal membranes has been shown to suppress phagocytic uptake, thereby prolonging half-life in circulation [43]. Engineered exosomes may also be stabilized through cross-linking strategies or by embedding stabilizing agents, such as cholesterol and sphingomyelin, to resist aggregation, enzymatic degradation, and membrane fusion events in vivo.

4.4.3 Controlled Drug Release Kinetics

One of the most significant translational advantages of exosome engineering is the ability to achieve temporally and spatially controlled release of therapeutic cargo. Unlike synthetic liposomes, native exosomes lack intrinsic mechanisms for controlled release. To address this, engineering efforts have focused on stimulus-responsive release systems, whereby exosomes are modified to respond to environmental cues such as pH, enzymatic activity, redox potential, or temperature. For instance, exosomes functionalized with acid-labile linkers can release their payload in acidic tumor microenvironments or within lysosomal compartments of target cells. Enzyme-sensitive peptides can be incorporated into exosome membranes or conjugated to drug molecules to enable cleavage and release in response to proteases like MMPs, which are overexpressed in inflammatory or fibrotic tissues [44]. In hybrid exosome–liposome systems, pH-sensitive lipids or polymer coatings further enable burst release kinetics under pathological conditions while maintaining payload stability in circulation.

4.4.4 Immune Evasion and Stealth Properties

As biological entities, exosomes naturally possess a degree of immune tolerance, particularly when derived from autologous or stem cell sources. However, engineering approaches can substantially enhance these properties, enabling stealth-like behavior even in allogeneic or xenogeneic contexts.

Surface modification with CD47, a transmembrane protein that inhibits phagocytosis by engaging SIRPα on macrophages, has been shown to reduce immunogenic clearance and prolong exosome retention dramatically [45]. Additionally, exosomes derived from mesenchymal stem cells (MSCs) or engineered to express PD-L1 can suppress immune activation and inflammatory cytokine production, making them suitable for applications in autoimmune disorders or transplant tolerance. Furthermore, minimizing the expression of immunogenic membrane proteins or using membrane cloaking techniques (e.g., erythrocyte membrane fusion) can convert exosomes into fully stealth nanocarriers, capable of systemic administration without eliciting adverse immune responses. Such strategies are pivotal for chronic disease therapy, often requiring repeated dosing and long-term exposure to therapeutic nanocarriers.

4.5 EXOSOME APPLICATIONS IN TARGETED DRUG DELIVERY

Exosomes represent a multifaceted therapeutic platform capable of delivering a wide array of bioactive molecules with high specificity and minimal toxicity. Exosome-based drug delivery systems are transforming therapeutic paradigms across diverse medical specialties. In oncology, they enable tumor-selective delivery while mitigating chemotherapy toxicity. For neurological disorders, they solve the decades-old BBB penetration challenge. Cardiovascular applications harness their innate regenerative capacity, while infectious disease strategies exploit their immunological functions [46]. Their adaptability across disease models highlights their potential as universal vectors for precision medicine. As clinical validation accelerates, exosome therapies promise to usher in a new era of precision medicine where treatments are simultaneously more effective and better tolerated. The diverse therapeutic applications of engineered exosomes across different disease areas are summarized in Figure 4.2. Future directions include the development of multifunctional exosomes that combine targeting, therapeutic, and diagnostic capabilities for truly personalized medicine [47].

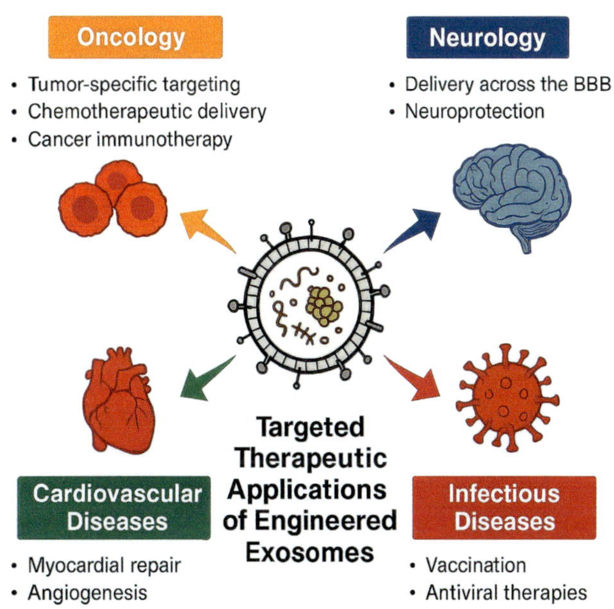

FIGURE 4.2 Targeted therapeutic applications of engineered exosomes. Credit: Biorender.com.

4.5.1 Oncology: Tumor-Targeted Chemotherapeutics and Immunomodulation

Exosome-based drug delivery has emerged as a revolutionary approach in cancer therapy, addressing critical challenges such as systemic toxicity, tumor resistance, and immune clearance. Their natural tropism for tumor tissues, combined with engineered targeting capabilities, makes them ideal vehicles for precision oncology.

- *Chemotherapeutic delivery:* Exosomes leverage their ability to deliver bioactive cargos to tumor tissues with high specificity while avoiding immune clearance. Studies demonstrate that tumor-derived exosomes inherently home to their parent tumor type, enabling a homotypic targeting strategy. Additionally, engineered exosomes can be decorated with tumor-specific ligands, such as GE11 for EGFR-positive cancers or iRGD for integrin-rich tumor vasculature, thereby enhancing selective uptake by malignant cells [48]. Chemotherapeutics such as paclitaxel, doxorubicin, and cisplatin have been successfully encapsulated in exosomes using techniques like electroporation and sonication. Exosomes loaded with doxorubicin or paclitaxel demonstrate superior tumor accumulation compared to free drugs or synthetic nanoparticles. In one notable study, macrophage-derived exosomes carrying paclitaxel significantly outperformed free drug formulations in murine models of pulmonary metastasis, achieving greater tumor accumulation and a superior therapeutic index [17]. Exosomes derived from mesenchymal stem cells (MSCs) loaded with paclitaxel exhibited 50-fold higher accumulation in lung tumors compared to conventional delivery in preclinical models. Furthermore, exosomes demonstrate improved drug retention and reduced cardiotoxicity compared to liposomal carriers.
- *Nucleic acid therapeutics:* RNA-based therapies have significantly benefited from exosome delivery platforms. Beyond small molecules, exosomes effectively deliver nucleic acid therapeutics, overcoming limitations of synthetic transfection reagents. siRNA against oncogenes packaged in exosomes silences target genes with high efficiency. Exosomes loaded with siRNA or miRNA targeting oncogenes, such as KRAS or MYC, have demonstrated strong gene silencing effects both in vitro and in vivo. In pancreatic cancer models, miRNA-122-loaded exosomes have been shown to sensitize drug-resistant tumor cells to chemotherapy [18]. CRISPR-Cas9 components delivered via exosomes enable precise genome editing in tumors, opening new avenues for personalized cancer therapy.
- *Immunotherapy applications:* Exosomes play a transformative role in cancer immunotherapy. On the immunological front, dendritic cell–derived exosomes (DEX) are under clinical investigation as cancer vaccines, functioning as "natural nanovaccines." These exosomes, enriched in MHC class I/II and co-stimulatory molecules, present tumor-associated antigens to naive T cells, initiating robust cytotoxic responses. Clinical trials (e.g., NCT01159288) have demonstrated that DEX loaded with tumor antigens increased the activation of tumor-specific CD8+ T cells and extended survival in patients with non-small cell lung cancer [6]. Engineered exosomes displaying checkpoint inhibitors reverse immunosuppression in the tumor microenvironment. Recent clinical trials have demonstrated that exosome-based immunotherapies induce durable responses in melanoma and non-small cell lung cancer, with minimal adverse effects.

4.5.2 Neurological Disorders: Crossing the Blood–Brain Barrier (BBB)

The blood–brain barrier presents a formidable obstacle to drug delivery in neurodegenerative diseases and brain tumors, with >98% of small molecules and nearly all biologics failing to reach the brain parenchyma. Exosomes, due to their nanometric size and endogenous origin, can cross the BBB via transcytosis, particularly those derived from brain endothelial cells or macrophages, which possess innate BBB-penetrating capabilities that synthetic carriers lack.

- *Targeting and delivery mechanisms:* Exosomes can be decorated with targeting ligands like rabies virus glycoprotein (RVG), transferrin, or apolipoprotein E (ApoE) to enhance BBB penetration. Recent advances include engineering exosomes with RVG peptides that specifically bind nicotinic acetylcholine receptors on neurons [20]. The unique lipid composition of neuronal exosomes facilitates fusion with target cells, enabling efficient delivery of neuroprotective factors.
- *Neurodegenerative disease applications:* Exosomes loaded with neuroprotective compounds such as curcumin, catalase, or brain-derived neurotrophic factor (BDNF) have demonstrated efficacy in reducing oxidative stress, inflammation, and neuronal apoptosis in models of Alzheimer's disease, Parkinson's disease, and ischemic stroke. In Alzheimer's disease, RVG-modified exosomes delivering BACE1 siRNA resulted in reduced amyloid beta accumulation and improved cognitive function in mouse models of Alzheimer's disease [19]. Following intravenous administration,

exosomes loaded with BACE1 siRNA reduce amyloid-β plaques by 60% in animal models. For Parkinson's disease, dopamine-loaded exosomes demonstrate targeted delivery to the nigrostriatal pathways, thereby improving motor function while minimizing peripheral side effects.

- *Gene therapy and alternative delivery routes:* Gene therapies also benefit from exosomal delivery. CRISPR-Cas9 components, antisense oligonucleotides, and therapeutic miRNAs can be packaged into exosomes for neuron-specific editing or silencing of neurotoxic genes. These targeted exosomes deliver antisense oligonucleotides to correct Huntington's disease mutations or deliver antioxidant enzymes to mitigate stroke damage [49]. Such approaches offer a safer and more efficient alternative to viral vectors. Notably, intranasal administration of exosome-based formulations bypasses the BBB entirely, achieving direct nose-to-brain delivery with rapid onset of action.

4.5.3 Cardiovascular Diseases: Repair and Regeneration

- *Cardiac repair and protection:* Exosome-mediated cardiac therapy focuses on promoting angiogenesis, reducing inflammation, and enhancing cardiomyocyte survival following myocardial infarction or ischemia-reperfusion injury. Cardiovascular applications leverage exosomes' natural cardioprotective and regenerative properties. Mesenchymal stem cell (MSC)-derived exosomes are particularly rich in reparative cargo, including VEGF, miR-21, miR-126, and HSP70. MSC-derived exosomes enriched with miR-21 and miR-210 reduce infarct size by 40% in myocardial infarction models through activation of pro-survival pathways [21]. These exosomes can be engineered to improve targeting toward damaged cardiac tissue by surface-display of peptides that bind to ischemic endothelium or infarct-specific integrins. When administered systemically or via intramyocardial injection, exosomes enhance neovascularization, suppress fibrosis, and restore left ventricular function. These exosomes inhibit fibrosis while promoting angiogenesis, outperforming stem cell therapies by eliminating the risks of arrhythmogenesis.
- *Vascular disease applications:* For atherosclerosis, macrophage exosomes engineered to carry statins, or anti-inflammatory miRNAs, selectively accumulate in plaques, reducing inflammation without systemic side effects. In peripheral artery disease, exosome hydrogels provide sustained release of pro-angiogenic factors, restoring blood flow in ischemic limbs [50].
- *Advanced delivery systems:* Cardioprotective exosomes have been utilized in conjunction with biomaterials, such as hydrogels or 3D-printed scaffolds, to create localized exosome delivery systems. These platforms provide sustained release and mechanical support, further improving myocardial repair and reducing infarct size. Emerging technologies include cardiac patch implants that incorporate exosome-releasing microspheres, which gradually repair damaged myocardium. These approaches address the short half-life of free exosomes in circulation, with some formulations maintaining therapeutic effects for weeks post-implantation [51].

4.5.4 Infectious Diseases: Exosome-Based Vaccines and Antiviral Therapies

- *Vaccine development:* Exosomes revolutionize infectious disease management through their dual role as vaccine platforms and antimicrobial delivery vehicles. Exosomes derived from infected cells contain pathogen-associated molecular patterns (PAMPs), viral antigens, and RNA fragments, making them suitable vehicles for immune priming. As natural carriers of PAMPs, exosomes from infected cells serve as potent vaccine antigens. Engineered exosomes expressing viral envelope proteins or peptide antigens have shown promise as acellular vaccines [52]. For instance, exosomes bearing SARS-CoV-2 spike proteins or influenza hemagglutinin have induced neutralizing antibody responses and T-cell activation in preclinical studies. Dendritic cell exosomes presenting the SARS-CoV-2 spike protein induce neutralizing antibody titers ten times higher than protein subunit vaccines in primate studies. Such vaccines eliminate the need for adjuvants and reduce the risk of infection associated with attenuated or inactivated viruses.
- *Antiviral therapies:* Beyond vaccination, exosomes are being explored as a means of delivering antiviral drugs. Antiviral siRNAs, CRISPR components targeting viral genomes, and interferon-stimulating RNAs have been successfully delivered using exosomal carriers. For antiviral therapy, exosomes deliver CRISPR-Cas9 ribonucleoproteins to excise the latent HIV provirus or deliver broad-spectrum antiviral miRNAs [14]. In HIV and hepatitis B models, exosome-mediated therapy has suppressed viral replication and reduced systemic viral load. In hepatitis B, exosomes that transport siRNA against viral transcripts achieve a 99% reduction in viral load without triggering interferon responses.
- *Novel therapeutic strategies:* Novel strategies include "decoy exosomes" that display high-affinity viral receptor mimics, neutralizing circulating pathogens.

During the COVID-19 pandemic, clinical trials demonstrated that inhaled exosomes carrying ACE2 receptors reduced viral load in severe patients by 90% within 72 hours [53].

4.6 APPLICATIONS OF ENGINEERED EXOSOMES IN CELL THERAPY

Engineered exosomes have emerged as revolutionary tools in regenerative medicine, immunotherapy, and targeted drug delivery. The inherent biological properties of parent cells are reflected in their derived exosomes, providing natural targeting and therapeutic capabilities that can be further enhanced through engineering approaches. Leveraging the innate biological characteristics and cargo-carrying potential of exosomes derived from diverse cell types, engineered exosomes offer a promising cell-free therapeutic approach with enhanced safety, scalability, and specificity.

4.6.1 Mesenchymal Stem Cell–Derived Exosomes

Mesenchymal stem cells (MSCs) have gained prominence as a versatile source of therapeutic exosomes, owing to their inherent immunomodulatory, anti-inflammatory, and regenerative potential. Rather than acting solely through direct differentiation, MSCs exert a majority of their therapeutic influence via paracrine signaling, in which exosomes play a pivotal role. These nanovesicles, typically 30–150 nm in diameter, encapsulate a rich cargo of proteins, lipids, and nucleic acids that mimic many of the parent cell's functions [54]. Unlike MSC transplantation, which poses risks of tumorigenesis, immune incompatibility, and heterogeneity in differentiation outcomes, MSC-derived exosomes offer a safer, more controllable, and storage-friendly alternative for cell-free therapy.

4.6.1.1 Biological Functions and Mechanisms of Action

The therapeutic potential of MSC exosomes spans multiple biological systems and disease contexts, facilitated by their complex and bioactive cargo. These exosomes influence recipient cells through the horizontal transfer of regulatory miRNAs (e.g., miR-21, miR-146a, miR-155), mRNAs, and proteins that modulate inflammation, fibrosis, apoptosis, and angiogenesis. For instance, miR-126 and miR-210 are well-documented for their pro-angiogenic properties, enhancing endothelial cell proliferation and tube formation. Moreover, MSC exosomes facilitate macrophage polarization toward the M2 anti-inflammatory phenotype and suppress T-cell activation by delivering immunosuppressive cytokines such as TGF-β and IL-10. MSC-derived exosomes also play a central role in tissue regeneration. They promote the proliferation of fibroblasts and keratinocytes, enhance extracellular matrix remodeling, and support the survival of resident progenitor cells under stress conditions, such as ischemia or oxidative injury. Importantly, these actions occur without invoking the immunogenic responses often associated with xenogeneic or allogeneic MSCs, making exosomes a more clinically feasible alternative [55].

4.6.1.2 Engineering Strategies to Enhance Therapeutic Efficacy

Several engineering approaches have been developed to enhance the therapeutic efficacy of MSC exosomes. Among the most extensively studied is preconditioning, where MSCs are exposed to specific environmental stimuli to optimize their exosomal output. Hypoxic preconditioning (1–5% O_2), which mimics the physiological stem cell niche, significantly alters exosome composition, leading to increased expression of HIF-1α, VEGF, and miR-210. These vesicles exhibit enhanced angiogenic activity and superior outcomes in myocardial infarction models, resulting in infarct size reductions of up to threefold compared to normoxic controls [56]. Another effective strategy is inflammatory preconditioning, which involves exposing MSCs to cytokines such as IFN-γ, TNF-α, or IL-1β. This approach augments the immunomodulatory capacity of exosomes by enhancing the expression of PD-L1, galectins, and other immunoregulatory molecules. In experimental autoimmune encephalomyelitis and colitis models, inflammatory-primed exosomes exhibit superior suppression of pathogenic T cell responses.

Genetic engineering offers even more targeted modifications. For example, overexpression of miR-146a in MSCs results in exosomes that more effectively inhibit NF-κB signaling, a critical pathway in inflammatory disorders. Similarly, exosomes enriched with miR-29 suppress fibrotic gene expression in hepatic and pulmonary fibrosis models by targeting collagen-encoding transcripts. Protein-based engineering is also promising; MSCs transfected with genes encoding neurotrophic factors, such as GDNF, or antiapoptotic proteins, like Bcl-2, produce exosomes that significantly improve neuronal survival in neurodegenerative disease models.

4.6.1.3 Surface Modification and Targeting Strategies

To increase the biodistribution specificity and reduce off-target effects, MSC exosomes can be surface-modified using diverse biochemical strategies. Peptide ligation, such as conjugation with the RGD peptide, enhances homing to ischemic tissues via integrin-mediated binding [57]. Antibody conjugation enables the selective delivery of antibodies to inflammation sites; for instance, anti-VCAM labeling promotes localization to inflamed vasculature. Additionally, glycan modification – particularly through the incorporation of sialic acid – has been shown to enhance BBB penetration, thereby facilitating therapeutic delivery to the brain. Innovative approaches that involve the embedding of superparamagnetic iron oxide nanoparticles into exosome membranes enable magnetically guided delivery, enhancing local accumulation at target sites by up to fivefold in murine models. These surface engineering strategies must be carefully balanced to preserve the native bioactivity and tropism of exosomes while improving disease-site selectivity. Over-modification may compromise membrane integrity or interfere with receptor-mediated uptake, underscoring the importance of optimization in clinical applications.

4.6.1.4 Disease Applications and Translational Impact

MSC-derived exosomes have shown therapeutic promise across a wide range of disease models. In cardiovascular diseases, particularly myocardial infarction and heart failure, they reduce infarct size, promote angiogenesis, and improve cardiac function metrics such as ejection fraction. Studies indicate a two- to threefold increase in capillary density and a 10–15% improvement in cardiac output following treatment with MSC exosomes. In the neurological domain, exosome therapy has led to significant improvements in preclinical models of Alzheimer's, Parkinson's, stroke, and multiple sclerosis. These effects are primarily attributed to their delivery of neurotrophic factors, anti-apoptotic proteins, and miRNAs that modulate neuroinflammation and synaptic plasticity. Alzheimer's models treated with MSC exosomes exhibit a 30–40% reduction in amyloid plaque burden, accompanied by improved spatial memory and cognitive performance [22].

MSC exosomes have also excelled in wound healing applications. During the inflammatory phase of tissue repair, they attenuate the expression of pro-inflammatory cytokines and promote the polarization of M2 macrophages. During the proliferative phase, they stimulate the proliferation of fibroblasts and keratinocytes and facilitate angiogenesis, resulting in faster granulation tissue formation. In the remodeling phase, they regulate collagen deposition and matrix metalloproteinase activity, minimizing scar formation. Clinical studies in diabetic ulcers and burns report up to 50% faster wound closure rates compared to conventional therapy [58]. Furthermore, in immunological disorders such as graft-versus-host disease, rheumatoid arthritis, and systemic lupus erythematosus, MSC-derived exosomes attenuate pathological immune responses without causing broad immunosuppression. By expanding regulatory T cells and inhibiting Th17 differentiation, these exosomes promote immune tolerance. In transplantation models, they reduce graft rejection and prolong graft survival, supporting their application in solid organ transplantation.

4.6.1.5 Future Perspectives and Clinical Translation

Despite their promising preclinical success, several challenges remain in translating MSC-derived exosomes into routine clinical practice. These include standardizing exosome isolation techniques, scaling up production while maintaining bioactivity, and ensuring consistent cargo profiles across batches. Advances in bioreactor systems, tangential flow filtration, and label-free purity assays are beginning to address these manufacturing bottlenecks. Regulatory frameworks are also evolving to accommodate the unique nature of exosome-based therapeutics. Importantly, ongoing clinical trials, such as those evaluating exosome therapy in myocardial infarction, osteoarthritis, and inflammatory bowel disease, are likely to provide the necessary safety and efficacy data to guide regulatory approval. As these platforms mature, engineered MSC-derived exosomes are expected to play a pivotal role in regenerative medicine, offering potent, targeted, and personalized treatment strategies across a wide array of diseases.

4.6.2 Immune Cell–Derived Exosomes

Exosomes derived from immune cells represent a specialized class of extracellular vesicles endowed with innate immunological functionality. These vesicles inherit many of the immunoregulatory, antigen-presenting, and cytotoxic properties of their parent cells, including dendritic cells (DCs), natural killer (NK) cells, and T lymphocytes. Unlike whole-cell therapies, immune cell–derived exosomes offer the advantage of reduced immunogenicity, improved safety profiles, and enhanced stability during storage and delivery [59]. Moreover, their small size allows efficient tissue penetration and traversal of biological barriers, making them highly suitable for both local and systemic immunotherapeutic applications. A comprehensive comparison of different cell sources for therapeutic exosomes is presented in Table 4.3.

4.6.2.1 Dendritic Cell–Derived Exosomes (Dex)

Dendritic cells, as professional antigen-presenting cells, play a central role in orchestrating adaptive immune responses. Exosomes secreted by these cells – termed dendritic cell–derived exosomes (Dex) – are enriched in major histocompatibility complex (MHC) class I and II molecules, costimulatory ligands, and adhesion proteins. These vesicles retain the capacity to present processed antigens to both CD4+ and CD8+ T cells through multiple mechanisms. Dex can directly engage memory T cells via surface peptide–MHC complexes or indirectly modulate immunity by transferring peptide–MHC complexes to recipient DCs in a process known as "cross-dressing." Additionally, intact antigens enclosed within Dex can be processed and presented by neighboring antigen-presenting cells, contributing to epitope spreading and broader immune activation [60].

The immunostimulatory efficacy of Dex is highly dependent on the maturation state of the parent dendritic cell and the method of antigen loading. Mature DCs produce exosomes with higher expression of MHC molecules, CD80/CD86, and ICAM-1, resulting in superior T-cell activation. These costimulatory signals are crucial for complete T-cell priming, and their absence can lead to T-cell anergy or tolerance [61]. Modulation of Dex content through exposure to pro-inflammatory stimuli or via genetic engineering to

TABLE 4.3
Cell Sources for Therapeutic Exosomes

Cell Source	Key Advantages	Engineering Approaches	Main Applications	Representative Cargo	References
MSCs	Immunomodulatory, regenerative, high yield	Preconditioning (hypoxia, inflammation), miRNA overexpression	Tissue repair, inflammation, and immunomodulation	IL-10, TGF-β, miR-146a, VEGF	[54, 55]
Dendritic cells	Antigen presentation, immune activation	Antigen loading (endogenous/exogenous), co-stimulatory ligand expression	Cancer vaccines, immune modulation	MHC–peptide complexes, CD80/CD86, ICAM-1	[60, 61]
Neural cells	BBB crossing, neuroprotection	Neurotrophic loading, surface targeting (e.g., RVG peptide)	Neurodegeneration, stroke, TBI	BDNF, NGF, miR-124, miR-132	[64, 65]
CAR-T cells	Tumor-specific cytotoxicity, lower toxicity	CAR display, cytotoxic protein/miRNA loading	Cancer therapy (especially solid tumors)	CAR constructs, granzyme B, and perforin	[66, 67]
Macrophages	Inflammation resolution, tissue penetration	M2 polarization, drug loading (chemo, siRNA)	Autoimmune/inflammatory diseases	IL-10, TGF-β, TNF-α inhibitors, drugs	[17]
NK cells	Natural cytotoxicity, antiviral potential	Cytotoxic protein enrichment, surface engineering	Cancer, viral infections	Perforin, granzyme A/B, IFN-γ	[59]
Endothelial cells	Vascular targeting, angiogenesis	Angiogenic cargo enrichment, peptide targeting	Cardiovascular regeneration, wound healing	VEGF, FGF2, angiopoietin-1, miR-126	[68]
Hepatocytes	Liver tropism, metabolic enzyme content	Enzyme/gene editing, hepatocyte-targeting peptides	Liver disease, metabolic disorders	CYP enzymes, urea cycle enzymes, miR-122	[32]

overexpress immune-enhancing ligands has been shown to improve T-cell proliferation and cytokine release by two- to fourfold compared to unmodified exosomes.

4.6.2.2 Cancer Vaccine Applications

One of the most promising applications of Dex lies in cancer immunotherapy. Several early-phase clinical trials have demonstrated the feasibility and safety of using autologous Dex loaded with tumor-associated antigens as cancer vaccines. These trials report partial immune responses, with some patients exhibiting tumor stabilization or delayed progression. Unlike dendritic cell–based vaccines, Dex offers advantages in standardization, long-term storage, and batch quality control – features that are critical for large-scale clinical deployment [61]. Recent developments have focused on enhancing the immunogenicity of Dex through the use of combinatorial strategies. For instance, co-loading Dex with toll-like receptor (TLR) ligands or STING agonists augments innate immune activation, while decorating their surface with antibodies or integrins facilitates targeted delivery to lymph nodes and secondary lymphoid organs. The incorporation of natural killer cell-activating ligands onto Dex further amplifies cytotoxic immune responses. These engineered Dex platforms have demonstrated up to fivefold improvement in antitumor immunity in murine models, surpassing the efficacy of conventional peptide vaccines or unmodified exosomes.

4.6.2.3 Broader Immunomodulatory Applications

Beyond oncology, Dex exhibits significant potential in the management of infectious diseases, chronic inflammation, and autoimmunity. For example, Dex enriched in tolerogenic signals can induce regulatory T cells and suppress Th17-mediated pathology, offering a novel strategy for treating diseases like rheumatoid arthritis and multiple sclerosis. Similarly, antigen-loaded Dex can be used to reestablish tolerance in allergy or organ transplantation settings, avoiding the need for systemic immunosuppression. Preclinical models have shown that antigen-specific Dex can prevent graft rejection and prolong allograft survival, likely through modulation of dendritic cell and T cell cross talk. The adaptability of Dex to diverse immunological contexts, combined with its potential for off-the-shelf production, makes it uniquely suited for both therapeutic and prophylactic immunomodulation. Continued advances in exosome biomanufacturing, cargo loading, and surface functionalization will be key to unlocking the full clinical potential of immune cell–derived exosomes [62, 63].

BBB: blood–brain barrier; BDNF: brain-derived neurotrophic factor; CAR: chimeric antigen receptor; IFN: interferon; MSCs: mesenchymal stem cells; NGF: nerve growth factor; NK: natural killer; VEGF: vascular endothelial growth factor.

4.6.3 CAR-T-Cell-Derived Exosomes

Chimeric antigen receptor (CAR)-T-cell therapy has revolutionized the treatment of hematologic malignancies, providing targeted tumor eradication through engineered T cells that express synthetic receptors recognizing specific tumor-associated antigens. However, the clinical translation of CAR-T-cell therapy into solid tumors and broader disease categories has been hindered by significant limitations, including cytokine release syndrome (CRS), neurotoxicity, limited tumor infiltration, and manufacturing complexity. In this context, exosomes derived from CAR-T cells have emerged as a promising acellular alternative that retains the antigen-specific targeting capacity of CAR-T cells while offering improved safety, scalability, and delivery potential [69].

CAR-T-cell-derived exosomes are naturally secreted extracellular vesicles that incorporate functional CAR proteins on their surface, with surface densities ranging between 50 and 200 CAR molecules per vesicle, depending on the design and expression system employed in the parental T cells. These exosomal CARs preserve antigen specificity and binding affinity, and upon interaction with target antigens, they can initiate signal transduction pathways and induce antitumor responses. Mechanistically, CAR-T exosomes mediate tumor cell apoptosis by delivering cytotoxic effector molecules, such as perforin and granzymes, and by interfering with oncogenic signaling pathways, thereby sensitizing tumor cells to chemotherapeutic or immunomodulatory agents. These functions closely mirror the cytotoxic activity of intact CAR-T cells but in a more refined, dose-controllable, and safer format.

One of the most compelling advantages of CAR-T exosomes lies in their superior safety profile. Unlike living CAR-T cells, exosomes are nonreplicative and lack the potential for autonomous cytokine production and proliferation. As a result, they eliminate the risk of cytokine storm and neurotoxicity, two major complications that have limited the application of CAR-T in broader clinical settings. Preclinical studies have demonstrated a reduction of up to 80–90% in adverse events with CAR-T exosome therapy compared to their cellular counterparts. Furthermore, the transient nature of exosome–cell interactions minimizes off-target, on-tissue toxicity while still allowing potent and specific tumor cell engagement [66]. This is particularly advantageous in targeting antigens with low expression in healthy tissues, where overactivation by CAR-T cells could result in autoimmune-like toxicity.

From a biophysical standpoint, the small size of CAR-T exosomes (typically 30–150 nm) facilitates deep penetration into dense tumor matrices and enhances biodistribution within poorly perfused tissues – a notable limitation for cellular therapies. These vesicles demonstrate enhanced tumor infiltration, particularly in solid tumor models where stromal or endothelial barriers limit T-cell access. Additionally, their membrane composition provides inherent protection to therapeutic cargo from enzymatic degradation, prolonging circulation half-life and increasing delivery efficiency to target tissues.

Recent research has also explored combinatorial strategies that further potentiate the utility of CAR-T exosomes. These include co-loading exosomes with immune checkpoint inhibitors (e.g., anti-PD-1 or anti-CTLA-4 antibodies) to overcome the immunosuppressive tumor microenvironment and incorporating chemotherapeutic agents such as doxorubicin or paclitaxel for dual cytotoxic activity. CAR exosomes have also been combined with oncolytic viruses and tumor-penetrating peptides to amplify antitumor efficacy through multi-modal mechanisms. Such approaches have demonstrated 60–80% greater tumor regression in preclinical models than monotherapy approaches alone, highlighting the potential versatility of the engineered CAR exosome platform. In parallel, advances in manufacturing techniques – such as stable lentiviral transduction of T cells, scalable bioreactor systems, and tangential flow filtration – have made it feasible to isolate CAR exosomes at clinically relevant scales. Their defined, cell-free nature also simplifies regulatory approval and quality control, compared to complex autologous cell therapies. Ongoing studies are evaluating the integration of CAR exosomes into next-generation immunotherapy pipelines, including "off-the-shelf" allogeneic strategies that eliminate the need for patient-specific manufacturing. In conclusion, CAR-T-cell-derived exosomes represent a novel and highly adaptable platform in cancer immunotherapy. They combine the tumor specificity of CAR engineering with the advantages of exosome-mediated delivery, offering a compelling approach to overcome the challenges associated with current cellular therapies [67]. As preclinical efficacy and safety data continue to accumulate, CAR exosomes are poised to advance toward early-phase clinical trials.

4.6.4 Neural Cell–Derived Exosomes

Exosomes derived from neural cells – comprising neurons, astrocytes, oligodendrocytes, and neural stem/progenitor cells – offer a unique opportunity for targeted therapy within the central nervous system (CNS). These vesicles not only reflect the molecular signature of their neural origin but also exhibit inherent abilities to navigate complex CNS environments, cross the blood–brain barrier (BBB), and modulate neural and glial cell functions. Given the high prevalence of neurodegenerative disorders, stroke, and traumatic brain injuries, many of which lack curative therapies, neural cell–derived exosomes have gained momentum as promising candidates in regenerative neurology and neuropharmacology. Neural exosomes are enriched with bioactive molecules that are highly relevant to neuronal survival, synaptic plasticity, and the regulation of neuroinflammation. These include neurotrophic factors such as brain-derived neurotrophic factor (BDNF), nerve growth factor (NGF), and glial cell line–derived neurotrophic factor (GDNF), as well as small regulatory RNAs like miR-124, miR-132, and miR-219. These components modulate key molecular pathways involved in neurogenesis, myelination, oxidative stress response, and microglial activation. For example, exosomes from neurons or neural progenitor cells containing miR-132 have been shown to enhance axonal outgrowth, promote synaptic remodeling, and suppress the release of pro-inflammatory cytokines from activated microglia [64].

One of the most compelling features of neural cell–derived exosomes is their ability to cross the BBB, a significant obstacle in neurotherapeutic delivery. Their endogenous lipid composition and expression of surface markers such as L1CAM and CD63 facilitate transcytosis and preferential uptake by neural cells [65]. This tropism is particularly beneficial in conditions such as Parkinson's disease, Alzheimer's disease, and multiple sclerosis, where deep tissue penetration and specific targeting of neurons, oligodendrocytes, or astrocytes is critical for efficacy. In preclinical models of Alzheimer's disease, exosomes derived from neural stem cells or iPSC--derived neurons have been shown to reduce amyloid-β aggregation, improve cognitive performance, and restore synaptic integrity by delivering antiapoptotic miRNAs and exosomal enzymes that degrade pathogenic protein aggregates.

In the context of ischemic stroke, neural exosomes play a crucial role in neurovascular remodeling by promoting angiogenesis, neurogenesis, and axonal regeneration. Astrocyte-derived exosomes, for instance, have demonstrated the ability to protect neurons against glutamate-induced excitotoxicity and oxidative stress by delivering antioxidant enzymes and stress-responsive miRNAs such as miR-181c. Similarly, oligodendrocyte-derived exosomes facilitate remyelination by transferring lipids and myelin-associated proteins to demyelinated axons, providing a therapeutic approach for diseases such as multiple sclerosis [70]. Engineering of neural exosomes further expands their therapeutic applications. For example, surface display of rabies virus glycoprotein (RVG) peptide has been used to enhance exosome delivery to cholinergic neurons. Moreover, loading exosomes with exogenous cargo, such as siRNAs targeting BACE1 or alpha-synuclein mRNA, has enabled gene silencing approaches in models of Alzheimer's and Parkinson's diseases, respectively. These strategies not only improve the specificity and efficacy of the exosomes but also allow their application as precision gene and drug delivery vehicles within the CNS. The clinical translation of neural exosomes is supported by their favorable immunological profile, low toxicity, and potential for both autologous and allogeneic applications. However, challenges remain, particularly in standardizing isolation techniques, controlling cargo heterogeneity, and achieving scalable manufacturing.

4.6.5 Emerging Cell Sources and Future Directions

While mesenchymal stem cells, immune cells, and neural cells currently dominate the field of exosome-based therapies, several emerging cell sources are rapidly gaining attention due to their unique biological advantages, enhanced scalability, and disease-specific targeting potential [71]. These novel sources, ranging from induced pluripotent stem cells (iPSCs) to placental and progenitor cells, promise to broaden the therapeutic repertoire of engineered exosomes and overcome existing limitations related to cargo heterogeneity, manufacturing consistency, and tissue specificity.

4.6.5.1 Induced Pluripotent Stem Cell (iPSC)-Derived Exosomes

iPSCs offer a renewable and ethically acceptable platform for generating lineage-specific exosomes with patient-specific compatibility. Unlike primary cells, iPSCs can be differentiated into virtually any somatic lineage, allowing them to produce exosomes with tailored biological functions. Moreover, their proliferative capacity and adaptability to defined culture conditions support the development of scalable, reproducible exosome production pipelines for clinical-grade applications. Exosomes derived from induced pluripotent stem cells (iPSCs) have demonstrated therapeutic potential in various preclinical models, particularly in neurodegenerative and cardiovascular diseases. For instance, iPSC-derived neural exosomes engineered to carry neuroprotective cargo, such as miR-124 or BDNF, significantly enhance neuronal survival and functional recovery in models of spinal cord injury and Parkinson's disease. Additionally, cardiomyocyte-like cells differentiated from iPSCs secrete exosomes enriched with cardiac-specific miRNAs (e.g., miR-1, miR-133, miR-208) that improve cardiac remodeling and reduce fibrosis following myocardial infarction. The significant advantage of iPSC-derived exosomes lies in their potential for personalized medicine. Autologous iPSC generation from patient-derived cells minimizes the risk of immune rejection, while gene editing technologies, such as CRISPR-Cas9, allow for the precise manipulation of exosomal cargo to address specific genetic or epigenetic abnormalities. However, issues related to the tumorigenic potential of residual undifferentiated cells, high manufacturing costs, and regulatory complexity must be addressed before the widespread clinical implementation of this technology [72, 73].

4.6.5.2 Cardiac and Tissue-Specific Progenitor Cell Exosomes

Exosomes derived from tissue-specific progenitor cells, such as cardiac progenitor cells (CPCs), endothelial progenitor cells (EPCs), and hepatic progenitors, possess a unique cargo signature that closely reflects their origin in the respective organ. CPC-derived exosomes are particularly enriched in miRNAs essential for myocardial survival and repair, including miR-1, miR-133a, and miR–208. These vesicles improve cardiomyocyte contractility, reduce infarct size, and enhance neovascularization following ischemic injury. Similarly, endothelial progenitor cell exosomes promote vascular regeneration and reduce oxidative injury in ischemic limbs and cerebrovascular disease models. Their high expression of angiogenic proteins (e.g., VEGF, FGF2) and endothelial markers enhances vascular integrity and functional recovery. The specificity and safety of progenitor-derived exosomes make them well-suited for precision regenerative therapies tailored to cardiovascular, hepatic, and renal injuries [68].

4.6.5.3 Placental and Amniotic Fluid–Derived Exosomes

The placenta and amniotic fluid represent ethically favorable and immunoprivileged sources of exosomes, rich in anti-inflammatory, pro-regenerative factors. These exosomes are obtained noninvasively from postpartum tissues typically discarded as medical waste, offering a sustainable and abundant source for therapeutic use. Placental exosomes are known to carry HLA-G, TGF-β, and IL-10, facilitating immune tolerance and tissue repair. In animal models, they have been shown to accelerate wound healing, reduce inflammation in colitis and arthritis, and support neurogenesis [74]. Amniotic fluid–derived exosomes similarly promote epithelialization, fibroblast migration, and angiogenesis, making them attractive candidates for treating chronic wounds, burns, and fibrotic diseases. Their low immunogenicity enables allogeneic application without the need for immunosuppression, and their cargo can be further enhanced through maternal or fetal preconditioning during gestation or collection.

4.6.5.4 Immortalized and Engineered Cell Lines for Exosome Biomanufacturing

To address the pressing need for consistent, high-yield exosome production, researchers have developed immortalized and genetically engineered cell lines optimized for exosome biomanufacturing. These cell lines provide predictable exosome output with stable cargo profiles across production batches, enabling adherence to good manufacturing practices (GMP) required for pharmaceutical development. Engineered cell lines can be designed to overexpress therapeutic payloads, targeting peptides, or membrane ligands that enhance exosome biodistribution, stability, and functional specificity [75]. For example, HEK293 or CHO cells stably transfected with therapeutic genes can produce exosomes loaded with antitumor siRNAs or neuroprotective proteins. These vesicles can be harvested using scalable culture systems, such as hollow-fiber bioreactors or microfluidic platforms, and then purified through advanced methods, including tangential flow filtration and size-exclusion chromatography. These systems enable large-scale, clinical-grade exosome manufacturing with enhanced reproducibility, reduced costs, and compliance with regulatory standards [76].

4.6.5.5 Hybrid and Smart Exosomes for Next-Generation Therapies

An exciting frontier in exosome engineering involves the development of hybrid or responsive exosomes that integrate features from multiple cell types or respond dynamically to the disease microenvironment. Hybrid exosomes, formed by fusing vesicles from different sources or combining exosomes with liposomes, enable multifunctional delivery platforms capable of simultaneous targeting, immune modulation, and regeneration. For instance, MSC–neuron hybrid exosomes may combine immunomodulatory and neurotrophic properties to address complex pathologies such as stroke or neuroinflammation. Smart exosomes, meanwhile, incorporate responsive elements – such as pH-sensitive lipids, redox-sensitive peptides, or enzyme-cleavable linkers – that enable controlled cargo release in response to specific pathological cues. This level of precision improves therapeutic index, reduces systemic toxicity, and enhances treatment outcomes. Concurrently, the development of predictive biomarkers to match patients with specific exosome-based therapies will be crucial in advancing personalized medicine [77].

4.7 CONCLUSION

Exosome engineering represents a rapidly evolving and transformative frontier at the intersection of cell biology, nanotechnology, and therapeutic innovation. The natural properties of exosomes, including their low immunogenicity, intrinsic targeting capabilities, and diverse molecular cargo capacity, provide significant advantages over traditional drug delivery methods. Advances in cargo-loading techniques, sophisticated surface modifications, and hybrid exosome–nanoparticle systems have broadened the therapeutic scope, facilitating targeted and personalized interventions across oncology, neurology, cardiovascular diseases, autoimmune conditions, and regenerative medicine. Despite substantial progress, several critical hurdles must be overcome, including heterogeneity of exosomes, challenges in large-scale manufacturing, regulatory complexities, and the need for standardized quality control measures. Addressing these barriers will require collaborative efforts among scientists, clinicians, biotechnology experts, and regulatory bodies. Recent technological innovations, such as CRISPR-based cell programming, AI-driven design methodologies, and scalable bioprocessing technologies, are emerging as pivotal solutions to these challenges.

Clinically, initial exosome therapies will likely target local administration applications, such as rheumatoid arthritis or wound healing, due to minimal systemic exposure risks. Subsequent advances will enable systemic treatments for conditions such as cancer and neurodegenerative diseases, as our understanding of exosome biodistribution and off-target effects improves. Establishing standardized methods for isolation, characterization, and potency assessment is critical for ensuring reproducibility and facilitating regulatory approval. The future of exosome engineering also involves increasingly sophisticated delivery platforms, including multiple targeting ligands, combination therapeutic payloads, and hybrid or fully synthetic exosome-mimetic structures. The integration of synthetic biology, microfluidics, and machine learning holds promise for accelerating personalized therapeutic development, further refining targeting specificity and treatment efficacy. Ultimately, engineered exosomes represent more than merely advanced drug carriers – they are transformative tools poised to redefine treatment paradigms for complex chronic diseases. By synergizing scientific innovation, practical manufacturing solutions, and responsive regulatory frameworks, exosome-based therapies will likely emerge as cornerstone modalities in next-generation therapeutics, significantly enhancing patient outcomes and quality of life.

AI DISCLOSURE STATEMENT

During the preparation of this work, the authors used Grammarly, QuillBot, and ChatGPT in order to enhance grammatical structure and narrative coherence. After using this tool/service, the authors reviewed and edited the content as needed and take full responsibility for the content of the publication.

CONFLICT OF INTEREST

None.

REFERENCES

[1] Y. Wang, H. Ma, X. Zhang, X. Xiao, Z. Yang, The increasing diagnostic role of exosomes in inflammatory diseases to leverage the therapeutic biomarkers, *J Inflamm Res* 17 (2024) 5005–5024 https://doi.org/10.2147/JIR.S475102.

[2] A. Vucemilovic, Exosomes: Intriguing mediators of intercellular communication in the organism's response to noxious agents, *Arh Hig Rada Toksikol* 75 (2024) 228–239 https://doi.org/10.2478/aiht-2024-75-3923.

[3] H.B. Koh, H.J. Kim, S.W. Kang, T.H. Yoo, Exosome-based drug delivery: Translation from bench to clinic, *Pharmaceutics* 15 (2023) https://doi.org/10.3390/pharmaceutics15082042.

[4] Q. Yang, S. Li, H. Ou, Y. Zhang, G. Zhu, S. Li, L. Lei, Exosome-based delivery strategies for tumor therapy: An update on modification, loading, and clinical application, *J Nanobiotechnol* 22 (2024) 41 https://doi.org/10.1186/s12951-024-02298-7.

[5] L. Bai, L. Yu, M. Ran, X. Zhong, M. Sun, M. Xu, Y. Wang, X. Yan, R.J. Lee, Y. Tang, J. Xie, Harnessing the potential of exosomes in therapeutic interventions for brain disorders, *Int J Mol Sci* 26 (2025) 2491 https://doi.org/10.3390/ijms26062491.

[6] G. Schwarz, X. Ren, W. Xie, H. Guo, Y. Jiang, J. Zhang, Engineered exosomes: A promising drug delivery platform with therapeutic potential, *Front Mol Biosci* 12 (2025) 1583992 https://doi.org/10.3389/fmolb.2025.1583992.

[7] X. Jiang, L. Shi, H. Feng, Y. Zhang, J. Dong, Z. Shen, Engineered exosomes loaded with triptolide: An innovative approach to enhance therapeutic efficacy in rheumatoid arthritis, *Int Immunopharmacol* 129 (2024) 111677 https://doi.org/10.1016/j.intimp.2024.111677.

[8] M. Deng, S. Wu, P. Huang, Y. Liu, C. Li, J. Zheng, Engineered exosomes-based theranostic strategy for tumor metastasis and recurrence, *Asian J Pharm Sci* 18 (2023) 100870 https://doi.org/10.1016/j.ajps.2023.100870.

[9] Y. Zhang, J. Bi, J. Huang, Y. Tang, S. Du, P. Li, Exosome: A review of its classification, isolation techniques, storage, diagnostic and targeted therapy applications, *Int J Nanomed* 15 (2020) 6917–6934 https://doi.org/10.2147/IJN.S264498.

[10] M. Bahadorani, M. Nasiri, K. Dellinger, S. Aravamudhan, R. Zadegan, Engineering exosomes for therapeutic applications: Decoding biogenesis, content modification, and cargo loading strategies, *Int J Nanomed* 19 (2024) 7137–7164 https://doi.org/10.2147/IJN.S464249.

[11] D. Ferreira, J.N. Moreira, L.R. Rodrigues, New advances in exosome-based targeted drug delivery systems, *Crit Rev Oncol Hematol* 172 (2022) 103628 https://doi.org/10.1016/j.critrevonc.2022.103628.

[12] Y. Lu, K. Godbout, G. Lamothe, J.P. Tremblay, CRISPR-Cas9 delivery strategies with engineered extracellular vesicles, *Mol Ther Nucleic Acids* 34 (2023) 102040 https://doi.org/10.1016/j.omtn.2023.102040.

[13] D. Bauer, M.A. Cornejo, T.T. Hoang, J.S. Lewis, B.M. Zeglis, Click chemistry and radiochemistry: An update, *Bioconjug Chem* 34 (2023) 1925–1950 https://doi.org/10.1021/acs.bioconjchem.3c00286.

[14] V.P. Chavda, A. Pandya, L. Kumar, N. Raval, L.K. Vora, S. Pulakkat, V. Patravale, Salwa, Y. Duo, B.Z. Tang, Exosome nanovesicles: A potential carrier for therapeutic delivery, *Nano Today* 49 (2023) 101771 https://doi.org/10.1016/j.nantod.2023.101771.

[15] C.Y. Huang, Clinical applications of exosomes. N.T. Liem, N.R. Forsyth, M. Heke, *Cell Therapy*, Springer Nature Singapore, Singapore, 2025, pp. 387–434 https://doi.org/10.1007/978-981-96-1261-1_23.

[16] D. Ho, S.R. Quake, E.R.B. McCabe, W.J. Chng, E.K. Chow, X. Ding, B.D. Gelb, G.S. Ginsburg, J. Hassenstab, C.M. Ho, W.C. Mobley, G.P. Nolan, S.T. Rosen, P. Tan, Y. Yen, A. Zarrinpar, Enabling technologies for personalized and precision medicine, *Trends Biotechnol* 38 (2020) 497–518 https://doi.org/10.1016/j.tibtech.2019.12.021.

[17] M.S. Kim, M.J. Haney, Y. Zhao, D. Yuan, I. Deygen, N.L. Klyachko, A.V. Kabanov, E.V. Batrakova, Engineering macrophage-derived exosomes for targeted paclitaxel delivery to pulmonary metastases: In vitro and in vivo evaluations, *Nanomedicine* 14 (2018) 195–204 https://doi.org/10.1016/j.nano.2017.09.011.

[18] P.K. Tiwari, P. Shanmugam, V. Karn, S. Gupta, R. Mishra, S. Rustagi, M. Chouhan, D. Verma, N.K. Jha, S. Kumar, Extracellular vesicular miRNA in pancreatic cancer: From lab to therapy, *Cancers (Basel)* 16 (2024) https://doi.org/10.3390/cancers16122179.

[19] Q. Si, L. Wu, D. Pang, P. Jiang, Exosomes in brain diseases: Pathogenesis and therapeutic targets, *MedComm* (2020) 4 (2023) e287 https://doi.org/10.1002/mco2.287.

[20] K. Haroon, H. Zheng, S. Wu, Z. Liu, Y. Tang, G.Y. Yang, Y. Liu, Z. Zhang, Engineered exosomes mediated targeted delivery of neuroprotective peptide NR2B9c for the treatment of traumatic brain injury, *Int J Pharm* 649 (2024) 123656 https://doi.org/10.1016/j.ijpharm.2023.123656.

[21] C.W. Chae, G. Choi, T. Yoon, Y.W. Kwon, Exosome-based therapy in cardiovascular diseases: A new frontier in cardiovascular disease treatment, *Korean Circ J* 55 (2025) 461–480 https://doi.org/10.4070/kcj.2025.0022.

[22] E. Suzuki, D. Fujita, M. Takahashi, S. Oba, H. Nishimatsu, Stem cell-derived exosomes as a therapeutic tool for

cardiovascular disease, *World J Stem Cells* 8 (2016) 297–305 https://doi.org/10.4252/wjsc.v8.i9.297.
[23] O. Martinez-Arroyo, A. Ortega, M.J. Forner, R. Cortes, Mesenchymal stem cell-derived extracellular vesicles as non-coding RNA therapeutic vehicles in autoimmune diseases, *Pharmaceutics* 14 (2022) https://doi.org/10.3390/pharmaceutics14040733.
[24] O.G. de Jong, S.A.A. Kooijmans, D.E. Murphy, L. Jiang, M.J.W. Evers, J.P.G. Sluijter, P. Vader, R.M. Schiffelers, Drug delivery with extracellular vesicles: From imagination to innovation, *Acc Chem Res* 52 (2019) 1761–1770 https://doi.org/10.1021/acs.accounts.9b00109.
[25] S. Gurung, D. Perocheau, L. Touramanidou, J. Baruteau, The exosome journey: From biogenesis to uptake and intracellular signalling, *Cell Commun Signal* 19 (2021) 47 https://doi.org/10.1186/s12964-021-00730-1.
[26] K.S. Makarova, V. Tobiasson, Y.I. Wolf, Z. Lu, Y. Liu, S. Zhang, M. Krupovic, M. Li, E.V. Koonin, Diversity, origin, and evolution of the ESCRT systems, *mBio* 15 (2024) e0033524 https://doi.org/10.1128/mbio.00335-24.
[27] Q.F. Han, W.J. Li, K.S. Hu, J. Gao, W.L. Zhai, J.H. Yang, S.J. Zhang, Exosome biogenesis: Machinery, regulation, and therapeutic implications in cancer, *Mol Cancer* 21 (2022) 207 https://doi.org/10.1186/s12943-022-01671-0.
[28] M. Liu, Z. Wen, T. Zhang, L. Zhang, X. Liu, M. Wang, The role of exosomal molecular cargo in exosome biogenesis and disease diagnosis, *Front Immunol* 15 (2024) 1417758 https://doi.org/10.3389/fimmu.2024.1417758.
[29] S. Hallal, A. Tuzesi, G.E. Grau, M.E. Buckland, K.L. Alexander, Understanding the extracellular vesicle surface for clinical molecular biology, *J Extracell Vesicles* 11 (2022) e12260 https://doi.org/10.1002/jev2.12260.
[30] H. Valadi, K. Ekstrom, A. Bossios, M. Sjostrand, J.J. Lee, J.O. Lotvall, Exosome-mediated transfer of mRNAs and microRNAs is a novel mechanism of genetic exchange between cells, *Nat Cell Biol* 9 (2007) 654–659 https://doi.org/10.1038/ncb1596.
[31] R. Kalluri, V.S. LeBleu, The biology, function, and biomedical applications of exosomes, *Science* 367 (2020) https://doi.org/10.1126/science.aau6977.
[32] Y. Ma, S. Brocchini, G.R. Williams, Extracellular vesicle-embedded materials, *J Control Release* 361 (2023) 280–296 https://doi.org/10.1016/j.jconrel.2023.07.059.
[33] D. Shi, D. Beasock, A. Fessler, J. Szebeni, J.Y. Ljubimova, K.A. Afonin, M.A. Dobrovolskaia, To PEGylate or not to PEGylate: Immunological properties of nanomedicine's most popular component, polyethylene glycol and its alternatives, *Adv Drug Deliv Rev* 180 (2022) 114079 https://doi.org/10.1016/j.addr.2021.114079.
[34] H. Javid, M.A. Oryani, N. Rezagholinejad, A. Esparham, M. Tajaldini, M. Karimi-Shahri, RGD peptide in cancer targeting: Benefits, challenges, solutions, and possible integrin-RGD interactions, *Cancer Med* 13 (2024) e6800 https://doi.org/10.1002/cam4.6800.
[35] A.E. Basyoni, A. Atta, M.M. Salem, T.M. Mohamed, Harnessing exosomes for targeted drug delivery systems to combat brain cancer, *Cancer Cell Int* 25 (2025) 150 https://doi.org/10.1186/s12935-025-03731-z.
[36] M. Xu, Q. Yang, X. Sun, Y. Wang, Recent advancements in the loading and modification of therapeutic exosomes, *Front Bioeng Biotechnol* 8 (2020) 586130 https://doi.org/10.3389/fbioe.2020.586130.
[37] H. Zeng, S. Guo, X. Ren, Z. Wu, S. Liu, X. Yao, Current strategies for exosome cargo loading and targeting delivery, *Cells* 12 (2023) 1416 https://doi.org/10.3390/cells12101416.
[38] Y.T. Sato, K. Umezaki, S. Sawada, S.A. Mukai, Y. Sasaki, N. Harada, H. Shiku, K. Akiyoshi, Engineering hybrid exosomes by membrane fusion with liposomes, *Sci Rep* 6 (2016) 21933 https://doi.org/10.1038/srep21933.
[39] J. Cao, G. Lv, F. Wei, Engineering exosomes to reshape the immune microenvironment in breast cancer: Molecular insights and therapeutic opportunities, *Clin Transl Med* 14 (2024) e1645 https://doi.org/10.1002/ctm2.1645.
[40] J. He, W. Ren, W. Wang, W. Han, L. Jiang, D. Zhang, M. Guo, Exosomal targeting and its potential clinical application, *Drug Deliv Transl Res* 12 (2022) 2385–2402 https://doi.org/10.1007/s13346-021-01087-1.
[41] Y. Yu, W. Li, L. Mao, W. Peng, D. Long, D. Li, R. Zhou, X. Dang, Genetically engineered exosomes display RVG peptide and selectively enrich a neprilysin variant: A potential formulation for the treatment of Alzheimer's disease, *J Drug Target* 29 (2021) 1128–1138 https://doi.org/10.1080/1061186X.2021.1929257.
[42] S.A.A. Kooijmans, L.A.L. Fliervoet, R. van der Meel, M. Fens, H.F.G. Heijnen, P.M.P. van Bergen En Henegouwen, P. Vader, R.M. Schiffelers, PEGylated and targeted extracellular vesicles display enhanced cell specificity and circulation time, *J Control Release* 224 (2016) 77–85 https://doi.org/10.1016/j.jconrel.2016.01.009.
[43] H. Kim, H. Jang, H. Cho, J. Choi, K.Y. Hwang, Y. Choi, S.H. Kim, Y. Yang, Recent advances in exosome-based drug delivery for cancer therapy, *Cancers (Basel)* 13 (2021) 4435 https://doi.org/10.3390/cancers13174435.
[44] S. Kamerkar, V.S. LeBleu, H. Sugimoto, S. Yang, C.F. Ruivo, S.A. Melo, J.J. Lee, R. Kalluri, Exosomes facilitate therapeutic targeting of oncogenic KRAS in pancreatic cancer, *Nature* 546 (2017) 498–503 https://doi.org/10.1038/nature22341.
[45] S.E.L. Andaloussi, I. Mager, X.O. Breakefield, M.J. Wood, Extracellular vesicles: Biology and emerging therapeutic opportunities, *Nat Rev Drug Discov* 12 (2013) 347–357 https://doi.org/10.1038/nrd3978.
[46] J. Li, J. Wang, Z. Chen, Emerging role of exosomes in cancer therapy: Progress and challenges, *Mol Cancer* 24 (2025) 13 https://doi.org/10.1186/s12943-024-02215-4.
[47] R. Tenchov, K.J. Hughes, M. Ganesan, K.A. Iyer, K. Ralhan, L.M. Lotti Diaz, R.E. Bird, J.M. Ivanov, Q.A. Zhou, Transforming medicine: Cutting-edge applications of nanoscale materials in drug delivery, *ACS Nano* 19 (2025) 4011–4038 https://doi.org/10.1021/acsnano.4c09566.
[48] Z. Wang, Q. Wang, F. Qin, J. Chen, Exosomes: A promising avenue for cancer diagnosis beyond treatment, *Front Cell Dev Biol* 12 (2024) 1344705 https://doi.org/10.3389/fcell.2024.1344705.
[49] E. Akyuz, F.S. Aslan, E. Gokce, O. Ilmaz, F. Topcu, S. Kakac, Extracellular vesicle and CRISPR gene therapy: Current applications in Alzheimer's disease, Parkinson's disease, amyotrophic lateral sclerosis, and Huntington's disease, *Eur J Neurosci* 60 (2024) 6057–6090 https://doi.org/10.1111/ejn.16541.
[50] L. Zisser, C.J. Binder, Extracellular vesicles as mediators in atherosclerotic cardiovascular disease, *J Lipid Atheroscler* 13 (2024) 232–261 https://doi.org/10.12997/jla.2024.13.3.232.
[51] H. Rayat Pisheh, M. Sani, Mesenchymal stem cells derived exosomes: A new era in cardiac regeneration, *Stem Cell Res Ther* 16 (2025) 16 https://doi.org/10.1186/s13287-024-04123-2.
[52] M.N. Huda, M. Nurunnabi, Potential application of exosomes in vaccine development and delivery, *Pharm Res* 39 (2022) 2635–2671 https://doi.org/10.1007/s11095-021-03143-4.

[53] D. Yang, Application of nanotechnology in the COVID-19 pandemic, *Int J Nanomed* 16 (2021) 623–649 https://doi.org/10.2147/IJN.S296383.

[54] B. Wei, M. Wei, H. Huang, T. Fan, Z. Zhang, X. Song, Mesenchymal stem cell-derived exosomes: A promising therapeutic strategy for age-related diseases, *Cell Prolif* 58 (2025) e13795 https://doi.org/10.1111/cpr.13795.

[55] Z. Shen, W. Huang, J. Liu, J. Tian, S. Wang, K. Rui, Effects of mesenchymal stem cell-derived exosomes on autoimmune diseases, *Front Immunol* 12 (2021) 749192 https://doi.org/10.3389/fimmu.2021.749192.

[56] T. Premchandani, A. Tatode, J. Taksande, M. Umekar, M. Qutub, U.M. Hussain, P. Singanwad, Engineered exosomes as smart drug carriers: Overcoming biological barriers in CNS and cancer therapy, *Drugs Drug Candidates* 4 (2025) 19 https://doi.org/10.3390/drugsdrugcandidates4010019.

[57] M. Qiu, J. Zou, Z. Yang, D. Yang, R. Wang, H. Guo, Strategies for targeting peptide-modified exosomes and their applications in the lungs, *Int J Nanomedicine* 19 (2024) 8175–8188 https://doi.org/10.2147/IJN.S472038.

[58] X. Yu, P. Liu, Z. Li, Z. Zhang, Function and mechanism of mesenchymal stem cells in the healing of diabetic foot wounds, *Front Endocrinol (Lausanne)* 14 (2023) 1099310 https://doi.org/10.3389/fendo.2023.1099310.

[59] J.M. Essola, M. Zhang, H. Yang, F. Li, B. Xia, J.F. Mavoungou, A. Hussain, Y. Huang, Exosome regulation of immune response mechanism: Pros and cons in immunotherapy, *Bioact Mater* 32 (2024) 124–146 https://doi.org/10.1016/j.bioactmat.2023.09.018.

[60] J.M. Pitt, F. Andre, S. Amigorena, J.C. Soria, A. Eggermont, G. Kroemer, L. Zitvogel, Dendritic cell-derived exosomes for cancer therapy, *J Clin Invest* 126 (2016) 1224–1232 https://doi.org/10.1172/JCI81137.

[61] Z. Du, Z. Huang, X. Chen, G. Jiang, Y. Peng, W. Feng, N. Huang, Modified dendritic cell-derived exosomes activate both NK cells and T cells through the NKG2D/NKG2D-L pathway to kill CML cells with or without T315I mutation, *Exp Hematol Oncol* 11 (2022) 36 https://doi.org/10.1186/s40164-022-00289-8.

[62] M.Y. Zamanian, H. Zafari, M.K. Osminina, A.A. Skakodub, R.F.A. Al-Aouadi, M. Golmohammadi, N. Nikbakht, I. Fatemi, Improving dexamethasone drug loading and efficacy in treating rheumatoid arthritis via liposome: Focusing on inflammation and molecular mechanisms, *Anim Model Exp Med* 8 (2025) 5–19 https://doi.org/10.1002/ame2.12518.

[63] Y. Huang, X. Guo, Y. Wu, X. Chen, L. Feng, N. Xie, G. Shen, Nanotechnology's frontier in combatting infectious and inflammatory diseases: Prevention and treatment, *Signal Transduct Target Ther* 9 (2024) 34 https://doi.org/10.1038/s41392-024-01745-z.

[64] Y. Li, B. Fang, Neural stem cell-derived extracellular vesicles: The light of central nervous system diseases, *Biomed Pharmacother* 165 (2023) 115092 https://doi.org/10.1016/j.biopha.2023.115092.

[65] N. Sanadgol, M. Abedi, M. Hashemzaei, Z. Kamran, R. Khalseh, C. Beyer, C. Voelz, Exosomes as nanocarriers for brain-targeted delivery of therapeutic nucleic acids: Advances and challenges, *J Nanobiotechnol* 23 (2025) 453 https://doi.org/10.1186/s12951-025-03528-2.

[66] J.S. Wang, S.J. Schellenberg, A. Demeros, A.Y. Lin, Exosomes in review: A new frontier in CAR-T cell therapies, *Neoplasia* 62 (2025) 101147 https://doi.org/10.1016/j.neo.2025.101147.

[67] H.W. Song, M. Prochazkova, L. Shao, R. Traynor, S. Underwood, M. Black, V. Fellowes, R. Shi, M. Pouzolles, H.C. Chou, A.T. Cheuk, N. Taylor, P. Jin, R.P. Somerville, D.F. Stroncek, J. Khan, S.L. Highfill, CAR-T cell expansion platforms yield distinct T cell differentiation states, *Cytotherapy* 26 (2024) 757–768 https://doi.org/10.1016/j.jcyt.2024.03.003.

[68] R. Farahzadi, E. Fathi, B. Valipour, S. Ghaffary, Stem cells-derived exosomes as cardiac regenerative agents, *Int J Cardiol Heart Vasc* 52 (2024) 101399 https://doi.org/10.1016/j.ijcha.2024.101399.

[69] R.C. Sterner, R.M. Sterner, CAR-T cell therapy: Current limitations and potential strategies, *Blood Cancer J* 11 (2021) 69 https://doi.org/10.1038/s41408-021-00459-7.

[70] Y. Sun, G. Wan, X. Bao, Extracellular vesicles as a potential therapy for stroke, *Int J Mol Sci* 26 (2025) 3130 https://doi.org/10.3390/ijms26073130.

[71] H. Yari, M.V. Mikhailova, M. Mardasi, M. Jafarzadehgharehziaaddin, S. Shahrokh, L. Thangavelu, H. Ahmadi, N. Shomali, Y. Yaghoubi, M. Zamani, M. Akbari, S. Alesaeidi, Emerging role of mesenchymal stromal cells (MSCs)-derived exosome in neurodegeneration-associated conditions: A groundbreaking cell-free approach, *Stem Cell Res Ther* 13 (2022) 423 https://doi.org/10.1186/s13287-022-03122-5.

[72] M. Chehelgerdi, F. Behdarvand Dehkordi, M. Chehelgerdi, H. Kabiri, H. Salehian-Dehkordi, M. Abdolvand, S. Salmanizadeh, M. Rashidi, A. Niazmand, S. Ahmadi, S. Feizbakhshan, S. Kabiri, N. Vatandoost, T. Ranjbarnejad, Exploring the promising potential of induced pluripotent stem cells in cancer research and therapy, *Mol Cancer* 22 (2023) 189 https://doi.org/10.1186/s12943-023-01873-0.

[73] S. Yamanaka, Induced pluripotent stem cells: Past, present, and future, *Cell Stem Cell* 10 (2012) 678–684 https://doi.org/10.1016/j.stem.2012.05.005.

[74] Y. Zhang, J. Yan, Y. Liu, Z. Chen, X. Li, L. Tang, J. Li, M. Duan, G. Zhang, Human amniotic fluid stem cell-derived exosomes as a novel cell-free therapy for cutaneous regeneration, *Front Cell Dev Biol* 9 (2021) 685873 https://doi.org/10.3389/fcell.2021.685873.

[75] S.H. Ahn, S.W. Ryu, H. Choi, S. You, J. Park, C. Choi, Manufacturing therapeutic exosomes: From bench to industry, *Mol Cells* 45 (2022) 284–290 https://doi.org/10.14348/molcells.2022.2033.

[76] E. Tan, C.S.H. Chin, Z.F.S. Lim, S.K. Ng, HEK293 cell line as a platform to produce recombinant proteins and viral vectors, *Front Bioeng Biotechnol* 9 (2021) 796991 https://doi.org/10.3389/fbioe.2021.796991.

[77] J. Mondal, S. Pillarisetti, V. Junnuthula, M. Saha, S.R. Hwang, I.K. Park, Y.K. Lee, Hybrid exosomes, exosome-like nanovesicles and engineered exosomes for therapeutic applications, *J Control Release* 353 (2023) 1127–1149 https://doi.org/10.1016/j.jconrel.2022.12.027.

5 Exosome-Based Therapeutics in Neurological Disorders

Abhipsa Priyadarshinee[1], *Akshita Jain*[1], *Etikala Amulya*[1], *Vivek Phatale*[1], *Tejaswini Kolipaka*[1], *Anupama Sikder*[1], *Mayank Gangwar*[2], and *Saurabh Srivastava*[1]

[1]Pharmaceutical Innovation and Translational Research Lab (PITRL), Department of Pharmaceutics, National Institute of Pharmaceutical Education and Research (NIPER), Hyderabad, Telangana, India

[2]Clinical Studies and Trials Unit, Division of Development, Indian Council of Medical Research, Department of Health Research, Ministry of Health & Family Welfare, New Delhi, India

5.1 INTRODUCTION

Neurological disorders represent a wide array of the most prevalent pathological conditions among severe diseases globally. Common neurological disorders such as Alzheimer's disease (AD), amyotrophic lateral sclerosis (ALS), Parkinson's disease (PD), and glioblastoma multiforme (GBM) pose significant health hurdles due to their complexity and limited therapeutic options [1]. They show a significant global burden on public health due to their high mortality and morbidity rates. These disorders, characterized by irreversible neuronal damage and a progressive decline in motor function, pose major challenges to effective treatment. The relentless nature of neurodegeneration, coupled with the increasing rate of disease progression despite therapeutic interventions, has hindered the development of successful countermeasures. Among these conditions, AD and PD stand as the most prevalent neurodegenerative disorders worldwide, followed by Huntington's disease (HD) and ALS [2]. A key pathological hallmark shared by these diseases is the abnormal accumulation of misfolded proteins, which triggers neuronal degeneration in specific brain regions, leading to profound physiological and functional impairments. The pathological protein aggregation may result from either a toxic gain of function, where mutated proteins form harmful aggregates, or a loss of essential cellular functions, rendering key proteins nonfunctional. Each neurodegenerative disease is associated with unique protein aggregates: AD is characterized by amyloid β (Aβ)-peptide and tau (τ) deposits, PD is marked by Parkin and α-synuclein accumulation, and HD involves Huntingtin aggregates within the nucleus and cytoplasm of affected neurons. Understanding these proteinopathies is crucial for developing targeted therapeutic strategies to combat these debilitating disorders.

The major challenges associated with the treatment of neurological disorders include (a) lack of curative treatment, (b) lack of early detection techniques, and (c) high cost of novel therapeutics that significantly contribute to the global neurological burden. Moreover, treating neurological disorders faces a formidable challenge owing to the presence of the blood–brain barrier (BBB) and blood–cerebrospinal fluid barrier (B-CSF) [3]. These highly selective barriers block over 98% of therapeutics and nearly 100% of nucleic acids, peptides, and proteins from reaching the CNS parenchyma, drastically limiting the effectiveness of conventional treatments. The restricted drug penetration into the brain necessitates higher doses and frequent administration, increasing the risk of neurotoxicity and unwanted side effects. To circumvent these limitations, researchers must delve into an in-depth understanding of BBB architecture, cellular transport mechanisms, and strategic drug selection when designing an optimal drug delivery system (DDS). Advances in nanotechnology, receptor-mediated transport, and targeted formulations are paving the way for innovative solutions that can effectively bypass these biological barriers, and enhance therapeutic efficacy while minimizing systemic toxicity.

Exosome (EXO)-based drug delivery and biomarker detection is an innovative and rapidly advancing field in neurodegenerative disease. These vesicles are secreted by all cell types and are naturally found in breast milk, serum, plasma, CSF, saliva, urine, and other bodily secretions. EXOs play a key role in cell-to-cell communication, synaptic plasticity, and nerve regeneration, primarily through EXO-mediated signaling cross talks. In the central nervous system (CNS), their release is regulated by glutaminergic neurons, where calcium influx via AMPA and NMDA receptors modulates their secretion.

One of the most remarkable attributes of EXOs is their innate ability to traverse the BBB, which is a major hurdle in CNS drug delivery. With their unique capacity to traverse biological barriers, protect therapeutic payloads, and facilitate targeted delivery, EXOs are emerging as a powerful nanocarrier for next-generation CNS therapeutics.

DOI: 10.1201/9781003623632-5

The use of EXOs in neurological disorders offers several advantages, including their capability to transport neuroprotective agents, modulate neuroinflammation, and promote neural regeneration [4]. Furthermore, engineered EXOs can be functionalized with therapeutic molecules, which enhances their efficacy and specificity in targeting diseased neural tissues. Their ability to function as natural nanocarriers, along with the progress in bioengineering methods, has paved the way for innovative developments. In the last few years, EXO-derived nanoarchitectonics has emerged as a potential solution for overcoming the challenges attributed to drug delivery in neurological disorders. Their biocompatibility, low immunogenicity, and ability to be engineered for targeted delivery make them highly attractive for CNS therapeutics. Their potential to deliver drugs, siRNA, miRNA, and neuroprotective agents directly to the brain opens new frontiers in the treatment of neurodegenerative conditions. Harnessing EXOs as a natural, efficient, and minimally invasive drug delivery system holds the promise of revolutionizing neurotherapeutics, paving the way for more effective and precise treatments for neurological disorders [5]. This chapter emphasizes the source from which EXOs are derived, and their potential role in various neurological disorders. Moreover, EXO-based nanoarchitectonics for the management of neurological ailments have been deliberated. Further, we have shed light on the clinical perspectives of EXOs-based therapeutics.

5.2 HARNESSING EXOS FROM DIVERSE SOURCES FOR VARIOUS NEUROLOGICAL DISORDERS

5.2.1 Bone Mesenchymal Stem Cell–Derived EXOs (BMSC-EXOs)

BMSC-EXOs have shown considerable interest in treating various neurological disorders, owing to their neuroprotective and regenerative capabilities. Wen and his team confirmed that BMSC-EXOs facilitated the shift of activated BV2 microglia toward an anti-inflammatory phenotype, suppressed pro-inflammatory cytokine expression, and enhanced anti-inflammatory cytokines in vitro. Additionally, in vivo studies indicate the neuroinflammation-reducing effects of BMSC-derived EXOs following TBI, likely by regulating the IL-10/STAT3 pathway through the action of miR-181b [6]. Another study depicts that BM-MSCs-EXOs alleviate cognitive impairment in an AD model by enhancing miR-146a expression in the hippocampus. This increase in miR-146a reduces neuroinflammation by suppressing NF-κB signaling in astrocytes. Improved astrocytic function promotes synaptogenesis, contributing to cognitive recovery. These findings emphasize the intriguing potential of BM-MSC-EXOs as a possible treatment option for AD [7]. Cui and his team investigated the effects of rabies viral glycoprotein (RVG)-tagged MSCs-derived EXOs in an AD model. These EXOs portrayed enhanced transit to the hippocampus and cortex, leading to reduced astrocyte activation, plaque deposition, and Aβ accumulation. EXO-treated mice exhibited significant improvements in learning and memory. Furthermore, the levels of anti-inflammatory cytokines (IL-10, IL-4, and IL-13) were found to be elevated, and the pro-inflammatory cytokines (TNF-α, IL-β, and IL-6) were reduced [8]. BM-MSCs-derived EXOs subsided the Aβ1–40, BACE1, Aβ1–42, and PS1 levels and endorsed the expression of neprilysin in APP/PS1 mice, thereby confirming that BMSC-EXOs reduce Aβ and promote cognitive function recovery by triggering SphK/S1P pathway [9]. A study by Mendes-Pinheiro and his colleagues highlights the therapeutic potential of human BM-MSCs (hBMSCs) secretome in a 6-OHDA rat model of PD. Compared to direct hBMSC transplantation, secretome injection led to better motor recovery, enhanced dopaminergic neuron survival, and promoted in vitro neuronal differentiation [10]. Likewise, BM-MSCs have been shown to suppress the Sp1 pathway, resulting in reduced neuronal loss and a diminished inflammatory response in an MPTP model [11]. A study by Wang et al. highlighted that exosomal miR-193b-5p derived from BM-MSCs reduced infarct volume and the expression of pyroptosis-related molecules in ischemic stroke by inhibiting the AIM2 pathway [12]. A study by Safakheil M and his team highlighted the role of combination therapy, that is, BM-MSCs with rosuvastatin in an ischemic stroke model. This approach reduced the size of infarct volume while downregulating the NLRP1 and NLRP3, indicating decreased cell death. Additionally, this therapy also reduced the number of GFAP-positive cells and MDA levels, reinforcing the overall therapeutic benefit of the exosomal strategy in the stroke model [13].

5.2.2 Immune Cell–Derived EXOs

Immune cell–derived EXOs have emerged as key mediators in the management of various neurological disorders. These EVs secreted by various immune cells, including microglia, dendritic cells, and macrophages, play a vital role in neuroprotection, neuroinflammation, and immune regulation.

Shakespear and his team investigated the role of astrocyte-derived exosomal miR-200a-3p in preventing MPP+-induced apoptosis. EXOs released by astrocytes have been shown to protect primary mesencephalic dopaminergic neurons and SH-SY5Y cells from MPP+-induced toxicity by decreasing cell death and suppressing the expression of MKK4, a key upstream regulator in the JNK-mediated apoptotic pathway [14]. M2 microglia-derived EXOs (M2-EXOs) portrayed the decrease in cell death, restored the mitochondrial membrane potential, and alleviated the levels of ROS in vitro. These EXOs were found to reduce the Aβ oligomers in APP/PS1 mice. Overall findings from the study depict that M2-EXOs play a crucial role in AD management by ameliorating PINK1/Parkin-mediated mitophagy [15]. Song and his team assessed the role of M2-EXOs in the stroke model. The M2-EXOs reduced the apoptosis of the oxygen-glucose-deprived (OGD) neuronal cells. The in vivo studies confirmed that M2-EXOs improved the behavioral outcomes and reduced the infarct volume [16]. Collectively, these studies underscore the promising role of immune cell–derived EXOs in neuroprotection and disease modulation.

5.2.3 Adipose-Derived Mesenchymal Stem Cell (ADMSCs)

ADMSCs exhibit significant potential in tissue repair mechanisms. The ADMSCs are highly preferred due to their high renewability and lower isolation costs, making them particularly advantageous for EXO production. This addresses major concerns associated with other techniques, such as BMSC [17, 18]. In neurodegenerative research, they have extensively explored their therapeutic applications in managing these disorders.

In AD patients, lower levels of neprilysin contribute to the accumulation of Aβ plaques. ADMSCs-derived EXOs play a crucial role in decreasing Aβ plaque formation [19, 20]. Kastuda et al. evaluated the neprilysin levels in an in vitro cell culture model using the N2a cell line. The results demonstrated enhanced neprilysin activity in comparison to control (phosphate buffer), with a progressive increase in activity over incubation time. Subsequently, the authors introduced thiorphan, a metallopeptidase inhibitor, which led to a decrease in neprilysin levels. This study confirms that EXOs release neprilysin, which is enzymatically degraded by thiorphan. Further, a comparative analysis was conducted using EXO derived from BMSC. The findings revealed that ADMSCs exhibited twice the neprilysin activity in comparison to BMSCs after 60 minutes of incubation. The study also revealed a significant decrease in Aβ plaque levels ($p = 0.0073$). Additionally, the authors have validated EXO transfer from the fluorescent-labeled ADMSCs to N2a cells. The results showed a unidirectional movement of EXO toward N2a cells, with no observable transfer to ADMSCs [21]. Moving toward PD, the stimulation of inflammasome (NLRP3) and cell division protein kinase (CDK-5) that mediates the autophagy, has a significant impact on the PD. So, targeting the gene level could be an effective strategy to prevent autophagy, which is responsible for PD. Qiang Li and team investigated the activity of miR-188-3p in PD at in vitro and in vivo levels. Initially, RT-PCR studies were performed on PD patients to understand miR-188-3p levels, revealing a significant decrease compared to the control ($p < 0.001$). Subsequently, the release of miR-188-3p from ADMSCs and their EXOs was analyzed, showing ~1.66-fold higher release from EXOs. Furthermore, key findings from the in vivo study using an MPTP model demonstrated a significant decrease in CDK-5 and NLRP-3 levels [22]. The results show effective treatment for PD patients. The overall findings in AD and PD reveal the significance of ADMSCs in neurological diseases.

5.2.4 Bacterial Extracellular Vesicles (BEVs)

BEVs are nanosized vesicles released by bacteria that facilitate communication between the gut bacteria and the brain. These vesicles contain various bioactive constituents, including nucleic acids, lipoglycans, quorum sensing, and proteins [23]. The formation of BEVs differs in Gram-positive and Gram-negative bacteria. In case of Gram-negative bacteria, vesicles are formed through protrusions on the outer membrane, which subsequently detach and are released through cell lysis. Conversely, Gram-positive bacteria produce vesicles from the cytoplasmic membrane via an endocytosis mechanism. These vesicles can regulate gene expression and play a pivotal role in neurodegenerative disorders and neurological diseases, AD and PD [24]. Khadijeh Sheikh and team have used BEVs derived from *Lactobacillus acidophilus* (Gram-positive) to evaluate their effect in PD. The results demonstrated that these vesicles could cross both the intestinal barrier and the BBB, resulting in enhanced 5-HT levels in the substantia nigra, a region affected in PD [25]. Another study by Pan et al. explored vesicles derived from the Gram-negative bacteria, which were hitchhiked on neutrophils. The authors loaded the vesicles with pioglitazone, a therapeutic agent for its anti-inflammatory properties. The treatment exhibited inhibitory effects on NLRP3 inflammasome activation, reducing reperfusion injury and demonstrating neuroprotective activity [26]. The gut–brain axis offers a promising route for the treatment of neurological diseases. Leveraging BEVs as a delivery system could enhance the therapeutic potential of neuroprotective agents.

5.2.5 Miscellaneous

EXOs derived from various MSCs have gained substantial attention due to their low immunogenicity and broad therapeutic potential. These stem cell–derived EXOs contain bioactive molecules like microRNAs (miRNAs), proteins, and lipids, which contribute to their regenerative and immunomodulatory effects, thereby exhibiting a vital role in neurological disorders. Yi-An Chen and colleagues explored the therapeutic role of Wharton's jelly MSCs-derived EXOs in the AD model. Results demonstrated that the Aβ expression was downregulated and the neuronal memory/synaptic plasticity-related gene expression was restored. The treatment of AD transgenic mice with EXOs resulted in improved cognitive function and brain glucose metabolism. Additionally, the neuronal and astrocyte activity is also found to be regulated in vivo [27]. In another study, Zhang and his coworkers unveiled that exosomal miR-146a-5p from umbilical cord MSCs alleviates microglia-driven neuroinflammation following OGD in vitro. Additionally, the behavioral deficits were found to be improved, and microglial activation was modulated in vivo by targeting the IRAK1/TRAF6 signaling pathway in the ischemic stroke model [28].

5.3 ROLE OF EXOSOMES IN VARIOUS NEUROLOGICAL DISORDERS

EXOs play a decisive role in neurodegenerative diseases by acting as natural carriers for intercellular communication.

Their ability to carry bioactive molecules makes them both pathological mediators and potential therapeutic tools. Further, EXOs derived from neurons or glial cells contain disease-specific proteins (e.g., α-synuclein in PD, τ in AD) that can serve as diagnostic biomarkers and can mediate the transmission of misfolded proteins, facilitating disease progression in conditions like AD, ALS, and PD. Moreover, it holds a great promise in revolutionizing the diagnosis and treatment of neurodegenerative disorders by serving as both biomarkers and therapeutic delivery systems.

5.3.1 Alzheimer's Disease

AD is the most widespread disorder resulting in dementia with neuronal degeneration. The hallmark neuropathological characteristics of AD include the accumulation of Aβ plaques external to the brain cells and the formation of neurofibrillary tangles (NFTs) within the brain cells [29]. The International Alzheimer's Organization projects that the worldwide number of people affected by AD will reach nearly 150 million by 2050. Numerous studies have highlighted the dual mechanism of EXOs in AD. EXOs are capable of transmitting pathogenic Aβ and hyperphosphorylated τ between the cells, possibly inducing apoptosis, which contributes to neuronal degeneration. Conversely, EXOs can mitigate Aβ accumulation in the brain with the help of microglial uptake and are also recognized for transferring neuroprotective factors across cells. Recently, there has been a notable increase in the progress of EXO-derived drug delivery systems, owing to their capacity to effectively traverse the BBB as well as other biological barriers. These properties, along with several other attributes, render EXOs highly intriguing for the enhancement of innovative therapeutic approaches.

EXOs play a dual role in AD pathogenesis, acting as both drivers of pathology and potential protective agents. These EVs facilitate the spread of toxic proteins, mediate neuroinflammation, and influence disease progression through complex mechanisms. Aβ plaques are transported from the extracellular region into the intracellular space in AD with the help of EXOs. Additionally, EXOs facilitate the intercellular transmission of Aβ, allowing it to spread from a single cell to another. Further, EXOs carry γ-secretase and β-secretase (BACE1), accelerating Aβ aggregation and plaque formation. They act as "seeds" for Aβ oligomerization, similar to prions, enabling transcellular spread between neurons [30]. EXOs in AD play an additional role; although they are linked to the pathophysiology of AD, they also show protective effects by reducing cellular burden by enabling the removal of accumulated toxins like Aβ and α-synuclein [31]. EXOs enhance the removal of Aβ through multiple pathways. First, neuron-derived EXOs transport Aβ to microglial lysosomes, where insulin-degrading enzyme (IDE) and other proteases break it down [32]. Glycosphingolipid-enriched EXOs trap Aβ and shuttle it to microglia, reducing plaque burden in AD mouse models. Further, EXOs accelerate Aβ fibril formation, which reduces toxic oligomers and makes Aβ easier for microglia to phagocytose [33]. Additionally, microglial EXOs express triggering receptor expressed on myeloid cells-2 (TREM2), which binds Aβ and promotes its clearance. TREM2 also inhibits τ hyperphosphorylation and spreads by modulating the inflammatory response. Again, blocking the P2X purinoceptor 7 (P2RX7) in microglia reduces EXO secretion, decreasing τ accumulation and improving memory in tauopathy mice [34].

5.3.2 Parkinson's Disease

PD is a chronic, nerve-degenerating disorder marked by dysfunction of motor neurons and neuronal inflammation [35]. Additionally, there will be a depletion of dopaminergic neurons within the substantia nigra. The accumulation of neuronal α-synuclein is a prominent feature in PD [36]. As a result, scientists are concentrating on innovative drug delivery methods to address neurological disorders. Numerous researchers highlighted the role of innovative carrier systems like EXOs. However, EXOs play a fundamental role as natural carriers for delivering neuroprotective factors that help to maintain dopaminergic neuron survival. These vesicles play an insignificant role in the delivery of different therapeutic molecules, such as growth factors and antioxidants, which aid in mitigating oxidative stress and inhibiting apoptosis, which is a key contributor to neuronal degeneration in PD. Additionally, EXOs possess surface markers that enable them to interact with specific cells, including neurons, allowing for efficient BBB penetration, a major hurdle in PD management. Moreover, EXOs can deliver genetic material, such as mRNAs, which helps in the regulation of gene expression as well as signaling pathways involved in the pathology of PD. Protecting brain neurons is a crucial goal in developing effective therapies for degenerative diseases.

EXOs can be engineered to carry antioxidants like catalase, which reduces oxidative stress in PD models. Catalase-loaded EXOs (ExoCAT) have displayed significant neuroprotective properties by preserving dopaminergic neurons and alleviating oxidative damage in both in vitro and in vivo PD models. As a potential drug delivery system, these EXOs effectively cross the BBB and target neuronal cells [37]. Moreover, blood-derived EXOs from healthy donors have shown promise in restoring the body's natural balance against oxidative stress in mice models of PD induced by MPTP. By helping to rebalance oxidative homeostasis, these EXOs reduce the oxidative stress that often leads to neuronal damage, one of the main drivers behind the development of PD [38]. Further, the presence of anti-inflammatory molecules, like specific microRNAs (e.g., miR-133b), mitigates the formation of pro-inflammatory cytokines. Hence, EXOs help in protecting the dopaminergic neurons by reducing the neuroinflammation [36]. Ting Sun and team investigated that in MPTP-treated mice, EXO treatment alleviated motor impairments, rescued dopaminergic neurons in the substantia nigra and striatum, and reduced apoptosis. These findings highlight their potential as a therapeutic tool for PD [38]. Further, MSC-derived EXOs containing miR-133b promote neurite outgrowth and regulate tyrosine hydroxylase production, which is essential for dopamine synthesis. This enhances dopaminergic function and supports neuronal repair in PD models [39]. Moreover, EXOs have been discovered to act as transporters of the PD-related protein α-synuclein into neural cells. EXOs from α-synuclein-overexpressing SH-SY-5Y cells were found to be capable of transmitting α-synuclein to viable SH-SY-5Y cells in an in vitro investigation. The results presented clearly imply that EXOs are essential to the development of PD [40]. EXOs are important mediators of neuroinflammation in the brain. It has been estimated that peripheral EXOs contribute to neuroinflammation by controlling astrogliosis and microgliosis. In contrast, Zhang et al. showed that IL-1β-stimulated EXOs can be used as a targeted strategy to treat neuroinflammation. The scientists created an injectable, thermosensitive supramolecular HDU/SF1 hydrogel that efficiently improves the adherence of EXOs stimulated IL-1β. Neurons are retrieved as a result of the hydrogel's retention of EXOs, which prevents neuroinflammation [41].

5.3.3 Amyotrophic Lateral Sclerosis

ALS, commonly referred to as Lou Gehrig's disease, is a progressive and severe neurological disease characterized by the gradual degeneration of motor neurons in the spinal cord and brain. This progressive condition leads to impaired muscle control, resulting in weakness, atrophy, and eventual paralysis. Exosomes offer a valuable strategy for targeted delivery of therapeutics to motor neurons and glial cells, where they help modulate the inflammatory environment, alleviate oxidative stress, and improve intercellular communication. A key neuropathological feature of ALS is the accumulation of mutant superoxide dismutase 1 (SOD-1). Additionally, the risk of immune reactions reduces due to their natural biocompatibility.

An in vitro study discovered that EXOs helped with the fostering of SOD1 expression. Within the study, it was suggested that EXOs, which were set free from cells damaged by ALS, could potentially allow SOD1 to be passed on to adjacent cells and thus aid in the proliferation of the disease. These outcomes suggest that EXOs may profoundly impact the onset and progression of ALS [42]. Furthermore, EXOs display neuroprotective effects in ALS by modulating apoptosis, oxidative stress, protein aggregation, and mitochondrial injuries. EXOs defend neuronal cells from oxidative injury by transferring protective agents and altering cellular processes. For instance, EXOs are reported to lower the levels of reactive oxygen species (ROS) produced in ALS models, maintaining the integrity of neurons [43]. EXO treatment improves mitochondrial function by bringing the diseased phospho-CREB/CREB ratio and PGC-1α expression levels in ALS neuronal cells to normal values. These modifications increase energy production and mitigate the decline in function of mitochondria, which underlie the degeneration of motor neurons in ALS [44].

5.3.4 Peripheral Nerve Injury

Peripheral nerve injury (PNI) is one of the neurodegenerative conditions that impairs both the CNS sensory and motor functions. The most prominent causes of PNI are trauma and iatrogenic procedures that result in functional impairment and loss of structural integrity [45]. EXOs show a crucial role in the repair and regeneration of peripheral nerves following injury. Their protective functions are linked to their ability to create favorable microenvironments, enhance axon regeneration, and promote angiogenesis. NTN1 EC-EXOs are derived from endothelial cells overexpressing Netrin-1 (NTN1), a molecule known for its role in angiogenesis and nerve regeneration. The EXOs help establish a vascular niche that fosters early nerve repair by promoting angiogenesis, axon guidance, and remyelination. Multi-omics analysis revealed that NTN1 EC-EXOs activate key signaling pathways such as PI3K-AKT, mTOR, focal adhesion, and axon guidance. They also downregulate let7a-5p, which is associated with enhanced regenerative capacity [46]. Furthermore, the pattern of endocytosis in nerve regeneration in the case of PNI was discovered by Shi Guidong et al. During nerve regeneration, the authors found that genes like ARF1, ARF6, RAB7A, and VPS45 were upregulated, while genes like RAB11A, DNM3, and NEDD4 were downregulated [47]. Few studies have documented how EXOs regulate the progression of PNI, underscoring their critical function in peripheral nerve preservation and restoration. In case of PNI, Din Yi et al. created platelet-enriched EXOs derived from plasma (PRP-EXOs) that support nerve regeneration and repair [48, 49].

5.3.5 CNS Infection

CNS infections, which include encephalitis, abscesses, and meningitis, are the most common type of infectious disease. Of all these, the most prevalent one is Japanese encephalitis, a CNS infection that primarily affects South Asian nations [50]. The regulative function of EXOs in CNS infections, especially meningitis and Japanese encephalitis, has been thoroughly investigated. Scientists found that meningitis patients had higher levels of CD93 expression than healthy people [51]. Furthermore, it was discovered that inflammatory factors like TNF-α, IL-4, and IFN-γ, as well as serum EXOs CD93 levels, play a role in HIV-associated cryptococcal meningitis [52]. The immune cells of the CNS, that is, microglia, play a crucial role in encephalitis. EXOs released by infected microglia cause nearby cells to start getting inflamed. EXOs, however, can also negatively affect CNS infections. For example, the pathophysiology of the virus associated with Japanese encephalitis is promoted, and the severity of the disease is increased by the exosomal secretion of let-7a/b miRNA [53, 54]. EXOs derived from infected cells can carry proteins and RNAs that influence immune cell behavior. They can enhance immune responses by inducing the activation of immune cells. EXOs can also stimulate monocyte-macrophages, natural killer (NK) cells, B cells, T cells, and natural killer (NK) cells, enhancing their functionality against pathogens. They can present antigens to immune cells, facilitating the recognition and targeting of pathogens within the CNS [55]. EXOs can also directly inhibit the growth and spread of pathogens. They contain bioactive molecules that may disrupt pathogen replication or interfere with their capability to infect host cells. Additionally, EXOs have become useful agents in targeted therapies for TB-related CNS infections. Rifampicin-loaded Angiopep-2-modified EXOs have demonstrated a promising role in treating CNS infections linked to tuberculosis. These altered EXOs improve the therapeutic efficacy of rifampicin by facilitating its targeted delivery [56].

5.3.6 Spinal Cord Injury

Serious spinal cord damage, usually brought on by mechanical trauma or neurogenic impairment, results in SCI. SCI is linked to an extremely painful condition that may worsen if treatment is not received. For doctors and biomedical scientists, treating SCI is a difficult task. Researchers are currently focusing their efforts on understanding the full pathophysiology of SCI by examining the regulatory functions of particular proteins, enzymes, miRNAs, and other molecular aspects. Additionally, EXOs are crucial for the regulation and treatment of SCI. As natural drug carriers, EXOs help medications cross the BBB to treat neurodegenerative diseases and spinal cord injuries [57]. According to the research conducted by Shu-Qin Ding and colleagues, patients with SCI had changes in their serum exosomal miRNAs, including miR-152–3p, miR-125b, and miR-200c-3p [58]. The authors came to the conclusion that serum exosomal miRNAs might be useful biomarkers for spinal cord injury. Diverse cellular sources of EXOs show significant therapeutic promise in the treatment of SCI. In a murine model, EXOs generated by MSCs functionalized with circZFHX3 are specifically linked to the restorative processes of SCI by modifying mir-16–5p/IGF-1 signaling [59]. Furthermore, by modifying the miR-124–3p/MYH9 axis, the synergistic use of miRNAs and neuron-derived EXOs shows a protective effect in SCI [60]. Similarly, EXOs containing miR-181c promote protection against changes brought on by SCI by blocking PTEN and NF-KB signaling. By directly inhibiting the AIM2/ASC/Caspase-1 signaling cascade, EXOs carrying miR-672–5p also reduce neuronal pyroptosis and enhance neurofunctional restoration in SCI [61, 62].

5.3.7 CNS Injury and Stroke

Stroke is a multifaceted neurovascular disorder arising through two principal mechanisms: ischemia (blocked cerebral blood flow) and hemorrhage (ruptured blood vessel bleeding). Ischemic strokes occur when an arterial blockage restricts blood flow to the brain, leading to the death of neuronal cells due to oxygen depletion and the deprivation of nutrients. In contrast, hemorrhagic strokes result from blood vessels rupturing, causing blood to leak into brain tissue, increasing pressure, and damaging surrounding cells. Both processes trigger neuroinflammation and disrupt neurotransmission, leading to a cascade of inflammation, oxidative stress, and excitotoxicity.

The two main causes of death and morbidity worldwide are CNS injuries and stroke. Ischemic stroke and hemorrhagic stroke are the two general classifications into which stroke falls. An ischemic stroke arises when the blood flow to a specific part of the brain is not adequate, whereas a hemorrhagic stroke occurs when blood vessels burst, leading to internal bleeding. In the pathogenesis of CNS injuries and stroke, EXOs play a vital role. Their important role in these conditions has been shown by a few preclinical investigations. It is interesting to note that by modifying their target proteins and particular miRNAs, EXOs produced from cerebral endothelial cells stimulate axonal growth [63]. EXOs that contain miRNAs are essential for both therapeutic and diagnostic purposes. Another study offered strong evidence of the protective role of miRNA-26a derived from EXOs in the setting of ischemic brain injury, which is facilitated by microglial polarization modulation. Additionally, by modifying the notch1 protein, exosomal miRNA-137 derived from microglia has a neuroprotective effect [64]. Jinju Wang and team examined the function of EXOs made from endothelial progenitor cells modified by miRNA-126. They found that it had a neuroprotective effect by encouraging angiogenesis, boosting cerebral blood flow, and downregulating caspase 3 before upregulating VEGFR2 [65]. Furthermore, in the context of neuroinflammation, the cerebro-protective action of EXOs containing miR-181c-3p was noted, in which they restrict CXCL1 in ischemic brain injury [66].

5.3.8 Glioblastoma Multiforme

GBM is an aggressive and intricate disease classified as a grade IV tumor by the WHO. Its heterogeneous nature presents considerable challenges in treatment [67]. Recently, the role of EXOs in glioma progression has garnered increasing attention among cancer researchers. EXOs play a crucial role in GBM treatment by serving as a natural nanocarrier system for drug delivery, enabling targeted therapy with minimal toxicity. EXOs can traverse the BBB, thereby making them an effective vehicle for delivering anticancer drugs as well as genes that will help in the management of tumor growth and combat tumor resistance [68]. EXOs-based therapies can improve immune responses against GBM by carrying tumor-associated antigens to activate immune cells. EXOs mediate intercellular communication, making them valuable in real-time monitoring of treatment efficacy [69]. Although they possess several advantages, certain bottlenecks, like off-target effects and scale-up challenges, need to be addressed before clinical translation.

Glioma-derived EXOs (GDEs) play a major role in aiding tumor cell growth, proliferation, and tumor angiogenesis. GDEs also promote initiation, progression, and resistance to therapies [70]. EXOs regulate the survival, invasion, and proliferation of tumor cells. GDEs carry various RNA molecules, both coding and noncoding, which further help in tumor cell survival and invasion [71].

5.4 EXOSOME-BASED NANOARCHITECTONICS FOR THE MANAGEMENT OF NEUROLOGICAL DISORDERS

The effective delivery of drugs to the brain remains a significant challenge due to the presence of the protective barrier known as the BBB, which restricts the entry of most therapeutics, including small molecules, proteins, and gene-based drugs [72]. Moreover, a wide range of diseases are affecting the CNS, which further complicates the development of effective drug delivery strategies. Additionally, conventional treatment approaches often suffer from many drawbacks, like poor bioavailability, systemic toxicity, and off-target effects [73]. This scenario represents the pivotal need for the development of innovative and efficient drug delivery strategies.

Recently, numerous researchers highlighted the emerging role of EXO-based nanoarchitectonics as a promising approach for overcoming conventional treatment challenges. Owing to their intrinsic biocompatibility, potential to permeate the BBB, and ability for targeted drug delivery, EXOs have garnered significant attention in the management of neurological conditions [74]. In this section, we explore the influence of various neurological conditions on the neurological microenvironment. It especially highlights the potential of EXOs as a therapeutic agent, analyzing data from multiple sources and their diverse effects on improving microenvironmental conditions. Moreover, further comprehensive research is necessary to determine the particular therapeutic value of EXOs, which is derived from different types of cells or modified EXOs. The goal is to explore better strategies that can effectively alleviate the existing challenges in the realm of neurological disorders.

5.4.1 Exosome-Based Nanoarchitectonics in PD

Conventional drug delivery methods for PD face major obstacles, including poor permeability across the BBB, low bioavailability, and systemic side effects. In this regard, nanotechnology holds great promise and has the potential to revolutionize the treatment of neurological diseases in the coming days [75]. A range of nanotechnological strategies is being investigated for drug delivery that targets neurological conditions. These consist of nanoscale carriers such as dendrimers, mesosomes, EXOs, and nanoparticles (NPs), which are commonly used to transport medications to the affected region of the brain.

Human umbilical cord MSCs (hucMSCs) are known due to their neuroprotective properties. However, recent research by Chen et al. suggests that stem cells communicate with other cells through EXOs. Additionally, the authors confirmed whether hucMSC-derived EXOs could serve as an alternative to hucMSCs. To support the research findings, hucMSCs were successfully isolated and identified to demonstrate their ability to promote the proliferation of SH-SY5Y cells. Moreover, hucMSCs were exposed to 6-hydroxydopamine (6-OHDA), a neurotoxic agent for PD modeling. Similarly, EXOs derived from hucMSCs also show excellent proliferation of SH-SY5Y cells. Notably, authors revealed that hucMSCs activated 6 OHDA and helped to stimulate SH-SY5Y cell proliferation ($P < 0.01$).

Further, in vivo experiments in the rat model of PD showed that EXOs crossed the BBB and reached the substantia nigra. Moreover, they alleviated apomorphine-induced rotational asymmetry, reduced dopaminergic neuron depletion and apoptosis, and increased dopamine levels in the striatum. These findings highlight the potential of hucMSC-derived EXOs as a novel therapeutic approach for PD [76].

Furthermore, we report another innovative approach based on synthetic biology-inspired control devices known as the transfer of EXOs into Cells (EXOtic) devices. These devices optimize key steps in mRNA delivery of EXOs, eliminating the necessity for EXO concentration. Researchers have explored whether engineered EXOs carrying catalase mRNA, which is produced by EXOs-generating cells consisting of the EXOtic devices, help to mitigate the cell death of neurons, which is induced by 6-hydroxydopamine (6-OHDA). As a neurotoxin, 6-OHDA is primarily utilized to model experimental PD due to its ability to damage neurons, although not entirely, by generating a cytotoxic stage of ROS. In another innovative research, they explored the use of HEK293 cells, which are transduced with plasmids that are designed to produce catalase mRNA-loaded EXOs, specifically targeting the brain for the treatment of PD. A promising strategy for directing EXOs toward the brain involves genetically transforming EXO-producing cells by introducing genes encoding target molecules such as peptides or receptors that associate with EXO membrane constituents like tetra-spanins, lysosomal membrane-associated protein 2B (LAMP-2B), or domain C1C2. For instance, scientists modified cells to produce a fusion protein combined with rabies virus glycoprotein (RVG) and LAMP2B, resulting in EXOs containing RVG embedded in their membranes. These RVG-expressing EXOs exhibited enhanced localization to the brain, facilitated by the interaction of RVG with neuronal and glial surface receptors. Additionally, a suite of EXOtic devices has been developed to enable the effective and customizable generation of engineered EXOs in genetically modified mammal cells. These encoded devices enhance EXOs production, targeted loading of mRNA, and delivery of mRNA into recipient cells without requiring EXOs concentration. Furthermore, transformed producer cells implanted in living organisms can reliably transport cargo mRNA to the brain. In both in vitro and in vivo experimental models of PD, the administration of mRNA catalase through engineered EXOs demonstrated neuroprotective effects by reducing neurotoxicity and inflammation of neurons. This research explored the potential of EXOtic devices as a promising platform for RNA-based therapeutic applications (Figure 5.1) [77].

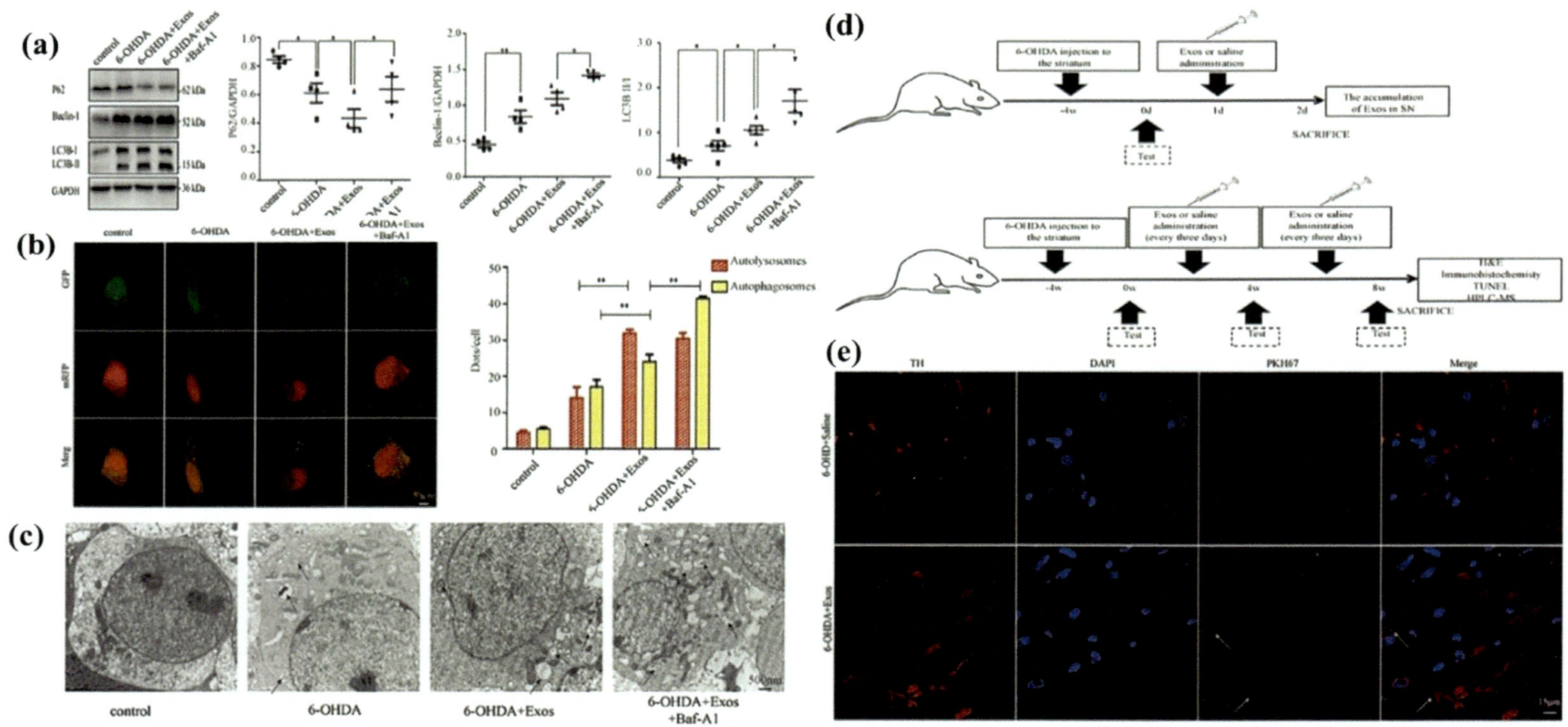

FIGURE 5.1 Representation of the (a) 6-OHDA-induced alterations in autophagy proteins was detected by Western blotting. (b) The tandem fluorescent GFP-RFG-LC3 assay detected autolysosomes and autophagosomes. (c) TEM evaluated autophagosomes in four groups. (D) An in vivo experimental technique. (e) Rats were given intravenous injections (IV) of PKH67-labeled EXOs. The figure shows the distribution of EXOs in brain tissue. *Source:* Adapted with permission from Ref. [77]. Copyright 2018, published under a Creative Commons CC BY License (http://creativecommons.org/licenses/by/4.0/).

5.4.2 Exosome-Based Nanoarchitectonics in AD

Nanotechnology-based EXOs offer significant advantages in AD treatment, particularly in overcoming challenges like BBB penetration and targeted drug delivery. Further, it helps in multi-therapeutic cargo capacity, targeted delivery, and sustained release, and has diagnostic potential. Ongoing research aims to optimize their drug-loading capacity and scalability for clinical use.

Multi-targeted treatment has arisen as a novel approach that is aimed at detecting and mitigating the pathogenic processes of AD. However, the present strategies remain restricted by suboptimal efficacy. Zhang and coworkers have developed activated EXOs (MP@Cur-MEXOs) derived from neutrophils to enhance the function of mitochondria in neurons by targeting and mitigating neurotoxicity induced by Aβ. MP@Cur-MEXOs are EXOs resulting from IL-8-activated neutrophils, functionalized with targeting mitochondria and Aβ-targeting ligands, and transformed with superparamagnetic iron oxide nanoparticles (SPION). Furthermore, EXOs that are engineered can be cleaved into matrix metalloproteinase-2, which are overexpressed in the AD. Consequently, the released SPION and engineered EXOs loaded with curcumin work together to protect neuronal cells from Aβ-induced mitochondrial dysfunction. Additionally, MP@Cur-MEXOs efficiently aggregated in the affected areas of the AD brain at an initial phase, enabling early detection of AD through bimodal (MRI/IVIS) imaging. Significantly, in an initial-phase AD mouse model, IV injection of MP@Cur-MEXOs recovered the function of mitochondria and mitigated mitochondrial damage induced by Aβ, also helping in slowing AD progression. From this, it can be concluded that it could be a promising delivery system for the prevention of mitochondrial damage and further progression of AD.

Sheykhhasan et al. explored the neuroprotective potential of adipose-derived stem cell EXOs (ADSCs-EXOs) loaded with coenzyme Q10 (CoQ10) in a streptozotocin-induced AD rat model. The outcomes of the study mainly focus on cognitive enhancement, memory retention, and neuronal proliferation. Their study builds on prior evidence showing CoQ10's ability to reduce Aβ pathology and oxidative stress while improving behavioral outcomes in AD models. The EXO-based delivery method may enhance CoQ10's bioavailability, aligning with research highlighting its role in mitigating mitochondrial dysfunction and supporting neuronal health in neurodegenerative conditions (Figure 5.2) [78].

Many treatment strategies for AD have failed due to the complex pathology of AD, as well as the challenges posed by the BBB and other biological barriers. Numerous effective compounds demonstrate limited therapeutic improvement due to inadequate penetration at the site of action, primarily caused by the BBB. One such compound is silibinin, which has shown higher efficacy in AD treatment in preclinical studies but exhibits reduced in vivo effectiveness due to poor brain penetration. Huo et al. developed macrophage-derived EXOs (EXOs-Slb) loaded with silibinin to enhance their targeted delivery in AD models. In SD rats, PKH26-labeled EXOs-Slb accumulated preferentially in hippocampal regions (CA1, CA3, dentate gyrus) and cortical areas, with fluorescence co-localization observed with GFAP-positive astrocytes, indicating efficient astrocyte-mediated uptake. This exosomal delivery doubled silibinin's hippocampal concentration compared to free administration, improving its bioavailability. Mechanistically, EXOs-Slb disrupted Aβ pathology by binding $A\beta_{1-42}$ monomers and suppressing fibril aggregation, aligning with silibinin's known anti-amyloid effects. The formulation also attenuated astrocyte-driven neuroinflammation in Aβ-exposed SH-SY5Y cells, reducing cytokine-mediated neuronal damage. These dual actions – Aβ modulation and astrocyte deactivation – synergistically improved spatial memory and cognitive performance in AD model rats, as evidenced by behavioral assays. The study highlights EXO-mediated delivery as a strategy to amplify silibinin's neuroprotective effects, leveraging its ability to cross biological barriers and target glial cells involved in AD progression [79].

5.4.3 Exosome-Based Nanoarchitectonics in ALS

As of now, an efficient cure for ALS remains elusive. Nevertheless, there is an increasing attention on the utilization of stem cells for managing neurodegenerative disorders. Conventional approaches struggle with various disadvantages, prompting researchers to explore nano-based strategies like EXOs with unique advantages. Researchers have demonstrated that stem cells possess an advantageous paracrine effect, primarily attributed to the secretion of EVs that serve as key mediators in cellular communication.

In a pioneering study by Bonafede et al., researchers explored an innovative free-of-cell therapeutic technique using EXOs obtained from murine stromal cells derived from adipose. The study examined the effects of these EXOs on motor neurons like NSC-34 cells, including those expressing ALS-associated mutations, under in vitro conditions. The researchers especially analyzed the impact of EXOs on both untreated NSC-34 cells and those overexpressing human SOD-1(G937R), SOD-1(G93A), or SOD-1(A4V) mutations when exposed to oxidative stress. Findings revealed that EXOs enhanced cell viability and protected against oxidative damage, a key factor in ALS progression. These results underscore the promise of stem cell–derived EXOs as a foreboding medical innovative technique for ALS treatment (Figure 5.3) [80].

5.4.4 Exosome-Based Nanoarchitectonics in Multiple Sclerosis

Multiple sclerosis (MS) is a persistent and debilitating autoimmune neurological disease. The immune system targets the CNS, triggering inflammation, demyelination, and disrupted signaling of nerve. The administration of EXOs allows

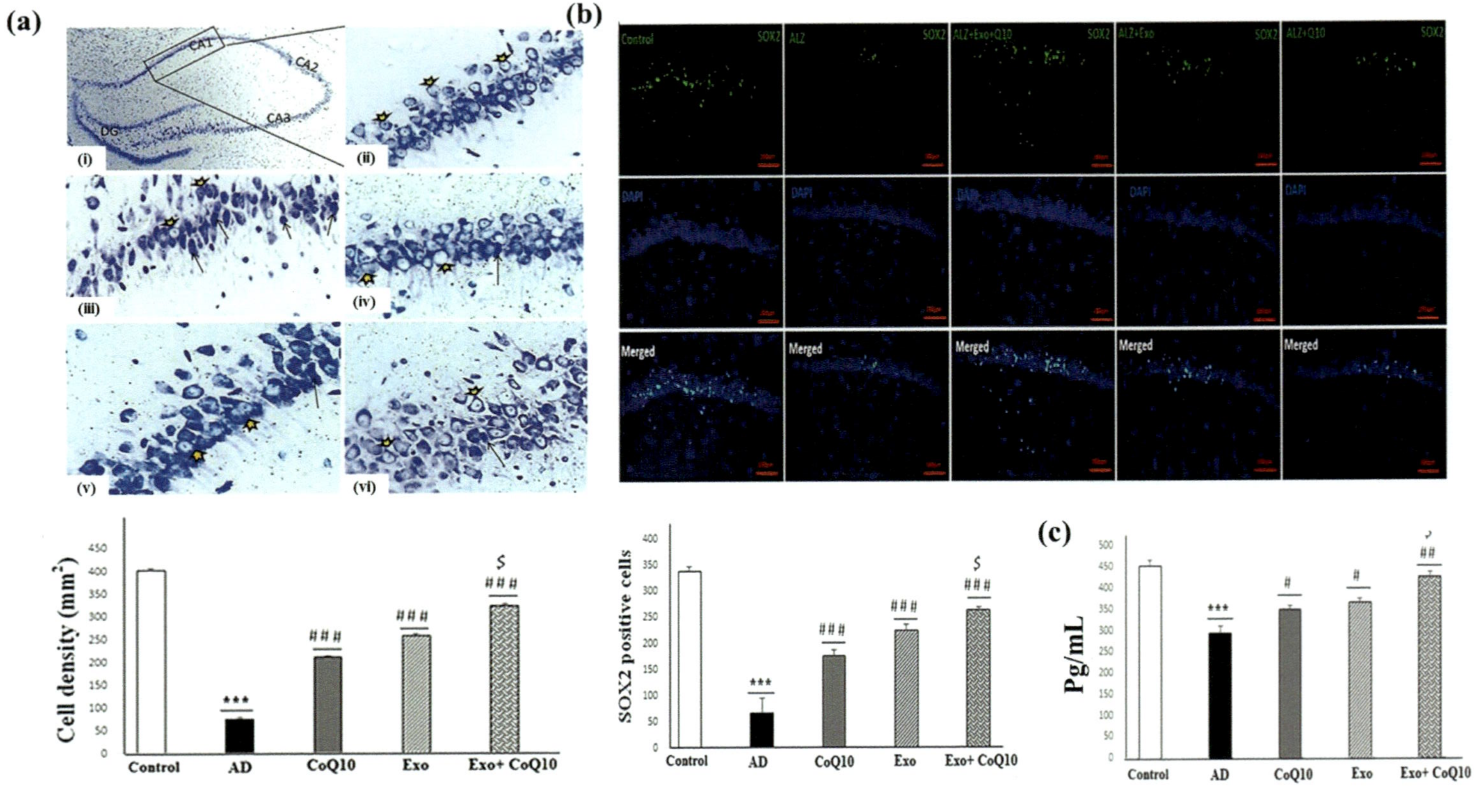

FIGURE 5.2 (a) Effects of ADSCs-EXO loaded with CoQ10 on the hippocampal CA1 area. (b) Impact of EXOs on the AD rat hippocampal SOX-2 protein. (c) Effectiveness of EXOs on AD rat hippocampal BDNF levels. *Source:* Adapted with permission from Ref. [78]. Copyright 2020, published under a Creative Commons CC BY-NC-ND License (https://creativecommons.org/licenses/by-nc-nd/4.0/).

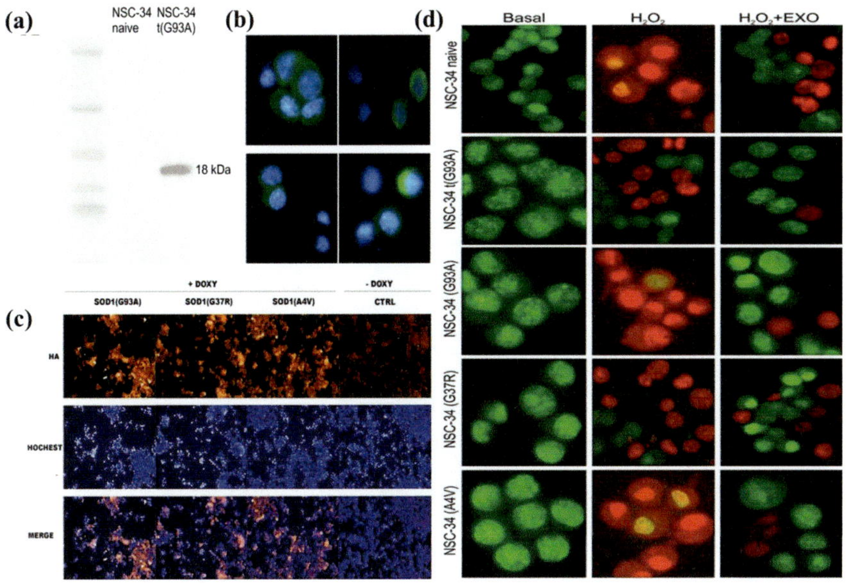

FIGURE 5.3 Representative images of NSC-34 cell line stable and transient transfection. (a) Western blot examination of transiently transfected NSC-34 cells' (NSC-34 t(G93A)) expression of HIS-HA-SOD1(G93A) (18 kDa). (b) Detecting the presence of HIS-HA-SOD1(G93A) protein in transiently transfected NSC-34 cells using immunofluorescence labeling with Anti-His Tag antibody (green). DAPI (blue) was used to counterstain the nuclei. (c) Alexa Fluor 594 immunofluorescence study of NSC34 Tet-To determine the sub-population of cells overexpressing mutant SOD-1 genes. (d) The trend of cell survival following treatments and AO/PI double labeling on several NSC-34 cell lines. *Source:* Adapted with permission from Ref. [80]. Copyright 2015, published by Elsevier.

interaction with immune cells, delivering therapeutic molecules that suppress pro-inflammatory responses and facilitate the immune-tolerant state. EXO-based nanotherapeutics are gaining recognition as a groundbreaking strategy for MS, offering solutions for key challenges such as modulating neuroinflammation, promoting tissue repair and regeneration, and aiding in the restoration of damaged myelin and neurons.

Recent research has explored the therapeutic promise of small EVs, particularly EXOs obtained from MSCs, administered intranasally (IN) in an experimental autoimmune encephalomyelitis (EAE) model of mice. The results indicated that MSC-derived small EVs (MSC-SEVs) were more efficient in reducing clinical scores than MSC treatment alone. Moreover, the reduction in clinical symptoms was accompanied by enhanced immunomodulatory responses, including an increased presence of Foxp3+ CD25+ T-cell regulators. While both MSC-SEV and MSC treatments elevated TGF-β levels, only MSC treatment led to an increase in IL-10 levels. Ultimately, the study showed that the IN administration of MSC-SEVs significantly decreased clinical scores and CNS histological lesions, emphasizing their potential as a promising therapeutic approach for MS (Figure 5.4) [81].

5.4.5 Exosome-Based Nanoarchitectonics in Stroke

Conventional therapies for ischemic stroke face noteworthy challenges, including narrow therapeutic windows, adverse effects, and limited efficacy, highlighting the requirement for ongoing research into innovative and more effective treatments. EXO-based nanotherapeutics are emerging as a promising strategy for stroke treatment, offering potential solutions for neuroprotection, neural repair, and functional recovery. Further, these nanomedicine offers a targeted approach for delivering neuroprotective and anti-inflammatory agents to ischemic brain regions [82].

A groundbreaking concept suggests that treating stroke with EXOs derived from multipotent MSCs can enhance neurovascular remodeling and promote functional recovery. The researchers demonstrated that administering MSC-derived EXOs to stroke-induced rats resulted in improved recovery, with increased axonal density and synaptophysin levels in the ischemic border zone compared to the PBS-treated group. Additionally, EXO's treatment significantly elevated the expression of von Willebrand factor and doublecortin, key markers for endothelial cells and neuroblasts, respectively. These remarkable findings indicate that the cell-free IV delivery of MSC-derived EXOs may facilitate neurite remodeling, angiogenesis, and neurogenesis after a stroke, which presents a highly promising therapeutic approach [83].

Neural stem cells (NSCs) possess the ability to home to ischemic regions, differentiate, and regenerate, which renders them a promising therapy for ischemic stroke. However, their effectiveness is hindered by the harsh microenvironment of ischemic areas, which are rich in ROS. The enhancement of NSCs with brain-derived neurotrophic factor (BDNF) through gene transfection can improve their therapeutic potential, but this typically requires viral vectors due to the inefficiency of current nonviral methods. To address this challenge, Jiang and colleagues developed the first effective polymer carrier, ROS-responsive charge-reversal poly(2-acryloyl)ethyl(*p*-boronic acid benzyl)diethylammonium bromide] (B-PDEA) EXOs, which helps to efficiently

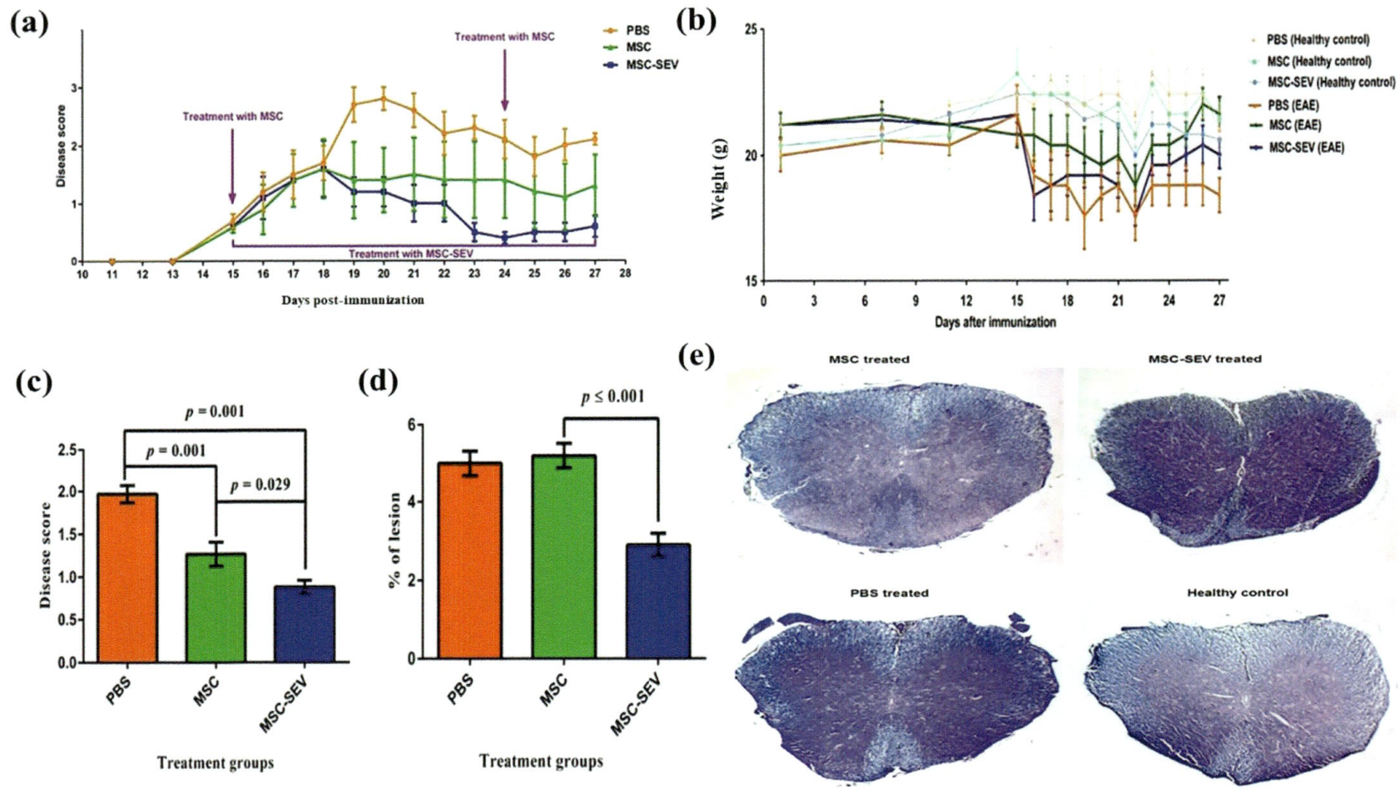

FIGURE 5.4 (a) Clinical scores of treatment groups. (b) The variations in the EAE and weight of healthy mice after treatment. (c) The mean indications for each group during the course of treatment are shown in the figure. (d) The reduction in histological lesions in EAE mice treated with MSC-SEV, MSC, as compared to PBS. (e) To assess the lesions in EAE mice treated with MSC, MSC-SEV, and PBS, luxol fast blue cresyl violet (LFB-CV) was used to stain the lumbar region of the mice's spinal tissue (caudal). *Source:* Adapted with permission from Ref. [81]. Copyright 2020, published by Elsevier.

gene transfection in NSCs. This innovation enhances BDNF expression while simultaneously reducing ROS stress by facilitating rapid plasmid release in NSCs within ischemic stroke regions. When injected IV into ischemic stroke mice, the modified NSCs (BDNF-NSCs) exhibited the same homing efficiency as unmodified NSCs but produced significantly higher BDNF levels. As a result, the survival rate of treated mice increased to 60%, compared to 0% in untreated mice and approximately 20% in those receiving unmodified NSCs. Additionally, BDNF-NSCs promoted faster and more effective functional recovery, highlighting their potential as a powerful stroke therapy [84].

5.4.6 Exosome-Based Nanoarchitectonics in GBM

One of the biggest challenges in treating GBM is the presence of the BBB, which severely confines the ability of conventional therapeutics to reach the tumor site. To overcome this, encapsulating drugs in nanocarrier systems like EXOs will offer a promising solution. Nanocarriers provide several advantages, such as bypassing biological barriers, enhancing drug delivery, and improving therapeutic efficacy. They also increase drug solubility and stability while ensuring better retention at the tumor site through the enhanced permeability and retention (EPR) effect. Nanocarriers can transport drugs across the BBB through passive diffusion, receptor-mediated uptake, or temporary disruption of the barrier. By increasing drug effectiveness while minimizing side effects, nanocarriers are emerging as a powerful tool in the fight against GBM.

Lee et al. utilized selumetinib-loaded EXOs derived from U87MG (U87-Selu EXO) and determined their efficacy for the management of GBM. The U87-Selu EXOs were developed using the electroporation method. The size of the formed EXOs was found to be 170 nm and 134 nm for the blank and drug-loaded EXOs, respectively, with a globular morphology. The drug loading efficiency of selumetinib-loaded U87MG EXOs was found to be 80–90%. In vitro EXO-targeting experiments utilized DiI-labeled EXO, and the results portrayed a 1.8-fold higher fluorescence intensity of DiI-U87-Selu EXO than DiI-A549-Selu EXO. In vivo U87MG nude mice model depicted 3.1-fold of tumor growth inhibition in the group treated with U87-Selu EXO than the saline group. Although the studies proved the potential anticancer activity of the prepared EXO, further *in vivo* homing effect of the glioma-derived EXO still needs to be performed for effective clinical translation [85].

In another study, Zhu et al. fabricated a paclitaxel (PTX) tumor-targeted human embryonic stem cells–derived EXOs modified with (cRGDyk) peptide (cRGD-Exo-PTX) and confirmed its inhibitory efficiency both in vitro and in vivo. The peptide was attached to the Exo by the post-insertion method, and PTX was loaded via coincubation. TEM results of both cRGD-Exo-PTX and PTX-loaded EXOs (Exo-PTX) portrayed a size of 125 nm and 107 nm, respectively. U87 and U251 cells assessed the targeting capacity of cRGD-Exo-PTX. Results from fluorescence microscopy revealed a stronger fluorescence signal of cRGD-Exo-PTX than Exo-PTX *in vitro,* suggesting better targeting capability of cRGD-Exo-PTX. The results from the antiproliferation assay depicted that cRGD-Exo-PTX more prominently reduced the cell viability of U87 and U251 cells than Exo-PTX, free PTX, and PBS. In vivo orthotopic xenograft mice model portrayed a significantly higher fluorescence intensity of DiR loaded cRGD-Exo-PTX than Exo-PTX in the tumor region, suggesting that the cRGD-Exo-PTX could gather in the glioma site more proficiently than Exo-PTX, depicting the tumor-targeting ability of cRGD-Exo-PTX. In vivo antitumor activity studies revealed a 2.9-fold reduction in tumor volume in the cRGD-Exo-PTX group compared to the Exo-PTX group. The formed Exo was found to be safe with no significant toxicity in the animal model. The developed Exos could act as a novel platform for brain tumor therapy (Figure 5.5) [86].

5.5 CLINICAL PERSPECTIVES OF EXOSOME-BASED THERAPEUTICS

Over the past decade, the nanomedicine research field has increasingly turned its attention to EXOs and their therapeutic potential. These nanosized, lipid-based vesicles play an emerging role in cell-to-cell communication by transferring essential biomolecules like nucleic acids, lipids, and amino acids. As researchers explore the complexities of EXOs' biology, their promise for medical applications has become a major focus. Clinical trials serve as a critical link between preclinical research and real-world experiments. Furthermore, clinical trials are key to understanding the safety, effectiveness, and feasibility of EXO-based techniques. This section takes a closer look at clinical trials that are either continuing or completed, giving a glance into the progress being made in this rapidly evolving field. By highlighting key findings, we seek to emphasize the challenges, breakthroughs, and potential of EXOs in medicine. Several current studies are exploring their safety, effectiveness, and diverse applications, bringing us closer to unlocking their full therapeutic potential. The clinical trials investigating the role of EXO-based therapies have been listed in Table 5.1.

5.6 CONCLUSION AND FUTURE DIRECTIONS

Neurological disorders pose a profound burden on global health due to their complex, progressive, and often irreversible nature. The BBB remains a significant impediment to efficient drug delivery, constraining the therapeutic potential of conventional approaches. EXOs have emerged as intriguing natural carriers that can cross the BBB and transport bioactive chemicals straight to the CNS. Their ability to modulate critical pathological pathways such as apoptosis, inflammation, and oxidative stress emphasizes their potential as next-generation therapies for neurological disorders such as AD, MS, stroke, PD, and ALS.

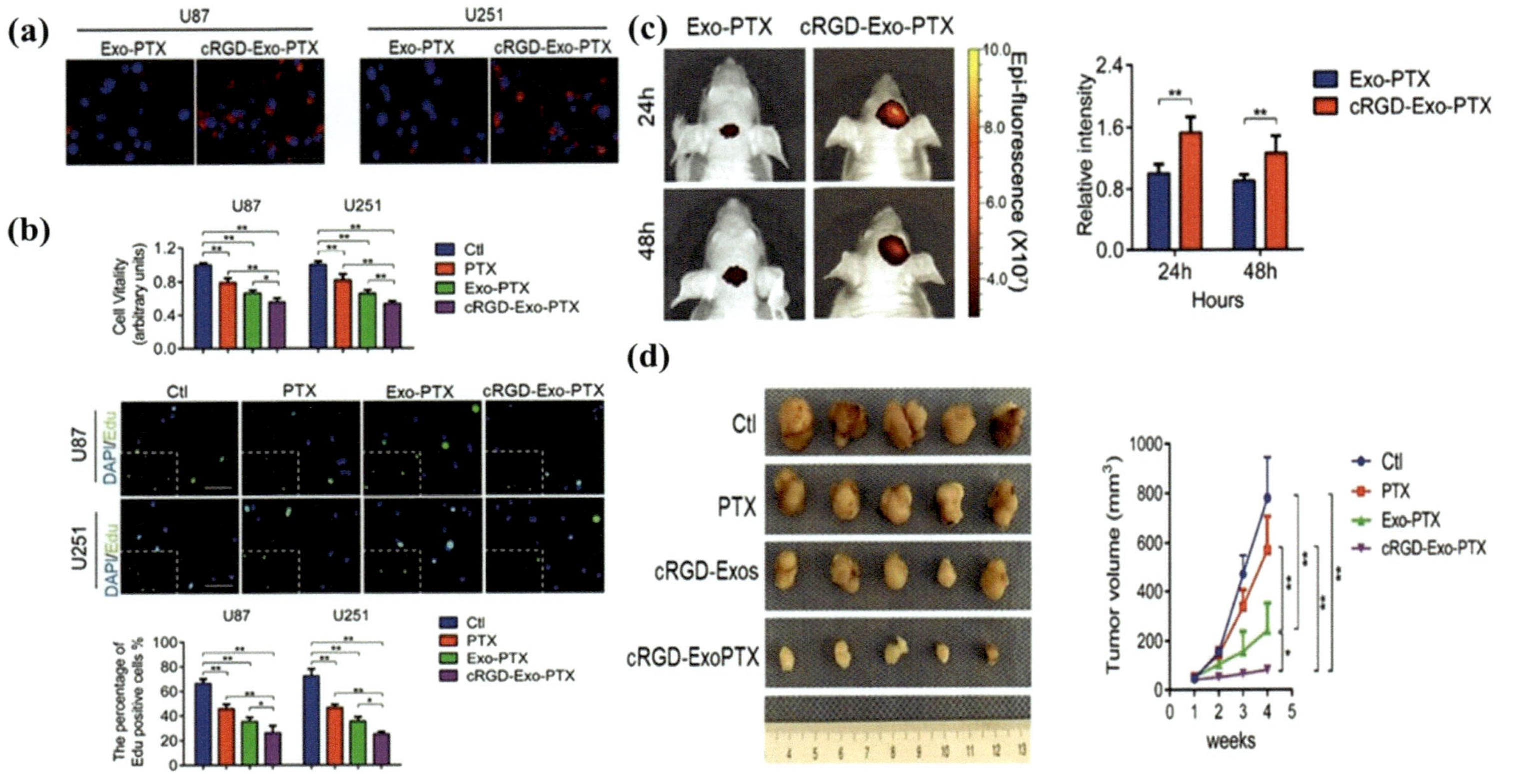

FIGURE 5.5 (a) Pictographic representation of internalization of DiI-labeled Exo-PTX and cRGD-Exo-PTX in U87 and U251 cells. (b) Cell viability and antiproliferation assay. (c) *In vivo* fluorescence images of administered Exo-PTX and the cRGD-Exo-PTX. (d) Tumor volume measurement. *Source:* Adapted with permission from Ref. [86]. Copyright 2019, published under a Creative Commons CC BY License (http://creativecommons.org/licenses/by/4.0/).

TABLE 5.1
Clinical Trials Related to Exosome-Based Therapeutics in Various Neurological Disorders

Clinical Trial ID	Pathological Condition	Study Title	Study Phase	Intervention	Current Status	Sponsor
NCT06598202	ALS	Exploring nasal drop therapy with small extracellular vesicles for ALS (de-ALS)	II	EXOs derived from hucMSCs	Recruiting	Xuanwu Hospital, Beijing
NCT06319742	Stroke	Extracellular vesicle surface markers in acute cerebrovascular syndromes	Not applicable	EV analysis	Recruiting	Ospedale Civico, Lugano
NCT06138210	Stroke	The Effect of GD-iExo-003 in Acute Ischemic Stroke	I	EXOs derived from human-induced pluripotent stem cells for injection	Recruiting	Xuanwu Hospital, Beijing
NCT01860118	PD	LRRK2 and other novel exosome proteins in Parkinson's disease	N/A	N/A	Completed	University of Alabama at Birmingham
NCT04202770	Neurodegenerative diseases	Focused ultrasound and exosomes to treat depression, anxiety, and dementias		EXOs	Suspended	Neurological Associates of West Los Angeles
NCT05490173	Preterm intraventricular hemorrhage	the pilot experimental study of the neuroprotective effects of EXOs in extremely low birth weight infants		EXOs derived from mesenchymal stromal cells	Not yet recruiting	Federal State Budget Institution Research Center for Obstetrics, Gynecology and Perinatology, Ministry of Healthcare
NCT05370105	Stroke	Extracellular vesicles as stroke biomarkers		–	Completed	Fondazione Don Carlo Gnocchi Onlus
NCT04388982	AD	The safety and efficacy evaluation of allogenic adipose MSC-Exos in patients with Alzheimer's disease	II	MSC-EXOs administered for nasal drip	Unknown status	Ruijin Hospital

Source: ClinicalTrials.gov (https://clinicaltrials.gov/).

This chapter focused on EXOs' biogenesis, composition, and functional significance in neurotherapeutic strategies, revealing their critical role in improving therapeutic outcomes. Despite promising preclinical results, several challenges must be addressed to fully realize the clinical potential of EXO-based therapies.

Standardization in EXO separation, purification, and characterization is critical to ensure reproducibility, safety, and efficacy. Furthermore, the mass production of EXOs while retaining biological activity remains a technical hurdle. Future studies should concentrate on designing EXOs for targeted delivery, increasing loading capacity, and reducing immunogenicity. EXO nanoarchitectonics advancements, such as the production of hybrid and biomimetic vesicles, provide great potential for improving therapeutic effectiveness. Furthermore, well-designed clinical trials are crucial for proving the safety and long-term advantages of EXO-based therapies in neurological diseases. Overall, integrating EXOs technology into neurotherapeutics is a game-changing frontier with the potential to drastically enhance patient outcomes in the next years.

ABBREVIATIONS

AD	Alzheimer's disease
ALS	amyotrophic lateral sclerosis
EV	extracellular vesicle
EXOs/EXO	exosomes
GD-iExo-003	Good Doctor induced Exosome 003 (investigational exosome product name used in the cited trial)
hucMSCs	human umbilical cord–derived mesenchymal stem cells
LRRK2	leucine-rich repeat kinase 2
MSC-Exos/ MSC-EXOs	mesenchymal stem cell–derived exosomes
PD	Parkinson's disease
SOD-1	superoxide dismutase 1

ACKNOWLEDGMENTS

The authors would like to acknowledge the support extended by the Department of Pharmaceuticals (DoP), Ministry of Chemicals and Fertilizers, Government of India, to Pharmaceutical Innovation and Translational Research Lab (PITRL), Department of Pharmaceutics, National Institute of Pharmaceutical Education and Research (NIPER), Hyderabad, India.

AI DISCLOSURE STATEMENT

During the preparation of this work, the authors used Grammarly, QuillBot, and ChatGPT in order to enhance grammatical structure and narrative coherence. After using this tool/service, the authors reviewed and edited the content as needed and take full responsibility for the content of the publication.

CONFLICT OF INTEREST

Reported none.

REFERENCES

[1] W. Liao, Y. Du, C. Zhang, F. Pan, Y. Yao, T. Zhang, Q. Peng, Exosomes: The next generation of endogenous nanomaterials for advanced drug delivery and therapy, *Acta Biomater.* 86 (2019) 1–14. https://doi.org/10.1016/j.actbio.2018.12.045.

[2] E. Amulya, D. Bahuguna, M. Negi, V. Phatale, A. Sikder, G. Vambhurkar, C.B. Katta, M.P. Dandekar, J. Madan, S. Srivastava, Lipid engineered nanomaterials: A novel paradigm shift for combating stroke, *Appl. Mater. Today.* 38 (2024). https://doi.org/10.1016/j.apmt.2024.102194.

[3] E. Amulya, A. Sikder, G. Vambhurkar, S. Shah, D.K. Khatri, R.S. Raghuvanshi, S.B. Singh, S. Srivastava, Nanomedicine based strategies for oligonucleotide traversion across the blood–brain barrier, *J. Control. Release.* 354 (2023) 554–571. https://doi.org/10.1016/j.jconrel.2023.01.031.

[4] R. Kalluri, V.S. LeBleu, The biology, function, and biomedical applications of exosomes, *Science* (80-.). 367 (2020). https://doi.org/10.1126/science.aau6977.

[5] V. Phatale, T. Kolipaka, P. Khairnar, G. Pandey, M. Negi, S. Saraf, S. Srivastava, Nano-biomaterial-assisted targeted therapy of central nervous system disorders, *Biomater. Nanomedicines Target. Ther.* (2024) 305–328. https://doi.org/10.1007/978-981-97-3925-7_11.

[6] L. Wen, Y.D. Wang, D.F. Shen, P.D. Zheng, M. Di Tu, W.D. You, Y.R. Zhu, H. Wang, J.F. Feng, X.F. Yang, Exosomes derived from bone marrow mesenchymal stem cells inhibit neuroinflammation after traumatic brain injury, *Neural Regen. Res.* 17 (2022) 2717–2724. https://doi.org/10.4103/1673-5374.339489.

[7] M. Nakano, K. Kubota, E. Kobayashi, T.S. Chikenji, Y. Saito, N. Konari, M. Fujimiya, Bone marrow-derived mesenchymal stem cells improve cognitive impairment in an Alzheimer's disease model by increasing the expression of microRNA-146a in hippocampus, *Sci. Rep.* 10 (2020). https://doi.org/10.1038/s41598-020-67460-1.

[8] G.H. Cui, H.D. Guo, H. Li, Y. Zhai, Z. Bin Gong, J. Wu, J.S. Liu, Y.R. Dong, S.X. Hou, J.R. Liu, RVG-modified exosomes derived from mesenchymal stem cells rescue memory deficits by regulating inflammatory responses in a mouse model of Alzheimer's disease, *Immun. Ageing.* 16 (2019). https://doi.org/10.1186/s12979-019-0150-2.

[9] X. Wang, G. Yang, Bone marrow mesenchymal stem cells-derived exosomes reduce Aβ deposition and improve cognitive function recovery in mice with Alzheimer's disease by activating sphingosine kinase/sphingosine-1-phosphate signaling pathway, *Cell Biol. Int.* 45 (2021) 775–784. https://doi.org/10.1002/cbin.11522.

[10] B. Mendes-Pinheiro, S.I. Anjo, B. Manadas, J.D. Da Silva, A. Marote, L.A. Behie, F.G. Teixeira, A.J. Salgado, Bone marrow mesenchymal stem cells' secretome exerts neuroprotective effects in a Parkinson's disease rat model, *Front. Bioeng. Biotechnol.* 7 (2019). https://doi.org/10.3389/fbioe.2019.00294.

[11] Y. Cai, M.M. Zhang, M. Wang, Z.H. Jiang, Z.G. Tan, Bone marrow-derived mesenchymal stem cell-derived exosomes containing Gli1 alleviate microglial activation and neuronal apoptosis in vitro and in a mouse Parkinson disease model by direct inhibition of Sp1 signaling, *J. Neuropathol. Exp. Neurol.* 81 (2022) 522–534. https://doi.org/10.1093/jnen/nlac037.

[12] Y. Wang, H. Chen, X. Fan, C. Xu, M. Li, H. Sun, J. Song, F. Jia, W. Wei, F. Jiang, G. Li, D. Zhong, Bone marrow mesenchymal stem cell-derived exosomal miR-193b-5p reduces pyroptosis after ischemic stroke by targeting AIM2, *J. Stroke Cerebrovasc. Dis.* 32 (2023). https://doi.org/10.1016/j.jstrokecerebrovasdis.2023.107235.

[13] M. Safakheil, H. Safakheil, The effect of exosomes derived from bone marrow stem cells in combination with Rosuvastatin on functional recovery and neuroprotection in rats after ischemic stroke, *J. Mol. Neurosci.* 70 (2020) 724–737. https://doi.org/10.1007/s12031-020-01483-1.

[14] N. Shakespear, M. Ogura, J. Yamaki, Y. Homma, Astrocyte-derived exosomal microRNA miR-200a-3p prevents MPP+-induced apoptotic cell death through down-regulation of MKK4, *Neurochem. Res.* 45 (2020) 1020–1033. https://doi.org/10.1007/s11064-020-02977-5.

[15] N. Li, J. Shu, X. Yang, W. Wei, A. Yan, Exosomes derived from M2 microglia cells attenuates neuronal impairment and mitochondrial dysfunction in Alzheimer's disease through the PINK1/Parkin pathway, *Front. Cell. Neurosci.* 16 (2022). https://doi.org/10.3389/fncel.2022.874102.

[16] Y. Song, Z. Li, T. He, M. Qu, L. Jiang, W. Li, X. Shi, J. Pan, L. Zhang, Y. Wang, Z. Zhang, Y. Tang, G.Y. Yang, M2 microglia-derived exosomes protect the mouse brain from ischemia-reperfusion injury via exosomal miR-124, *Theranostics.* 9 (2019) 2910–2923. https://doi.org/10.7150/thno.30879.

[17] A. Tang, Q. Shu, S. Jia, Z. Lai, J. Tian, Adipose mesenchymal stem cell-derived exosomes as nanocarriers for treating musculoskeletal disorders, *Int. J. Nanomedicine.* 19 (2024) 13547–13562. https://doi.org/10.2147/IJN.S486622.

[18] G. Yahao, W. Xinjia, The role and mechanism of exosomes from umbilical cord mesenchymal stem cells in inducing osteogenesis and preventing osteoporosis, *Cell Transplant.* 30 (2021). https://doi.org/10.1177/09636897211057465.

[19] T. Katsuda, K. Oki, T. Ochiya, Potential application of extracellular vesicles of human adipose tissue-derived mesenchymal stem cells in Alzheimer's disease therapeutics, *Methods Mol. Biol.* 1212 (2015) 171–181. https://doi.org/10.1007/7651_2014_98.

[20] A.M. Gorabi, N. Kiaie, G.E. Barreto, M.I. Read, H.A. Tafti, A. Sahebkar, The therapeutic potential of mesenchymal stem cell–derived exosomes in treatment of neurodegenerative diseases, *Mol. Neurobiol.* 56 (2019) 8157–8167. https://doi.org/10.1007/s12035-019-01663-0.
[21] T. Katsuda, R. Tsuchiya, N. Kosaka, Y. Yoshioka, K. Takagaki, K. Oki, F. Takeshita, Y. Sakai, M. Kuroda, T. Ochiya, Human adipose tissue-derived mesenchymal stem cells secrete functional neprilysin-bound exosomes, *Sci. Rep.* 3 (2013). https://doi.org/10.1038/srep01197.
[22] Q. Li, Z. Wang, H. Xing, Y. Wang, Y. Guo, Exosomes derived from miR-188-3p-modified adipose-derived mesenchymal stem cells protect Parkinson's disease, *Mol. Ther. Nucleic Acids.* 23 (2021) 1334–1344. https://doi.org/10.1016/j.omtn.2021.01.022.
[23] A. Chronopoulos, R. Kalluri, Emerging role of bacterial extracellular vesicles in cancer, *Oncogene.* 39 (2020) 6951–6960. https://doi.org/10.1038/s41388-020-01509-3.
[24] N.H. Pirolli, W.E. Bentley, S.M. Jay, Bacterial extracellular vesicles and the gut-microbiota brain axis: Emerging roles in communication and potential as therapeutics, *Adv. Biol.* 5 (2021). https://doi.org/10.1002/adbi.202000540.
[25] K. Sheikh, J. Arasteh, M. Tajabadi Ebrahimi, A. Hesampour, Membrane vesicles from lactobacillus acidophilus reduce intestinal inflammation and increase 5-HT in the substantia nigra of rats with parkinson's disease, *Arch. Med. Res.* 56 (2025). https://doi.org/10.1016/j.arcmed.2024.103143.
[26] J. Pan, Z. Wang, X. Huang, J. Xue, S. Zhang, X. Guo, S. Zhou, Bacteria-derived outer-membrane vesicles hitchhike neutrophils to enhance ischemic stroke therapy, *Adv. Mater.* 35 (2023). https://doi.org/10.1002/adma.202301779.
[27] Y.A. Chen, C.H. Lu, C.C. Ke, S.J. Chiu, F.S. Jeng, C.W. Chang, B.H. Yang, R.S. Liu, Mesenchymal stem cell-derived exosomes ameliorate alzheimer's disease pathology and improve cognitive deficits, *Biomedicines.* 9 (2021). https://doi.org/10.3390/biomedicines9060594.
[28] Z. Zhang, X. Zou, R. Zhang, Y. Xie, Z. Feng, F. Li, J. Han, H. Sun, Q. Ouyang, S. Hua, B. Lv, T. Hua, Z. Liu, Y. Cai, Y. Zou, Y. Tang, X. Jiang, Human umbilical cord mesenchymal stem cell-derived exosomal miR-146a-5p reduces microglial-mediated neuroinflammation via suppression of the IRAK1/TRAF6 signaling pathway after ischemic stroke, *Aging (Albany, NY).* 13 (2021) 3060–3079. https://doi.org/10.18632/aging.202466.
[29] M. Negi, E. Amulya, V. Phatale, N. Abraham, A. Hedaoo, D.A. Srinivasarao, S. Srivastava, Surface engineered nano architectonics: An evolving paradigm for tackling Alzheimer's disease, *Life Sci.* 358 (2024). https://doi.org/10.1016/j.lfs.2024.123155.
[30] M. Sun, Z. Chen, Unveiling the complex role of exosomes in Alzheimer's disease, *J. Inflamm. Res.* 17 (2024) 3921–3948. https://doi.org/10.2147/JIR.S466821.
[31] R. Perez-Gonzalez, S.A. Gauthier, A. Kumar, E. Levy, The exosome secretory pathway transports amyloid precursor protein carboxyl-terminal fragments from the cell into the brain extracellular space, *J. Biol. Chem.* 287 (2012) 43108–43115. https://doi.org/10.1074/jbc.M112.404467.
[32] T. Liang, Z. Wu, J. Li, S. Wu, W. Shi, L. Wang, The emerging double-edged sword role of exosomes in Alzheimer's disease, *Front. Aging Neurosci.* 15 (2023). https://doi.org/10.3389/fnagi.2023.1209115.
[33] Z. Song, Y. Xu, W. Deng, L. Zhang, H. Zhu, P. Yu, Y. Qu, W. Zhao, Y. Han, C. Qin, Brain derived exosomes are a double-edged sword in Alzheimer's disease, *Front. Mol. Neurosci.* 13 (2020). https://doi.org/10.3389/fnmol.2020.00079.
[34] N. Annadurai, J.B. De Sanctis, M. Hajdúch, V. Das, Tau secretion and propagation: Perspectives for potential preventive interventions in Alzheimer's disease and other tauopathies, *Exp. Neurol.* 343 (2021). https://doi.org/10.1016/j.expneurol.2021.113756.
[35] Q. Shi, W. Kang, Z. Liu, X. Zhu, The role of exosomes in the diagnosis of Parkinson's disease, *Heliyon.* 9 (2023) e20595. https://doi.org/10.1016/j.heliyon.2023.e20595.
[36] H. Yu, T. Sun, J. An, L. Wen, F. Liu, Z. Bu, Y. Cui, J. Feng, Potential roles of exosomes in Parkinson's disease: From pathogenesis, diagnosis, and treatment to prognosis, *Front. Cell Dev. Biol.* 8 (2020). https://doi.org/10.3389/fcell.2020.00086.
[37] M.J. Haney, N.L. Klyachko, Y. Zhao, R. Gupta, E.G. Plotnikova, Z. He, T. Patel, A. Piroyan, M. Sokolsky, A. V. Kabanov, E. V. Batrakova, Exosomes as drug delivery vehicles for Parkinson's disease therapy, *J. Control. Release.* 207 (2015) 18–30. https://doi.org/10.1016/j.jconrel.2015.03.033.
[38] T. Sun, Z.X. Ding, X. Luo, Q.S. Liu, Y. Cheng, Blood exosomes have neuroprotective effects in a mouse model of Parkinson's disease, *Oxid. Med. Cell. Longev.* 2020 (2020). https://doi.org/10.1155/2020/3807476.
[39] H. Xin, Y. Li, B. Buller, M. Katakowski, Y. Zhang, X. Wang, X. Shang, Z.G. Zhang, M. Chopp, Exosome-mediated transfer of miR-133b from multipotent mesenchymal stromal cells to neural cells contributes to neurite outgrowth, *Stem Cells.* 30 (2012) 1556–1564. https://doi.org/10.1002/stem.1129.
[40] L. Alvarez-Erviti, J. Cooper, C. Gardiner, I.L. Sargent, A.H. Schapira, Y. Seow, M.J.A. Wood, Lysosomal dysfunction increases exosome-mediated alpha-synuclein release and transmission, *Neurobiol. Dis.* 42 (2011) 360–367. https://doi.org/10.1016/j.nbd.2011.01.029.
[41] M. Zhang, R. Zhang, H. Chen, X. Zhang, Y. Zhang, H. Liu, C. Li, Y. Chen, Q. Zeng, G. Huang, Injectable supramolecular hybrid hydrogel delivers IL-1β-stimulated exosomes to target neuroinflammation, *ACS Appl. Mater. Interfaces.* 15 (2023) 6486–6498. https://doi.org/10.1021/acsami.2c19997.
[42] J.M. Silverman, D. Christy, C.C. Shyu, K.M. Moon, S. Fernando, Z. Gidden, C.M. Cowan, Y. Ban, R. Greg Stacey, L.I. Grad, L. McAlary, I.R. Mackenzie, L.J. Foster, N.R. Cashman, CNS-derived extracellular vesicles from superoxide dismutase 1 (SOD1)G93A ALS mice originate from astrocytes and neurons and carry misfolded SOD1, *J. Biol. Chem.* 294 (2019) 3744–3759. https://doi.org/10.1074/jbc.RA118.004825.
[43] M. Lee, J.J. Ban, K.Y. Kim, G.S. Jeon, W. Im, J.J. Sung, M. Kim, Adipose-derived stem cell exosomes alleviate pathology of amyotrophic lateral sclerosis in vitro, *Biochem. Biophys. Res. Commun.* 479 (2016) 434–439. https://doi.org/10.1016/j.bbrc.2016.09.069.
[44] T. Gschwendtberger, N. Thau-Habermann, J. von der Ohe, T. Luo, R. Hass, S. Petri, Protective effects of EVs/exosomes derived from permanently growing human MSC on primary murine ALS motor neurons, *Neurosci. Lett.* 816 (2023). https://doi.org/10.1016/j.neulet.2023.137493.
[45] B. Lopes, P. Sousa, R. Alvites, M. Branquinho, A.C. Sousa, C. Mendonça, L.M. Atayde, A.L. Luís, A.S.P. Varejão, A.C. Maurício, Peripheral nerve injury treatments and advances: One health perspective, *Int. J. Mol. Sci.* 23 (2022). https://doi.org/10.3390/ijms23020918.
[46] J. Huang, J. Li, S. Li, X. Yang, N. Huo, Q. Chen, W. Wang, N. Yang, Y. Wang, N. Zhou, Netrin-1–engineered endothelial cell exosomes induce the formation of pre-regenerative

niche to accelerate peripheral nerve repair, *Sci. Adv.* 10 (2024). https://doi.org/10.1126/sciadv.adm8454.
[47] G. Shi, D. Hao, L. Zhang, J. Qin, G. Tian, B. Ma, X. Zhou, Endocytosis-associated patterns in nerve regeneration after peripheral nerve injury, *J. Orthop. Transl.* 31 (2021) 10–19. https://doi.org/10.1016/j.jot.2021.09.004.
[48] J.P. Bischoff, A. Schulz, H. Morrison, The role of exosomes in intercellular and inter-organ communication of the peripheral nervous system, *FEBS Lett.* 596 (2022) 655–664. https://doi.org/10.1002/1873-3468.14274.
[49] D. Yi, Y. Zhang, M. Li, J. Chen, X. Chen, L. Wang, G. Xing, S. Chen, Y. Zhu, Y. Wang, Ultrasound-targeted microbubble destruction assisted delivery of platelet-rich plasma-derived exosomes promoting peripheral nerve regeneration, *Tissue Eng. – Part A.* 29 (2023) 645–662. https://doi.org/10.1089/ten.tea.2023.0133.
[50] H. Wang, G. Liang, Epidemiology of Japanese encephalitis: Past, present, and future prospects, *Ther. Clin. Risk Manag.* 11 (2015) 435–448. https://doi.org/10.2147/TCRM.S51168.
[51] G. Singh, A. Mehra, S. Arora, D. Gugulothu, L.K. Vora, R. Prasad, D.K. Khatri, Exosome-mediated delivery and regulation in neurological disease progression, *Int. J. Biol. Macromol.* 264 (2024). https://doi.org/10.1016/j.ijbiomac.2024.130728.
[52] D. Nie, Y. Tan, B. Chen, T. Jiang, X. Sun, Effect of amphotericin B combined with flucytosine on treatment of HIV-associated cryptococcal meningitis and changes of serum exosomes CD93 and inflammatory factors, *Chin. J. Nosocomiol.* 29 (2019) 3390–3414.
[53] N. Luong, J.K. Olson, Exosomes secreted by microglia during virus infection in the central nervous system activate an inflammatory response in bystander cells, *Front. Cell Dev. Biol.* 9 (2021). https://doi.org/10.3389/fcell.2021.661935.
[54] S. Mukherjee, I. Akbar, B. Kumari, S. Vrati, A. Basu, A. Banerjee, Japanese encephalitis virus-induced let-7a/b interacted with the NOTCH-TLR7 pathway in microglia and facilitated neuronal death via caspase activation, *J. Neurochem.* 149 (2019) 518–534. https://doi.org/10.1111/jnc.14645.
[55] W. Zhang, X. Jiang, J. Bao, Y. Wang, H. Liu, L. Tang, Exosomes in pathogen infections: A bridge to deliver molecules and link functions, *Front. Immunol.* 9 (2018). https:/ /doi.org/10.3389/fimmu.2018.00090.
[56] H. Li, Y. Ding, J. Huang, Y. Zhao, W. Chen, R. Chen, C. Hu, Q. Tang, Y. An, Angiopep-2 modified exosomes load rifampicin with potential for treating central nervous system tuberculosis, *Int. J. Nanomed.* 18 (2023) 489–503. https://doi.org/10.2147/IJN.S395246.
[57] J. Wang, L. Yang, The role of exosomes in central nervous system tissue regeneration and repair, *Biomed. Mater.* 18 (2023). https://doi.org/10.1088/1748-605X/ace39c.
[58] S.Q. Ding, Y.Q. Chen, J. Chen, S.N. Wang, F.X. Duan, Y.J. Shi, J.G. Hu, H.Z. Lü, Serum exosomal microRNA transcriptome profiling in subacute spinal cord injured rats, *Genomics.* 112 (2020) 5086–5100. https://doi.org/10.1016/j.ygeno.2019.09.021.
[59] F. Tian, J. Yang, R. Xia, Exosomes secreted from circ-ZFHX3-modified mesenchymal stem cells repaired spinal cord injury through mir-16-5p/IGF-1 in mice, *Neurochem. Res.* 47 (2022) 2076–2089. https://doi.org/10.1007/s11064-022-03607-y.
[60] D. Jiang, F. Gong, X. Ge, C. Lv, C. Huang, S. Feng, Z. Zhou, Y. Rong, J. Wang, C. Ji, J. Chen, W. Zhao, J. Fan, W. Liu, W. Cai, Neuron-derived exosomes-transmitted miR-124-3p protect traumatically injured spinal cord by suppressing the activation of neurotoxic microglia and astrocytes, *J. Nanobiotechnol.* 18 (2020). https://doi.org/10.1186/s12951-020-00665-8.
[61] M. Zhang, L. Wang, S. Huang, X. He, Exosomes with high level of miR-181c from bone marrow-derived mesenchymal stem cells inhibit inflammation and apoptosis to alleviate spinal cord injury, *J. Mol. Histol.* 52 (2021) 301–311. https:/ /doi.org/10.1007/s10735-020-09950-0.
[62] Z. Zhou, C. Li, T. Bao, X. Zhao, W. Xiong, C. Luo, G. Yin, J. Fan, Exosome-shuttled miR-672-5p from anti-inflammatory microglia repair traumatic spinal cord injury by inhibiting AIM2/ASC/Caspase-1 signaling pathway mediated neuronal pyroptosis, *J. Neurotrauma.* 39 (2022) 1057–1074. https://doi.org/10.1089/neu.2021.0464.
[63] Y. Zhang, Y. Qin, M. Chopp, C. Li, A. Kemper, X. Liu, X. Wang, L. Zhang, Z.G. Zhang, Ischemic cerebral endothelial cell-derived exosomes promote axonal growth, *Stroke.* 51 (2020) 3701–3712. https://doi.org/10.1161/STROKEAHA.120.031728.
[64] D. Zhang, G. Cai, K. Liu, Z. Zhuang, K. Jia, S. Pei, X. Wang, H. Wang, S. Xu, C. Cui, M. Sun, S. Guo, W. Song, G. Cai, Microglia exosomal miRNA-137 attenuates ischemic brain injury through targeting Notch1, *Aging (Albany. NY).* 13 (2021) 4079–4095. https://doi.org/10.18632/aging.202373.
[65] J. Wang, S. Chen, W. Zhang, Y. Chen, J.C. Bihl, Exosomes from miRNA-126-modified endothelial progenitor cells alleviate brain injury and promote functional recovery after stroke, *CNS Neurosci. Ther.* 26 (2020) 1255–1265. https://doi.org/10.1111/cns.13455.
[66] H. Song, X. Zhang, R. Chen, J. Miao, L. Wang, L. Cui, H. Ji, Y. Liu, Cortical neuron-derived exosomal microRNA-181c-3p inhibits neuroinflammation by downregulating CXCL1 in astrocytes of a rat model with ischemic brain injury, *Neuroimmunomodulation.* 26 (2019) 217–233. https://doi.org/10.1159/000502694.
[67] S.S.K. Yalamarty, N. Filipczak, X. Li, M.A. Subhan, F. Parveen, J.A. Ataide, B.A. Rajmalani, V.P. Torchilin, Mechanisms of resistance and current treatment options for glioblastoma multiforme (GBM), *Cancers (Basel).* 15 (2023). https://doi.org/10.3390/cancers15072116.
[68] J.P. Dardet, N. Serrano, I.E. András, M. Toborek, Overcoming blood-brain barrier resistance: Implications for extracellular vesicle-mediated drug brain delivery, *Front. Drug Deliv.* 2 (2022). https://doi.org/10.3389/fddev.2022.855017.
[69] M. Karami Fath, J. Azami, A. Masoudi, R. Mosaddeghi Heris, E. Rahmani, F. Alavi, A. Alagheband Bahrami, Z. Payandeh, B. Khalesi, M. Dadkhah, N. Pourzardosht, V. Tarhriz, Exosome-based strategies for diagnosis and therapy of glioma cancer, *Cancer Cell Int.* 22 (2022). https://doi.org/10.1186/s12935-022-02642-7.
[70] S. Yang, Y. Sun, W. Liu, Y. Zhang, G. Sun, B. Xiang, J. Yang, Exosomes in glioma: Unraveling their roles in progression, diagnosis, and therapy, *Cancers (Basel).* 16 (2024). https://doi.org/10.3390/cancers16040823.
[71] J. Shi, Y. Zhang, B. Yao, P. Sun, Y. Hao, H. Piao, X. Zhao, Role of exosomes in the progression, diagnosis, and treatment of gliomas, *Med. Sci. Monit.* 26 (2020). https://doi.org/10.12659/MSM.924023.
[72] C. Villarroya-Beltri, F. Baixauli, C. Gutiérrez-Vázquez, F. Sánchez-Madrid, M. Mittelbrunn, Sorting it out: Regulation

of exosome loading, *Semin. Cancer Biol.* 28 (2014) 3–13. https://doi.org/10.1016/j.semcancer.2014.04.009.

[73] T.T. Nguyen, T.T. Dung Nguyen, T.K. Vo, N.-M.-A. Tran, M.K. Nguyen, T. Van Vo, G. Van Vo, Nanotechnology-based drug delivery for central nervous system disorders, *Biomed. Pharmacother.* 143 (2021) 112117. https://doi.org/10.1016/j.biopha.2021.112117.

[74] L.A. Bors, F. Erdő, Overcoming the blood–brain barrier. Challenges and tricks for CNS drug delivery, *Sci. Pharm.* 87 (2019) 6. https://doi.org/10.3390/scipharm87010006.

[75] J. Huang, X. Yuan, L. Chen, B. Hu, H. Wang, W. Huang, The biology, pathological roles of exosomes and their clinical application in Parkinson's disease, *Neuroscience*. 531 (2023) 24–38. https://doi.org/10.1016/j.neuroscience.2023.09.001.

[76] H.X. Chen, F.C. Liang, P. Gu, B.L. Xu, H.J. Xu, W.T. Wang, J.Y. Hou, D.X. Xie, X.Q. Chai, S.J. An, Exosomes derived from mesenchymal stem cells repair a Parkinson's disease model by inducing autophagy, *Cell Death Dis.* 11 (2020). https://doi.org/10.1038/s41419-020-2473-5.

[77] R. Kojima, D. Bojar, G. Rizzi, G.C. El Hamri, M.D. El-Baba, P. Saxena, S. Ausländer, K.R. Tan, M. Fussenegger, Designer exosomes produced by implanted cells intracerebrally deliver therapeutic cargo for Parkinson's disease treatment, *Nat. Commun.* 9 (2018). https://doi.org/10.1038/s41467-018-03733-8.

[78] M. Sheykhhasan, R. Amini, S. Soleimani Asl, M. Saidijam, S.M. Hashemi, R. Najafi, Neuroprotective effects of coenzyme Q10-loaded exosomes obtained from adipose-derived stem cells in a rat model of Alzheimer's disease, *Biomed. Pharmacother.* 152 (2022) 113224. https://doi.org/10.1016/j.biopha.2022.113224.

[79] Q. Huo, Y. Shi, Y. Qi, L. Huang, H. Sui, L. Zhao, Biomimetic silibinin-loaded macrophage-derived exosomes induce dual inhibition of Aβ aggregation and astrocyte activation to alleviate cognitive impairment in a model of Alzheimer's disease, *Mater. Sci. Eng. C.* 129 (2021) 112365. https://doi.org/10.1016/j.msec.2021.112365.

[80] R. Bonafede, I. Scambi, D. Peroni, V. Potrich, F. Boschi, D. Benati, B. Bonetti, R. Mariotti, Exosome derived from murine adipose-derived stromal cells: Neuroprotective effect on in vitro model of amyotrophic lateral sclerosis, *Exp. Cell Res.* 340 (2016) 150–158. https://doi.org/10.1016/j.yexcr.2015.12.009.

[81] A. Fathollahi, S.M. Hashemi, M. Haji Molla Hoseini, S. Tavakoli, E. Farahani, F. Yeganeh, Intranasal administration of small extracellular vesicles derived from mesenchymal stem cells ameliorated the experimental autoimmune encephalomyelitis, *Int. Immunopharmacol.* 90 (2021) 107207. https://doi.org/10.1016/j.intimp.2020.107207.

[82] H. Xin, Y. Li, Y. Cui, J.J. Yang, Z.G. Zhang, M. Chopp, Systemic administration of exosomes released from mesenchymal stromal cells promote functional recovery and neurovascular plasticity after stroke in rats, *J. Cereb. Blood Flow Metab.* 33 (2013) 1711–1715. https://doi.org/10.1038/jcbfm.2013.152.

[83] H. Xin, M. Katakowski, F. Wang, J.-Y. Qian, X.S. Liu, M.M. Ali, B. Buller, Z.G. Zhang, M. Chopp, MicroRNA-17-92 cluster in exosomes enhance neuroplasticity and functional recovery after stroke in rats, *Stroke*. 48 (2017) 747–753.https://doi.org/10.1161/STROKEAHA.116.015204.

[84] X. Jiang, J. Xiang, H. Wu, T. Zhang, D. Zhang, Q. Xu, X. Huang, X. Kong, J. Sun, Y. Hu, K. Li, Y. Tabata, Y. Shen, J. Gao, Neural stem cells transfected with reactive oxygen species–responsive polyplexes for effective treatment of ischemic stroke, *Adv. Mater.* 31 (2019). https://doi.org/10.1002/adma.201807591.

[85] H. Lee, K. Bae, A.R. Baek, E. Bin Kwon, Y.H. Kim, S.W. Nam, G.H. Lee, Y. Chang, Glioblastoma-derived exosomes as nanopharmaceutics for improved glioma treatment, *Pharmaceutics*. 14 (2022). https://doi.org/10.3390/pharmaceutics14051002.

[86] Q. Zhu, X. Ling, Y. Yang, J. Zhang, Q. Li, X. Niu, G. Hu, B. Chen, H. Li, Y. Wang, Z. Deng, Embryonic stem cells-derived exosomes endowed with targeting properties as chemotherapeutics delivery vehicles for glioblastoma therapy, *Adv. Sci.* 6 (2019). https://doi.org/10.1002/advs.201801899.

6 Innovative Exosomes Therapy in Cardiovascular Disease Management

Nitin Deshmukh[1], *Nilesh Mandloi*[1], *Mohini Patidar*[1], *Mukesh Sharma*[3,4], *Kushagra Nagori*[3,4], *and Sunita Minz*[2]
[1]GRY Institute of Pharmacy, Borawan, Khargone, Madhya Pradesh, India
[2]Department of Pharmacy, Indira Gandhi National Tribal University (a Central University), Anuppur, Madhya Pradesh, India
[3]Rungta College of Pharmaceutical Sciences and Research, Bhilai, Chhattisgarh, India
[4]School of Pharmacy, Rungta International Skills University, Bhilai, Chhattisgarh, India

6.1 INTRODUCTION

While advances in medical therapies and surgical techniques have improved cardiovascular care, heart disease remains a major global health challenge. The need for innovative treatments to repair damaged heart tissue highlights the complexity and urgency of this issue. Stem cell–based therapies have demonstrated significant potential for tissue regeneration, but their clinical application is limited by challenges such as immune rejection and tumor risk. Exosomes, small membrane-bound vesicles secreted by most cell types, have emerged as promising mediators of tissue repair. They carry a variety of bioactive molecules that reflect the state of their originating cells and play key roles in intercellular communication [1].

Exosomes are natural nanocarriers that can carry biological information in humans. Their origin, structural properties, and functions make them interesting objects for the diagnosis of various rare diseases, like cancer, and also as innovative tools for the drug delivery process.

Exosomes play a crucial role in cardiovascular disease (CVD) by influencing its development, progression, and potential treatment. As CVD – including coronary artery disease, heart failure, and myocardial infarction – remains a leading global health challenge, understanding how exosomes contribute offers promising avenues for new biomarkers and therapies.

Various studies have revealed that exosomes are involved in several key cardiovascular processes:

- Fibrosis
- Angiogenesis
- Apoptosis
- Cardiac remodeling
- Inflammation

Exosome contents, particularly microRNAs, have been identified as promising biomarkers for early detection and prognosis of cardiovascular disease (CVD) due to their stability in bodily fluids and disease-specific expression profiles. In addition, exosome-based delivery systems are under investigation for the targeted transport of drugs, RNA molecules, or protective proteins to injured heart tissue. This approach represents a novel and potentially less invasive treatment strategy [2].

6.1.1 Overview of Exosomes

Intercellular communication is a prerequisite for the transmission of various signals. The regular function of a multicellular organism depends on intercellular contacts or the transfer of secreted molecules using exocytosis. In the last two decades, the transfer of molecules using extracellular vesicles (EVs) has been discovered as a critical intercellular communication mechanism [3].

EVs are released by all cells, prokaryotes, and eukaryotes, and regulate intercellular communication in health and disease. Based on the release pathway, function, biogenesis, size, and content, EVs are divided into four classes: exosomes (30–200 nm), microvesicles (MVs) (100–1,000 nm), apoptotic bodies (>1,000 nm), and retrovirus-like vesicles [4].

6.1.1.1 Structure and Composition of Exosomes

Exosomes were initially described by Johnstone as uniquely secreted by reticulocytes. In 1987, the term "exosomes" was proposed to designate the extracellularly released intraendosomal vesicles. In the 1980s, exosomes were suggested to serve as a means to eliminate unwanted components such as plasma membrane receptors during the maturation of reticulocytes. A more complex function of exosomes emerged in the late 1990s. Here, exosomes secreted by antigen-presenting cells were shown to activate T cells, hence inducing an immune response. Exosomes are produced by a number of cell types. Exosome secretion was described to originate from hematopoietic cells such as reticulocytes, B lymphocytes and T cells, platelets, mast cells, dendritic cells, and macrophages. Exosomes are

DOI: 10.1201/9781003623632-6

also produced by cells of non-hematopoietic origin, like epithelial cells (intestinal epithelial cells), Schwann cells, astrocytes, neurons, melanocytes, mesothelioma cells, adipocytes, fibroblasts, and tumor cells [5].

Exosomes have been described in numerous biological fluids, such as sperm, blood (serum and/or plasma), breast milk, urine, amniotic fluid, saliva, cerebrospinal fluid, nasal secretions, bronchoalveolar lavage, bile, synovial fluid, malignant effusions, pleural effusions, and ascites. Common characteristics of exosomes include structure (vesicles limited by a lipid bilayer), size (50–100 nm in diameter), density (flotation at 1.13–1.21 g/mL on a sucrose gradient), and an endocytic origin (enrichment in HSP70, tetraspanins, Tsg101, Alix, MHC molecules). A typical structure and components of cell-derived exosomes have been provided in Figure 6.1.

Exosomes are composed of a distinct and well-organized collection of biological molecules, including proteins, lipids, and other cellular components, which originate from the parent cell that produces them. Their protein content features a variety of transmembrane proteins, notably tetraspanins (such as CD9, CD63, and CD81), integrins, members of the immunoglobulin superfamily, and membrane-associated peptidases. These surface proteins play a crucial role in guiding exosomes to specific target cells and mediating their interaction with the recipient cell's surface.

In terms of lipid composition, exosomes are particularly enriched in structural and signaling lipids, including cholesterol, phosphatidylserine, and phosphatidylcholine. These lipids not only contribute to the stability and rigidity of the exosomal membrane but also play roles in cell recognition and membrane fusion events.

Overall, the specific molecular profile of exosomes mirrors their origin from the endocytic pathway and highlights their important role in intercellular communication, as they transfer functional biomolecules between cells [6].

6.1.1.2 Biogenesis of Exosomes

Exosomes are generated from late endosomes via inward budding of the limiting membrane of multivesicular bodies (MVBs), which produces small vesicles within MVBs called intraluminal vesicles (ILVs). During this process, specific membrane proteins become incorporated into the invaginating membrane, and cytosolic components are sequestered within the developing ILVs. Upon fusion of MVBs with the plasma membrane, ILVs are released into the extracellular space and are referred to as exosomes. In contrast, fusion of MVBs with lysosomes leads to degradation

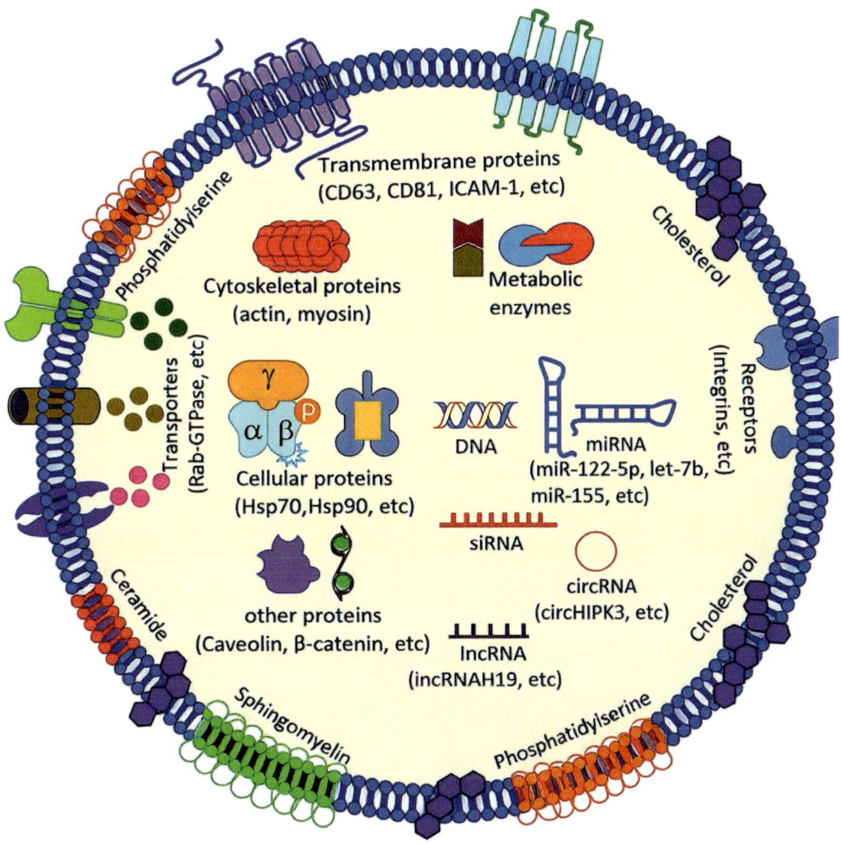

FIGURE 6.1 Representation of the structure of exosomes. Exosomes encapsulating molecular cargo of variety of lipids, metabolites, proteins, and RNAs, DNAs, receptors, cytoskeletal proteins, cellular proteins, and membrane transporters, and it also contains miRNA, IncRNA, circRNA, and metabolites surrounded by a lipid bilayer. *Source:* Adapted from Ref. [5], under Creative Commons 4.0.

of their contents. The following section describes the structural characteristics of exosomes [7].

Alternatively, multivesicular bodies (MVBs) may fuse with lysosomes, leading to the degradation of their contents. Exosomes exhibit a biconcave or cup-shaped morphology when dried for imaging. In suspension, transmission electron microscopy reveals a spherical shape. Their buoyant density ranges from 1.13 g/mL in B cells to 1.19 g/mL in epithelial cells, as determined by sucrose gradient centrifugation [8].

The generation of intraluminal vesicles (ILVs) primarily depends on the endosomal sorting complex required for transport (ESCRT) machinery, which consists of four protein complexes: ESCRT-0, ESCRT-I, ESCRT-II, and ESCRT-III. These complexes function sequentially to mediate cargo sorting, vesicle budding, and multivesicular body (MVB) formation. ESCRT-0 initiates the process by identifying and clustering ubiquitinated proteins at endosomal membrane domains. Interactions with ESCRT-I and ESCRT-II promote further assembly, followed by ESCRT-III, which drives vesicle budding. After budding, the ATPase Vps4 dissociates the ESCRT-III complex from the membrane. Although the requirement for ESCRT in exosome release remains under investigation, numerous ESCRT-associated proteins and ubiquitinated molecules have been consistently identified in isolated exosomes. For example, the protein Alix interacts with ESCRT components such as TSG101 and CHMP4 and participates in both membrane budding and selective cargo packaging, particularly through its association with syndecan [9]. Figure 6.2 depicts the biogenesis and release of exosomes.

6.1.1.3 Biological Functions of Exosomes

After budding, the ATPase Vps4 dissociates the ESCRT-III complex from the membrane. Although the requirement for ESCRT in exosome release remains under investigation, numerous ESCRT-associated proteins and ubiquitinated molecules have been consistently identified in isolated exosomes. For example, the protein Alix interacts with ESCRT components such as TSG101 and CHMP4 and participates in both membrane budding and selective cargo packaging, particularly through its association with syndecan.

Exosomes serve as essential mediators of intercellular communication by transporting proteins, lipids, and RNAs between cells. They interact with recipient cells through mechanisms such as membrane fusion, receptor binding, or endocytosis. These interactions modulate immune responses, antigen presentation, and cellular signaling, particularly within the contexts of immunity, cancer, and cell regulation [10].

Exosomes facilitate intercellular communication by transferring genetic material such as messenger RNA (mRNA) and microRNA (miRNA) between cells. This transfer alters gene expression and modulates recipient cell behavior, affecting immune regulation, gene expression control, and disease progression. Although the mechanisms underlying nucleic acid packaging remain under investigation, exosomes are established as critical mediators of genetic signaling in physiological and pathological contexts [11].

Exosomes regulate immune function by transporting specific proteins and RNA between cells. Exosomes secreted by immune cells amplify immune responses by enhancing antigen presentation and activating T cells. Conversely, tumor-derived exosomes can suppress immune activity, enabling cancer cells to evade detection. Through these opposing actions – either enhancing or inhibiting immunity – exosomes are key mediators of immune modulation and are being evaluated for therapeutic use in immunotherapy [10].

Exosomes carry a range of biologically active molecules and reflect the internal state of their parent cells. Research shows that exosomes from the central nervous system are found in cerebrospinal fluid and peripheral bodily fluids. Their contents vary with disease progression. Exosomes

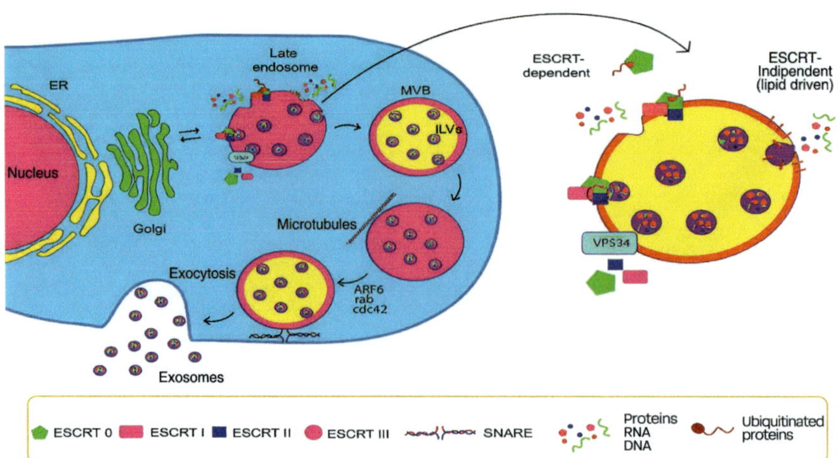

FIGURE 6.2 Biogenesis and release of exosomes. MVB: multivesicular body; ILVs: intraluminal vesicles; ESCRT: endosomal sorting complex required for transport. *Source:* Adapted from Ref. [7], under Creative Commons 4.0.

can cross the blood–brain barrier and remain stable in circulation, preserving disease-specific molecules. This stability makes them promising as biomarkers in neurological disorders. They support early detection, tracking of disease progression, and improvement of treatment strategies [12].

Exosomes have emerged as promising biomarkers, as accumulating evidence demonstrates that they transport proteins and nucleic acids associated with a range of diseases, including cancer, hepatic and renal disorders, neurodegenerative diseases, infectious diseases, and metabolic syndromes. These extracellular vesicles can be isolated from multiple biological sources, such as blood, urine, saliva, cerebrospinal fluid, amniotic fluid, and cultured cells. The molecular cargo of exosomes, which includes lipids, messenger RNA (mRNA), nucleic acids, and proteins involved in cell signaling and membrane transport, reflects the physiological state and cellular origin of the parent cells [13].

Exosomes are emerging as innovative therapeutic tools due to their natural ability to deliver molecules between cells. They can be engineered to carry drugs, RNA, or proteins to specific target cells, enhancing treatment precision and minimizing side effects. This makes them promising vehicles for drug delivery, gene therapy, and regenerative medicine. Exosomes from immune cells or tumors are also being studied as components of cancer immunotherapy to boost the body's defense against tumors. Their biocompatibility and ability to cross biological barriers position exosomes as strong candidates for next-generation therapeutics [14].

6.1.2 Current Landscape of Cardiovascular Diseases

Cardiovascular disease (CVD) continues to be a leading contributor to early deaths and increasing healthcare expenses. A combination of metabolic conditions, lifestyle choices, environmental exposures, and social determinants significantly influences the risk of developing CVD. To effectively inform policy and support strategic decision-making, it is crucial to carry out standardized and reliable evaluations of long-term trends and global patterns in cardiovascular health [15].

The current landscape of cardiovascular disease (CVD) reflects a growing global health concern, with its remaining the leading cause of death worldwide. Projections suggest that the global crude prevalence of CVD will nearly double between 2025 and 2050, rising from 598 million to approximately 1.14 billion cases – equivalent to an average annual growth rate of 3.6%. During the same period, worldwide cardiovascular-related deaths are expected to increase from 20.5 million to 35.6 million, marking a 73.4% rise, or about 2.9% per year. The highest number of CVD deaths by 2050 is anticipated in Southeast Asia, East Asia, and Oceania, with an estimated 15.9 million deaths, followed by South Asia with around 7.78 million fatalities [16].

Conditions such as hypertension, diabetes, obesity, and high cholesterol, alongside habits like poor diet, physical inactivity, tobacco use, and insufficient sleep, are key contributors. In its 2025 edition of the *World Heart Report*, the World Heart Federation (WHF) draws attention to the increasing influence of obesity on cardiovascular health worldwide. The report emphasizes that the global surge in obesity is a key contributor to the rising burden of preventable heart disease. Notably, middle- and low-income countries are experiencing a disproportionate burden, often due to limited access to preventive care and treatment. Advances in medical technology and risk prediction tools offer potential for earlier diagnosis and better management, but emphasis on public health strategies – such as lifestyle modification, early screening, and education – remains crucial to curbing the growing CVD epidemic [17].

6.1.3 Limitations of Conventional CVD Therapies

While conventional approaches for treating cardiovascular disease – like statins, beta-blockers, ACE inhibitors, and anticoagulants – have greatly enhanced survival rates and improved quality of life, they possess significant drawbacks. First, these treatments typically aim to manage symptoms or slow the progression of the disease instead of reversing the underlying damage. Many patients also face side effects or develop resistance, which can hinder long-term adherence and effectiveness. Moreover, standard therapies may not adequately consider the intricate relationships among genetic, metabolic, and lifestyle-related risk factors that contribute to CVD. There is also a gap in access to these therapies, particularly in low- and middle-income areas, where issues of affordability and healthcare infrastructure present challenges.

Research and clinical trials have associated statin use with various potential side effects. These can consist of muscle-related issues, liver impairment, an increased risk of diabetes, cognitive deterioration, and, on occasion, hemorrhagic stroke. One especially alarming condition is statin-induced myopathy, which manifests as unexplained muscle pain or weakness and can greatly impact a patient's quality of life [18].

Beta blockers can be linked to side effects like tiredness and diminished sexual performance. However, these potential issues should be weighed against the established benefits of beta blockers in treating cardiovascular ailments [19]. ACE inhibitors, which are frequently used to manage high blood pressure, may result in side effects, including a persistent dry cough, elevated potassium levels in the bloodstream, low blood pressure, and reduced kidney function. While angiotensin receptor blockers (ARBs) might produce similar negative effects, they are generally linked to a decreased likelihood of causing a cough. Calcium channel blockers could result in side effects such as swelling of the legs and ankles, constipation, dizziness, and an increased heart rate. Diuretics result in dehydration, frequent urination, and imbalances in electrolytes. Antiplatelet drugs like aspirin are associated with the risk of gastrointestinal bleeding, stomach ulcers, and allergic responses. Anticoagulants, like warfarin, are associated with a risk of excessive bleeding, which can sometimes be life-threatening. Medications such as nitrates can cause headaches, dizziness, and an increased heart rate. Aldosterone antagonists may lead to elevated potassium levels, kidney impairment, and, in some cases, breast enlargement in men (gynecomastia). Additionally, SGLT2 inhibitors – originally designed to manage type 2 diabetes and now also used in heart failure treatment – may increase the risk of urinary tract and genital infections [20].

Ranolazine, used in the management of chronic angina, is contraindicated in patients with cirrhosis, advanced kidney dysfunction, or a history of prolonged QTc intervals. It should also be avoided in patients taking medications that either extend the QTc interval or inhibit the CYP3A4 enzyme, due to potential drug interactions and safety concerns [21].

6.2 EXOSOMES AS THERAPEUTIC AGENTS IN CVD

6.2.1 Mechanism of Exosomes' Action in Cardiac Repair

6.2.1.1 Anti-Inflammatory

Research has shown that mesenchymal stem cells (MSCs) have extensive anti-inflammatory properties, which include influencing macrophage polarization, inhibiting the activation of effector T cells, and dampening cytokine secretion by B cells. Sun et al. [22] found that exosomes originating from MSCs can encourage the transformation of macrophages from a pro-inflammatory (M1) phenotype to an anti-inflammatory (M2) phenotype. Zhao et al. [23] discovered that MSC-derived exosomes, which are rich in miR-182, aided in the transition of M1 macrophages to the M2 phenotype in the peri-infarct area after myocardial injury. Likewise, Wang et al. [24] reported that miR-233-enriched exosomes from MSCs, upon being taken up by macrophages, reduced the expression of SEMA3A and STAT3, leading to a decrease in inflammation driven by macrophages. Beyond influencing macrophage function, exosomes from MSCs have been shown to affect adaptive immune responses by inhibiting the proliferation of effector T cells, promoting apoptosis in activated T cells, and diminishing B cell activity. Furthermore, miR-181c, which is delivered via MSC-derived exosomes, has been recognized as a crucial factor in regulating T-cell function by blocking the inflammatory signaling pathway mediated by Toll-like receptor 4 (TLR4) [25]. This immunomodulatory role further aids in decreasing infarct injury and ensuring plaque stability [26].

6.2.1.2 Angiogenesis Promotion

Stem cell–derived extracellular vesicles, especially those from mesenchymal stem cells and cardiac progenitor cells, are being studied for their ability to help repair heart tissue. These vesicles carry important molecules like growth factors, antiapoptotic proteins, and microRNAs that aid in new blood vessel formation and tissue healing [27].

Following internalization by cardiac cells, stem cell–derived extracellular vesicles (EVs) facilitate reparative processes. These include enhancing cardiomyocyte survival, suppressing apoptosis, and promoting neovascularization. Collectively, these mechanisms contribute to the restoration of myocardial function after injury.

Exosomes enriched with microRNA-21 (miR-21) have been extensively studied. miR-21 modulates signaling pathways associated with programmed cell death and inflammatory responses. Consequently, miR-21-loaded EVs reduce myocardial fibrosis and improve cardiac remodeling, demonstrating therapeutic potential in preclinical models of myocardial infarction (MI) [28].

In addition to proteins, the RNA cargo of extracellular vesicles (EVs), particularly noncoding RNAs such as microRNAs (miRNAs) and long noncoding RNAs (lncRNAs), plays a critical role in regulating cardiovascular pathology. Recent multi-omics studies have identified distinct miRNA profiles within EVs that modulate key signaling pathways involved in cardiac hypertrophy, fibrosis, apoptosis, and angiogenesis.

Certain EV-associated miRNAs serve as antifibrotic agents by inhibiting key molecules in profibrotic signaling, thereby reducing extracellular matrix deposition and slowing cardiac fibrosis. Liu et al. reported that low levels of exosomal miR-425 and miR-744 in heart failure patients are strongly associated with increased fibrotic remodeling, indicating their diagnostic and therapeutic potential. In contrast, other EV-enclosed miRNAs support tissue regeneration. Some promote angiogenesis by downregulating antiangiogenic factors, which improves blood flow and oxygenation in ischemic myocardial tissue. For example, miR-146a in cardiosphere-derived EVs suppresses inflammation and apoptosis while promoting cardiomyocyte proliferation and capillary formation in murine myocardial infarction models [29].

6.2.1.3 Myocardial Regeneration

Exosomes have emerged as promising paracrine agents in myocardial regeneration. They transport various biological molecules and serve as vehicles for therapeutics, such as proteins and small interfering RNA (siRNA). Research on engineered CD34+ stem cells (SCs) shows that these cells can release exosomes loaded with pro-angiogenic factors. When given to mice with myocardial infarction, these exosomes reduced infarct size, enhanced new blood vessel formation, and improved long-term cardiac repair [30].

6.2.1.4 Antifibrotic Properties

Exosomes enriched with miR-22, from mesenchymal stem cells (MSCs), can help reduce fibrotic progression in cardiomyocytes after acute myocardial infarction (AMI). This effect partly involves interactions with methylated

CpG-binding protein 2, which also helps lessen myocardial necrosis and fibrosis. When miR-29 is suppressed, inhibition of collagen mRNA expression is lifted, leading to a stronger fibrotic response. Inhibition of miR-15b during long-term cardiac pressure overload in vivo has led to worse cardiac hypertrophy and fibrosis. Lower levels of miR-21, miR-34, miR-130, and miR-378 in exosomes from bone marrow MSCs have been linked to reduced myocardial fibrosis in mice. In contrast, exosomal miR-208a appears to play a harmful role, as it is involved in both the beginning and worsening of cardiac fibrosis [31].

6.2.2 Sources of Therapeutic Exosomes

Exosomes are small extracellular vesicles that facilitate intercellular communication. They transfer bioactive molecules such as proteins, lipids, and nucleic acids between cells. Exosomes are secreted by a wide range of cell types, including mesenchymal stem cells (MSCs), cardiac progenitor cells (CPCs), induced pluripotent stem cells (IPSCs), and other cell sources like dendritic cells, macrophages, blood and bone marrow cells, and cancer cells. They also originate from nonhuman sources, such as milk, saliva, plants, and microbes [32]. Numerous studies have successfully isolated exosomes from these diverse origins and explored their potential roles in CVD. Exosomes are present in many body fluids, including blood, saliva, urine, breast milk, amniotic fluid, pleural effusions, and ascitic fluid. They can also be harvested from the culture media of nearly all cell types. These conventional sources have been instrumental in advancing our understanding of exosome biology. This research sheds light on their roles in various physiological processes and highlights their potential applications in biomedical research and therapy [33].

6.2.2.1 Mesenchymal Stem Cells–Derived Exosomes (MSC-Exos)

Mesenchymal stem cells (MSCs) are multipotent stem cells capable of self-renewal and differentiation into multiple cell lineages. MSCs exert therapeutic effects via autocrine, paracrine, and endocrine mechanisms. Paracrine signaling is primarily mediated by exosomes, which are small extracellular vesicles that transport bioactive molecules, including lipids, proteins, transfer RNAs (tRNAs), microRNAs (miRNAs), cytokines, and chemokines. MSC-derived exosomes (MSC-Exos) display specific surface markers such as CD9, CD63, and CD81, as well as proteins associated with stress response (HSP60, HSP70, HSP90), membrane trafficking (Alix, Tsg101), and cell adhesion (CD29, CD44, CD73). In addition, MSC-Exos are enriched with growth-promoting factors, including vascular endothelial growth factor A (VEGF-A), fibroblast growth factor 2 (FGF-2), hepatocyte growth factor (HGF), and platelet-derived growth factor BB (PDGF-BB). These factors play critical roles in tissue regeneration and are particularly effective in promoting skin wound healing [34].

Exosomes (Exo) derived from mesenchymal stem cells (MSCs), referred to as MSCs-Exo, contain a high concentration of bioactive molecules and mediate intercellular communication in both physiological and pathological contexts. Mesenchymal stem cells are among the most extensively studied cell types for cardiac repair and regeneration. These multipotent adult stem cells can be isolated from bone marrow, adipose tissue, and umbilical cord sources. Substantial preclinical and clinical evidence demonstrates the therapeutic potential of MSCs in reducing myocardial injury. Nevertheless, clinical application of MSC-based therapies has yielded only limited improvements in patient outcomes [35, 36].

Recent studies demonstrate that exosomes derived from mesenchymal stem cells (MSC-Exo) localize to ischemic myocardial tissue and modulate essential cellular processes, including proliferation, apoptosis, autophagy, inflammation, and angiogenesis during cardiac tissue repair. In addition to their inherent therapeutic properties, MSC-Exos have been engineered as targeted drug delivery vehicles by encapsulating specific proteins or microRNAs to enhance their effects on damaged myocardium. These exosomes regulate the expression of key proteins and influence multiple signaling pathways in cardiomyocytes and macrophages, which contributes to decreased myocardial injury and improved cardiac function [37].

6.2.2.2 Cardiac Progenitor Cell–Derived Exosomes (CPC-Exos)

Cardiac progenitor cells (CPCs) were isolated from atrial appendage tissue obtained during heart surgeries for valvular or ischemic conditions. Exosomes were extracted from the conditioned media of these cultured CPCs. Proteomic analysis indicated that these exosomes contained proteins involved in redox regulation. In contrast, Wu et al. demonstrated that exosomes derived from gastric cancer cells promote tumor cell proliferation by activating the nuclear factor-κB (NF-κB) signaling pathway. Furthermore, a separate study found that fibroblasts internalized exosomes released by human amniotic epithelial cells (hAECs-Exo), which enhanced fibroblast migration and proliferation through activation of matrix metalloproteinase-1 (MMP-1). Multiple studies have reported that transplanted cardiac progenitor cells (CPCs) secrete paracrine factors that reduce tissue damage and facilitate the repair and regeneration of injured tissue [38].

6.2.2.3 Induced Pluripotent Stem Cell–Derived Exosomes (iPSC-Exos)

Since the pioneering work by Yamanaka et al., a new era in regenerative medicine began. Their research showed that introducing transcription factors Oct3/4, Sox2, c-Myc, and Klf4 into somatic fibroblasts via retroviral delivery could reprogram these cells into pluripotent stem cells. These are now known as induced pluripotent stem cells (iPSCs). This groundbreaking discovery was recognized with the Nobel Prize in Physiology or Medicine in 2012 [39].

Induced pluripotent stem cell (iPSC) technology reprograms fully differentiated somatic cells back into a pluripotent state. This is done through the introduction of specific transcription factors. This process reverses cellular differentiation and restores the ability of the cells to give rise to various cell types. These cells become similar to embryonic stem cells. Unlike conventional embryonic stem cell approaches, iPSC technology does not require the use of embryos or oocytes, which are often linked to ethical concerns. Additionally, iPSCs can be generated from a patient's own somatic cells. They offer the advantage of producing personalized stem cells with a lower risk of immune rejection [40].

In recent years, stem cell–derived exosomes have gained attention in regenerative medicine as a promising alternative to conventional cell-based therapies.

Building on this, iPSC-derived exosomes (iPSC-Exos) have been identified as key players in cardiovascular diseases, functioning as therapeutic agents, signaling messengers, and, in some cases, contributors to disease progression. Previous research has shown that exosomes derived from embryonic stem cells (ESCs) can enhance endogenous cardiac repair mechanisms and improve heart function after myocardial infarction (MI). More recently, growing evidence supports the therapeutic potential of iPSC-Exos in MI models. In one study, Wang et al. isolated exosomes from mouse iPSCs and delivered them via intramyocardial injection into a murine model of ischemic heart injury. The results demonstrated that miPSC-Exos provided cardioprotection against myocardial ischemia/reperfusion (MIR) injury. Complementary in vitro experiments further revealed that these exosomes exert antiapoptotic and antioxidant effects – for instance, shielding H9C2 cardiomyoblasts from oxidative damage induced by hydrogen peroxide (H_2O_2) by suppressing caspase 3/7 activation.

Extracellular vesicles (EVs) from mouse iPSCs, human iPSC-derived cardiomyocytes, iPSC-derived cardiovascular progenitors, and hESC-derived cardiovascular progenitors have demonstrated significant potential in promoting cardiac repair and regeneration [41, 42].

6.2.2.4 Other Cell Sources

In addition to stem and progenitor cells, exosomes can also be extracted from immune cells such as dendritic and T cells, endothelial cells, neural precursors, and some tumor cells. The specific molecular signatures present in these exosomes vary based on their source and can be utilized for personalized therapeutic approaches in different diseases [28].

6.2.2.4.1 Mast Cell–Derived Exosomes

Mast cells that originate from the bone marrow are innate immune cells derived from hematopoietic tissue. They can release exosomes through either a continuous (constitutive) pathway or in response to certain triggers via regulated exocytosis. The exosomes produced by mast cells (mast-Exos) generally range from 60 nm to 100 nm in diameter and feature a unique cup-shaped morphology when viewed under electron microscopy. Early studies demonstrated that exosomes secreted by mast cells are abundant in immune-related proteins such as MHC class II, CD86, LFA-1, and ICAM-1. These molecular elements have been found to activate and enhance the proliferation of B and T lymphocytes, in addition to encouraging the secretion [43].

6.2.2.4.2 Dendritic Cell–Derived Exosomes

Dendritic cells (DCs) release exosomes with different functional properties depending on their level of activation and the cytokine environment they are in. When DCs are activated under inflammatory conditions, the exosomes they produce can boost innate immune responses by increasing the expression of ligands for NKG2D receptors located on natural killer (NK) cells, including MIC-A, MIC-B, and ULBP-1. These exosomes also contribute to adaptive immunity by activating antigen-specific T cells through both direct and indirect mechanisms. In the direct pathway, dendritic cell–derived exosomes (DEXs) display MHC class II–peptide complexes and co-stimulatory molecules, enabling them to directly engage with T-cell receptors. In the indirect pathway, these exosomes can transfer MHC II–peptide complexes to other DCs in a process known as "MHC dressing." Additionally, exosomes promote the transfer of antigenic peptides from activated DCs to their inactive counterparts, significantly enhancing the surface expression of MHC II–peptide complexes and boosting T-cell activation [44].

6.2.2.4.3 Macrophage-Derived Exosomes

Macrophages can be categorized into different activation states, including the pro-inflammatory M1 type and the anti-inflammatory M2 type, with each type producing exosomes that have distinct compositions and functions. M1 macrophages and their exosomes are mainly linked to the promotion of inflammation, which can impede tissue regeneration. On the other hand, M2 macrophages are well-known for their role in terminating inflammation and aiding in tissue repair. Exosomes released by M2 macrophages (M2-Exos) replicate many of the advantageous functions of their parent cells and demonstrate significant potential for immune modulation. This section examines the therapeutic potential of M2-Exos within the realm of regenerative medicine [45].

6.2.2.4.3 T-Cell-Derived Exosomes

T cells, which are essential for regulating the immune response, comprise several functional subtypes. Notably, exosomes produced by regulatory T cells (Treg-Exos) are recognized for their potent immunosuppressive effects and have been extensively researched in regenerative medicine. A significant area of research involves their application in preventing transplant rejection. For example, in a humanized mouse model of skin transplantation, Treg-Exos derived from humans considerably decreased immune cell infiltration and lessened immune-mediated tissue damage, thus prolonging the survival of the skin graft [37].

6.3 APPLICATION OF EXOSOME THERAPY IN SPECIFIC CVDS

6.3.1 Myocardial Infarction (MI) and Ischemic Heart Disease

Exosomes have gained recognition as potential therapeutic agents and biomarkers in heart-related conditions, including myocardial infarction and ischemic heart disease. These tiny vesicles, released by different cell types, transport proteins, lipids, and nucleic acids that mirror the physiological state of their originating cells.

In the case of myocardial infarction, exosomes from stem cells (such as mesenchymal stem cells or cardiac progenitor cells) have shown promise in facilitating cardiac repair by promoting blood vessel formation, decreasing cell death, and modulating immune responses in the damaged tissue. They can convey cardioprotective microRNAs (like miR-21, miR-126, and miR-210), which aid in alleviating inflammation and fibrosis, while enhancing heart function. Likewise, in ischemic heart disease, exosomes can enhance blood flow to areas affected by ischemia by encouraging the formation of new blood vessels and maintaining endothelial function. Their role in communication between cells also makes them valuable for delivering therapeutic substances and for diagnostic purposes, as their content indicates early molecular alterations in the heart muscle. Overall, therapies based on exosomes present a noncellular, low-immunogenic option for heart regeneration, showing increasing potential in tailored cardiovascular medicine [46].

6.3.2 Heart Failure (HF)

Exosomes are attracting interest as new regenerative agents because of their capacity to deliver growth factors, microRNAs, and various other biologically active substances directly to damaged heart tissue. This precise delivery aids in tissue repair, reduces inflammatory responses, and engages the body's own repair mechanisms. In comparison to traditional stem cell therapies, exosomes present numerous practical benefits – which include simpler manufacturing techniques, easier storage, and a lower risk of immune rejection – making them more appropriate for clinical applications. Furthermore, they can be administered via minimally invasive techniques such as intravenous, subcutaneous, or even intranasal routes, improving patient safety and compliance [47].

Significant uses include exosomes as diagnostic biomarkers that facilitate accurate disease identification and patient categorization. In therapeutic contexts, exosomes can be modified to transport specific agents, like necroptosis inhibitors (e.g., Nec-1) or cardioprotective microRNAs, which assist in reducing oxidative stress, maintaining mitochondrial health, and preventing fibrotic remodeling in damaged heart tissue. Moreover, exosomes sourced from mesenchymal stem cells (MSCs) or induced pluripotent stem cells (iPSCs) have demonstrated potential in transmitting healing signals, encouraging angiogenesis, and curbing programmed cell death within myocardial tissue.

Overall, strategies utilizing exosomes indicate a promising transformation in heart failure management – providing personalized, regenerative, and minimally invasive treatment options [20].

6.3.3 Atherosclerosis

Exosomes are vital in addressing atherosclerosis, presenting both diagnostic and therapeutic opportunities. These small vesicles transport molecules linked to disease – such as microRNAs, proteins, and lipids – that indicate ongoing changes in the vascular system, making them excellent candidates for stable, noninvasive biomarkers found in bodily fluids. Certain exosomal microRNAs, like miR-122–5p, miR-92a-3p, and miR-155, have indicated promise in diagnosing atherosclerosis, tracking its development, and forecasting clinical outcomes. On the therapeutic side, exosomes can be modified to deliver active substances, such as siRNAs, medications, or regulatory RNAs, directly to target cells like endothelial cells or macrophages, thereby influencing crucial pathological processes. Exosomes that originate from mesenchymal stem cells or immune cells inherently exhibit anti-inflammatory and vascular-protective characteristics. For example, miR-342-5p derived from adipose tissue exosomes has been shown to inhibit plaque formation by preserving endothelial function. Furthermore, regulatory microRNAs such as miR-155, when conveyed through exosomes, can diminish inflammation driven by immune cells and restore vascular health. With progress in exosome engineering for targeted delivery and their ability to reflect the molecular profile of donor cells, these vesicles are leading the way toward customizable and precise cardiovascular treatments [26].

6.3.4 Pulmonary Hypertension

Exosome-derived microRNAs (miRNAs) released by different cell types play unique roles depending on the progression or treatment of pulmonary hypertension (PH). An earlier investigation by Caruso et al. found that levels of miR-22, miR-30, and miR-let-7f were significantly diminished during the development of PH. At the same time, Fagan and colleagues revealed in two widely recognized rodent models – chronic hypoxia and monocrotaline (MCT)-induced injury – that both sustained and intermittent low-oxygen conditions initiate abnormal remodeling of the pulmonary arteries, which is a crucial factor leading to PH and possible lung failure. Notably, in these animal models, the levels of miR-22, miR-30, and miR-let-7f not only increased but also showed heightened expression of miR-322 and miR-451. The importance of miRNAs in the pathology of PH is increasingly apparent, largely due to their ability to control multiple gene targets at once. Numerous investigations have recognized several miRNAs as being elevated in both experimental models and patients suffering from pulmonary arterial hypertension (PAH), while others are consistently observed to be reduced, highlighting a complex regulatory network implicated in the progression of the disease [20].

6.4 ENGINEERING AND OPTIMIZATION OF EXOSOMES FOR CVD THERAPY

Developments in bioengineering have greatly improved the therapeutic potential of exosomes for treating cardiovascular disease (CVD). Exosomes from natural sources, especially those obtained from mesenchymal stem cells (MSCs) and induced pluripotent stem cells (iPSCs), demonstrate encouraging regenerative and protective effects on the heart. Nonetheless, to enhance their clinical efficacy, exosomes are increasingly being modified and optimized for targeted delivery, effective cargo loading, and better therapeutic outcomes in cardiac repair [28].

6.4.1 Exosome Isolation and Purification Techniques

The effective and standardized separation of exosomes is essential for their therapeutic application. The following are the commonly used methods:

- Differential ultracentrifugation (considered the gold standard, but it is labor-intensive)
- Size-exclusion chromatography (SEC)
- Ultrafiltration
- Precipitation-based techniques (such as ExoQuick)

Innovative methods like microfluidics and tangential flow filtration (TFF) are more scalable and reproducible, providing improved purity and yield for the production of clinical-grade exosomes [43].

6.4.2 Cargo-Loading and Genetic Engineering of Exosomes

6.4.2.1 miRNA and siRNA Delivery

Exosomes naturally contain a high concentration of regulatory RNAs. Modifying exosomes to deliver specific miRNAs or siRNAs can influence gene expression in targeted cardiac cells. For instance:

- *miR-21 and miR-126:* Encourage angiogenesis and decrease apoptosis.
- *siRNAs:* Target and silence harmful genes associated with fibrosis or inflammation.

The cargo can be loaded through methods such as electroporation, transfection of donor cells, or co-incubation techniques [44].

6.4.2.2 Protein and Drug Delivery

Exosomes can be modified to transport cardioprotective proteins (such as HSPs or VEGF) or small-molecule medications like anti-inflammatory or antifibrotic drugs. Methods including chemical conjugation, fusion protein expression, and lipid insertion are used to effectively incorporate therapeutic agents into or onto the exosome's surface [45].

6.4.3 Targeting Strategies for Enhanced Efficacy

Targeted delivery to the injured myocardium enhances therapeutic outcomes and reduces off-target effects. Strategies include the following:

- Surface modification with targeting ligands (e.g., cardiac homing peptides, antibodies)
- Genetic engineering of parent cells to express exosome surface proteins like Lamp2b fused with tissue-specific ligands
- Magnetic guidance or ultrasound-triggered release for site-specific delivery

Such enhancements improve exosome biodistribution and retention in cardiac tissues [46].

6.4.4 Biocompatibility and Immunogenicity Considerations

Exosomes have low immunogenicity, particularly when sourced from autologous or allogeneic MSCs. Nevertheless, exosomes that are engineered or loaded with specific cargo may trigger immune responses depending on their surface characteristics or the materials they contain. It is crucial to conduct preclinical research to assess the following:

- Compatibility with cardiac tissues
- Potential for immune activation
- Distribution and clearance rates within the body

Utilizing xeno-free, serum-free culture systems alongside stringent quality control measures can enhance safety and promote clinical acceptance [47].

6.5 PRECLINICAL AND CLINICAL PROGRESS

6.5.1 In Vitro Studies and Animal Models

Mesenchymal stem cells (MSCs) and induced pluripotent stem cells (iPSCs) have become significant resources in regenerative medicine. Their therapeutic potential is increasingly recognized not just for their ability to replace cells but also due to paracrine signaling, which is mainly facilitated by extracellular vesicles (EVs), such as exosomes. These nanosized vesicles, ranging from 30 nm to 150 nm, are released by stem cells and are essential in modifying the microenvironment, aiding tissue repair, and managing immune responses. Preclinical models that utilize exosomes derived from MSCs and iPSCs offer a critical understanding of their functional roles and therapeutic effectiveness in diseases affecting the heart, nervous system, liver, and kidneys, and in inflammatory conditions. This section reviews the common in vitro and in vivo models used to study exosomes derived from MSCs and iPSCs [48].

6.5.1.1 MSC-Derived Exosome Models

(A) Exosomes: Source and Characteristics

- Mesenchymal stem cells (MSCs) can be obtained from various sources, including bone marrow, adipose tissue, umbilical cord, dental pulp, and placenta.
- The exosomes derived from MSCs contain anti-inflammatory factors, promote angiogenesis, and prevent apoptosis:
  - *microRNAs:* miR-21, miR-126, miR-146a
  - *Proteins:* TSG101, Alix, HSP70, VEGF, TGF-β
  - *Lipids and enzymes:* play a role in signal transduction

(B) In Vitro Models of MSC-Derived Exosomes

- *Cardiac cells (such as H9c2 and primary cardiomyocytes):*
  - Evaluate protective effects against hypoxia-reoxygenation injury
  - Assess levels of apoptosis, reactive oxygen species (ROS) production, and the integrity of mitochondria.
- *Endothelial cells (such as HUVECs):*
  - Utilized in the investigation of angiogenesis through tube formation assays.
  - Assays for migration and proliferation.
- *Immune cells (such as macrophages and peripheral blood mononuclear cells [PBMCs]):*
  - Examine immunomodulatory effects and secretion of cytokines.
- *Fibroblasts (such as human dermal fibroblasts):*
  - Study processes related to wound healing and the remodeling of the extracellular matrix (ECM) [49, 50].

(C) In Vivo MSC–Exosome Models

In vivo MSC–exosome models have gained significant attention in recent years due to their potential therapeutic applications. Mesenchymal stem cells (MSCs) are known for their robust immunomodulatory and regenerative properties, which are partly attributed to their paracrine activity. Exosomes, small extracellular vesicles secreted by MSCs, play a crucial role in this process. In vivo MSC–exosome models are shown in Table 6.1.

6.5.1.2 iPSC-Derived Exosome Models

(A) Exosomes: Source and Characteristics

- iPSCs are somatic cells reprogrammed into a pluripotent state using Yamanaka factors.
- Exosomes can be harvested from the following:
  - Undifferentiated iPSCs
  - iPSC-derived cell types, such as cardiomyocytes, neurons, or MSCs
- Cargo includes the following:
  - miRNAs: miR-294, miR-199a, miR-210
  - Pluripotency-associated proteins
  - Regenerative molecules promoting cell cycle reentry, survival, and angiogenesis

(B) In Vitro iPSC–Exosome Models

- *Neuronal cells (e.g., SH-SY5Y):* Used to test neuroprotective effects under oxidative or excitotoxic stress
- *Cardiomyocytes (e.g., iPSC-CMs):* Assess contractility, calcium handling, and cell survival

TABLE 6.1
In Vivo MSC–Exosome Models

Disease Area	Model	Outcome of MSC–Exosome Therapy	References
Myocardial infarction	LAD ligation in mice/rats	Reduced infarct size, improved LVEF, and increased angiogenesis	[51]
Stroke	MCAO (middle cerebral artery occlusion)	Decreased neuroinflammation, enhanced neurogenesis	[52]
Kidney injury	Cisplatin-induced or IRI model	Protection of renal function, reduced apoptosis	[53]
Liver fibrosis	CCl_4-induced liver injury	Antifibrotic, hepatoprotective effects	[54]
Wound healing	Full-thickness skin wound	Accelerated re-epithelialization, angiogenesis	[55]
Inflammatory diseases	Colitis, arthritis models	Suppression of cytokines (e.g., IL-6, TNF-α), M2 macrophage polarization	[55]

- *Endothelial cells:* Examine promotion of angiogenesis and migration
- *Fibrotic models:* Test antifibrotic responses to TGF-β stimulation

(C) In Vivo iPSC–Exosome Models

In vivo models have been used to study the regenerative and therapeutic capabilities of exosomes derived from induced pluripotent stem cells (iPSCs). These tiny vesicles contain bioactive substances like proteins, lipids, and noncoding RNAs, which can influence cellular signaling and repair mechanisms. There are some in vivo iPSC exosome models shown in Table 6.2.

(D) Mechanisms of Action in Preclinical Models

MSC–Exosome Effects

- *Antiapoptotic:* Inhibition of caspase-3, upregulation of Bcl-2
- *Immunomodulatory:* M2 macrophage polarization, inhibition of T-cell activation
- *Pro-angiogenic:* Induction of VEGF, bFGF, and miR-126-mediated endothelial repair
- *Antifibrotic:* Suppression of TGF-β/Smad signaling

iPSC–Exosome Effects

- *Regeneration:* Induction of cardiomyocyte proliferation via miR-199a and YAP/TAZ activation
- *Neuroprotection:* Reduction of oxidative stress and inflammation
- *Pluripotency signaling:* Maintenance of repair potential in recipient cells
- *BBB penetration:* Effective brain delivery after systemic or intranasal administration [61]

6.5.2 Current Clinical Trials and Their Outcomes

Several early-phase clinical trials have evaluated exosome-based therapies, primarily derived from mesenchymal stem cells. Indications under investigation include cancer, COVID-19-related ARDS, chronic wounds, and neurological disorders. Early outcomes suggest that exosomes are generally safe and well-tolerated, with encouraging signs of immune modulation, tissue regeneration, and functional improvement. However, most trials remain in phase I or II, and further studies are needed to establish efficacy and optimize dosing [62].

6.5.3 Regulatory Pathways and Challenges

Regulatory classification of exosomes remains a key challenge due to their hybrid nature – they may be considered biologics, drugs, or combination products depending on their source and intended use.**56** Agencies like the FDA and EMA require rigorous data on identity, purity, potency, and safety for clinical approval. Current barriers include the lack of standardized isolation protocols, variability in product quality, and unclear guidelines specific to exosome therapeutics. Harmonization of regulatory pathways and development of quality control standards are essential for clinical success [63].

6.6 CURRENT CLINICAL TRIALS AND THEIR OUTCOMES

6.6.1 Standardization of Production and Quality Control

A major challenge in exosome-based therapeutics is the *lack of standardized protocols* for production, purification, and quality control. Variations in cell source, culture conditions, isolation techniques, and storage methods lead to inconsistencies in exosome yield, purity, and functional activity. Quality control parameters – such as size, concentration, surface markers, and bioactivity – must be rigorously defined and validated according to good manufacturing practice (GMP) standards to ensure reproducibility, safety, and regulatory compliance [63].

6.6.2 Scaling Up for Clinical Translation

Exosome production at laboratory scale is often inefficient and labor-intensive, making clinical-scale manufacturing a significant hurdle. Current isolation methods, such as

TABLE 6.2
In Vivo iPSC–Exosome Models

Disease Area	Model	Outcome of iPSC–Exosome Therapy	References
Heart failure/MI	Ischemia via LAD ligation	Increased cardiomyocyte survival, neovascularization, and reduced fibrosis	[56]
Parkinson's disease	6-OHDA-lesioned rat	Motor improvement, dopaminergic neuron preservation	[57]
Alzheimer's disease	APP/PS1 mouse model	Cognitive improvement, reduced Aβ plaques	[58]
Liver regeneration	Partial hepatectomy	Enhanced hepatocyte proliferation	[59]
Lung injury	Bleomycin-induced fibrosis	Reduced collagen deposition, improved lung function	[60]

ultracentrifugation and size-exclusion chromatography, are not easily scalable. Therefore, there is a pressing need for automated, scalable, and cost-effective bioprocessing platforms – including tangential flow filtration, bioreactor systems, and microfluidic technologies – that can produce high-yield, high-purity exosomes suitable for clinical use [64].

6.6.3 Long-Term Safety and Efficacy

While short-term studies show promise, the long-term safety and efficacy of exosome-based therapies remain unclear. Potential risks include immunogenicity, off-target effects, tumorigenicity, and bioaccumulation in nontarget organs. Comprehensive toxicological studies, biodistribution analyses, and long-term follow-up in both preclinical and clinical settings are necessary to establish safety profiles. In addition, understanding the durability of therapeutic effects and optimal dosing strategies will be crucial for sustained clinical success [33].

6.6.4 Combination Therapies with Exosomes

Combining exosomes with other therapeutic modalities, such as drugs, small molecules, or gene therapies, offers synergistic benefits.**58** Exosomes can act as natural delivery vehicles, enhancing the targeted transport and bioavailability of co-administered agents. This strategy is especially promising in cancer and regenerative medicine, where exosomes may boost drug efficacy while reducing systemic toxicity. However, challenges remain in co-loading strategies, release kinetics, and regulatory classification of such combination products [26].

6.6.5 Personalized Exosome Therapy

Personalized exosome therapy aims to utilize patient-derived or patient-matched exosomes, potentially minimizing immune rejection and enhancing therapeutic relevance. These autologous exosomes can be engineered to carry disease-specific cargos (e.g., miRNAs or proteins) tailored to the individual's genetic or pathological profile. However, the approach is currently limited by cost, complexity of production, and quality control issues, which must be addressed to make it clinically viable [16].

6.6.6 Nanotechnology and Exosome Delivery

The integration of nanotechnology with exosome platforms has opened new frontiers in precision medicine. Techniques such as surface engineering, magnetic tagging, and stimuli-responsive coatings enable controlled, targeted, and trackable delivery of exosome cargo. Hybrid exosome–nanoparticle systems combine the biocompatibility of exosomes with the functional versatility of synthetic nanomaterials, improving therapeutic efficacy in oncology, neurology, and beyond. The field is advancing rapidly, but demands thorough evaluation of biosafety and scalability before clinical adoption [65].

6.7 CONCLUSION

Exosomes have emerged as a transformative class of extracellular vesicles with immense potential in diagnostics, therapeutics, and regenerative medicine. Derived from various cell types, including mesenchymal stem cells (MSCs) and induced pluripotent stem cells (iPSCs), exosomes are natural carriers of bioactive molecules – such as proteins, lipids, and RNAs – that enable intercellular communication and tissue modulation. Their ability to cross biological barriers, low immunogenicity, and intrinsic targeting capacity make them promising candidates for noncellular therapies in a wide range of diseases, including cancer, cardiovascular disorders, neurological conditions, and inflammatory diseases.

Despite their promise, significant challenges remain. These include the need for standardized protocols for exosome isolation, characterization, and quantification; scalable and GMP-compliant production platforms; robust potency assays; and thorough evaluation of long-term safety and efficacy. Regulatory ambiguity and variability in product quality continue to hinder clinical translation.

Looking ahead, advances in bioengineering, nanotechnology, and personalized medicine are expected to enhance the specificity, efficacy, and safety of exosome-based applications. With growing interdisciplinary collaboration and regulatory alignment, exosomes hold great potential to become the next generation of precision therapeutics and diagnostics, bridging the gap between biology and medicine.

ACKNOWLEDGMENT

All the authors want to credit their affiliated institutes.

AI DISCLOSURE STATEMENT

During the preparation of this work, the authors used Grammarly, QuillBot, and ChatGPT in order to enhance grammatical structure and narrative coherence. After using this tool/service, the authors reviewed and edited the content as needed and take full responsibility for the content of the publication.

CONFLICT OF INTEREST

None.

REFERENCES

[1] Hossein RP, Maha S, Mesenchymal stem cells derived exosomes: A new era in cardiac regeneration, *Stem Cell Research & Therapy*, 16, **16** (2025). https://doi.org/10.1186/s13287-024-04123-2

[2] Milosevits G, Szebeni J, Krol S, Exosomes: Potential model for complement-stealth delivery systems, *European Journal of Nanomedicine*, 7, **7** (2015) 207–218. https://doi.org/10.1515/ejnm-2015-0005

[3] Ameneh J, Amirhesam B, Meghdad AA, Nayebali A, Mostafa RT, Exosomes and cancer: From molecular mechanisms to clinical applications, *Medical Oncology*, 38, **45** (2021). https://doi.org/10.1007/s12032-021-01491-0
[4] Gurung S, Perocheau D, Touramanidou L, Bartuteau J, The exosome journey: From biogenesis to uptake and intracellular signaling, *Cell Communication and Signaling*, 19, **47** (2021). https://doi.org/10.1186/s12964-021-00730-1
[5] Tasleem AJ, Safikur R, Raied B, Eun JL, Ehab HM, Elrashdy MR, Inho C, Expedition into exosome biology: A perspective of progress from discovery to therapeutic development, *Cancers*, 13, **5** (2021) 1157. https://doi.org/10.3390/cancers13051157
[6] Clotilde T, Laurence Z, Sebastian A, Exosomes: Composition, biogenesis and function, *Nature Reviews Immunology*, **2** (2002) 569–579. https://doi.org/10.1038/nri855
[7] Zarà M, Amadio P, Campodonico J, Sandrini L, Barbieri SS, Exosomes in cardiovascular diseases, *Diagnostics*, **10** (2020) 943. https://doi.org/10.3390/diagnostics10110943
[8] Xu M, Ji J, Jin D, Wu Y, Wu T, Lin R, Zhu S, Jiang F, Ji Y, Bao B, Li M, Xu W, Xiao M, The biogenesis and secretion of exosomes and multivesicular bodies (MVBs): Intercellular shuttles and implications in human diseases, *Genes Disease*, 22, **10** (2022) 1894–1907. https://doi.org/10.1016/j.gendis.2022.03.021
[9] Villarroya-Beltri C, Baixauli F, Gutierrez-Vazquez C, Sanchez-Madrid F, Mittelbrunn M, Sorting it out: Regulation of exosome loading, *Seminars in Cancer Biology*, **28** (2014) 3–13. https://doi.org/10.1016/j.semcancer.2014.04.009
[10] Rahmati S, Shojaei F, Shojaeian A, Rezakhani L, Dehkordi MB, An overview of current knowledge in biological functions and potential theragnostic applications of exosomes, *ChemPhys Lipids*, **226** (2020) 104836. https://doi.org/10.1016/j.chemphyslip.2019
[11] Kalluri R, LeBleu VS, The biology, function, and biomedical applications of exosomes. *Science*, **7** (2020). https://doi.org/10.1126/science.aau6977
[12] Liu W, Bai X, Zhang A, Huang J, Xu S, Zhang J, Role of exosomes in central nervous system diseases, *Frontiers in Molecular Neuroscience*, **12** (2019) 240. https://doi.org/10.3389/fnmol.2019.00240
[13] Nurul Huda M, Nafiujjaman M, Deaguero IG, Okonkwo J, Hill ML, Kim T, Nurunnabi M, Potential use of exosomes as diagnostic biomarkers and in targeted drug delivery: Progress in clinical and preclinical applications, *ACS Biomater SciEng*, **7** (2021) 2106–2149. https://doi.org/10.1021/acsbiomaterials.1c00217
[14] Anguluri NVL, Subramanian S, Ramakrishna S, Therapeutic applications of exosomes in various diseases: A review, *Biomaterials Advances*, **134** (2022) 112579. https://doi.org/10.1016/j.msec.2021.112579
[15] Roth GA, Mensah GA, Johnson CO, Global burden of cardiovascular diseases and risk factors, 1990–2019, *Update From the GBD Study*, **76** (2019) 2982–3021. https://doi.org/10.1016/j.jacc.2020.11.010
[16] Chong B, Jayabaskaran J, Jauhari SM, Chan SP, Goh R, WahKueh MT, Global burden of cardiovascular diseases: Projections from 2025 to 2050, *European Journal of Preventive Cardiology*, **281** (2024). https://doi.org/10.1093/eurjpc/zwae281
[17] Cesare MD, Perel P, Taylor S, Kabudula C, Bixby H, Gaziano TA, McGhie DV, The heart of the world, *Global Heart*, **19** (2024), 11. https://doi.org/10.5334/gh.1288
[18] Khatiwada N, Hong Z, Potential benefits and risks associated with the use of statins, *Pharmaceutics*, **16** (2024) 214. https://doi.org/10.3390/pharmaceutics16020214
[19] Dennis TK, Hebert PR, Coffey CS, Sedrakyan A, Curtis JP, Krumholz HM, Beta-blocker therapy and symptoms of depression, fatigue, and sexual dysfunction, *JAMA*, **288** (2020) 351–357. https://doi.org/10.1001/jama.288.3.351
[20] Netala VR, Teertam SK, Li H, Zhang Z, A comprehensive review of cardiovascular disease management: Cardiac biomarkers, imaging modalities, pharmacotherapy, surgical interventions, and herbal remedies, *Cells*, **13** (2024) 1471. https://doi.org/10.3390/cells13171471.
[21] Aldakkak M, Stowe DF, Camara AKS, Safety and efficacy of ranolazine for the treatment of chronic angina pectoris, *Clinical Medicine Insights. Therapeutics*, **5** (2013) 1–14. https://doi.org/10.4137/CMT.S7824.
[22] Sun X, Shan A, Wei Z, Xu B, Intravenous mesenchymal stem cell-derived exosomes ameliorate myocardial inflammation in the dilated cardiomyopathy, *Biochemical and Biophysical Research Communications*, 503, **4** (2018) 2611–2618. https://doi.org/10.1016/j.bbrc.2018.08.012.
[23] Zhao J, Li X, Hu J, Chen F, Qiao S, Sun X, Gao L, Xie, Xu B, Mesenchymal stromal cell-derived exosomes attenuate myocardial ischaemia-reperfusion injury through miR-182-regulated macrophage polarization, *Cardiovascular Research*, 115, **7** (2019), 1205–1216. https://doi.org/10.1093/cvr/cvz040
[24] Wang X, Gu H, Qin D, Yang L, Huang W, Essandoh K, Wang Y, Caldwell CC, Exosomal miR-223 contributes to mesenchymal stem cell-elicited cardioprotection in polymicrobial sepsis, *Scientific Reports*, **5** (2015) 13721. https://doi.org/10.1038/srep13721
[25] Li X, Liu L, Yang J, Yu Y, Chai J, Wang L, Ma L, Yin H, Exosome derived from human umbilical cord mesenchymal stem cell mediates MiR-181c attenuating burn-induced excessive inflammation, *EBioMedicine*, **8** (2016) 72–82. https://doi.org/10.1016/j.ebiom.2016.04.030.
[26] Xu Y, Wan W, Zeng H, Xiang Z, Li M, Yao Y, Li Y, Bortolanza M, Wu J, Exosomes and their derivatives as biomarkers and therapeutic delivery agents for cardiovascular diseases: Situations and challenges, *Journal of Translational Internal Medicine*, 11, **4** (2023) 341–354. https://doi.org/10.2478/jtim-2023-0124.
[27] Farahzadi R, Fathi E, Valipour B, Ghaffary S. Stem cells-derived exosomes as cardiac regenerative agents, *International Journal of Cardiology. Heart & Vasculature*, **52** (2024) 101399. https://doi.org/10.1016/j.ijcha.2024.101399.
[28] Barile L, Vassalli G, Exosomes: Therapy delivery tools and biomarkers of diseases, *Pharmacology & Therapeutics*, **174** (2017) 63–78. https://doi.org/10.1016/j.pharmthera.2017.02.020
[29] Fu E, Pan K, Li Z, Engineering extracellular vesicles for targeted therapeutics in cardiovascular disease, *Frontiers in Cardiovascular Medicine*, **11** (2024) 1503830. https://doi.org/10.3389/fcvm.2024.1503830
[30] Adamiak M, Sahoo S, Exosomes in myocardial repair: Advances and challenges in the development of next-generation therapeutics, *Molecular Therapy: The Journal of the American Society of Gene Therapy*, 26, **7** (2018) 1635–1643. https://doi.org/10.1016/j.ymthe.2018.04.024
[31] Tang T, Leng M, Tang W, Cai Z, Yang L, Wang L, Zhang Y, Guo J, The roles of exosome-derived microRNAs in cardiac

fibrosis, *Molecules (Basel, Switzerland)*, 29, **6** (2024) 1199. https://doi.org/10.3390/molecules29061199

[32] Panigrahi AR, Srinivas L, Panda J, Exosomes: Insights and therapeutic applications in cancer, *Translational Oncology*, **21** (2022) 101439. https://doi.org/10.1016/j.tranon.2022.101439

[33] Janouskova O, Herma R, Semeradtova A, Poustka D, Liegertova M, Auer MH, Maly J, Conventional and nonconventional sources of exosomes-isolation methods and influence on their downstream biomedical application, *Frontiers in Molecular Biosciences*, 2, **9** (2022) 846650. https://doi.org/10.3389/fmolb.2022.846650

[34] Xu F, Zhang Q, Liu Y, Tang R, Li H, Yang H, Lin L, The role of exosomes derived from various sources in facilitating the healing of chronic refractory wounds, *Pharmacological Research*, **216** (2025) 107753. https://doi.org/10.1016/j.phrs.2025.107753

[35] Liu L, Chen Y, Li X, Wang J, Yang L, Therapeutic potential: The role of mesenchymal stem cells from diverse sources and their derived exosomes in diabetic nephropathy, *Biomedicine & Pharmacotherapy*, **175** (2024) 116672. https://doi.org/10.1016/j.biopha.2024.116672.

[36] Ozaki Tan SJ, Floriano JF, Nicastro L, Emanueli C, Catapano F, Novel applications of mesenchymal stem cell-derived exosomes for myocardial infarction therapeutics, *Biomolecules*, 10, **5** (2020) 707. https://doi.org/10.3390/biom10050707

[37] Kundu D, Shin SY, Chilian WM, Dong F, The potential of mesenchymal stem cell-derived exosomes in cardiac repair, *International Journal of Molecular Sciences*, **25** (2024) 13494. https://doi.org/10.3390/ijms252413494

[38] Xiao J, Pan Y, Li XH, Yang XY, Feng YL, Tan HH, Jiang L, Feng J, Yu XY, Cardiac progenitor cell-derived exosomes prevent cardiomyocytes apoptosis through exosomal miR-21 by targeting PDCD4, *Cell Death & Disease*, 7, **6** (2020) 2277. https://doi.org/10.1038/cddis.2016.181.

[39] Wang AYL, Human induced pluripotent stem cell-derived exosomes as a new therapeutic strategy for various diseases, *International Journal of Molecular Sciences*, 22, **4** (2021) 1769. https://doi.org/10.3390/ijms22041769.

[40] Aldali F, Yang Y, Deng C, Li X, Cao X, Xu J, Li Y, Ding J, Chen H, Induced pluripotent stem cell-derived exosomes promote peripheral nerve regeneration in a rat sciatic nerve crush injury model: A safety and efficacy study, *Cells*, 14, **7** (2025) 529. https://doi.org/10.3390/cells14070529

[41] Abdul Malik SZ, Muhilan Y, Nordin F, Hwei Ng M, Xian Law J, Imran M, Mohd Idris I, Jun Ty G, Stem cell derived exosome trilogy: An epic comparison of human MSCs, ESCs and iPSCs, *Stem Cell Research & Therapy*, **16** (2025) 318. https://doi.org/10.1186/s13287-025-04440-0

[42] Wang Y, Zhang L, Li Y, Chen L, Wang X, Guo W, Exosomes/microvesicles from induced pluripotent stem cells deliver cardioprotective miRNAs and prevent cardiomyocyte apoptosis in the ischemic myocardium, *International Journal of Cardiology*, **192** (2015) 61–69. https://doi.org/10.1016/j.ijcard.2015.05.020

[43] Lai RC, Yeo RW, Tan KH, Lim SK, Exosomes for drug delivery – a novel application for the mesenchymal stem cell, *Biotechnology Advances*, 31, **5** (2013) 543–551. https://doi.org/10.1016/j.biotechadv.2012.08.001

[44] Ibrahim AG, Cheng K, Marbán E, Exosomes as critical agents of cardiac regeneration triggered by cell therapy, *Stem Cell Reports*, 2, **5** (2014) 606–619. https://doi.org/10.1016/j.stemcr.2014.04.006

[45] Vader P, Mol EA, Pasterkamp G, Schiffelers RM, Extracellular vesicles for drug delivery, *Advanced Drug Delivery Reviews*, **106** (2016) 148–156. https://doi.org/10.1016/j.addr.2016.02.006

[46] Tkach M, Théry C, Communication by extracellular vesicles: Where we are and where we need to go, *Cell*, 164, **6** (2016) 1226–1232. https://doi.org/10.1016/j.cell.2016.01.043

[47] Xu R, Greening DW, Zhu HJ, Takahashi N, Simpson RJ, Extracellular vesicle isolation and characterization: Toward clinical application, *The Journal of Clinical Investigation*, 126, **4** (2016) 1152–1162. https://doi.org/10.1172/JCI81129

[48] Colletti M, Galardi A, De Santis M, Guidelli GM, Di Giannatale A, Di Luigi L, Antinozzi C, Exosome in cardiovascular diseases: A new therapeutic strategy, *International Journal of Molecular Sciences*, 20, **18** (2019) 4337. https://doi.org/10.3390/ijms20184337

[49] Andaloussi EI, Mäger S, Breakefield IXO, Wood, MJA, Extracellular vesicles: Biology and emerging therapeutic opportunities, *Nature Reviews Drug Discovery*, 12, **5** (2013) 347–357. https://doi.org/10.1038/nrd3978

[50] Kim MS, Haney MJ, Zhao Y, Mahajan V, Deygen I, Klyachko NL, Inskoe E, Piroyan A, Sokolsky M, Okolie O, Hingtgen SD, Kabanov AV, Batrakova EV, Development of exosome-encapsulated paclitaxel to overcome MDR in cancer cells, *Nanomedicine: Nanotechnology, Biology and Medicine*, 12, **3** (2016) 655–664. https://doi.org/10.1016/j.nano.2015.10.012

[51] Christopher JC, Renee RL, Teresa Y, Stephane MIM, Ruenn CL, Dominique PV, Sai KL, Mark R, 1,3 Systemic mesenchymal stem cell-derived exosomes reduce myocardial infarct size: Characterization with MRI in a porcine model, *Frontiers in Cardiovascular Medicine*, **7** (2020) 1–12. https://doi.org/10.3389/fcvm.2020.601990

[52] Larson A, Natera RDE, Crane ALD, Low WC, Grande AW, Lee J, Emerging roles of exosomes in stroke therapy, *International Journal of Molecular Sciences*, 25, **12** (2024) 6507. https://doi.org/10.3390/ijms25126507

[53] Li L, Wang R, Jia Y, Rong R, Xu M, Zhu T, Exosomes derived from mesenchymal stem cells ameliorate renal ischemic-reperfusion injury through inhibiting inflammation and cell apoptosis, *Frontiers in Medicine*, **6** (2019) 269. https://doi.org/10.3389/fmed.2019.00269

[54] Sabry D, Mohamed A, Monir M, Ibrahim HA, The effect of mesenchymal stem cells derived microvesicles on the treatment of experimental CCL4 induced liver fibrosis in rats, *International Journal of Stem Cells*, 12, **3** (2019) 400–409. https://doi.org/10.15283/ijsc18143.

[55] Wiklander OPB, Brennan MA, Lötvall J, Breakefield XO, El Andaloussi S, Advances in therapeutic applications of extracellular vesicles, *Science Translational Medicine*, 11, **492** (2019) 8521. https://doi.org/10.1126/scitranslmed.aav8521

[56] Chen J, Liang X, Han Q, He H, Huang X, Shen Y, Qiu J, Lin F, Mai C, Li Z, Ma K, Hu B, Li X, Zhang Y, Exosomal miR-202-5p derived from iPSC-MSCs protects against myocardial infarction through inhibition of cardiomyocyte pyroptosis, *Stem Cell Research & Therapy*, 16, **1** (2025) 282. https://doi.org/10.1186/s13287-025-04390-7

[57] Pinnell JR, Cui M, Tieu K, Exosomes in Parkinson disease, *Journal of Neurochemistry*, 157, **3** (2021) 413–428. https://doi.org/10.1111/jnc.15288

[58] Cao Z, Kong F, Ding J, Chen C, He F, Deng W, Promoting Alzheimer's disease research and therapy with stem cell technology, *Stem Cell Research & Therapy*, 15, **136** (2024). https://doi.org/10.1186/s13287-024-03737-w

[59] Nojima H, Freeman CM, Schuster RM, Japtok L, Kleuser B, Edwards MJ, Gulbins E, Lentsch AB, Hepatocyte exosomes mediate liver repair and regeneration via sphingosine-1-phosphate, *Journal of Hepatology*, 64, **1** (2016) 60–68. https://doi.org/10.1016/j.jhep.2015.07.030

[60] Zhou Y, Gao Y, Zhang W, Chen Y, Jin M, Yang Z, Exosomes derived from induced pluripotent stem cells suppresses M2-type macrophages during pulmonary fibrosis via miR-302a-3p/TET1 axis, *International Immunopharmacology*, **99** (2021) 108075. https://doi.org/10.1016/j.intimp.2021.108075

[61] Kalluri R, LeBleu VS, The biology, function, and biomedical applications of exosomes. *Science*, 367, **6478** (2020) 6977. https://doi.org/10.1126/science.aau6977

[62] Liang Y, Duan, L, Lu J, Xia J, Engineering exosomes for targeted drug delivery, *Theranostics*, 11, **7** (2021) 3183–3195. https://doi.org/10.7150/thno.53744

[63] Lener T, Gimona M, Aigner L, Börger V, Buzas E, Camussi G, Chaput N, Chatterjee D, Court FA, Del Portillo HA, O'Driscoll L, Fais S, Falcon-Perez JM, Felderhoff-Mueser U, Fraile L, Gho YS, Görgens A, Gupta RC, Hendrix A, Hermann DM, Hill AF, Hochberg F, Horn PA, de Kleijn D, Kordelas L, Kramer BW, Krämer-Albers EM, Laner-Plamberger S, Laitinen S, Leonardi T, Lorenowicz MJ, Lim SK, Lötvall J, Maguire CA, Marcilla A, Nazarenko I, Ochiya T, Patel T, Pedersen S, Pocsfalvi G, Pluchino S, Quesenberry P, Reischl IG, Rivera FJ, Sanzenbacher R, Schallmoser K, Slaper-Cortenbach I, Strunk D, Tonn T, Vader P, van Balkom BWM, Wauben MHM, El Andaloussi S, Théry C, Rohde E, Giebel B, Applying extracellular vesicles based therapeutics in clinical trials – an ISEV position paper, *Journal of Extracellular Vesicles*, 4, **1** (2015) 30087. https://doi.org/10.3402/jev.v4.30087

[64] Mendt M, Kamerkar S, Sugimoto H, McAndrews KM, Wu CC, Gagea M, Yang S, Blanko EVR, Peng Q, Ma X, Marszalek JR, Maitra A, Yee C, Rezvani K, Shpall EJ, LeBleu V, Kalluri R, Generation and testing of clinical-grade exosomes for pancreatic cancer, *JCI Insight*, 3, **8** (2018) 99263. https://doi.org/10.1172/jci.insight.99263

[65] Lötvall J, Hill AF, Hochberg F, Buzás EI, Di Vizio D, Gardiner C, Gho YS, Kurochkin IV, Mathivanan S, Quesenberry P, Sahoo S, Tahara H, Wauben MH, Witwer KW, Théry C, Minimal experimental requirements for definition of extracellular vesicles and their functions: A position statement from the International Society for Extracellular Vesicles, *Journal of Extracellular Vesicles*, 3, **1** (2014) 26913. https://doi.org/10.3402/jev.v3.26913

7 Exosome-Based Targeted Drug Delivery in Cancer Treatment

Ravi Prakash Shukla[1]
[1]Icahn School of Medicine at Mount Sinai, New York, NY

7.1 INTRODUCTION

Cancer is the second leading cause of death worldwide. Cancer cells are different from other cells because they can divide quickly and make lumps and tumor which can spread to other parts of the body [1, 2]. Additionally, their ability to trigger and switch between different molecular pathways plays a big role in drug resistance, which makes therapy much harder to work. To deal with these issues, different treatment plans have been created to slow the growth of tumors. These include nucleic acid–based medicines and targeted anticancer drugs. Along with these treatments, better delivery methods like nanoparticles have been developed to enable getting medicines to cancer cells more precise and effective [3, 4].

In recent years, more attention has been paid to the role of extracellular vesicles, also known as exosome/EVs in cancer biology. EVs are nano- and microvesicles that are released by most bacterial and eukaryotic cells via longstanding pathway. EVs play a key role in how cells communicate with each other and can transport a wide range of useful chemicals, such as proteins, lipids, and nucleic acids. EVs come in different types based on their size, biogenesis, and release processes. These types include EVs, microparticles, shedding vesicles, apoptotic bodies, tolerosomes, proteasomes, and prominosomes. While some EVs emerge directly from the plasma membrane, others are generated by exocytosis of multivesicular bodies, demonstrating the complexity of their biogenesis [5, 6].

EVs, ranging in size from 30 nm to 150 nm, have piqued researchers' interest because of their critical roles in cancer progression and therapeutic potential. Furthermore, they transport mRNAs and microRNAs to recipient cells, where they regulate gene expression and genomic stability [7, 8]. Their intrinsic biocompatibility, stability, and ability to cross physiological barriers, such as the blood–brain barrier (BBB), make them excellent candidates for drug delivery applications. Recent advances in nanotechnology and bioengineering have led to the development of customized EVs designed to improve targeted delivery efficiency and therapeutic effects in cancer treatment [9].

In addition to acting as delivery vehicles, EVs play a crucial role in the dynamic interplay of the tumor microenvironment (TME). The TME is a complex network of cellular components that includes cancer-associated fibroblasts (CAFs), endothelial cells, immune cells, and nontransformed stromal cells, as well as noncellular elements such as cytokines, growth hormones, and extracellular matrix. EVs enhance local and systemic communication by carrying bioactive cargo that affects tumor angiogenesis, proliferation, invasion, metastasis, and drug resistance. Notably, EVs produced by both tumor cells and TME components can speed up tumor growth and induce treatment resistance through a variety of mechanisms, including the activation of oncogenic signaling pathways and immune responses [10, 11].

EVs, with their diverse functions, are emerging as diagnostic and therapeutic targets in oncology, as well as potential drug carriers. The goal of this study is to thoroughly examine the ever-changing landscape of EVs research in cancer. To gain a solid knowledge, we first investigate the EVs discovery, constitution, and biogenesis methods. Thereafter we investigate how EVs influence the TME and promote important cancer processes such as proliferation, migration, and treatment resistance [12]. We also examine the significance of exosomal noncoding RNAs (ncRNAs) in tumor behavior and the primary molecular signaling pathways influenced by EVs. Finally, we examine the clinical potential of tumor-derived EVs, including their use as diagnostic tools and therapeutic agents, and provide recommendations for future translational and clinical research work. Given their potential in targeted delivery and intercellular communication, EV-based treatments, including drug delivery, antigen presentation, and cancer vaccines, are being explored as promising therapeutic approaches for various human diseases, including cancer (Figure 7.1) [13, 14].

7.1.1 Biology and Mechanism of EVs

7.1.1.1 Mechanism of Biogenesis and Release of EVs

EVs are released by almost all types of human cells. They are very important for signaling between cells. The process of making them is tightly controlled and has several steps, such as starting the process, taking in substances, making multivesicular bodies (MVBs), and finally releasing them. At the start of this process, the plasma membrane buds inward, creating MVBs with many intraluminal vesicles (ILVs). These MVBs can go in two main ways: they either go to lysosomes to be broken down or they join with the

DOI: 10.1201/9781003623632-7

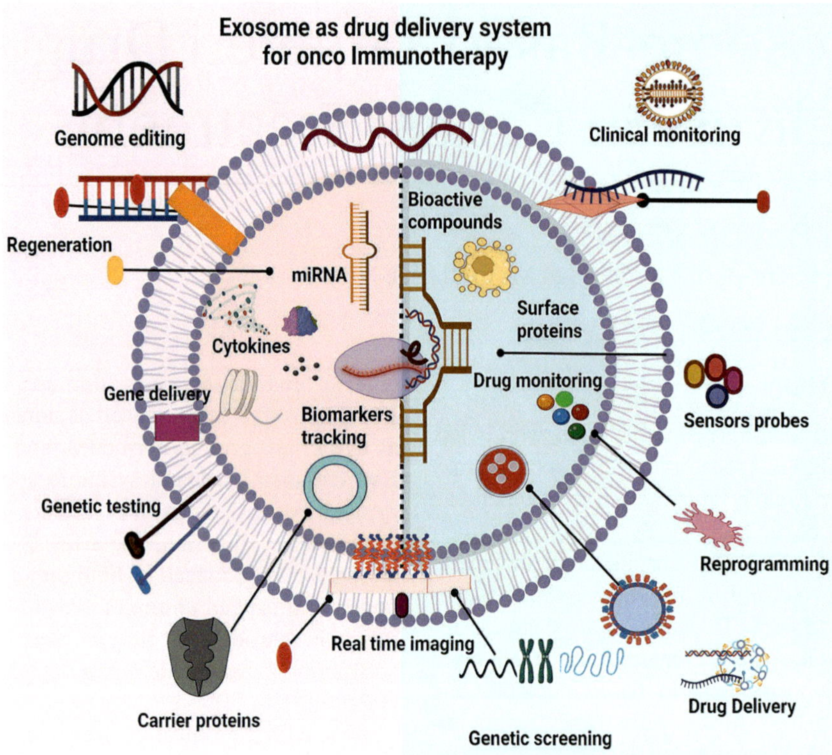

FIGURE 7.1 EVs as drug delivery vehicle. *Source:* Adapted with permission from Ref. [9].

plasma membrane, which lets ILVs move into the extracellular space as EVs [15].

At the molecular level, both ESCRT-dependent and ESCRT-independent pathways work together to sort cargo into ILVs within MVBs. There are four main parts that make up the ESCRT gear: ESCRT-0, ESCRT-I, ESCRT-II, and ESCRT-III. ESCRT-0 and ESCRT-I are mostly in charge of finding and grabbing ubiquitinated protein cargo. ESCRT-II and ESCRT-III, on the other hand, cause membrane splitting and scission to finish forming vesicles. This process is controlled even more by helper proteins like Alix, Syntenin, and VPS4 [16, 17].

On the other hand, EVs can be created without the ESCRT parts, thanks to microdomains that are rich in ceramide and tetraspanin. Ceramide, which is made by sphingomyelinases, helps the membrane curve and ILVs form. Problems with these enzymes may stop EV production. It has been shown that tetraspanins like CD9, CD81, and CD82 affect both the choice of cargo and the budding of vesicles in this route.

EVs are released into the external environment when MVBs are moved and fused with the plasma membrane. This process is controlled by small GTPases such as RAB11, RAB27, RAB35, and RAB27B. Interestingly, cells that divide quickly, like mesenchymal stem cells (MSCs), tend to send out more EVs. Scientists have found that overexpressing the MYC oncogene in MSCs greatly increases the release of EVs without changing their biological features. Conditions inside cells, like pH and calcium ion levels, can also change how EVs are released. For example, using the Na+/H+ exchanger drug monensin raises the amount of Ca^{2+} inside cells, which makes EVs release more. The acidic microenvironment that forms in tumor cores because of poor blood flow and lack of nutrients makes EVs release and be taken up even more. When EVs are released, they can either work directly on the cells that made them (autocrine signaling) or on cells and areas far away (paracrine signaling) [18]. Figuring out the finer points of EVs biogenesis, cargo sorting, and release not only helps us learn more about how cells communicate with each other, but it could also lead to the creation of new cancer treatments.

7.1.1.2 Uptake of EVs by Recipient Cells

EVs are taken up by recipient cells in a very controlled and specific way, not by chance. This happens in three main ways: (i) EVs directly fuse with the plasma membrane and release their cargo into the cytoplasm. (ii) EVs enter the cell through endocytosis, where the membrane bulges and the EVs are taken inside. (iii) Receptor-mediated internalization, where proteins on the EV surfaces interact with receptors on the receiving cell in specific ways [19, 20].

Transmembrane proteins on EVs, whose makeup changes depending on the type of cell that releases them and the conditions under which they are released, play a key part in receptor-mediated uptake. Researchers have found that tetraspanins on EVs and integrins on target cells work together to make binding and internalization more effective. Also, when there is a lot of inflammation, B-cell-derived EVs that contain ICAM-1 can help target cells stick to each other better. Bioactive lipid receptors, like the TIM family receptors and the G2A receptor, also help with uptake by recognizing lyso phosphatidyl choline (LPC) and phosphatidylserine (PS) on the surface of the EVs [21]. Also, heparan sulfate proteoglycans on the surface of cells have been found to play a big role in how EVs bind and enter cells. This shows how complicated and multifaceted the interactions between EVs and cells are.

7.2 EVS AS TARGETED DRUG DELIVERY VEHICLES

EVs are natural transporters of bioactive compounds, making them ideal for drug delivery systems. They have several advantages for therapeutic delivery, including the ability to cross biological barriers, safety in clinical settings like graft-versus-host disease, expression of the "don't eat me" marker CD47, which helps them evade phagocytosis and inherent anti-inflammatory and regenerative properties that enhance therapeutic outcomes [22] (Table 7.1).

7.2.1 Parental Cell Engineering

Parental cell engineering is a technique for altering exosomes by modifying the parent cells that generate them, resulting in the secretion of exosomes with targeted therapeutic content or surface alterations. Strategies encompass the genetic modification of cells using viral vectors or plasmids to express therapeutic proteins or nucleic acids, as well as the physical preconditioning of cells using pharmaceuticals or other agents. This "pre-isolation" method facilitates the generation of exosomes with improved targeting, augmented therapeutic efficacy, and the transport of compounds that cannot be incorporated into exosomes post-isolation.

TABLE 7.1
EVs for Targeted Drug Delivery in Cancer Therapy

Cargo/Agent	Target/Disease Context	Function/Therapeutic Outcome	Surface Ligand/Strategy	References
Curcumin + SPION	Glioma	Imaging and therapy via dual targeting	Neuropilin-1-targeting peptide	[25]
Elastin + folate	Breast cancer	Induction of ferroptosis selectively	DSPE-PEG-folate	[53]
Survivin siRNA	Breast, prostate, colorectal cancers	Tumor-selective RNAi delivery	RNA aptamers (PSMA, EGFR, folate)	[53]
let-7 miRNA + VEGF siRNA	Nucleolin-expressing cancer cells	Tumor-specific RNA therapy	AS1411 aptamer	[53]
Dox	Leukemia cells	Targeted chemotherapeutic delivery	sgc8 aptamer + diacyl lipid(PEG)$_2$	[54]
Paclitaxel	Breast cancer	Aptamer-guided targeted chemotherapy	AS1411 + Chol-PEG2000	[55]
Methotrexate + KLA peptide	Glioblastoma	Brain tumor-specific cytotoxicity	ApoA-1 mimetic peptide + lipid conjugate	[38]
Photosensitizer + NLS peptide	Carcinoma, colorectal cancer (CT26)	Dual-targeted photodynamic therapy	C16-conjugated NLS	[56]
Mannosamine + RGD	HUVEC (angiogenesis model)	Targeted imaging and pro-angiogenic activity	DSPE-PEG-RGD	[57]
KRAS siRNA	Lung adenocarcinoma cells (A549)	Silencing mutant KRAS oncogene	LAMP-2B + iRGD peptide	[58]
DOX	Breast cancer	Tumor-specific drug delivery	LAMP-2B + αv-integrin targeting iRGD peptide	[58]
SOX2 siRNA	Non-small cell lung cancer stem cells	Gene suppression for cancer therapy	LAMP-2B + tLyP-1 peptide	[31]
Imatinib + BCR-ABL siRNA	Chronic myelogenous leukemia	Dual inhibition of oncogene expression and enhanced drug accumulation	LAMP-2B + IL-3 ligand	[27]
5-FU + anti-miR-21	Colorectal cancer	Reversal of chemoresistance	LAMP-2B + zHER affibody	[59]
TPD50 siRNA	HER2+ breast cancer	RNAi-mediated gene silencing	LAMP-2B + HER2-targeting DARPin	[60]
miRNA-let-7a	EGFR-expressing tumors	Gene regulation and tumor suppression	LAMP-2B + GE11 peptide	[21]
Bispecific αCD3/αEGFR	Breast cancer + T-cell activation	T cell-based immunotherapy via dual targeting	Smart-EVs strategy	[61]
miRNA-26a	Hepatocellular carcinoma	Inhibits tumor proliferation and migration	CD63 + ApoA-1 peptide	[62]
OVA antigen	CD8+ T cells	Boosts immunogenic response for vaccine purposes	CD63 modification	[63]

7.2.2 Small Molecular Weight Medicines

Small-molecule medicines often have inadequate absorption and limited therapeutic index due to complicated pharmacokinetic and pharmacodynamic characteristics [23]. EVs are a promising alternative for enhancing delivery efficiency. Paclitaxel-loaded MSCs produced EVs with higher antiproliferative effects against pancreatic cancer compared to free PTX [24]. Curcumin treatment in breast cancer cells improved exosomal protein ubiquitination and NK cell function [25]. Improved medication loading by combining EVs and fusogenic liposomes enhanced delivery of hydrophobic substances.

7.2.3 Nucleic Acids

Oligonucleotides such as siRNAs, shRNAs, miRNAs, and ASOs can efficiently control gene expression [26]. However, successfully delivering these drugs, especially to non-hepatic areas, remains a significant translational challenge. EVs are a promising delivery technique because of their inherent targeting capability and biocompatibility.

HEK293T cells generate EVs with anti-c-Met siRNA, which targets the overexpressed oncogene in stomach cancer [27]. These siRNA-loaded EVs suppressed tumor growth in mouse xenograft models and increased cisplatin sensitivity in gastric cancer cells in vitro. Source cells can be transfected with expression plasmids to generate EVs encoding therapeutic mRNAs, miRNA precursors, shRNAs, or proteins [28]. EVs from cells overexpressing miR-134 were found to inhibit cancer cell migration and invasion while increasing susceptibility to anti-Hsp90 medicines [29].

This approach also allows for the administration of larger therapeutic proteins, which are typically clinically limited due to poor pharmacokinetics. Tumor necrosis factor-related apoptosis-inducing ligand (TRAIL) was successfully packed into EVs following source cell transfection [30]. Yang et al. employed electrical nanoporation to improve EVs production and mRNA loading, resulting in a 50-fold increase in EVs yield and a more than 1,000-fold increase in exosomal mRNA compared to standard methods. Using this approach, large amounts of PTEN mRNA-loaded EVs were produced and distributed systemically, greatly increasing survival in PTEN-deficient glioma mouse models [31, 32].

7.2.4 Protein

Yim et al. developed a new optogenetic device called EXPLORs that loads proteins into EVs. They exploited blue light-controlled reversible protein–protein interactions to efficiently insert cargo proteins into EVs during their natural biogenesis. This approach enabled the successful delivery of functional proteins into target cells' cytoplasm, significantly increasing intracellular protein levels both in vitro and in vivo [33].

7.2.5 Viral Vectors

Viral vectors are widely used for gene delivery because to their high transfection efficiency, but they face challenges such as immunogenicity, neutralizing antibodies, and limited targeting capabilities [34]. Garofalo et al. employed EVs to administer oncolytic viruses (OVs) to specific malignancies by intravenous injection [35]. Labani-Motlagh et al. revealed that OV-infected melanoma cells produce EVs harboring viral transgenes, which activate immune responses at tumor sites [36].

7.3 EVS SURFACE ENGINEERING

EVs have potential in cancer therapy, although their natural form lacks targeting capabilities, resulting in low drug accumulation at tumor sites. Phagocytosis is critical in the rapid clearance of these particles from the bloodstream [37]. Researchers employed CD47, an anti-phagocytic protein that interacts with SIRPα, to inhibit immune absorption [38]. Research has shown that EVs with higher CD47 levels have longer circulation times and lesser clearance [39] (Figure 7.2).

7.3.1 EVs Biology and Immunology

Biological modification is a typical strategy for increasing EVs targeting, which involves painting their membranes with specific peptides or aptamers. Cyclic RGD peptides, like c(RGDyK), enhance EVs affinity for $\alpha v\beta 3$ integrin receptors on tumor cells and vasculature, leading to better drug delivery in malignancies like osteosarcoma [40]. Other peptides, such as tLyp-1, target NRP1 and NRP2 receptors in lung cancer, whereas c-Met-targeted EVs show enhanced absorption in triple-negative breast cancer. Aptamers, which are short oligonucleotides or peptides with unique structures, enhance specificity. For example, E3-aptamer-modified EVs harboring SIRT6 siRNA significantly suppressed prostate cancer growth, whereas AS1411 aptamer-modified EVs increased transport across the blood–brain barrier in glioblastoma mice [41].

Immunological modification seeks to boost antitumor immunity. EVs expressing fibroblast activation protein-α (FAP) operate as tumor vaccines, triggering cytotoxic T-lymphocyte responses against tumors and FAP-positive stromal cells [42]. This alters the TME in malignancies such colorectal and melanoma. EVs containing antibodies, such as anti-CD3, anti-EGFR, or anti-HER2, can also direct T cells toward cancer cells, improving therapeutic precision. Incorporating high-affinity PD-1 variants into EVs reduces PD-L1-mediated T-cell suppression, leading to stronger immune responses [43].

To avoid quick clearance by the mononuclear phagocyte system, EVs can be programmed to overexpress CD47, a "don't eat me" signal that reduces phagocytosis and hence increases circulation time. These findings establish modified EVs as promising techniques for combining immunotherapy

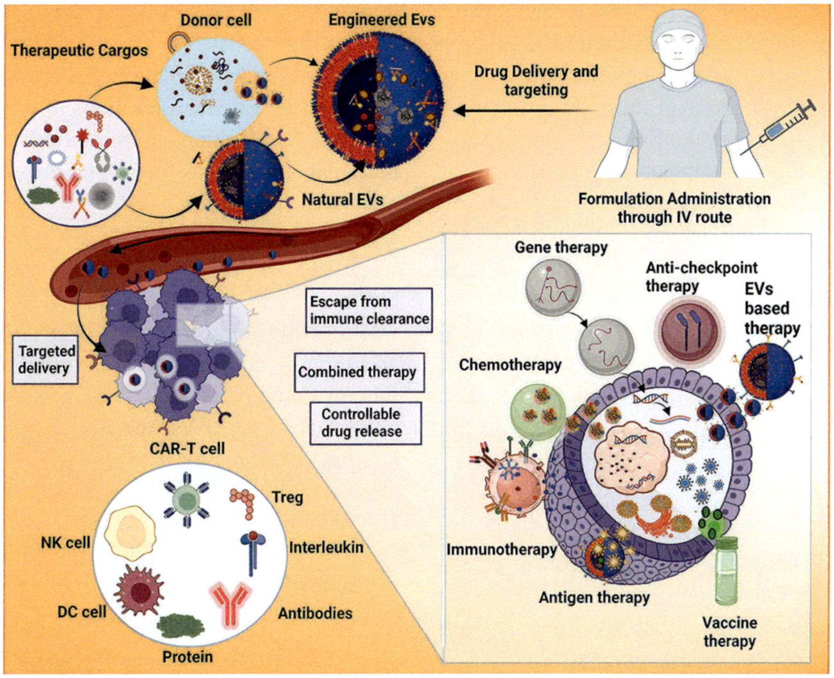

FIGURE 7.2 Schematic representation of engineered EVs for targeted cancer therapy. EVs derived from donor cells are loaded with therapeutic cargos and modified to enhance targeting, immune evasion, and controllable drug release. *Source:* Adapted with permission from Ref. [9].

and targeted drug delivery by enhancing specificity, retention, and immune activation in the TME [38].

7.3.2 Chemical Alteration

The acidic nature of tumor microenvironments (TMEs), which is primarily caused by increased glycolysis and lactate production, presents an opportunity to develop pH-responsive modified EVs. EVs designed with i-motifs, a cytosine-rich, pH-sensitive DNA sequence, can, for example, preferentially release doxorubicin (DOX) in the acidic environment of breast cancer tissue. Loading EVs with sodium bicarbonate ($NaHCO_3$) produces CO_2 bubbles inside cancer cells after endocytosis, leading to enhanced paclitaxel release [44]. Aside from that, researchers developed pH- and temperature-sensitive polymeric adhesives that preferentially adhere to tumor cells at pH 6.8 but are inactive at normal physiological pH (7.4). These smart modifications exploit the acidic TME to regulate drug release, revealing a possible pathway for therapeutically relevant EVs-based delivery systems [45].

7.3.3 External Physical Methods to Enhance Targeting

When engineered EVs are paired with external physical interventions such as magnetic fields, laser irradiation, and ultrasound (US), tumor targeting can improve. For example, EVs from neutrophils have been functionalized with superparamagnetic iron oxide nanoparticles (SPIONs), allowing them to aggregate selectively at tumor locations when subjected to an external magnetic field. EVs containing TNF-α and SPIONs improve tumor targeting and reduce toxicity in in vitro and in vivo models when guided by magnetic fields [46, 47].

Beyond magnetic targeting, laser-based techniques such as photothermal treatment (PTT) have also been investigated [47]. Drug-loaded thermosensitive liposomes can be coupled with EVs to form hybrid vesicles that accumulate mostly in malignancies. Under laser irradiation, photothermal agents within these vesicles generate localized heat, resulting in efficient tumor ablation. Near-infrared (NIR) light can also cause the regulated release of chemotherapy medicines in glioma models [48]. Despite accurate targeting, the clinical use of lasers is constrained by difficulties such as tissue damage, expensive prices, and poor reproducibility.

In comparison, ultrasonography provides a safer and more repeatable option. For example, researchers produced EVs containing chlorin e6 (Ce6) and coated them with CP05-TK-mPEG [49]. When exposed to ultrasound, Ce6 produces reactive oxygen species (ROS), which disrupt the thioketal (TK) linkers, resulting in controlled drug release.

7.3.4 Improved Pharmacokinetics of EVs

EVs have been shown to have improved pharmacokinetics. On top of that, surface integrins play a significant part in the process of EVs biodistribution. In their study published in 2015, tumor-derived EVs, which contained either α6β4 or αVβ5 integrins, accumulated in the lungs and liver to varying degrees. Reducing the activity of these integrins led to a reduction in organ-specific uptake [50].

Based on these findings, it appears that modifying the proteins that are found on the surface of EVs can improve pharmacokinetics and targeted delivery.

7.3.5 EVs that Have Been Engineered for Spatial Targeting

EVs that include targeting molecules, such as peptides, have the ability to transport therapeutic cargo to tumor sites with greater concentration and accuracy. EVs that were loaded with miR-HER2 and decorated with HER2-binding ligands, for instance, were selectively picked up by HER2-positive laryngeal carcinoma cells and caused stronger anticancer effects than EVs that were not targeted. In a similar manner, it has been demonstrated that EVs-like nanovesicles (ELNVs) derived from medicinal herbs like *Asparagus cochinchinensis* (ACNVs) can prevent hepatocellular carcinoma by triggering apoptosis [51]. On the other hand, the rapid clearance of tumors by the mononuclear phagocyte system (MPS) takes place in organs, including the liver and the spleen, which limits the growth of tumors. PEGylation, also known as DSPE-PEG2000, was used by researchers to modify ACNVs in order to better this situation. This modification expanded the circulation length and boosted the suppression of tumor growth while simultaneously minimizing the negative effects. In spite of these developments, it is still challenging to identify optimum targeting ligands that boost tumor selectivity while simultaneously restricting uptake by healthy tissues.

7.3.6 Engineered EVs for Temporal Targeting

While targeting ligands aids EVs in reaching tumors, regulated release at the tumor site remains limited. To address this, researchers created hybrid nanovesicles (hGLVs) by mixing EVs and drug-loaded thermosensitive liposomes [52]. In vitro investigations revealed that these hybrids released pharmaceuticals more effectively after laser irradiation, presenting a mechanism for precise, stimulus-responsive medication delivery into tumors.

7.4 EVS DELIVERY TECHNIQUES

7.4.1 Natural EVs

Natural EVs are nanosized vesicles released by different cell types and found in bodily fluids or cell culture media. They have gained popularity as drug delivery vehicles due to their biocompatible lipid bilayer, minimal immunogenicity, and capacity to transport proteins, lipids, and nucleic acids directly to target cells – even across barriers such as the blood–brain barrier. This makes them promising in the treatment of cancer and neurological illnesses. However, hurdles remain, such as large-scale production, purification, and standardized isolation procedures. Researchers are investigating surface alterations to route EVs more directly to tumor cells, which could improve therapeutic efficacy in malignancies such as breast and lung cancers [5, 64].

7.4.2 Synthetic EVs

Synthetic EVs are bioengineered to mimic natural EVs while giving greater control over targeting, drug loading, and therapy administration. Synthetic EVs are utilized in therapy delivery. By doing so, they solve problems that are present in typical EVs, such as inconsistent production and low payload capacity. The first step in the process involves locating and modifying the source cells in order to produce EVs that appear to have specific surface markings. Using techniques such as ultracentrifugation or chromatography, surface engineering techniques, which include chemical conjugation, genetic modification, and physical methods such as electroporation, improve the ability to target and the stability of the product after it has been manufactured and isolated. It is possible to load drugs, nucleic acids, or proteins either during the process of biogenesis (known as endogenous loading) or directly into isolated EVs (known as exogenous loading) by the use of techniques such as sonication or electroporation, with varying degrees of success. In conclusion, synthetic EVs are subjected to a comprehensive examination using methods such as transmission electron microscopy (TEM), differential light scattering (DLS), and flow cytometry in order to ascertain their dimensions, composition, and cargo content. Considering these advancements, synthetic EVs have the potential to become a promising platform for precise and targeted distribution in the treatment of cancer and other illnesses [65, 66].

7.5 FUNCTIONAL STRATEGIES TO ENHANCE THERAPEUTIC EFFICACY

7.5.1 Leveraging the EPR Effect for Passive Tumor Targeting

Chemotherapy for human solid tumors in clinical practice is far from effective. Despite the discovery and synthesis of hundreds of thousands of anticancer compounds targeting various critical units in cancer cell proliferation and metabolism, the fundamental issue is the lack of selective delivery of these compounds into solid tumor tissue to maintain an effective concentration level for a certain period of time for drug–tumor interaction to carry out anticancer activities [67, 68]. The enhanced permeability and retention effect (EPR effect) is a universal pathophysiological phenomenon and mechanism in which macromolecular compounds such as albumin and other polymer-conjugated drugs larger than 40 kDa can gradually accumulate in the tumor-vascularized area, resulting in targeted delivery and retention of anticancer compounds into solid tumor tissue. Targeting therapy with the EPR effect in clinical practice is not always successful since the strength of the EPR effect varies depending on the type and location of tumors, the level of blood circulation in tumors, and the physical–chemical properties of macromolecular anticancer medicines. This chapter emphasizes the importance of theconcept and mechanism of EPR effect, as well as techniques for better using the EPR effect in constructing smarter macromolecular nanomedicine for successful clinical applications [69, 70].

7.5.2 CXCR4-Engineered EVs for Targeting Hypoxic Tumor Niches

Active targeting techniques improve tumor selectivity by loading EVs with receptors or ligands that identify tumor-associated signals. One interesting method is to modify EVs to overexpress CXCR4, a chemokine receptor associated with tumor invasion, migration, and hypoxia-driven metastasis. Hypoxic areas within tumors create large quantities of CXCL12 (also known as SDF-1), making them perfect targets for CXCR4-engineered EVs. Studies have demonstrated that these modified EVs achieve 3.8 times more absorption in hypoxic glioblastoma cells than controls, resulting in a considerable reduction in tumor volume (about 64%) in preclinical models.

In hepatocellular cancer, CXCR4-engineered EVs activated the SDF-1/CXCR4/CXCR7 pathway, increasing the metastatic potential by about 45%. In addition to increasing drug delivery, EVs can have an impact on the tumor microenvironment. For example, in melanoma models, CXCR4-overexpressing EVs promoted tumor cell migration to pre-metastatic bone niches, increasing tumor burden by around 62% via CXCL12 signaling. Additionally, combining CXCR4 targeting with immunotherapy has yielded promising results. One study found that EVs designed to contain both anti-PD-L1 antibodies and CXCR4 ligands increased T-cell infiltration into tumors, resulting in 72% tumor shrinkage in treated animals.

Overall, CXCR4-engineered EVs have a high promise for increasing tumor-specific delivery, altering the tumor environment, and supplementing immune-based treatments. While passive targeting through the EPR effect is still crucial, active targeting tactics such as CXCR4 alteration can dramatically improve the precision and efficacy of EVs-based cancer treatment, especially in aggressive, hypoxic tumors [71, 72].

7.5.3 Stimuli-Responsive Drug Release

Modern EVs-based drug delivery systems are increasingly employing tumor-specific triggers to provide regulated, localized drug release, increasing therapeutic efficacy while decreasing side effects. Among these, pH-sensitive and enzyme-responsive systems have shown significant promise in responding to circumstances specific to the TME [23, 73].

7.5.3.1 pH-Responsive Systems

The acidic nature of TMEs (pH ~6.2–6.8) relative to healthy tissues (pH ~7.4) has been utilized to construct EVs that release medications only in malignancies. For example, pH-sensitive peptide-modified EVs loaded with DOX selectively released the medication in acidic glioma tissues, lowering tumor volume by 68% in mice compared to traditional liposomal delivery [74]. Yamamoto et al. found that medication release from EVs increased significantly – up to 15-fold – in acidic tumor settings. Aside from DOX, similar pH-triggered devices have been studied for additional chemotherapeutics [75]. Chen et al. found that pH-sensitive DOX nanoparticles aggregated in endosomes and released medicines efficiently at late-endosomal pH (~5.0), enhancing intracellular delivery [40].

7.5.3.2 Enzyme-Responding Systems

EVs have also been designed to respond to tumor-overexpressed enzymes such as matrix metalloproteases (MMP-2 and MMP-9), which promote tumor invasion and metastasis. EVs are created with MMP-cleavable peptide linkers that contained DOX. In the presence of MMP-2, these EVs released the medication within 4 hours, resulting in significant tumor cell death and a 73% reduction in tumor volume in mice models. Importantly, these systems reduced off-target drug leakage, with insignificant release reported at normal enzyme levels, whereas tumor-associated MMP-9 levels caused a 5.3-fold increase in drug release. Combining enzyme- and pH-responsive mechanisms has demonstrated even more promise; for example, other developed dual-sensitive EVs, which obtained a 6.4-fold increase in medication concentration at metastatic locations, greatly reduced tumor recurrence when compared to nonresponsive carriers [76, 77].

To summarize, pH-sensitive and enzyme-cleavable EV devices provide tailored, on-demand medication delivery within malignancies. These novel approaches improve therapeutic precision and show potential for increasing the safety and efficacy of future cancer therapies.

7.6 OVERCOMING MULTIDRUG RESISTANCE (MDR)

Tumors frequently create drug efflux pumps, antiapoptotic proteins, and metabolic modifications to avoid chemotherapy, which is why MDR is still a big problem in cancer treatment. By combining the delivery of chemotherapeutic medicines with siRNA or by using CRISPR-Cas9 gene-editing techniques to silence resistance-related genes, EVs-based systems provide a potential answer to MDR [78].

7.6.1 Chemotherapeutic and siRNA Co-Delivery

Overexpressed proteins linked to drug resistance, such as P-glycoprotein (P-gp), BCL-2, and survivin, can be targeted in an effective manner. By delivering siRNAs against P-gp and MCL-1 simultaneously using lipid-modified polyethylenimine (PEI), Parang et al. demonstrated that breast cancer cells could be resensitized to DOX, leading to cell death rates of up to 90% in cells that had previously been resistant [79]. Navarro et al. also used a unique PEI-dioleoyl phosphatidyl ethanol amine combination to increase DOX cytotoxicity and drug retention in DOX-resistant breast cancer cells by delivering siRNA targeting P-gp by using liposomes loaded with siRNA [80].

7.6.2 EVs Loaded with CRISPR-Cas9

One use of CRISPR-Cas9 technology is the direct knockout of drug resistance genes; this technique also allows for precise gene editing. In animal models, Ganesh et al. showed that cisplatin sensitivity was enhanced by 30–60% when CRISPR-Cas9 targeted survivin and BCL-2 in lung cancer cells [81]. This resulted in a significant decrease in tumor size. Also, Jiang et al. increased intracellular DOX by 4.2 times and decreased tumor burden by targeting the HMGA2/mTOR/P-gp pathway in gastric cancer cells using CRISPR-Cas9 supplied by EVs. When taken as a whole, these methods demonstrate how modified EV systems can aid in overcoming MDR, increasing drug accumulation in tumors, and boosting the efficacy of treatment [82].

7.6.3 Immunomodulatory EVs

By manipulating immunosuppressive environments and taking advantage of checkpoint pathways, cancer cells are able to evade immune detection. Reactivating antitumor immunity can be achieved through the targeted delivery of immunotherapeutic drugs, such as tumor antigen vaccines and checkpoint inhibitors, through engineered EVs [83].

7.6.3.1 Checkpoint Inhibitor Delivery: Targeted Anti-PD-1/PD-L1 Therapy

Although systemic administration of immune checkpoint drugs that target the PD-1/PD-L1 pathway has revolutionized cancer treatment, it can also cause substantial off-target damage. One way to minimize adverse effects and localize the activity of these inhibitors is to use EVs for delivery. As an example, Yu et al. enhanced T-cell infiltration and tumor suppression by 4.5-fold using paclitaxel and P-gp inhibitors loaded into nanoliposomes coated with PD-L1 antibodies [84]. EVs treated with PD-L1 inhibitors decreased immune suppression by 98% in glioblastoma patients, which was associated with better antitumor responses.

7.6.3.2 EVs Vaccines: Activating Dendritic Cells with Tumor Antigens

Enhancement of dendritic cell activation can trigger cytotoxic T-cell responses against cancer when EVs containing tumor-specific antigens are introduced. Through the use of gp100 antigen-loaded EVs, Tobias et al. showed that anti-PD-1 therapy had a much stronger impact on melanoma, resulting in more lymphocyte infiltration and better survival [85]. EVs containing survivin peptides also enhanced antigen-specific T-cell responses and raised interferon-gamma levels by 72% in breast cancer models. Put together, these methods show how to modify immune responses efficiently and how to build EVs to evade drug resistance (by co-delivering medications and siRNA or CRISPR-Cas9 editing, for example). EVs have enormous promise as a new generation of individualized cancer therapies due to their ability to combine targeted medication delivery with immunotherapy [86].

7.7 CHALLENGES AND MITIGATION STRATEGIES

Despite promising developments, clinical translation of EVs-based medicines confronts a number of technological, regulatory, and financial challenges. Addressing these challenges through inventive innovation, rigorous characterization, and explicit legislation is critical to incorporating EVs into mainstream clinical practice.

7.7.1 Technical Obstacles

7.7.1.1 Off-Target Uptake and PEGylation

Rapid EVs clearance by the mononuclear phagocyte system, particularly liver Kupffer cells, poses a significant challenge. Surface PEGylation, which involves adding polyethylene glycol (PEG) chains, can increase circulation time while decreasing liver uptake. Researchers found that PEGylated, copper-64-labeled EVs had over 60% greater tumor accumulation than unmodified EVs [57]. However, recurrent PEGylation might cause the accelerated blood clearance (ABC) phenomenon, which is driven by anti-PEG IgM antibodies and lowers therapeutic efficacy.

7.7.1.2 Immunogenicity and Source Selection

The immunogenic character of EVs is strongly dependent on the cell from which they originate. Human platelet-derived or MSC-derived EVs have minimal immunological activation, whereas immune cell–derived EVs can elicit stronger immune responses [5]. Yan et al. found that plant-derived EVs, such as those from ginger and grapes, have anti-inflammatory properties and lower immune activation, indicating alternative sources for mitigating immune-related damage [87].

7.7.2 Clinical Translation Barriers

7.7.2.1 Scalability and Consistency

Producing EVs on a clinical scale with consistent quality is difficult. Current isolation procedures, such as ultracentrifugation and chromatography, result in fluctuation in yield and purity [88].

7.7.2.2 Heterogeneity and Characterization

EV's cargo varies depending on their biological origin and culture circumstances, complicating standardization. Better methods, such as high-throughput proteomics and single-vesicle analysis, are required to ensure therapeutic uniformity [89].

7.7.2.3 Systemic Clearance and Tumor Penetration

EVs frequently aggregate in the liver and spleen rather than tumors, reducing efficacy. Engineering methods, such as lipid modification or PEGylation, can increase circulation, but balancing targeting and safety remains challenging. Nanocarriers (such as EVs–liposome hybrids) increase targeting, and penetrating into dense tumor tissue is still limited [45].

7.7.3 Regulatory and Commercial Challenges

7.7.3.1 Regulatory Classification (FDA/EMA)

The regulatory frameworks differ: the FDA frequently defines EV therapies as biologics, whereas the EMA may regard them as sophisticated medical goods. Strict GMP requirements and variable product potency lead to low success rates beyond phase I, with just ~5% of IND applications moving forward [90].

7.7.3.2 Intellectual Property and Patents

Rapid growth in the EVs sector has resulted in patent battles, particularly over isolation methods. Over 600 patents have been submitted, with conflicts involving processes such as tangential flow filtration (TFF). While these methods produce higher purity, emerging approaches, such as nanoparticle-based isolation, may increase efficiency and reduce costs.

7.8 CONCLUSION AND FUTURE PERSPECTIVES

EVs are very useful and hopeful nanoscale platforms for delivering drugs to specific areas. They have many benefits: they are biocompatible in nature, able to cross biological barriers, and have the potential for custom surface engineering. EVs-based medicines can now deliver a wider range of cargos, such as small molecules, nucleic acids, proteins, and gene-editing tools like CRISPR-Cas9. This is possible due to progress in parental cell engineering, EV surface modifications, and stimuli-responsive release systems. Strategies that use the EPR effect, hypoxia-targeting changes like increasing CXCR4, and the addition of immunomodulatory agents have also made it possible for more specific and personalized cancer treatments.

Even with these amazing improvements, there are still big problems to solve before EV-based treatments can be widely used in patients. Technical problems like immune activation, off-target buildup, and limited scalability still make it hard to make sure the product works consistently and on a large scale. Different types of tumors and fast clearance from the body make it harder to know what treatments will work, and unclear regulations and intellectual property issues slow down commercialization.

Moving forward, the field will benefit from bringing together bioengineering, nanotechnology, and immunology to create standardized production methods, enhance characterization at the single-vesicle level, and create programmable EVs capable of dynamically targeting tumors. New technologies like AI and machine learning may also be able to predict how EVs will act in living things, which would make individual treatment plans even better. It will also be important to have clearer regulatory systems and fair intellectual property policies to protect patients and encourage new ideas.

In conclusion, EVs have the ability to completely change precision medicine by making therapeutic strategies safer, more targeted, and more useful in more than one way. To get past the problems today, researchers see the full medical potential of EVs-based drug delivery methods for cancer and other diseases, and hence need to keep researching and coming up with new ideas.

ACKNOWLEDGMENT

None.

AI DISCLOSURE STATEMENT

During the preparation of this work the authors used Grammarly, QuillBot, and ChatGPT in order to enhance grammatical structure and narrative coherence. After using this tool/service, the authors reviewed and edited the content as needed and take full responsibility for the content of the publication.

CONFLICT OF INTEREST

None.

REFERENCES

[1] M. Sharma, A.K. Bakshi, N. Mittapelly, S. Gautam, D. Marwaha, N. Rai, N. Singh, P. Tiwari, N. Agarwal, A. Kumar, P.R. Mishra, Recent updates on innovative approaches to overcome drug resistance for better outcomes in cancer, *J. Control. Release*. 346 (2022) 43–70. https://doi.org/10.1016/J.JCONREL.2022.04.007.

[2] S. Gautam, D. Marwaha, N. Singh, N. Rai, M. Sharma, P. Tiwari, S. Urandur, R. Shukla, V. Teja, P. Mishra, Self-assembled redox-sensitive polymeric nanostructures facilitate the intracellular delivery of paclitaxel for improved breast cancer therapy, *Mol. Pharm.* 20 (2023). https://doi.org/10.1021/acs.molpharmaceut.2c00673.

[3] D. Marwaha, S. Gautam, N. Singh, N. Rai, M. Sharma, P. Tiwari, R. Shukla, S. Urandur, V. Teja, M. Nilakanth, A. Kumar, P. Mishra, Synergistic delivery of Imatinib through multifunctional nano-crystalline capsules, in response to redox environment for improved breast cancer therapy, *Colloids Surf. B Biointerfaces*. 226 (2023) 113316. https://doi.org/10.1016/j.colsurfb.2023.113316.

[4] R.P. Shukla, S. Urandur, V.T. Banala, D. Marwaha, S. Gautam, N. Rai, N. Singh, P. Tiwari, P. Shukla, P.R. Mishra,

Development of putrescine anchored nano-crystalsomes bearing doxorubicin and oleanolic acid: Deciphering their role in inhibiting metastatic breast cancer, *Biomater. Sci.* 9 (2021) 1779–1794. https://doi.org/10.1039/D0BM01033B.

[5] P. Tiwari, K. Yadav, R.P. Shukla, S. Gautam, D. Marwaha, M. Sharma, P.R. Mishra, Surface modification strategies in translocating nano-vesicles across different barriers and the role of bio-vesicles in improving anticancer therapy, *J. Control. Release.* 363 (2023) 290–348. https://doi.org/10.1016/j.jconrel.2023.09.016.

[6] P. Tiwari, R.P. Shukla, K. Yadav, N. Singh, D. Marwaha, S. Gautam, A.K. Bakshi, N. Rai, A. Kumar, D. Sharma, P.R. Mishra, Dacarbazine-primed carbon quantum dots coated with breast cancer cell-derived exosomes for improved breast cancer therapy, *J. Control. Release Off. J. Control. Release Soc.* 365 (2023) 43–59. https://doi.org/10.1016/j.jconrel.2023.11.005.

[7] N. Singh, D. Marwaha, S. Gautam, N. Rai, P. Tiwari, M. Sharma, R.P. Shukla, M.N. Mugale, A. Kumar, P.R. Mishra, Surface-modified lyotropic crystalline nanoconstructs bearing doxorubicin and buparvaquone target sigma receptors through pH-sensitive charge conversion to improve breast cancer therapy, *Biomacromolecules.* 24 (2023) 5780–5796. https://doi.org/10.1021/acs.biomac.3c00795.

[8] M. Sharma, D. Panwar, P. Tiwari, A. Kumar, S. Gautam, D. Marwaha, N. Rai, N. Singh, A.K. Bakshi, N. Agarwal, N.K.C. Singh, K. Mitra, V.M. Prajapati, P.R. Mishra, Immobilized doxorubicin and ribociclib carbamate linkers encaged in surface modified cubosomes spatially target tumor reductive environment to enhance antitumor efficacy, *Biomater. Adv.* 155 (2023) 213672. https://doi.org/10.1016/j.bioadv.2023.213672.

[9] P. Tiwari, K. Yadav, R.P. Shukla, A.K. Bakshi, D. Panwar, S. Das, P.R. Mishra, Extracellular vesicles-powered immunotherapy: Unleashing the potential for safer and more effective cancer treatment, *Arch. Biochem. Biophys.* 756 (2024) 110022. https://doi.org/10.1016/j.abb.2024.110022.

[10] S. Urandur, V.T. Banala, R.P. Shukla, N. Mittapelly, G. Pandey, N. Kalleti, K. Mitra, S.K. Rath, R. Trivedi, P. Ramarao, P.R. Mishra, Anisamide-anchored lyotropic nano-liquid crystalline particles with AIE effect: A smart optical beacon for tumor imaging and therapy, *ACS Appl. Mater. Interfaces.* 10 (2018) 12960–12974. https://doi.org/10.1021/ACSAMI.7B19109/ASSET/IMAGES/LARGE/AM-2017-19109Y_0004.JPEG.

[11] V.T. Banala, S. Urandur, S. Sharma, M. Sharma, R.P. Shukla, D. Marwaha, S. Gautam, M. Dwivedi, P.R. Mishra, Targeted co-delivery of the aldose reductase inhibitor epalrestat and chemotherapeutic doxorubicin via a redox-sensitive prodrug approach promotes synergistic tumor suppression, *Biomater. Sci.* 7 (2019) 2889–2906. https://doi.org/10.1039/C9BM00221A.

[12] A. Gebeyehu, N. Kommineni, D.G.J. Meckes, M.S. Sachdeva, Role of exosomes for delivery of chemotherapeutic drugs, *Crit. Rev. Ther. Drug Carrier Syst.* 38 (2021) 53–97. https://doi.org/10.1615/CritRevTherDrugCarrierSyst.2021036301.

[13] T.-T. Tang, L.-L. Lv, H.-Y. Lan, B.-C. Liu, Extracellular vesicles: Opportunities and challenges for the treatment of renal diseases, *Front. Physiol.* 10 (2019) 226. https://doi.org/10.3389/fphys.2019.00226.

[14] R.P. Shukla, J. Dewangan, S. Urandur, V.T. Banala, M. Diwedi, S. Sharma, S. Agrawal, S.K. Rath, R. Trivedi, P.R. Mishra, Multifunctional hybrid nanoconstructs facilitate intracellular localization of doxorubicin and genistein to enhance apoptotic and anti-angiogenic efficacy in breast adenocarcinoma, *Biomater. Sci.* 8 (2020) 1298–1315. https://doi.org/10.1039/C9BM01246J.

[15] B. György, M.E. Hung, X.O. Breakefield, J.N. Leonard, Therapeutic applications of extracellular vesicles: Clinical promise and open questions, *Annu. Rev. Pharmacol. Toxicol.* 55 (2015) 439–464. https://doi.org/10.1146/annurev-pharmtox-010814-124630.

[16] U. Sterzenbach, U. Putz, L.-H. Low, J. Silke, S.-S. Tan, J. Howitt, Engineered exosomes as vehicles for biologically active proteins, *Mol. Ther.* 25 (2017) 1269–1278. https://doi.org/10.1016/j.ymthe.2017.03.030.

[17] T. Juan, M. Fürthauer, Biogenesis and function of ESCRT-dependent extracellular vesicles, *Semin. Cell Dev. Biol.* 74 (2018) 66–77. https://doi.org/10.1016/j.semcdb.2017.08.022.

[18] M. Grapp, A. Wrede, M. Schweizer, S. Hüwel, H.-J. Galla, N. Snaidero, M. Simons, J. Bückers, P.S. Low, H. Urlaub, J. Gärtner, R. Steinfeld, Choroid plexus transcytosis and exosome shuttling deliver folate into brain parenchyma, *Nat. Commun.* 4 (2013) 2123. https://doi.org/10.1038/ncomms3123.

[19] R.P. Shukla, Targeting of tumor microenvironment through nano-biomaterial-based chemotherapy, in: M. Pradhan, K. Yadav, N. Singh Chauhan (Eds.), *Biomaterial-Inspired Nanomedicines for Targeted Therapies*, Springer Nature Singapore, Singapore, 2024: pp. 147–182. https://doi.org/10.1007/978-981-97-3925-7_6.

[20] G. Pandey, N. Mittapelly, G.R. Valicherla, R.P. Shukla, S. Sharma, V.T. Banala, S. Urandur, A.K. Jajoriya, K. Mitra, D.P. Mishra, J.R. Gayen, P.R. Mishra, P-gp modulatory acetyl-11-keto-β-boswellic acid based nanoemulsified carrier system for augmented oral chemotherapy of docetaxel, *Colloids Surf. B Biointerfaces.* 155 (2017) 276–286. https://doi.org/10.1016/j.colsurfb.2017.04.028.

[21] G. Liang, S. Kan, Y. Zhu, S. Feng, W. Feng, S. Gao, Engineered exosome-mediated delivery of functionally active miR-26a and its enhanced suppression effect in HepG2 cells, *Int. J. Nanomed.* 13 (2018) 585–599. https://doi.org/10.2147/IJN.S154458.

[22] R. van der Meel, M.H.A.M. Fens, P. Vader, W.W. van Solinge, O. Eniola-Adefeso, R.M. Schiffelers, Extracellular vesicles as drug delivery systems: Lessons from the liposome field, *J. Control. Release.* 195 (2014) 72–85. https://doi.org/10.1016/j.jconrel.2014.07.049.

[23] D. Marwaha, N. Singh, S. Gautam, N. Rai, P. Tiwari, A.K. Bakshi, A. Kumar, N. Agarwal, M. Sharma, R.P. Shukla, P.R. Mishra, Pemetrexed-loaded supramolecular acetal-functionalized pH-responsive nanocarriers selectively induce apoptosis through biotin receptors to enhance antitumor efficacy, *Colloids Surf. B Biointerfaces.* 245 (2025) 114247. https://doi.org/10.1016/j.colsurfb.2024.114247.

[24] M.S. Kim, M.J. Haney, Y. Zhao, D. Yuan, I. Deygen, N.L. Klyachko, A.V. Kabanov, E. V Batrakova, Engineering macrophage-derived exosomes for targeted paclitaxel delivery to pulmonary metastases: In vitro and in vivo evaluations, *Nanomedicine.* 14 (2018) 195–204. https://doi.org/10.1016/j.nano.2017.09.011.

[25] G. Jia, Y. Han, Y. An, Y. Ding, C. He, X. Wang, Q. Tang, NRP-1 targeted and cargo-loaded exosomes facilitate simultaneous imaging and therapy of glioma in vitro and in vivo, *Biomaterials.* 178 (2018) 302–316. https://doi.org/10.1016/j.biomaterials.2018.06.029.

[26] Z. Naseri, R.K. Oskuee, M.R. Jaafari, M. Forouzandeh Moghadam, Exosome-mediated delivery of functionally

active miRNA-142-3p inhibitor reduces tumorigenicity of breast cancer in vitro and in vivo, *Int. J. Nanomed.* 13 (2018) 7727–7747. https://doi.org/10.2147/IJN.S182384.

[27] S.K. Limoni, M.F. Moghadam, S.M. Moazzeni, H. Gomari, F. Salimi, Engineered exosomes for targeted transfer of siRNA to HER2 positive breast cancer cells, *Appl. Biochem. Biotechnol.* 187 (2019) 352–364. https://doi.org/10.1007/s12010-018-2813-4.

[28] T.A. Shtam, R.A. Kovalev, E.Y. Varfolomeeva, E.M. Makarov, Y. V Kil, M. V Filatov, Exosomes are natural carriers of exogenous siRNA to human cells in vitro, *Cell Commun. Signal.* 11 (2013) 88. https://doi.org/10.1186/1478-811X-11-88.

[29] A. Taghikhani, F. Farzaneh, F. Sharifzad, S. Mardpour, M. Ebrahimi, Z.M. Hassan, Engineered tumor-derived extracellular vesicles: Potentials in cancer immunotherapy, *Front. Immunol.* 11 (2020). https://doi.org/10.3389/fimmu.2020.00221.

[30] L. Zhu, S. Kalimuthu, P. Gangadaran, J.M. Oh, H.W. Lee, S.H. Baek, S.Y. Jeong, S.-W. Lee, J. Lee, B.-C. Ahn, Exosomes derived from natural killer cells exert therapeutic effect in melanoma, *Theranostics.* 7 (2017) 2732–2745. https://doi.org/10.7150/thno.18752.

[31] G. Liang, Y. Zhu, D.J. Ali, T. Tian, H. Xu, K. Si, B. Sun, B. Chen, Z. Xiao, Engineered exosomes for targeted co-delivery of miR-21 inhibitor and chemotherapeutics to reverse drug resistance in colon cancer, *J. Nanobiotechnol.* 18 (2020) 10. https://doi.org/10.1186/s12951-019-0563-2.

[32] Z. Yang, J. Shi, J. Xie, Y. Wang, J. Sun, T. Liu, Y. Zhao, X. Zhao, X. Wang, Y. Ma, V. Malkoc, C. Chiang, W. Deng, Y. Chen, Y. Fu, K.J. Kwak, Y. Fan, C. Kang, C. Yin, J. Rhee, P. Bertani, J. Otero, W. Lu, K. Yun, A.S. Lee, W. Jiang, L. Teng, B.Y.S. Kim, L.J. Lee, Large-scale generation of functional mRNA-encapsulating exosomes via cellular nanoporation, *Nat. Biomed. Eng.* 4 (2020) 69–83. https://doi.org/10.1038/s41551-019-0485-1.

[33] N. Yim, S.-W. Ryu, K. Choi, K.R. Lee, S. Lee, H. Choi, J. Kim, M.R. Shaker, W. Sun, J.-H. Park, D. Kim, W. Do Heo, C. Choi, Exosome engineering for efficient intracellular delivery of soluble proteins using optically reversible protein–protein interaction module, *Nat. Commun.* 7 (2016) 12277. https://doi.org/10.1038/ncomms12277.

[34] S. Warashina, T. Nakamura, Y. Sato, Y. Fujiwara, M. Hyodo, H. Hatakeyama, H. Harashima, A lipid nanoparticle for the efficient delivery of siRNA to dendritic cells, *J. Control. Release.* 225 (2016) 183–191. https://doi.org/10.1016/J.JCONREL.2016.01.042.

[35] M. Garofalo, A. Villa, N. Rizzi, L. Kuryk, B. Rinner, V. Cerullo, M. Yliperttula, V. Mazzaferro, P. Ciana, Extracellular vesicles enhance the targeted delivery of immunogenic oncolytic adenovirus and paclitaxel in immunocompetent mice, *J. Control. Release Off. J. Control. Release Soc.* 294 (2019) 165–175. https://doi.org/10.1016/j.jconrel.2018.12.022.

[36] A. Labani-Motlagh, S. Naseri, J. Wenthe, E. Eriksson, A. Loskog, Systemic immunity upon local oncolytic virotherapy armed with immunostimulatory genes may be supported by tumor-derived exosomes, *Mol. Ther. Oncolytics.* 20 (2021) 508–518. https://doi.org/10.1016/j.omto.2021.02.007.

[37] M. Morishita, Y. Takahashi, A. Matsumoto, M. Nishikawa, Y. Takakura, Exosome-based tumor antigens-adjuvant co-delivery utilizing genetically engineered tumor cell-derived exosomes with immunostimulatory CpG DNA, *Biomaterials.* 111 (2016) 55–65. https://doi.org/10.1016/j.biomaterials.2016.09.031.

[38] E. Koh, E.J. Lee, G.-H. Nam, Y. Hong, E. Cho, Y. Yang, I.-S. Kim, Exosome-SIRPα, a CD47 blockade increases cancer cell phagocytosis, *Biomaterials.* 121 (2017) 121–129. https://doi.org/10.1016/j.biomaterials.2017.01.004.

[39] S. Kamerkar, V.S. LeBleu, H. Sugimoto, S. Yang, C.F. Ruivo, S.A. Melo, J.J. Lee, R. Kalluri, Exosomes facilitate therapeutic targeting of oncogenic KRAS in pancreatic cancer, *Nature.* 546 (2017) 498–503. https://doi.org/10.1038/nature22341.

[40] C. Chen, Z. Duan, Y. Yuan, R. Li, L. Pang, J. Liang, X. Xu, J. Wang, Peptide-22 and cyclic RGD functionalized liposomes for glioma targeting drug delivery overcoming BBB and BBTB, *ACS Appl. Mater. Interfaces.* 9 (2017) 5864–5873. https://doi.org/10.1021/acsami.6b15831.

[41] J. Zou, M. Shi, X. Liu, C. Jin, X. Xing, L. Qiu, W. Tan, Aptamer-functionalized exosomes: Elucidating the cellular uptake mechanism and the potential for cancer-targeted chemotherapy, *Anal. Chem.* 91 (2019) 2425–2430. https://doi.org/10.1021/acs.analchem.8b05204.

[42] A.M. Scott, J.D. Wolchok, L.J. Old, Antibody therapy of cancer, *Nat. Rev. Cancer.* 12 (2012) 278–287. https://doi.org/10.1038/nrc3236.

[43] G. Chen, A.C. Huang, W. Zhang, G. Zhang, M. Wu, W. Xu, Z. Yu, J. Yang, B. Wang, H. Sun, H. Xia, Q. Man, W. Zhong, L.F. Antelo, B. Wu, X. Xiong, X. Liu, L. Guan, T. Li, S. Liu, R. Yang, Y. Lu, L. Dong, S. McGettigan, R. Somasundaram, R. Radhakrishnan, G. Mills, Y. Lu, J. Kim, Y.H. Chen, H. Dong, Y. Zhao, G.C. Karakousis, T.C. Mitchell, L.M. Schuchter, M. Herlyn, E.J. Wherry, X. Xu, W. Guo, Exosomal PD-L1 contributes to immunosuppression and is associated with anti-PD-1 response, *Nature.* 560 (2018) 382–386. https://doi.org/10.1038/s41586-018-0392-8.

[44] H. Abumanhal-Masarweh, L. Koren, A. Zinger, Z. Yaari, N. Krinsky, G. Kaneti, N. Dahan, Y. Lupu-Haber, E. Suss-Toby, E. Weiss-Messer, M. Schlesinger-Laufer, J. Shainsky-Roitman, A. Schroeder, Sodium bicarbonate nanoparticles modulate the tumor pH and enhance the cellular uptake of doxorubicin, *J. Control. Release Off. J. Control. Release Soc.* 296 (2019) 1–13. https://doi.org/10.1016/j.jconrel.2019.01.004.

[45] P. Tiwari, R.P. Shukla, K. Yadav, M. Sharma, A.K. Bakshi, D. Panwar, N. Singh, N. Agarwal, M.N. Mugale, P.R. Mishra, YIGSR functionalized hybrid exosomes spatially target dasatinib to laminin receptors for precision therapy in breast cancer, *Adv. Healthc. Mater.* 14 (2025) 2402673. https://doi.org/10.1002/adhm.202402673.

[46] A. Niescioruk, D. Nieciecka, A.K. Puszko, A. Królikowska, P. Kosson, G.Y. Perret, P. Krysinski, A. Misicka, Physicochemical properties and in vitro cytotoxicity of iron oxide-based nanoparticles modified with antiangiogenic and antitumor peptide A7R, *J. Nanoparticle Res. Interdiscip. Forum Nanoscale Sci. Technol.* 19 (2017) 160. https://doi.org/10.1007/s11051-017-3859-x.

[47] R.J.C. Bose, S. Uday Kumar, Y. Zeng, R. Afjei, E. Robinson, K. Lau, A. Bermudez, F. Habte, S.J. Pitteri, R. Sinclair, J.K. Willmann, T.F. Massoud, S.S. Gambhir, R. Paulmurugan, Tumor cell-derived extracellular vesicle-coated nanocarriers: An efficient theranostic platform for the cancer-specific delivery of anti-miR-21 and imaging agents, *ACS Nano.* 12 (2018) 10817–10832. https://doi.org/10.1021/acsnano.8b02587.

[48] T. Dube, U.B. Kompella, J.J. Panda, Near infrared triggered chemo-PTT-PDT effect mediated by glioma directed twin functional-chimeric peptide-decorated gold nanoroses, *J. Photochem. Photobiol. B Biol.* 228 (2022) 112407. https://doi.org/10.1016/j.jphotobiol.2022.112407.
[49] S. Salunkhe, Dheeraj, M. Basak, D. Chitkara, A. Mittal, Surface functionalization of exosomes for target-specific delivery and in vivo imaging & tracking: Strategies and significance, *J. Control. Release.* 326 (2020) 599–614. https://doi.org/10.1016/j.jconrel.2020.07.042.
[50] B. Costa-Silva, N.M. Aiello, A.J. Ocean, S. Singh, H. Zhang, B.K. Thakur, A. Becker, A. Hoshino, M.T. Mark, H. Molina, J. Xiang, T. Zhang, T.-M. Theilen, G. García-Santos, C. Williams, Y. Ararso, Y. Huang, G. Rodrigues, T.-L. Shen, K.J. Labori, I.M.B. Lothe, E.H. Kure, J. Hernandez, A. Doussot, S.H. Ebbesen, P.M. Grandgenett, M.A. Hollingsworth, M. Jain, K. Mallya, S.K. Batra, W.R. Jarnagin, R.E. Schwartz, I. Matei, H. Peinado, B.Z. Stanger, J. Bromberg, D. Lyden, Pancreatic cancer exosomes initiate pre-metastatic niche formation in the liver, *Nat. Cell Biol.* 17 (2015) 816–826. https://doi.org/10.1038/ncb3169.
[51] L. Zhang, F. He, L. Gao, M. Cong, J. Sun, J. Xu, Y. Wang, Y. Hu, S. Asghar, L. Hu, H. Qiao, Engineering exosome-like nanovesicles derived from asparagus cochinchinensis can inhibit the proliferation of hepatocellular carcinoma cells with better safety profile, *Int. J. Nanomed.* 16 (2021) 1575–1586. https://doi.org/10.2147/IJN.S293067.
[52] Q. Lv, L. Cheng, Y. Lu, X. Zhang, Y. Wang, J. Deng, J. Zhou, B. Liu, J. Liu, Thermosensitive exosome-liposome hybrid nanoparticle-mediated chemoimmunotherapy for improved treatment of metastatic peritoneal cancer, *Adv. Sci. (Weinheim, Baden-Wurttemberg, Ger.)* 7 (2020) 2000515. https://doi.org/10.1002/advs.202000515.
[53] Y. Wan, L. Wang, C. Zhu, Q. Zheng, G. Wang, J. Tong, Y. Fang, Y. Xia, G. Cheng, X. He, S.-Y. Zheng, Aptamer-conjugated extracellular nanovesicles for targeted drug delivery, *Cancer Res.* 78 (2018) 798–808. https://doi.org/10.1158/0008-5472.CAN-17-2880.
[54] Z. Ye, T. Zhang, W. He, H. Jin, C. Liu, Z. Yang, J. Ren, Methotrexate-loaded extracellular vesicles functionalized with therapeutic and targeted peptides for the treatment of glioblastoma multiforme, *ACS Appl. Mater. Interfaces.* 10 (2018) 12341–12350. https://doi.org/10.1021/acsami.7b18135.
[55] H. Cheng, J.-H. Fan, L.-P. Zhao, G.-L. Fan, R.-R. Zheng, X.-Z. Qiu, X.-Y. Yu, S.-Y. Li, X.-Z. Zhang, Chimeric peptide engineered exosomes for dual-stage light guided plasma membrane and nucleus targeted photodynamic therapy, *Biomaterials.* 211 (2019) 14–24. https://doi.org/10.1016/j.biomaterials.2019.05.004.
[56] J. Wang, W. Li, Z. Lu, L. Zhang, Y. Hu, Q. Li, W. Du, X. Feng, H. Jia, B.-F. Liu, The use of RGD-engineered exosomes for enhanced targeting ability and synergistic therapy toward angiogenesis, *Nanoscale.* 9 (2017) 15598–15605. https://doi.org/10.1039/c7nr04425a.
[57] S. Shi, T. Li, X. Wen, S.Y. Wu, C. Xiong, J. Zhao, V.R. Lincha, D.S. Chow, Y. Liu, A.K. Sood, C. Li, Copper-64 labeled PEGylated exosomes for in vivo positron emission tomography and enhanced tumor retention, *Bioconjug. Chem.* 30 (2019) 2675–2683. https://doi.org/10.1021/acs.bioconjchem.9b00587.
[58] D. Bellavia, S. Raimondo, G. Calabrese, S. Forte, M. Cristaldi, A. Patinella, L. Memeo, M. Manno, S. Raccosta, P. Diana, G. Cirrincione, G. Giavaresi, F. Monteleone, S. Fontana, G. De Leo, R. Alessandro, Interleukin 3-receptor targeted exosomes inhibit in vitro and in vivo chronic myelogenous leukemia cell growth, *Theranostics.* 7 (2017) 1333–1345. https://doi.org/10.7150/thno.17092.
[59] S. Ohno, M. Takanashi, K. Sudo, S. Ueda, A. Ishikawa, N. Matsuyama, K. Fujita, T. Mizutani, T. Ohgi, T. Ochiya, N. Gotoh, M. Kuroda, Systemically injected exosomes targeted to EGFR deliver antitumor microRNA to breast cancer cells, *Mol. Ther.* 21 (2013) 185–191. https://doi.org/10.1038/mt.2012.180.
[60] Q. Cheng, X. Shi, M. Han, G. Smbatyan, H.-J. Lenz, Y. Zhang, Reprogramming exosomes as nanoscale controllers of cellular immunity, *J. Am. Chem. Soc.* 140 (2018) 16413–16417. https://doi.org/10.1021/jacs.8b10047.
[61] T. Kanuma, T. Yamamoto, K. Kobiyama, E. Moriishi, Y. Masuta, T. Kusakabe, K. Ozasa, E. Kuroda, N. Jounai, K.J. Ishii, CD63-mediated antigen delivery into extracellular vesicles via DNA vaccination results in robust CD8(+) T cell responses, *J. Immunol.* 198 (2017) 4707–4715. https://doi.org/10.4049/jimmunol.1600731.
[62] R.B. Rountree, S.J. Mandl, J.M. Nachtwey, K. Dalpozzo, L. Do, J.R. Lombardo, P.L. Schoonmaker, K. Brinkmann, U. Dirmeier, R. Laus, A. Delcayre, Exosome targeting of tumor antigens expressed by cancer vaccines can improve antigen immunogenicity and therapeutic efficacy, *Cancer Res.* 71 (2011) 5235–5244. https://doi.org/10.1158/0008-5472.CAN-10-4076.
[63] M. Wang, S. Altinoglu, Y.S. Takeda, Q. Xu, Integrating protein engineering and bioorthogonal click conjugation for extracellular vesicle modulation and intracellular delivery, *PLoS One.* 10 (2015) e0141860. https://doi.org/10.1371/journal.pone.0141860.
[64] K. Yadav, R. Vijayalakshmi, K. Kumar Sahu, P. Sure, K. Chahal, R. Yadav, Sucheta, A. Dubey, M. Jha, M. Pradhan, Exosome-based macromolecular neurotherapeutic drug delivery approaches in overcoming the blood-brain barrier for treating brain disorders, *Eur. J. Pharm. Biopharm.* 199 (2024) 1–23. https://doi.org/10.1016/j.ejpb.2024.114298.
[65] B. Vestad, A. Llorente, A. Neurauter, S. Phuyal, B. Kierulf, P. Kierulf, T. Skotland, K. Sandvig, K.B.F. Haug, R. Øvstebø, Size and concentration analyses of extracellular vesicles by nanoparticle tracking analysis: A variation study, *J. Extracell. Vesicles.* 6 (2017). https://doi.org/10.1080/20013078.2017.1344087.
[66] Y.-J. Li, J.-Y. Wu, J. Liu, W. Xu, X. Qiu, S. Huang, X.-B. Hu, D.-X. Xiang, Artificial exosomes for translational nanomedicine, *J. Nanobiotechnol.* 19 (2021) 242. https://doi.org/10.1186/s12951-021-00986-2.
[67] S. Urandur, V.T. Banala, R.P. Shukla, S. Gautam, D. Marwaha, N. Rai, M. Sharma, S. Sharma, P. Ramarao, P.R. Mishra, Theranostic lyotropic liquid crystalline nanostructures for selective breast cancer imaging and therapy, *Acta Biomater.* 113 (2020) 522–540. https://doi.org/10.1016/j.actbio.2020.06.023.
[68] V.T. Banala, S. Sharma, P. Barnwal, S. Urandur, R.P. Shukla, N. Ahmad, N. Mittapelly, G. Pandey, M. Dwivedi, N. Kalleti, K. Mitra, S.K. Rath, R. Trivedi, P.R. Mishra, Synchronized ratiometric codelivery of metformin and topotecan through engineered nanocarrier facilitates in vivo synergistic precision levels at tumor site, *Adv. Healthc. Mater.* 7 (2018). https://doi.org/10.1002/ADHM.201800300.
[69] P. Tiwari, R.P. Shukla, K. Yadav, D. Panwar, N. Agarwal, A. Kumar, N. Singh, A.K. Bakshi, D. Marwaha, S. Gautam, N. Rai, P.R. Mishra, Exploring nanocarriers as innovative

materials for advanced drug delivery strategies in onco-immunotherapies, *J. Mol. Graph. Model.* (2024) 108702. https://doi.org/10.1016/j.jmgm.2024.108702.
[70] S. Gautam, N. Singh, D. Marwaha, N. Rai, M. Sharma, P. Tiwari, S. Singh, A. Kumar Bakshi, A. Kumar, N. Agarwal, R. Prakash Shukla, P. Ranjan Mishra, Celastrol-loaded polymeric mixed micelles shows improved antitumor efficacy in 4 T1 bearing xenograft mouse model through spatial targeting, *Int. J. Pharm.* 659 (2024) 124234. https://doi.org/10.1016/j.ijpharm.2024.124234.
[71] S. Santagata, C. Ieranò, A.M. Trotta, A. Capiluongo, F. Auletta, G. Guardascione, S. Scala, CXCR4 and CXCR7 signaling pathways: A focus on the cross-talk between cancer cells and tumor microenvironment, *Front. Oncol.* 11 (2021) 591386. https://doi.org/10.3389/fonc.2021.591386.
[72] Y. Yang, J. Li, W. Lei, H. Wang, Y. Ni, Y. Liu, H. Yan, Y. Tian, Z. Wang, Z. Yang, S. Yang, Y. Yang, Q. Wang, CXCL12-CXCR4/CXCR7 axis in cancer: From mechanisms to clinical applications, *Int. J. Biol. Sci.* 19 (2023) 3341–3359. https://doi.org/10.7150/ijbs.82317.
[73] R.P. Shukla, P. Tiwari, A. Sardar, S. Urandur, S. Gautam, D. Marwaha, A.K. Tripathi, N. Rai, R. Trivedi, P.R. Mishra, Alendronate-functionalized porous nano-crystalsomes mitigate osteolysis and consequent inhibition of tumor growth in a tibia-induced metastasis model, *J. Control. Release.* 372 (2024) 331–346. https://doi.org/10.1016/j.jconrel.2024.06.009.
[74] Y. Tian, S. Li, J. Song, T. Ji, M. Zhu, G.J. Anderson, J. Wei, G. Nie, A doxorubicin delivery platform using engineered natural membrane vesicle exosomes for targeted tumor therapy, *Biomaterials.* 35 (2014) 2383–2390. https://doi.org/10.1016/j.biomaterials.2013.11.083.
[75] T.N. Yamamoto, R.J. Kishton, N.P. Restifo, Developing neoantigen-targeted T cell-based treatments for solid tumors, *Nat. Med.* 25 (2019) 1488.
[76] J. Wang, D. Chen, B. Li, J. He, D. Duan, D. Shao, M. Nie, Fe-MIL-101 exhibits selective cytotoxicity and inhibition of angiogenesis in ovarian cancer cells via downregulation of MMP, *Sci. Rep.* 6 (2016) 26126. https://doi.org/10.1038/srep26126.
[77] Z. Zou, X. He, D. He, K. Wang, Z. Qing, X. Yang, L. Wen, J. Xiong, L. Li, L. Cai, Programmed packaging of mesoporous silica nanocarriers for matrix metalloprotease 2-triggered tumor targeting and release, *Biomaterials.* 58 (2015) 35–45. https://doi.org/10.1016/j.biomaterials.2015.04.034.
[78] Y. Chen, L. Wang, M. Zheng, C. Zhu, G. Wang, Y. Xia, E.J. Blumenthal, W. Mao, Y. Wan, Engineered extracellular vesicles for concurrent Anti-PDL1 immunotherapy and chemotherapy, *Bioact. Mater.* 9 (2022) 251–265. https://doi.org/10.1016/j.bioactmat.2021.07.012.
[79] S. Mozaffari, D. Salehi, P. Mahdipoor, R. Beuttler, R. Tiwari, H.M. Aliabadi, K. Parang, Design and application of hybrid cyclic-linear peptide-doxorubicin conjugates as a strategy to overcome doxorubicin resistance and toxicity, *Eur. J. Med. Chem.* 226 (2021) 113836. https://doi.org/10.1016/j.ejmech.2021.113836.
[80] G. Navarro, R.R. Sawant, S. Biswas, S. Essex, C. Tros de Ilarduya, V.P. Torchilin, P-glycoprotein silencing with siRNA delivered by DOPE-modified PEI overcomes doxorubicin resistance in breast cancer cells, *Nanomedicine (Lond).* 7 (2012) 65–78. https://doi.org/10.2217/nnm.11.93.
[81] S. Ganesh, A.K. Iyer, J. Weiler, D. V Morrissey, M.M. Amiji, Combination of siRNA-directed gene silencing with cisplatin reverses drug resistance in human non-small cell lung cancer, *Mol. Ther. Nucleic Acids.* 2 (2013) e110. https://doi.org/10.1038/mtna.2013.29.
[82] L. Jiang, Y. Zhang, L. Guo, C. Liu, P. Wang, W. Ren, Exosomal microRNA-107 reverses chemotherapeutic drug resistance of gastric cancer cells through HMGA2/mTOR/P-gp pathway, *BMC Cancer.* 21 (2021) 1290. https://doi.org/10.1186/s12885-021-09020-y.
[83] N. Syn, L. Wang, G. Sethi, J.-P. Thiery, B.-C. Goh, Exosome-mediated metastasis: From epithelial-mesenchymal transition to escape from immunosurveillance, *Trends Pharmacol. Sci.* 37 (2016) 606–617. https://doi.org/10.1016/j.tips.2016.04.006.
[84] J. Yu, F. Hu, Q. Zhu, X. Li, H. Ren, S. Fan, B. Qian, B. Zhai, D. Yang, PD-L1 monoclonal antibody-decorated nanoliposomes loaded with Paclitaxel and P-gp transport inhibitor for the synergistic chemotherapy against multidrug resistant gastric cancers, *Nanoscale Res. Lett.* 15 (2020) 59. https://doi.org/10.1186/s11671-019-3228-z.
[85] E. Chatziioannou, J. Roßner, T.N. Aung, D.L. Rimm, H. Niessner, U. Keim, L.M. Serna-Higuita, I. Bonzheim, L. Kuhn Cuellar, D. Westphal, J. Steininger, F. Meier, O.T. Pop, S. Forchhammer, L. Flatz, T. Eigentler, C. Garbe, M. Röcken, T. Amaral, T. Sinnberg, Deep learning-based scoring of tumour-infiltrating lymphocytes is prognostic in primary melanoma and predictive to PD-1 checkpoint inhibition in melanoma metastases, *EBioMedicine.* 93 (2023) 104644. https://doi.org/10.1016/j.ebiom.2023.104644.
[86] B.T. Kreger, E.R. Johansen, R.A. Cerione, M.A. Antonyak, The enrichment of survivin in exosomes from breast cancer cells treated with paclitaxel promotes cell survival and chemoresistance, *Cancers (Basel).* 8 (2016). https://doi.org/10.3390/cancers8120111.
[87] Q. Yi, Z. Xu, A. Thakur, K. Zhang, Q. Liang, Y. Liu, Y. Yan, Current understanding of plant-derived exosome-like nanoparticles in regulating the inflammatory response and immune system microenvironment, *Pharmacol. Res.* 190 (2023) 106733. https://doi.org/10.1016/j.phrs.2023.106733.
[88] H. Chen, Q. Li, Recent advances in scalable exosome production: Challenges and innovations, *Chinese J. Plast. Reconstr. Surg.* (2025). https://doi.org/10.1016/j.cjprs.2025.05.001.
[89] N. Ludwig, T.L. Whiteside, T.E. Reichert, Challenges in exosome isolation and analysis in health and disease, *Int. J. Mol. Sci.* 20 (2019). https://doi.org/10.3390/ijms20194684.
[90] K. Cheng, R. Kalluri, Guidelines for clinical translation and commercialization of extracellular vesicles and exosomes based therapeutics, *Extracell. Vesicle.* 2 (2023) 100029. https://doi.org/10.1016/j.vesic.2023.100029.

8 Harnessing Exosomes for the Development of Next-Generation Cancer Vaccines

Prajakta Sahane[1], Niharika Puri[1], Pooja Khairnar[1], Swapnil Shinde[1], Shalini Shukla[1], Suraj Wagh[1], Mayank Gangwar[2], and Saurabh Srivastava[1]

[1]Pharmaceutical Innovation and Translational Research Lab (PITRL), Department of Pharmaceutics, National Institute of Pharmaceutical Education and Research (NIPER), Hyderabad, Telangana, India

[2]Clinical Studies and Trials Unit, Division of Development, Indian Council of Medical Research, Department of Health Research, Ministry of Health & Family Welfare, New Delhi, India

8.1 INTRODUCTION

Cancer has emerged as one of the most critical health issues in the current scenario, which is affecting millions of people globally. Due to this, it requires more research and advancements to develop an efficacious and targeted treatment option [1]. Several traditional approaches are accessible for cancer treatment, such as surgery, radiotherapy, and chemotherapy [2]. These therapies are highly resistant to the metastasized cancer as they possess several limitations, like imprecision, detrimental effects, multiple drug resistance, and nonselective distribution, which contribute to the toxicity [3, 4]. To circumvent all these challenges, cancer vaccines are one of the options available, which take advantage of the immune response for the precise and persistent tumor suppression [5].

The cancer vaccine takes advantage of the antigen presentation, cancer immunology, and immune evasion strategies to diminish the tumor cells [6]. The cancer vaccines are made to trigger the immune system by activating various cells, such as cluster of differentiation (CD4+, CD8+), regulatory T and B cells (Treg and Breg), to target and destroy tumor-specific antigens. The vaccine's goal is to provoke a significant and durable immune response by providing tumor-specific antigens (TSAs) or tumor-associated antigens (TAAs) to the immune system [7]. The various developments in the cancer vaccine field provide safe, efficient, and personalized cancer therapies. By utilizing all this previous knowledge, nowadays researchers are focusing on the development of vaccines that promote the tumor responses, improve the tumor detection capability, and elevate the therapeutic effectiveness [8]. Various types of cancer vaccines are available, like peptide-based vaccines, viral-based vaccines, and dendritic cell (DCs) vaccines [9, 10]. Cancer has a complex mechanism of immune detection, which requires breakthrough vaccines that circumvent the hurdles [11]. The cancer vaccines are based on the principle of the immunological defense mechanism through which the body continually tracks the immune system and eradicates the cancer cells [12]. Despite significant developments, limitations such as antigenic variability, low immunogenicity, and immune suppression in the tumor microenvironment (TME) continue to limit the clinical effectiveness of cancer vaccines [13].

One of the primary drawbacks of conventional cancer vaccines is their inability to accurately distribute antigens and adjuvants to antigen-presenting cells (APCs) in such a way that assures robust immune stimulation. This has encouraged researchers to investigate novel therapeutic strategies that improve the efficiency of cancer vaccines. In light of this, exosomes have generated a lot of interest as naturally occurring nanovesicles suitable for facilitating intercellular communication and assisting in antigen presentation [14]. Their unique properties make them superlative options for the development of next-generation cancer vaccines [15].

By considering all these issues of the cancer vaccines, exosomes have gained more interest due to their naturally occurring tendency, cell-to-cell communication, and antigen presentation. Exosomes possess a 30–150 nm diameter and are released by all mammalian cell types, such as tumor cells, stromal cells, and immune cells [16]. Exosomes are the crucial carriers for immune modulation in cancer therapy. Exosomes are classified into two types: natural-based exosomes and artificial exosomes based on whether they are tailored artificially or not. Exosomes exhibit special properties, including immunological inertness, biocompatibility, and stability under physiological conditions, making them a precise candidate in cancer-based vaccines, providing superior advantages over traditional delivery systems. Exosomes activate the anticancer immune action; along with this, they also attenuate the anticancer immune

DOI: 10.1201/9781003623632-8

responses based on their composition to facilitate the tumor development via their presentation of antigens and the intercellular communication. Tumor-derived exosomes (TEXs) are crucial mediators contributing to the transition of the immune cell function in cancer. TEXs are involved in initiating the immunosuppression, in several ways, such as by activating the CD4+ and CD8+ Treg cells, macrophage activation, and stimulating myeloid-derived suppressor cells (MDSCs), initiating the Breg cell [17]. Conversely, exosome-based vaccines inhibit anti-activity by the tumor response by restraining the CD4+, CD8+ cell activation, preventing the DC development and activation, and natural killer cells (NK)-induced cell death [18]. The dual application of the exosome as a vehicle of antigens and the controller of the immune action represents a new way for advanced investigation in cancer immunotherapy.

Furthermore, exosomes possess the lipids and the membrane proteins, which contribute significantly to the effective therapeutic transport and the successful interaction with the membrane of the target cells [19]. Exosomes act as the natural transporters of RNA that can be harnessed to convey the siRNA in gene silencing in the management of cancer [20]. Additionally, targeting the immunosuppressive gene through siRNA, coupled with the immune checkpoint inhibitors, holds an important application in the expansion of novel immunotherapies for malignancy treatment [21]. Hence, exosomes are the essential vehicles in the delivery of the immune-stimulating agents against cancer [22]. Thus, major scientific teams are actively focusing on developing effective and economically viable therapies leveraging the exosomes.

This chapter aims to assess the significance of the exosome-based cancer vaccines in immunotherapy. It presents a comprehensive understanding of the composition, mechanism, and functions of exosomes. Besides these, the isolation and characterization techniques have also been explored. Advanced strategies in designing exosome-based cancer vaccines further improve their effectiveness, particularly when incorporated with combination therapies such as chemotherapy and phototherapy and radiotherapy. Also, regulatory considerations and the clinical translation of exosome-derived cancer vaccines have been discussed, which offer valuable insights into their future potential in cancer management.

8.2 EXOSOMES: MECHANISM, COMPOSITION, AND FUNCTION

Exosomes are a predominant yet smallest subcategory of heterogeneous bilayer EVs derived from the intraluminal membranes. These are formed via the blending of the multivesicular bodies (MVB) with the cell membranes of immune or cancer cells finally expelled out of the cells [23]. With the usual diameter range of 30–150 nm, exosomes deliver an array of information about the genetic makeup of the originating cells, proving to be a carrier full of potential biomarkers of the cells of their origin, which widens the arena of their application in several diseases [24]. Exosomes contain a plethora of biomolecules concerned with various pathological and physiological processes, with intricate compositions that constitute transmembrane and cytosolic lipids as well as proteins that incorporate messenger ribonucleic acids (mRNAs), micro ribonucleic acids (miRNAs), and long noncoding DNAs and RNAs [25]. The exosomal membrane forms through a two-step invagination process. First, the plasma membrane folds inward, creating the late endosomal compartment. Then, the outer membrane of the endosome undergoes a controlled invagination, selectively incorporating specific membrane and cytosolic components. These invaginated vesicles bud into the endosomal lumen and are eventually released from the cell as exosomes [26]. The biogenetic pathways of exosomes may be either dependent or independent of endosomal sorting complexes required for transport (ESCRT). Though there is no clarity obtained on the detailed mechanism revolving around the ESCRT machinery, researchers have speculated the roles of multifaceted compounds involved in sorting and secreting the components into the exosomes by interacting with ubiquitinated membrane proteins [27]. ESCRT-0 sets off the MVB pathway, ESCRT-I and ESCRT-II aid in stabilizing the neck of budding vesicles, and ESCRT-III assists in membrane remodeling [28]. On the other hand, the ESCRT-independent mechanism involves the secretion of sphingolipids such as ceramides, which may be generated owing to cellular stress conditions. This pathway may be called the sphingomyelinase pathway and is characterized by tetraspanins (CD63), which may be supportive for the endosomal components sorting and secreting [29, 30].

The Exocarta database (version 4) has listed about 4,563 proteins, 194 lipids, 1,639 mRNAs, and 764 miRNAs, identified in different types of exosomes to date [28]. The composition of exosomes varies as per their origin and physiological state; however, a few sets of lipids, nucleic acids, and proteins are common to all. The lipidic subset is responsible for the stability of the exosomal membranes owing to the rigidity of these molecules, such as cholesterol, ceramides, saturated fatty acid chains, sphingolipids, and phosphoglycerides [31]. These molecules, in turn, impart rigidity and stability to the membrane. Additionally, in terms of proteins, various functional proteins are present in exosomes, such as antigen-presenting proteins (major histocompatibility complexes: MHC I, MHC II, and CD86), heat shock proteins (Hsp70, Hsp84, and Hsp90), signal transduction moieties (RhoA, Catenin), lipid raft-associated proteins (stomatin, cholesterol, LBPA), membrane fusion and transport proteins (annexins, GTPases, and flotillin), tetraspanins (CD9, CD37, CD63, CD81, and CD82), cytoskeletal proteins (vimentin, actin, and tubulin), and so on. Also, prominent ESCRT proteins such as TSG101 and Alix play a crucial role in cargo sorting for the exosomes [32, 33]. They are hence exclusive biomarkers present on the exosomes, setting them apart from the rest of the EVs, like apoptotic bodies and microvesicles. Exosomes are responsible for intracellular communications, leading to the regulation of various cellular processes. The exosomes release their cargo into the cells through a series of events: (i) exosome uptake by receptors, (ii) fusion of the plasma membrane with the exosome's membrane, triggering phagocytosis, and (iii) paracrine signaling leading to adherence of the cargo directly onto the surface of the recipient's cell [34]. Such a diverse composite of all the lipids and proteins proves to boost the stability of the exosomes toward the extracellular environment, and the anti-phagocytic protein CD47 prevents the uptake of exosomes by macrophagocytosis. Hence, these evade reticuloendothelial uptake and are highly biocompatible, making them suitable to be exploited as drug delivery vectors [35].

Exosomes can impact both the systemic and local environments, promoting cancer progression. At the same time, they can also trigger an antitumor response, hence bearing bimodal responsibilities in cancer. Exosome-facilitated communication with the cancerous and immune cells results in the recruitment of pro-tumorigenic factors. Several cancers display exosomes' role play evidently in tumorigenesis, angiogenesis, metastasis, and even resistance to therapeutics [36]. Hypoxic conditions in the TME result in the secretion of exosomes from the cancerous cells, which further release interleukins and miRNA. These subsequently trigger M2 polarization of macrophages by the single transducer and activator of transcription 3 (STAT3) signaling pathway, thereby promoting tumorigenesis [37]. Moreover, the release of growth factors, such as vascular endothelial growth factor (VEGF) and TGF-β, leads to angiogenesis, ensuring a rich supply of nutrients into the TME. The macrophage-derived exosomes produce matrix metalloproteinases (MMP-9), which degrade the extracellular matrix (ECM), making it suitable for invasion and metastasis. TEXs secrete miRNAs that downregulate E-cadherin and upregulate N-cadherin and vimentin, facilitating epithelial–mesenchymal transition (EMT) [38]. Exosomal integrins (α6β4, αVβ5) aid in determining the secondary tumor sites and, in turn, activate signaling cascades involved in the metastasis of the cancer stem cells. Additionally, miRNA (miRNA 21, miRNA 155) enhances tumorigenesis by downregulating tumor-suppressing genes. The hypoxic TME subsequently leads the exosomes to release miRNA-233, which in turn reduces PTEN expression. This further stimulates the PI3k/Akt signaling pathways, steering the tumor cells toward drug resistance [39, 40]. The exosomes originating from cancer-associated fibroblasts (CAF) advocate stemness in the tumor cells and impart them with radio-resistance in carcinoma cells.

Exosomes have functionality depending on their origin, which is apparent by the nature of exosomes generated by diverse immune cells present in and around TME. These exosomes may trigger the immune surveillance against tumors by TNF-α signaling [41]. Additionally, the exosomes derived from T-cells enhance the proliferative activity of cytotoxic T-cells, counteracting the immunosuppressive TME. DC-derived exosomes trigger the recruitment of natural killer cells to the tumor sites. TEXs comprise several antigens in the form of transmembrane proteins stemming from endosomal compartments of the donor cells (MART1, tyrosine-related proteins, gp100, and so on) [42]. These exosomes also aid in APCs, yielding an activated CD8+ T cell-mediated immune response, initiating an antitumor response. Exosomes' roles are shaped by the dynamic interplay between tumor-promoting and immune-activating signals in the TME, making them a double-edged sword for cancer (Figure 8.1a).

8.3 EXOSOME ISOLATION AND CHARACTERIZATION TECHNIQUES

It is indispensable to isolate exosomes and set out characterization techniques to harness their multifarious functions. Isolating exosomes involves separating them from a range of interfering cellular components while keeping up with the integrity of exosomes and producing a high yield of exosomes from versatile matrices [43]. Meanwhile, to record the size, size distribution, morphology, concentration, and biochemical composition of the exosomes, several characterization techniques are relevant to gauge their purity and integrity. Here are some of the isolation techniques explained briefly. The isolation and characterization techniques of exosomes are depicted in Figure 8.1b.

8.3.1 Isolation Techniques

8.3.1.1 Ultracentrifugation-Based Methods

Ultracentrifugation utilizes high centrifugal forces (up to 1,000,000 × *g*) at 4°C to sediment components based on their size and density, where larger and heavier particles settle first. While this technique is simple, predominantly employed (~56 % cases), and cost-effective, it can be time-consuming and low-yielding [44, 45]. In differential ultracentrifugation, the cellular elements are separated owing to their buoyant density according to different time cycles for pelletization and different centrifugal forces [46]. Yet another technique, density gradient centrifugation, is often employed, which is also categorized as isopycnic separation or moving zone density gradient centrifugation. Isopycnic separation relies on components remaining at their isodensity zone within prefilled tubes containing biocompatible media of varying densities, such as sucrose or iodixanol, under the centrifugal force. Meanwhile, the moving zone technique is a dynamic platform employing layering the sample on a low-density gradient to segregate exosomes based on size and shape differences, along with the densities [47].

8.3.1.2 Size-Based Methods

Size-based separation is less time-consuming than the previously described techniques. Meanwhile, they require larger volumes of samples and may induce the deterioration of the exosomes due to the shear stress involved [48]. These methods involve ultrafiltration, flow field-flow fractionation (AF4), and size-exclusion chromatography (SEC).

Ultrafiltration uses sequential cycles of various filters with predefined molecular weight (MW) cutoffs to separate exosomes by size or MW. A tandem syringe filter arrangement captures exosomes on the lower membrane (~100 nm cutoff) while retaining larger vesicles on the upper membrane (~500 kDa MW cutoff) [49]. In SEC, a porous stationary phase (agarose, polyacrylamide, allyl dextran, and other cross-linked dextrans) separates macromolecules by hydrodynamic size. Smaller components enter the pores and elute later, while larger ones, like exosomes, are excluded and elute earlier [50]. AF4, a relatively novel technique, separates particles in a chamber using cross and longitudinal flows. Parabolic crossflow pushes larger particles toward the chamber walls, while smaller ones stay in the center and elute first [51].

8.3.1.3 Immunoaffinity-Based Methods

The immunoaffinity-based interplay of abundant proteins (annexin, CD9, CD81, CD82, RAB5, and so on) on the exosomal surface and their receptors can be leveraged to separate the exosomes specifically [47]. Magnetic beads coated with specific antibodies bind to exosomal antigens, capturing the target exosomes while washing away other cellular matter. Anti-EpCAM-coated magnetic beads have shown efficiency comparable to ultrafiltration, requiring smaller sample volumes and achieving up to twice the yield of other methods. However, these techniques are costly and impractical for industrial applications [52].

8.3.1.4 Exosome Precipitation

Precipitation methods rely on certain water-excluding polymers (polyethylene glycol, lectins, etc.) to aggregate exosomes by altering their dispersibility and solubility and form a precipitate that can be further processed downstream. Several commercial kits (e.g., ExoPrep, ExoQuick, Exo-spin, Total Exosome Isolation Reagent, and Exosome Purification Kit) offer rapid isolation with minimal exosomal damage, making them widely used in clinical research [53]. Nonetheless, additional pre-purification steps may be required to avoid the co-precipitation of biopolymers and other proteinaceous components.

8.3.1.5 Microfluidics-Based Techniques

Unlike other noncontinuous isolation methods, microfluidics-based systems utilize both the biochemical and physical properties of exosomes at the microscale. Once introduced into miniaturized fluidic channels, exosomes can be separated through conventional size, immunoaffinity, and density-based techniques and advanced electromagnetic, electrophoretic, and acoustic manipulations, enhancing precision and efficiency [54]. Thus, the desired exosomes are efficiently directed into designated microchannels, while unwanted matter is expelled through the waste outlet. Recently, the ExoTIC device has gained recognition as a promising exosome isolation technique, offering high yield and purity. However, the limited sample capacity and complex setup restrict their broader application [55].

8.4 TUMOR-DERIVED EXOSOMES (TEXS): A KEY PLAYER IN CANCER PROGRESSION

TEXs are the exosomes obtained from the tumor cells. It can promote the activation of the immune system cells and decrease the tumor development. TEXs carry heterogeneous oncogenic components, including proteins, microRNAs, and mRNAs, which modulate the behavior of surrounding cells. These exosomes engage with stromal cells within the TME, inducing their transformation into CAFs that facilitate tumor progression. Several pieces of evidence showed that TEX functions as a carrier of the tumor-derived material, suggesting its potential as a diagnostic platform and a biomarker [56].

TEXs promote angiogenesis during cancer progression [57]. For instance, exosomes secreted by prostate cancer cells contain TGF-β, which may contribute to fibroblast activation, thereby promoting a tumor-supportive microenvironment [58]. In addition, exosomes originating from colorectal cancer have been reported to enhance the expression of α-SMA, a marker of CAFs [59]. The angiogenesis effect is likely attributed to the abundance of angiogenesis-related biomolecules, including miRNAs, mRNAs, and angiogenic proteins. These molecular components can induce phenotypic and functional alterations in stromal cells, enhancing their proliferative and migratory capacities, thereby contributing to tumor growth and metastasis [60]. Additionally, by transferring signaling molecules, exosomes constitute a critical factor in promoting angiogenesis, enhancing immune evasion, and developing pre-metastatic niches in distant organs. Skog et al. demonstrated that glioblastoma-derived exosomes, which carry mRNA, miRNA, and angiogenic proteins, can enhance angiogenesis upon uptake by noncancerous host cells, including brain microvascular endothelial cells [56]. Their findings revealed that TDEs contain angiogenic proteins that promote tubule formation through endothelial cells and promote the augmentation of a human glioma cell line, demonstrating a self-exalting mechanism that supports tumor growth and progression. Furthermore, their ability to enable intercellular communication over long distances suggests their significant role in facilitating metastasis [60, 61]. Figure 8.2 illustrates the role of exosomes in cancer therapy.

8.4.1 Role of Exosomes in Tumor Metastasis

Metastasis is initiated when tumor cells undergo EMT. Tumor cells can acquire migratory potential, enabling them to infiltrate the circulatory or lymphatic systems, disseminate, and establish metastatic lesions at distant sites. Exosomes influence the molecular pathways that govern cancer cell migration and invasion. EMT-driven metastasis is facilitated by the upregulation of N-cadherin and vimentin and downregulation of E-cadherin [62, 63]. EMT-inducing transcription factors such as ZEB1/2, Snail, TGF-β, Slug, and Twist

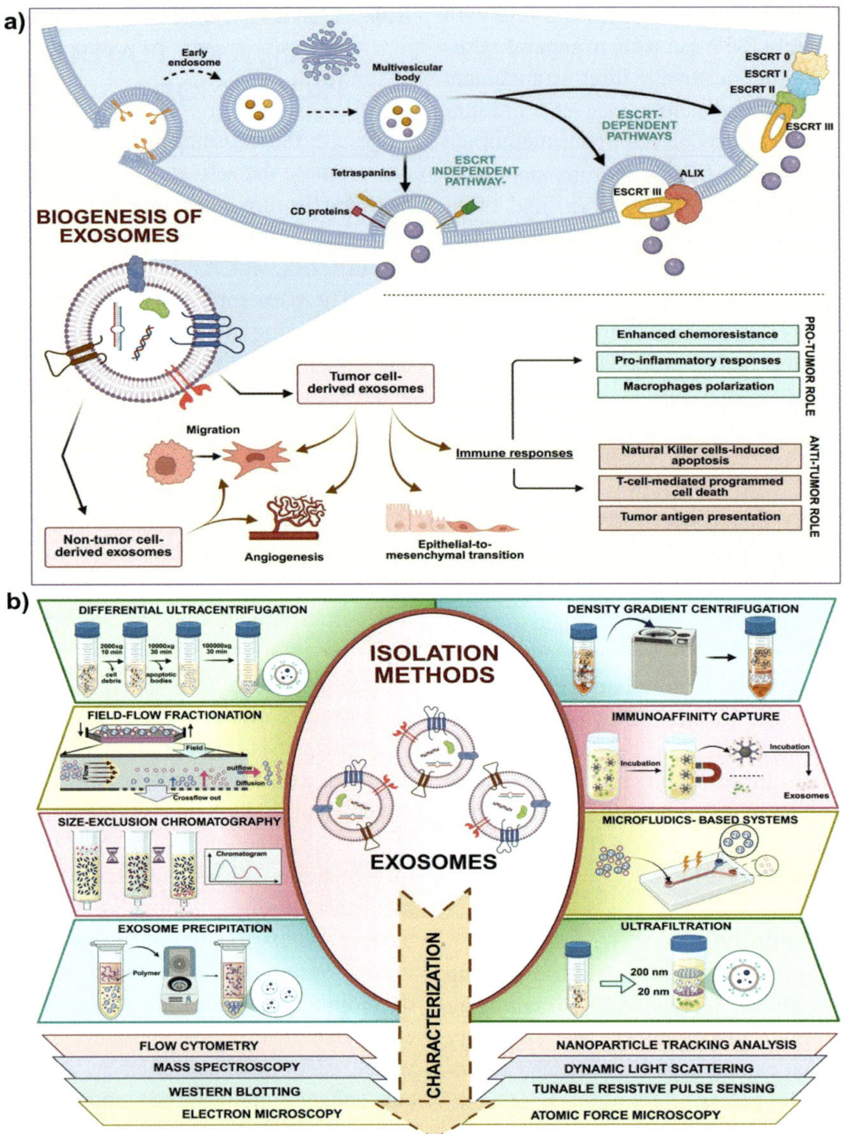

FIGURE 8.1 (a) Biogenesis and mechanism of action of exosomes. (b) Various techniques for the isolation and characterization of exosomes. *Source:* The illustration is created with BioRender.com.

stimulate EMT, leading to tumor invasion [64, 65]. Hypoxic conditions within the TME stimulate malignant cells to enhance exosome production. These hypoxia-induced exosomes are internalized by normal endothelial cells, where exosome cargo promotes angiogenesis [66]. Tumor angiogenesis is predominantly regulated by the hypoxia-inducible factor (HIF). Hypoxia stimulates the secretion of exosomes from cancer cells, driven by the HIF-mediated oxygen-sensing pathway [67, 68]. Wang et al. reported that oral squamous cell carcinoma (OSCC) cells secrete exosomes that contain miRNA-210-3p, which is upregulated in OSCC and functions as a tumor-promoting factor by enhancing the microvessel density. Exosomal miRNA-210-3p facilitates OSCC progression by downregulating Ephrin A3 expression, thereby activating the PI3K/Akt signaling pathway and inducing angiogenesis [69]. Further, cancer cells produce exosomes that exhibit a unique molecular composition, characterized by the EMT-associated proteins and key molecules involved in migration and invasion, such as $\alpha v\beta 6$ integrin present in exosomes secreted by prostate cancer and KIT in exosomes derived from gastrointestinal stromal tumors [70]. Exosomes secreted by tumor cells can migrate and engage with blood vessels, stromal components, and immune cells, facilitating the generation of a pre-metastatic niche [71]. Studies demonstrated that exosomes enriched with the integrin alpha 2 subunit (ITGA2) can promote EMT and significantly augment the metastatic potential of prostate cancer cells [72].

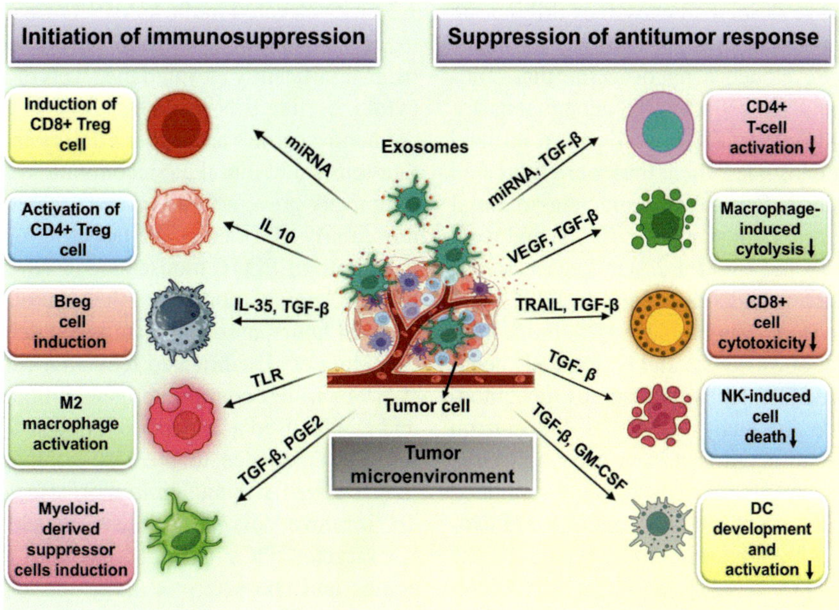

FIGURE 8.2 Role of exosomes in cancer therapy. *Source:* The illustration is created with BioRender.com.

8.4.2 Role of Exosomes in Immunosuppression

TEXs modulate immune responses by transmitting inhibitory signals to immune cells, thereby impairing their antitumor functions [73]. Suppressing the effect of immune cells contributes to tumor progression and immune evasion. These exosomes carry immunosuppressive molecules and bioactive factors that affect the differentiation and antitumor efficacy of immune cells [74]. TDEs regulate T-cell activity by suppressing signaling pathways, inhibiting tumor growth, and inducing apoptosis of CD8+ T cells [75]. Studies reported that tumors can alter the antitumor phenotype of cytotoxic CD8+ T cells, transforming cells into strong immunosuppressive functions [76, 77]. TDEs contain immunosuppressive molecules, including TGF-β, a key cytokine involved in cellular proliferation and differentiation [78]. Beyond its diverse roles in the TME, TGF-β has been shown to drive the development of CAF, which enhances ECM production and contributes to an immunosuppressive milieu [79].

8.5 EXOSOMES AS CANCER VACCINE DELIVERY VEHICLES

Exosomes are pivotal regulators of intercellular signaling, facilitating the transfer of proteins, lipids, RNA, and other biomolecules between cells [80]. Recently, exosomes have been increasingly acknowledged as promising carriers for the delivery of therapeutic and biomolecules, primarily due to their natural origin, which minimizes immunogenic responses [81]. Their inherent biocompatibility, high stability, ability to home to tumor tissues, and customizable targeting capabilities render exosomes a highly attractive strategy for cancer therapy [82].

In particular, exosomes have garnered significant attention as a delivery system for cancer vaccines, owing to their capacity to efficiently transport bioactive compounds to recipient cells, thereby enhancing antigen presentation and immune activation [83]. The unique structural and functional properties of exosomes, such as their nanoscale size, membrane composition, and ability to cross biological barriers, further augment their suitability for targeted therapeutic applications [84]. Despite these advancements, there remains a substantial discrepancy between our existing understanding of exosome biology and the full realization of their potential in clinical applications. Addressing this knowledge gap is critical for optimizing exosome-based delivery systems and translating their therapeutic potential into effective cancer immunotherapies.

Recently, nanocarriers have been proposed as a new approach toward vaccine administration. Exosomes, also known as biogenic nanocarriers, are microscopic membrane vesicles released by cells to transport proteins, RNAs, and lipids. More crucially, DEXs display tumor antigens, MHC molecules, and costimulatory molecules on their surfaces, thereby triggering the liberation of antigen-specific CD8+ and CD4+ T lymphocytes [85]. DEX's membrane structure prevents its substantial degradation and ensures biocompatibility and safety in vivo. DEXs can be held in vitro for longer periods, leading to lower production costs [86]. Preclinical studies indicate that they are more immunogenic and resistant to immunosuppressive effects than DC vaccines, resulting in superior antitumor effectiveness.

Li and coworkers have developed exosomes originating from DCs carrying neoantigens for personalized immunotherapies. They have developed a nanovaccine that comprises DEX carriers and patient-selective neoantigens for personalized immunotherapies. DC2.4 cells were utilized for the isolation of the exosomes; then, they were subjected to further characterization. Particle size and zeta-potential were assessed by dynamic light scattering (DLS) and were obtained as 186 ± 7.6 nm and $-(29.1 \pm 2.2)$ mV, respectively. To entrap exogenous proteins into exosomes, the BTX ECM 600 electroporation system was employed. Authors have studied the depot effect of exosomes and lymph node accumulation by injecting the DiR-labeled exosomes subcutaneously in C57BL/6 mice and analyzing the IVIS spectrum imaging system. They observed that even after 48 hours, the fluorescent signals for DiR-labeled exosomes remained close to the injection area, showing the depot effect of exosomes. Further, they studied the immune activation by the EXP-OVA in mouse models. They observed that IFN-γ secretion was highest after Exo-OVA therapy, indicating increased immunological stimulation. Following immunological activation, flow cytometry was utilized to assess the proportion of T-cell subpopulations following various treatments. They observed that following Exo-OVA vaccination, there was a substantial increase in CD4+ and CD8+ T lymphocytes in comparison to empty Exo and control groups. Further, they reported that mice vaccinated with Exo-OVA demonstrated the strongest cytotoxic T-lymphocyte (CTL) activity, resulting in the removal of around 83% of target cells. They investigated the antitumor efficacy in MC-38 tumors using a single-epitope neoantigen [Adpgk (ASMTNRELM→ASMTNMELM)]-based exovaccine. They reported that novavaccine increased the CD8/Treg and CD8/CD4 ratios in comparison to the control group, especially for Exo-Adpgk. These ratios of immune cells are excellent markers of a strong adaptive T-cell response [87]. Thus, we can conclude that the authors' research revealed superior approaches for cell-free vaccinations and proposed an exosome-derived nanoplatform for tailored nanotechnology and cancer immunotherapy.

Distinct type of cancers presents challenges in the delivery of the vaccines for immunotherapies owing to the low vascularity and intrinsic overexpressed stroma. One such type of cancer is pancreatic ductal adenocarcinoma (PDAC). It poses critical challenges in the delivery of vaccines. Zhou and coworkers have developed an adaptable approach for pancreatic cancer vaccination using exosomes produced from immunogenically dying tumor cells. They enhanced the antitumor immune response by surface-modifying exosomes with the MART-1 peptide (pMEXO) and electroporating MEXO (sMEXO) with CCL22 siRNA to disrupt the CCR4/CCL22 signaling pathway, which normally helps DCs recruit Tregs and attenuate the immune system function. Further, they checked the in vivo immune-activation effect of spMEXO, where they had used C57BL/6 mice. The number of mature DCs in the lymph nodes of mice was 1.2 times elevated in the spMEXO-treated group compared to the sMEXO-therapy group. Furthermore, spMEXO treatment significantly elevated the levels of immunostimulatory cytokines, like IFN-γ and IL-12p70, in the blood, indicating the nanovaccine's antitumor efficacy in vivo. In vivo investigations show that the concomitant system (spMEXO) can effectively prevent tumor growth when administered intramuscularly. Results of the targeting efficacy investigation in orthotopic PDAC mouse models (Bodipy-labeled MEXO (650/665 nm) demonstrated the accumulation of the MEXO after 24 hours, assuring its targeted delivery [88].

Zhang et al. fabricated leukemia cell-derived exosomes (LEXs), which contain antigens unique to leukemia cells, and can be used as a depot of antigen for DC vaccine loading. However, due to the low immunogenicity of LEXs, they showed limited antitumor effects in clinical trials. To improve this, researchers developed gene-modified LX-targeted DCs (LEX-8086) with costimulatory molecules and characterized them for antileukemia immune responses. Their results showed that LEX-8086 improved DCs' antigen-presenting and maturation capabilities.

Immunization with DCsLEX8086 enhanced the CTL response in vitro and markedly activated CD8+ T cells. More significantly, DCsLEX-8086 demonstrated antileukemia properties and efficiently inhibited tumor development in both preventive and therapeutic animal models [89]. Exosomes are a dynamic and intriguing delivery system for cancer vaccines that could transform cancer treatment. They are a top contender for next-generation cancer vaccines because of their capacity to effectively transport antigens, alter immune responses, and get beyond biological barriers. However, to fully exploit their therapeutic potential, more studies and clinical confirmation are necessary.

8.6 DESIGNING EXOSOME-BASED CANCER VACCINES

A variety of distinct sources can be utilized to isolate exosomes. Broadly, exosomes are classified into two categories depending on their source: (i) animal-sourced exosomes and (ii) plant-sourced exosomes [90]. Animal-sourced exosomes can be obtained from diverse sources, including cell cultures (immune cells, fibroblasts, mesenchymal stromal cells, immune cells, adipocytes, and tumor cells) and animal tissues (spleen, brain, and muscle) [91]. Exosomes are lipid bilayer-structured vesicles capable of encapsulating a wide range of compounds, which can be derived from either synthetic or biological origins [92]. Exosomes can be loaded by two techniques: (i) in vivo loading before isolation (preloading), or (ii) in vitro loading after isolation [93]. Preloading involves the integration of the cargo in the course of the exosome's synthesis, producing the exosomes that carry the cargo inside the donor cell before being secreted into the extracellular space. This is achieved through methods such as transfection, electroporation, and co-incubation.

Such a technique is preferred for high-molecular-weight RNA molecules. However, this method has the disadvantage of uncontrolled loading of the payload into exosomes. In post-loading, the substance is introduced after the exosomes have been isolated. This technique is more efficient than the method mentioned earlier [93].

Immune cells can identify exosomes through various surface proteins, including cell adhesion molecules (integrins), cytokines, and their associated receptors, T- and B-cell receptors, and lectins. Numerous in vitro and in vivo assessments have demonstrated that DEXs can activate both adaptive and innate immune action against tumor cells and overcome tumor-induced immunosuppression [94]. In general, exosomes can influence the immune system in different ways, stimulating, inhibiting, or inducing tolerance, depending on their source. Macrophages and certain immature myeloid cells can selectively uptake exosomes. Additionally, exosomes produced from the DCs and B cells, enriched with proteins or peptides with antigenic properties, have been observed to trigger antigen-specific responses in two main types of lymphocytes, T and B cells [94]. Given their immunomodulatory properties, exosomes hold potential for use in vaccines and other immunotherapeutic strategies.

While DEXs remain the most widely explored exosomes for immunotherapy, exosomes from other sources, like tumors or different immune cells, have the potential to also be harnessed for malignancy treatment. Utilizing DEXs or TEXs as potential cancer immunotherapy vaccines offers several key benefits, including convenience, cost-effectiveness, favorable safety, and lack of toxicity.

Furthermore, miRNAs are proposed as vaccine candidates for various diseases due to their role in regulating critical cellular processes such as growth, differentiation, vascularization, and progression [95]. Numerous clinical trial studies have been carried out to investigate TAA-related exosomes in cancer patients at the end stage [96]. Overall, exosomes utilized as cell-free vaccines demonstrated good endurance, although their therapeutic effectiveness remains limited. Exosome-derived cancer vaccines could be divided into three forms: cancer vaccines based on whole exosomes, DC-loaded exosome vaccines, and next-generation vaccines based on exosome membranes [97].

Current research on exosome-based cancer vaccines is largely concerned with their anticancer effectiveness and immunological responses. One investigation found that TEXs were found to be more successful than tumor cell lysates in increasing DCs' maturation and MHC molecule cross-presentation, resulting in a significant tumor-specific CTL response. Following TEX vaccinations, DCs expressed less programmed cell death ligand 1 (PD-L1), resulting in a smaller population of regulatory T cells in vitro [98]. These findings recommend that TEXs and DEXs have the potential to be versatile cancer vaccine candidates, warranting further investigation. Growing evidence suggests that TEXs can be altered, either nongenetically or genetically, to improve their antitumor properties and clinical potential. These modifications aim to enhance immune cell responses and reduce immunosuppression within the TME [99]. Lee et al. incorporated the class II transactivator gene into TEXs, leading to a significant increase in the expression of MHC class II and the induction of Th1-polarized immune activation [100]. Likewise, modified myeloma cell-based exosomes presenting membrane-bound heat shock protein 70 and P1A tumor antigen demonstrated enhanced anticancer immunity by inducing antitumor immunity of antigen-targeted $CD8^+$ CTL responses along with the type 1 $CD4^+$ helper T-cell responses [101].

DEXs are EVs produced by DCs and serve an essential role in presenting antigens and defensive immune system regulation. These exosomes contain important components such as co-stimulatory proteins, MHC class II and I, and cytokines, which can activate T cells and boost immunological responses. They are being investigated as promising candidates for cancer vaccines and immunotherapies because of their capacity to stimulate strong antitumor immunity [86].

DCs, which serve as specialized APCs, secrete exosomes containing key molecules for antigen presentation, utilizing MHC II and MHC I to showcase processed peptides. MHC II and MHC I on Dex have been shown to excite $CD4^+$ and $CD8^+$ T cells. Additionally, protein-based Dex can trigger CTL responses and suppress tumor progression by activating $CD4^+$ T cells and B cells. Dex can be incorporated with TAAs to trigger both T-cell and B-cellactivation, thereby promoting anticancer immune activity [102].

Zhu and coworkers constructed a compelling and biologically safe anticancer vaccine molecule by combining DEX with the MUC1 glycopeptide via a two-step biorthogonal cross-linking method. MUC1-Dex increased CD86 and CD80 expression on DCs, facilitating maturation and enhancement of MHC II and MHC I expression for better antigen presentation. In vivo, MUC1-Dex conjugates increased the formation of MUC1-targeted IgG antibodies, which strongly bind to tumor cells that express MUC1. In a therapeutic and preventative tumor-induced mouse model, the developed vaccinations suppressed tumor growth and produced a potent antibody-mediated immune response and CTLs specific to the antigens to destroy the tumor cells [103].

DEX can elicit both the T-cell and B-cell immune response. In addition to successfully activating adaptive immune cells like T and B cells, DEX has been demonstrated to alter innate immunological responses. Viaud et al. demonstrated that exosomes generated from immature DCs on their surface had ligands for the NKG2D receptor, which trigger the NK in vitro and in vivo, utilizing a mouse model. These ligands could interact directly with NK cells, activating them [104]. Figure 8.3 provides a stepwise illustration for synthesizing exosome-based cancer vaccines.

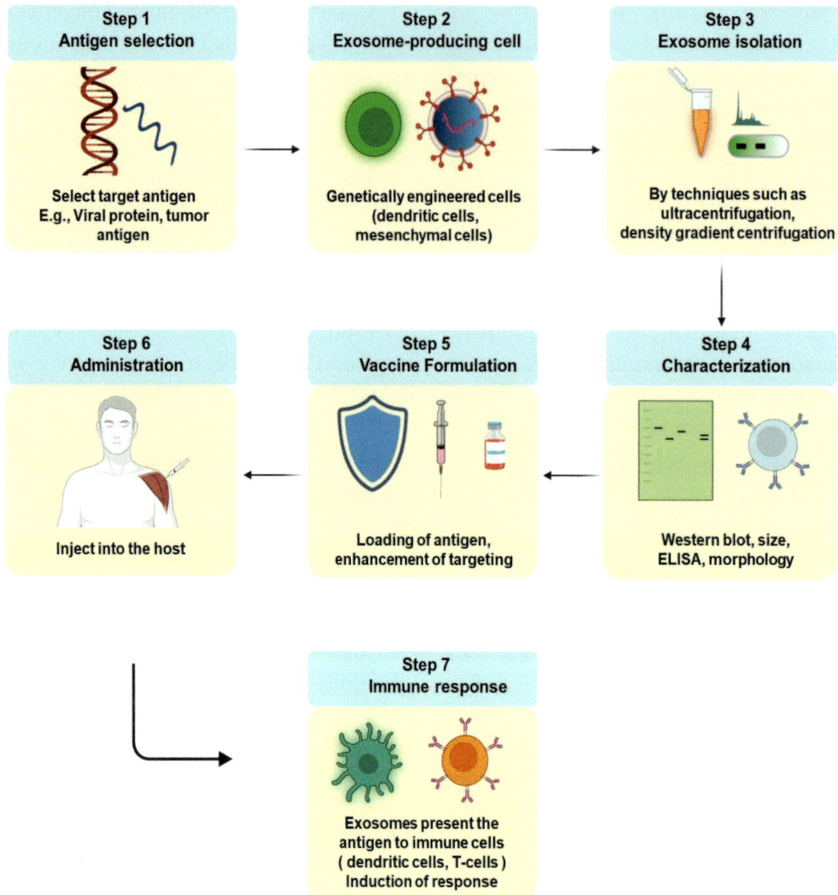

FIGURE 8.3 Step-by-step illustration of designing exosome-based cancer vaccines. *Source:* The illustration is created with BioRender.com.

8.7 EXOSOME-BASED CANCER VACCINES IN COMBINATION THERAPIES

8.7.1 Exosome Vaccines in Combination with Chemotherapy

Exosome-based vaccines, when used alongside chemotherapy, can enhance antitumor effects by inducing immunogenic cell death (ICD), facilitating tumor antigen release, and strengthening the immune response [105]. Chemotherapy not only helps to reduce tumor burden but also enhances the effectiveness of cancer vaccines by modifying the immune environment [106]. Recent findings suggest that the use of exosome-based vaccines has potential advantages when combined with chemotherapy. For instance, studies have shown that exosomes originating from DCs are capable of efficiently presenting tumor cell antigens, thereby triggering CTLs that destroy malignant cells [107].

In a study, Xiao et al. examined whether the therapeutic potential of tumor exosomes (TEX)-loaded DCs vaccination could be improved through the use of drugs that modulate MDSC. Vaccination with DC-TEX was administered, and the results depicted that vaccination resulted in the delayed tumor progression and the survival duration was significantly increased in the mice treated with DC-TEX. The mean tumor diameter was 0.7-fold less than the control group. To address MDSC-driven immunosuppression, researchers examined the efficacy of drugs that hinder MDSC growth, activation, and recruitment. C57BL/6 mice were administered an o.t. injection of UNKC cells to develop a tumor, along with all *trans*-retinoic acid (ATRA), which impairs MD maturation, Sunitinib (Sun) to inhibit MDSC differentiation, or Gemcitabine (GEM), to induce MDSC apoptosis. The influence of ATRA on T-cells facilitated CD8+ expansion and upregulated accessory molecules. However, in tumor-infiltrating lymphocytes (TIL), CD28 and CD25 expression decreased but was restored with simultaneous vaccination. Immunohistochemistry confirmed an increased level of CD4+ and CD8+ cells. ATRA treatment enhanced the proliferative activity of lymph node cells (LNC) and spleen cells (SC) in response to DC-TEX. LNC proliferation in ATRA-treated + DC-TEX-vaccinated mice surpassed that of vaccination alone. Sun treatment had a negligible effect on cytotoxic activity; however, Sun-treated, DC-TEX-vaccinated mice displayed a twofold increase in cytotoxicity against UNKC-breeding mice. Furthermore, the cytotoxic potential of GEM revealed that the combination of GEM and DC-TEX exhibited a 1.3-fold increase against LNC-UNKC and a 2.3-fold increase against SC-UNKC compared to vaccination alone, emphasizing the synergistic effect of the drug and vaccination. These findings indicated that the combined approach enhances both vaccination efficacy and cancer treatment [108].

8.7.2 Exosome Vaccines in Combination with Phototherapy

The phototherapies, such as photodynamic therapy and photothermal therapy, play a significant role in the tumor eradication. When these therapies are combined with exosome vaccines, they upgrade the therapeutic effectiveness of the formulation by improving the tumor-selective immune action and increasing the overall anticancer activity of phototherapies [109]. Phototherapy generates reactive oxygen species (ROS) due to the application of light, promoting the site-specific apoptosis of cancer cells [110]. This brings the release of the tumor antigens, which are employed by the exosome-based vaccines for improving the anticancer immunity by activating the immune cells [111]. Combining phototherapy with exosome vaccines improves therapeutic precision by utilizing exosomes' role in immune activation and photosensitizer localization [112].

By taking advantage of the combination therapy, Liu et al. fabricated the exosome-coated black phosphorus nanocarrier system (hEX@BP) for achieving effective action. To determine the effectiveness of the treatment, a subcutaneous lung carcinoma model in mice is used. To analyze the biocompatibility and the in vitro photothermal action of the hEX@BP, its capability for light absorption and the photothermal characteristics were assessed, and it showed that hEX@BP displayed a broad spectrum, concentration-dependent absorption in the NIR range. Researchers evaluated the in vivo prophylactic efficacy of the hEX@BP-based photonanovaccine using the LLC-induced lung cancer model in C57BL/6 mice. A significant delay in tumor growth (**$p <$ 0.01) was observed exclusively in the hEX@BP-immunized and BP-based PTT-treated groups. The antitumor immune mechanisms of the hEX@BP + PTT were evaluated, and the results depicted that compared to other groups, splenic cells from the hEX@BP+NIR-treated group exhibited significantly elevated levels of IL2/6 (**$P < 0.01$), IFN-γ (*$p < 0.05$), and TNF-α (**$P < 0.01$), demonstrating improved antitumor immunity. After performing several studies, it was revealed that the hEX@BP exhibited prolonged PTT efficacy, amplified thermal effect in the tumor tissue, and effective tumor-selective targeting in vivo. Therefore, these exosome-encapsulated BPQDs have a better biocompatibility, and the PTT enhances immune cell infiltration into tumors and strengthens the antitumor response of the hEX@BP-induced photo-nanovaccine [113].

8.7.3 Exosome Vaccines in Combination with Radiotherapy

Radiotherapy is a widely recognized standard of care for many cancers and holds significant potential for improving responses when paired with cancer vaccines. It directly damages tumor cells and triggers the liberation of tumor antigens, which is captured by exosome-based vaccines and presented to DCs, promoting a robust adaptive immune response [114]. In addition, exosome-based vaccines can help reprogram the TME by reducing immunosuppressive factors and increasing cytotoxic T-cell infiltration. This further leads to better tumor control and minimal recurrence [114]. The combination of radiotherapy with exosome-based vaccines enhances tumor cell destruction caused by radiation and stimulates a prolonged immune response, offering the potential to transform radiotherapy into a promising immunotherapy strategy [115].

In response to these advantages, Kim et al. confirmed the phenotypic and functional changes in DCs, which serve as a crucial link between adaptive and innate immunity after stimulation with exosomes derived from gamma-irradiated (G-exo) and nonirradiated (N-exo) melanoma cells. Additionally, researchers explored the potential of these activated DCs as therapeutic cancer vaccine agents. Bone marrow–derived DCs (BMDCs) viability was assessed to determine the cytotoxicity of G-exo and N-exo. At 5 and 25 μg/mL, both exosome types induced immune responses without causing significant cell death, as shown by Annexin V/PI staining. BMDCs internalized CFSE-labeled exosomes over 30–180 minutes. While uptake kinetics were comparable at 5 μg/mL, G-exo internalization was significantly higher at 25 μg/mL at 60 minutes (***$p < 0.001$) and 120 minutes (**$p < 0.01$), suggesting better binding affinity of G-exo to the DCs. G-exo stimulated DCs exhibited a dose-dependent enhancement in CD80, CD86, MHCI, and MHC II (***$p < 0.001$). N-exo induced similar or slightly higher MHC-I and MHC-II expression (**$p < 0.01$) at 25 μg/mL compared to controls. Further, G-exo-treated DCs also enhanced MHC-I and MHC-II antigen presentation (***$p < 0.001$), while N-exo-stimulated DCs showed no response. All finding concludes that gamma-irradiated tumor exosomes enhance anticancer T-cell immunity and improve DC-based cancer immunotherapy [116].

Further, the combination therapies utilizing exosome-based vaccines is summarized in Table 8.1.

8.8 CLINICAL TRANSLATION AND REGULATORY CONSIDERATIONS

Exosomes have gained more recognition due to their unique characteristics, which are essential during cancer management. In the sections mentioned earlier, we explained the potential of exosome-derived cancer vaccines in cancer therapy. To transfer the exosome-derived cancer vaccines from the lab to clinical application requires stringent rules and regulations to confirm the quality, efficacy, and safety. However, exosome-based cancer vaccines face several clinical translation hurdles such as classification, manufacturing, quality control and assurance, cost, and scalability, which must be resolved to enable the effective integration into clinical translation.

TABLE 8.1
Exosome-Based Vaccines in Combination with Various Therapies

Sr. No	Type of Exosome-Based Vaccine	Combination Therapy	Preclinical/ Clinical Study	Key Findings	References
1	DEXs	Exosome-based vaccines combined with cyclophosphamide in tumor models	Preclinical	This combination increased antitumor immune response. Cyclophosphamide reduced Treg, enhancing vaccine efficacy. This results in tumor inhibition	[117]
2	Melanoma cell-derived exosomes	Exosome vaccine combined with radiotherapy	Preclinical	The exosomes enhance maturation and activation of splenic DCs, as reflected in the elevated expression of CD80, CD86.PD-L1, and CD40	[118]
3	Tumor cell-derived reassembled exosome	Exosome-based vaccines combined with phototherapy	Preclinical	Tumor-derived reassembled exosomes containing Ce6 retained targeting capabilities. It enabled photoacoustic imaging-guided photodynamic therapy. Higher cytokine release from immune cells indicates its potential as an immunotherapeutic agent	[119]
4	Lung tumor–derived exosomes	Exosome vaccine combined with radiotherapy	Preclinical	Findings indicate that sEVs released from irradiated tumor cells enhance antitumor immune response by promoting CD8+ and CD4+ T-cell infiltration in primary tumors and lung metastases	[120]
5	Hepatocellular cancer-derived exosomes	Exosome vaccine combined with chemotherapy	Preclinical	Sorafenib treatment in mice bearing orthotropic hepatocellular carcinoma resulted in elevated levels of PD-L1 inhibitor and a rise in T-regulatory cells within the tumor tissues. The combination of sorafenib with DC-TEX and PD-1 antibody further improved the therapeutic effectiveness	[121]
6	DEXs	Exosome vaccine combined with chemotherapy	Clinical	Increased NKp30-driven NK cell activity was observed in a subset of NSCLC patients displaying dysfunctional NKp30 expression	[122]

The first regulatory barrier in exosome-based cancer vaccines is their classification under the regulatory frameworks. Regulatory authorities such as the European and US Food and Drug Administration (USFDA) [123] regulatory frameworks categorize exosomes as biological products, which designates them as a type of Advanced Therapy Medicinal Products (ATMPs) in the European Union and biologics under the FDA Center for Biologics Evaluation and Research (CBER) [124]. Exosome-based vaccines are generated from cell cultures, in contrast to conventional vaccines, which contain well-studied antigens and adjuvants. Hence, their classification is more intricate. These classifications require stringent regulations involving manufacturing, preclinical, and clinical studies, before going to market for authorization [125]. By modifying exosomes with the specific TAAs, they may be categorized as cellular or gene therapy products, requiring more regulatory controls. In addition, exosome-based vaccines generated from allogeneic or autologous origins may be evaluated at various levels. Autologous products may have fewer immunogenicity problems, but have more variability in production. This hurdle can be addressed by developing standardized regulatory frameworks for classifying exosome-based cancer vaccines with tailored guidelines for their unique characteristics [126].

Another hurdle is related to manufacturing and the standardization process. The manufacturing process of exosome-based cancer vaccines must align with the good manufacturing practice (GMP) guidelines to ensure the safety and consistency of the product. The lack of standardized protocols for exosome preparation, isolation, separation, and characterization poses a significant challenge. Current isolation methods, such as size-exclusion chromatography, immunoaffinity assay, and ultracentrifugation, differ in purity, yield, and scalability. Quality control parameters like concentration, size distribution, and marker expression on the surface and drug content must be accurately defined. Advanced analytical techniques such as DLS, NTA, and high-resolution microscopy are utilized to analyze the exosomes. Then, proteomic and transcriptomic profiling are essential to ensure that the exosomal cargo remains stable across the production batches. There is a need to define a standardized protocol and reproducible process that confirms batch-to-batch consistency [39].

The stability and potency of exosome-based cancer vaccines are another critical regulatory hurdle. Regulatory authorities require stability studies data to determine the appropriate storage temperatures, shelf life, and cryopreservation methods. Exosomes are susceptible to changes, which require optimized storage conditions to preserve their biological activity. In addition, potency assays must be established to determine the ability of the vaccine to provide an immune response, which is particularly challenging due to the heterogeneity of the exosomal payload. The functional assays, such as cytokine induction, T-cell activation, and antigen presentation efficacy, provide proof of immunogenic capability. To comply with the regulations, exosome-based vaccines must demonstrate reproducible biological activity in vitro and in vivo models. The batch release criteria must be developed to ensure integrity before clinical use. Such stability-related hurdles can be addressed by establishing optimized storage conditions and the development of robust potency assays [127].

The manufacturing cost of the exosome-based vaccine is high owing to its complex methods related to its production, such as separation and extraction methods like chromatography and ultracentrifugation. Hence, the cost of manufacturing and scaling up exosome-based vaccines remains a significant barrier to commercialization. The techniques used for manufacturing are time-consuming and result in minimal output. Moreover, the requirement of sophisticated evaluation tools such as NTA and RNA sequencing amplifies the cost of production. Scalability is a further hurdle, as preserving consistency in exosome output, integrity, and therapeutic efficacy becomes more challenging. Besides that, maintaining sterility and prolonged stability at a large-scale production requires advanced bioprocessing solutions, which heighten cost and regulatory burdens. Regulatory authorities require proof of economic viability and scalability for commercial-scale manufacturing [128].

Tackling all these hurdles will be essential to harnessing the full potential of exosome-based cancer vaccines in cancer therapy. Strategic partnerships among academic researchers, biotechnological organizations, and regulatory agencies promote the establishment of standardized production processes, expediting clinical translation.

8.9 CONCLUSION

The numerous cancer vaccines did not achieve the desired clinical effectiveness. This is because of the mechanism of cancer in the tumor evasion, which involves the accumulation of immunosuppressive cells and lower MHC in the TME. Exosome-based cancer vaccines mark an emerging breakthrough in oncology, establishing a pioneering avenue for cancer immunotherapy. Exosomes are natural carriers of several biomolecules, enabling them to modify the immune response. Exosomes have been employed as powerful candidates for the preparation of cancer vaccines owing to an in-depth knowledge of their biogenesis, composition, and mechanism of action. Recent developments in exosome isolation and characterization have made it practicable to generate high-purity and functionally active exosomes for effective immunotherapeutic purposes. The TEXs function in immunity evasion and metastases; they could be precisely modified to stimulate significant immune responses against tumors. Due to these advantages, TEXs are the cutting-edge platform in the progression of cancer vaccines. Exosome-derived cancer vaccines are being systematically explored with a focus on enhancing immunogenicity, providing precise targeted delivery.

Moreover, the traditional therapies developed resistance against cancer. Exosome-based vaccines unlock new avenues for combination approaches to address this limitation. The development of cancer vaccines is a complex process that requires strict compliance and meticulous regulatory frameworks to ensure effectiveness, safety, and consistency. The integration of machine learning (ML), artificial intelligence (AI), and 3D printing has the ability to reshape the production of exosome-derived cancer vaccines by increasing accuracy, scalability, and personalized therapy. Also, the advancements in bioinformatics, biomanufacturing, and immunotherapy greatly aid in expanding vaccine development. These innovations can reduce production costs, enhance vaccine potency, and efficacy. The evolution of next-generation cancer vaccines from lab to clinic translation is challenging; thus, continuous research, technological refinements, and a well-defined regulatory framework are required.

ABBREVIATIONS

BTX ECM 600	electroporation system model
C57BL/6	inbred laboratory mouse strain
CD28	T-cell costimulatory receptor
CD40	costimulatory receptor on antigen-presenting cells
CD47	integrin-associated "don't-eat-me" signal
CD80	B7-1 costimulatory molecule
CD86	B7-2 costimulatory molecule
CIITA	class II transactivator
DC2.4	murine dendritic cell line
DiR	near-infrared lipophilic fluorescent dye
ExoTIC	exosome total isolation chip
gp100	melanoma antigen (PMEL)
Hsp84	heat shock protein 84
Hsp90	heat shock protein 90
IVIS	in vivo imaging system
KIT	receptor tyrosine kinase (CD117)
LBPA	lysobisphosphatidic acid
LLC	Lewis lung carcinoma
MC-38	murine colon adenocarcinoma cell line
MUC1	mucin 1 glycoprotein
NKG2D	natural killer group 2D receptor
PD-L1	programmed death-ligand 1
PI3K	phosphoinositide 3-kinase
RAB5	small GTPase Rab5
RhoA	Ras homolog family member A
SEC	size-exclusion chromatography
TSG101	tumor susceptibility gene 101
UNKC	murine pancreatic cancer cell line (UNKC)
ZEB1/2	zinc finger E-box-binding homeobox 1/2
α6β4	integrin alpha-6 beta-4
αVβ5	integrin alpha-V beta-5
αvβ6	integrin alpha-V beta-6

ACKNOWLEDGMENTS

The authors would like to acknowledge the research funding support by the Department of Pharmaceuticals (DoP), Ministry of Chemicals and Fertilizers, Government of India, to "Pharmaceutical Innovation and Translational Research Lab" (PITRL), Department of Pharmaceutics, National Institute of Pharmaceutical Education and Research (NIPER), Hyderabad, India. We would like to acknowledge BioRender (BioRender.com) for providing the tools used to create the figures in this chapter.

AI DISCLOSURE STATEMENT

During the preparation of this work, the authors used Grammarly and QuillBot in order to enhance grammatical structure and narrative coherence. After using this tool/service, the authors reviewed and edited the content as needed and take full responsibility for the content of the publication.

CONFLICT OF INTEREST

None.

REFERENCES

[1] N.R. Raghani, M.R. Chorawala, M. Mahadik, R.B. Patel, B.G. Prajapati, P.S. Parekh, Revolutionizing cancer treatment: Comprehensive insights into immunotherapeutic strategies, *Medical Oncology* 41:2 (2024) 1–20. https://doi.org/10.1007/S12032-023-02280-7.

[2] B. Liu, H. Zhou, L. Tan, K.T.H. Siu, X.Y. Guan, Exploring treatment options in cancer: Tumor treatment strategies, *Signal Transduction and Targeted Therapy* 9:1 (2024) 1–44. https://doi.org/10.1038/s41392-024-01856-7.

[3] U. Anand, A. Dey, A.K.S. Chandel, R. Sanyal, A. Mishra, D.K. Pandey, V. De Falco, A. Upadhyay, R. Kandimalla, A. Chaudhary, J.K. Dhanjal, S. Dewanjee, J. Vallamkondu, J.M. Pérez de la Lastra, Cancer chemotherapy and beyond: Current status, drug candidates, associated risks and progress in targeted therapeutics, *Genes & Diseases* 10 (2023) 1367–1401. https://doi.org/10.1016/J.GENDIS.2022.02.007.

[4] M. Handa, P.S. Khairnar, R. Shukla, P. Kesharwani, Nanotechnology-mediated combinational drug delivery approach for cancer therapy, *Combination Drug Delivery Approach as an Effective Therapy for Various Diseases* (2022) 297–327. https://doi.org/10.1016/B978-0-323-85873-1.00008-3.

[5] D. Liu, X. Che, X. Wang, C. Ma, G. Wu, Tumor vaccines: Unleashing the power of the immune system to fight cancer, *Pharmaceuticals* 16 (2023) 1384. https://doi.org/10.3390/PH16101384.

[6] Y. Zhang, Z. Zhang, The history and advances in cancer immunotherapy: Understanding the characteristics of tumor-infiltrating immune cells and their therapeutic implications, *Cellular & Molecular Immunology* 17:8 (2020) 807–821. https://doi.org/10.1038/s41423-020-0488-6.

[7] A.T. Malacopol, P.J. Holst, Cancer vaccines: Recent insights and future directions, *International Journal of Molecular Sciences* 25 (2024) 11256. https://doi.org/10.3390/IJMS252011256.

[8] M. Saxena, S.H. van der Burg, C.J.M. Melief, N. Bhardwaj, Therapeutic cancer vaccines, *Nature Reviews Cancer* 21:6 (2021) 360–378. https://doi.org/10.1038/s41568-021-00346-0.

[9] M. Kaczmarek, J. Poznańska, F. Fechner, N. Michalska, S. Paszkowska, A. Napierała, A. Mackiewicz, Cancer vaccine therapeutics: Limitations and effectiveness – a literature review, *Cells* 12 (2023) 2159. https://doi.org/10.3390/CELLS12172159.

[10] P. Li, L. Jia, X. Bian, S. Tan, Application of engineered dendritic cell vaccines in cancer immunotherapy: Challenges and opportunities, *Current Treatment Options in Oncology* 24 (2023) 1703–1719. https://doi.org/10.1007/S11864-023-01143-7/METRICS.

[11] M. Gupta, A. Wahi, P. Sharma, R. Nagpal, N. Raina, M. Kaurav, J. Bhattacharya, S.M. Rodrigues Oliveira, K.G. Dolma, A.K. Paul, M. de Lourdes Pereira, P. Wilairatana, M. Rahmatullah, V. Nissapatorn, Recent advances in cancer vaccines: Challenges, achievements, and futuristic prospects, *Vaccines* 10 (2022) 2011. https://doi.org/10.3390/VACCINES10122011.

[12] A. Garg, P. Pandey, V. Singh, S. Ashique, Ethical considerations of cancer immunotherapy and patient perspectives on the benefits and risks of treatment, *Nanotechnology Based Strategies for Cancer Immunotherapy* (2024) 419–438. https://doi.org/10.1007/978-981-97-7022-9_16.

[13] S. Taefehshokr, A. Parhizkar, S. Hayati, M. Mousapour, A. Mahmoudpour, L. Eleid, D. Rahmanpour, S. Fattahi, H. Shabani, N. Taefehshokr, Cancer immunotherapy: Challenges and limitations, *Pathology – Research and Practice* 229 (2022) 153723. https://doi.org/10.1016/J.PRP.2021.153723.

[14] C. Feng, P. Tan, G. Nie, M. Zhu, Biomimetic and bioinspired nano-platforms for cancer vaccine development, *Exploration* 3 (2023) 20210263. https://doi.org/10.1002/EXP.20210263.

[15] M. Naseri, M. Bozorgmehr, M. Zöller, E. Ranaei Pirmardan, Z. Madjd, Tumor-derived exosomes: The next generation of promising cell-free vaccines in cancer immunotherapy, *Oncoimmunology* 9 (2020). https://doi.org/10.1080/2162402X.2020.1779991.

[16] R. Kalluri, V.S. LeBleu, The biology, function, and biomedical applications of exosomes, *Science* (1979) 367 (2020). https://doi.org/10.1126/SCIENCE.AAU6977/ASSET/0D942D3E-7BB0-410E-B762-FEC151E57E74/ASSETS/GRAPHIC/367_AAU6977_FA.JPEG.

[17] Q. Tang, S. Yang, G. He, H. Zheng, S. Zhang, J. Liu, S. Wei, Q. Fan, X. Peng, X. Li, D. Zhang, L. Yang, H. Li, Tumor-derived exosomes in the cancer immune microenvironment and cancer immunotherapy, *Cancer Letters* 548 (2022) 215823. https://doi.org/10.1016/J.CANLET.2022.215823.

[18] T. Zemanek, L. Danisovic, A. Nicodemou, Exosomes, their sources, and possible uses in cancer therapy in the era of personalized medicine, *Journal of Cancer Research and Clinical Oncology* 151:1 (2024) 1–14. https://doi.org/10.1007/S00432-024-06066-W.

[19] S. Dai, D. Wei, Z. Wu, X. Zhou, X. Wei, H. Huang, G. Li, Phase I clinical trial of autologous ascites-derived exosomes combined with GM-CSF for colorectal cancer, *Molecular Therapy* 16 (2008) 782–790. https://doi.org/10.1038/MT.2008.1.

[20] M. Lu, H. Xing, Z. Xun, T. Yang, P. Ding, C. Cai, D. Wang, X. Zhao, Exosome-based small RNA delivery: Progress and prospects, *Asian Journal of Pharmaceutical Sciences* 13 (2018) 1–11. https://doi.org/10.1016/J.AJPS.2017.07.008.

[21] P. Khairnar, T. Kolipaka, G. Pandey, V. Phatale, S. Shah, D.A. Srinivasarao, S. Saraf, S. Srivastava, Nanosponge-mediated oligonucleotide delivery: A cutting-edge technology towards cancer management, *Journal of Drug Delivery Science and Technology* (2023) 105226. https://doi.org/10.1016/J.JDDST.2023.105226.
[22] C. Giacobino, M. Canta, C. Fornaguera, S. Borrós, V. Cauda, Extracellular vesicles and their current role in cancer immunotherapy, *Cancers* 13 (2021) 2280. https://doi.org/10.3390/CANCERS13092280.
[23] L.M. Doyle, M.Z. Wang, Overview of extracellular vesicles, their origin, composition, purpose, and methods for exosome isolation and analysis, *Cells* 8 (2019) 727. https://doi.org/10.3390/CELLS8070727.
[24] K. Sidhom, P.O. Obi, A. Saleem, A review of exosomal isolation methods: Is size exclusion chromatography the best option? *International Journal of Molecular Sciences* 21 (2020) 6466. https://doi.org/10.3390/IJMS21186466.
[25] C. Rajagopal, K.B. Harikumar, The origin and functions of exosomes in cancer, *Frontiers in Oncology* 8 (2018) 313913. https://doi.org/10.3389/FONC.2018.00066/PDF.
[26] Q. Hu, H. Su, J. Li, C. Lyon, W. Tang, M. Wan, T. Ye Hu, Clinical applications of exosome membrane proteins, *Precision Clinical Medicine* 3 (2020) 54–66. https://doi.org/10.1093/PCMEDI/PBAA007.
[27] Y. Song, Y. Kim, S. Ha, S. Sheller-Miller, J. Yoo, C. Choi, C.H. Park, The emerging role of exosomes as novel therapeutics: Biology, technologies, clinical applications, and the next, *American Journal of Reproductive Immunology* 85 (2020) e13329. https://doi.org/10.1111/AJI.13329.
[28] M. Frydrychowicz, A. Kolecka-Bednarczyk, M. Madejczyk, S. Yasar, G. Dworacki, Exosomes – structure, biogenesis and biological role in non-small-cell lung cancer, *Scandinavian Journal of Immunology* 81 (2015) 2–10. https://doi.org/10.1111/SJI.12247.
[29] C. Subra, K. Laulagnier, B. Perret, M. Record, Exosome lipidomics unravels lipid sorting at the level of multivesicular bodies, *Biochimie* 89 (2007) 205–212. https://doi.org/10.1016/J.BIOCHI.2006.10.014.
[30] K. Trajkovic, C. Hsu, S. Chiantia, L. Rajendran, D. Wenzel, F. Wieland, P. Schwille, B. Brügger, M. Simons, Ceramide triggers budding of exosome vesicles into multivesicular endosomes, *Science* 319 (2008) 1244–1247. https://doi.org/10.1126/SCIENCE.1153124.
[31] C. Théry, L. Zitvogel, S. Amigorena, Exosomes: Composition, biogenesis and function, *Nature Reviews Immunology* 2:8 (2002) 569–579. https://doi.org/10.1038/nri855.
[32] J.S. Schorey, S. Bhatnagar, Exosome function: From tumor immunology to pathogen biology, *Traffic* 9 (2008) 871–881. https://doi.org/10.1111/J.1600-0854.2008.00734.X.
[33] S.P. Li, Z.X. Lin, X.Y. Jiang, X.Y. Yu, Exosomal cargo-loading and synthetic exosome-mimics as potential therapeutic tools, *Acta Pharmacologica Sinica* 39 (2018) 542. https://doi.org/10.1038/APS.2017.178.
[34] S. Samanta, S. Rajasingh, N. Drosos, Z. Zhou, B. Dawn, J. Rajasingh, Exosomes: New molecular targets of diseases, *Acta Pharmacologica Sinica* 39 (2018) 501–513. https://doi.org/10.1038/APS.2017.162.
[35] M. Ragusa, D. Barbagallo, M. Purrello, Exosomes: Nanoshuttles to the future of BioMedicine, *Cell Cycle* 14 (2015) 289–290. https://doi.org/10.1080/15384101.2015.1006535.
[36] Y. Liu, K. Shi, Y. Chen, X. Wu, Z. Chen, K. Cao, Y. Tao, X. Chen, J. Liao, J. Zhou, Exosomes and their role in cancer progression, *Frontiers in Oncology* 11 (2021) 639159. https://doi.org/10.3389/FONC.2021.639159/BIBTEX.
[37] W. Guo, T. Qiao, B. Dong, T. Li, Q. Liu, X. Xu, The effect of hypoxia-induced exosomes on anti-tumor immunity and its implication for immunotherapy, *Frontiers in Immunology* 13 (2022) 915985. https://doi.org/10.3389/FIMMU.2022.915985.
[38] B. Wang, Z. Tan, F. Guan, Tumor-derived exosomes mediate the instability of cadherins and promote tumor progression, *International Journal of Molecular Sciences* 20 (2019) 3652. https://doi.org/10.3390/IJMS20153652.
[39] Z. Xu, S. Zeng, Z. Gong, Y. Yan, Exosome-based immunotherapy: A promising approach for cancer treatment, *Molecular Cancer* 19:1 (2020) 1–16. https://doi.org/10.1186/S12943-020-01278-3.
[40] X. Li, A.L. Corbett, E. Taatizadeh, N. Tasnim, J.P. Little, C. Garnis, M. Daugaard, E. Guns, M. Hoorfar, I.T.S. Li, Challenges and opportunities in exosome research-Perspectives from biology, engineering, and cancer therapy, *APL Bioengineering* 3 (2019) 11503. https://doi.org/10.1063/1.5087122/22954.
[41] J. De Toro, L. Herschlik, C. Waldner, C. Mongini, Emerging roles of exosomes in normal and pathological conditions: New insights for diagnosis and therapeutic applications, *Frontiers in Immunology* 6 (2015) 129468. https://doi.org/10.3389/FIMMU.2015.00203/PDF.
[42] C. Théry, M. Ostrowski, E. Segura, Membrane vesicles as conveyors of immune responses, *Nature Reviews Immunology* 9:8 (2009) 581–593. https://doi.org/10.1038/nri2567.
[43] J. Gao, A. Li, J. Hu, L. Feng, L. Liu, Z. Shen, Recent developments in isolating methods for exosomes, *Frontiers in Bioengineering and Biotechnology* 10 (2023) 1100892. https://doi.org/10.3389/FBIOE.2022.1100892.
[44] P. Li, M. Kaslan, S.H. Lee, J. Yao, Z. Gao, Progress in exosome isolation techniques, *Theranostics* 7 (2017) 789. https://doi.org/10.7150/THNO.18133.
[45] I. Campoy, L. Lanau, T. Altadill, T. Sequeiros, S. Cabrera, M. Cubo-Abert, A. Pérez-Benavente, A. Garcia, S. Borrós, A. Santamaria, J. Ponce, X. Matias-Guiu, J. Reventós, A. Gil-Moreno, M. Rigau, E. Colas, Exosome-like vesicles in uterine aspirates: A comparison of ultracentrifugation-based isolation protocols, *Journal of Translational Medicine* 14 (2016) 1–12. https://doi.org/10.1186/S12967-016-0935-4/FIGURES/4.
[46] K.C. Miranda, D.T. Bond, J.Z. Levin, X. Adiconis, A. Sivachenko, C. Russ, D. Brown, C. Nusbaum, L.M. Russo, Massively parallel sequencing of human urinary exosome/microvesicle RNA reveals a predominance of non-coding RNA, *PLoS One* 9 (2014) e96094. https://doi.org/10.1371/JOURNAL.PONE.0096094.
[47] L. Zhu, H.T. Sun, S. Wang, S.L. Huang, Y. Zheng, C.Q. Wang, B.Y. Hu, W. Qin, T.T. Zou, Y. Fu, X.T. Shen, W.W. Zhu, Y. Geng, L. Lu, H. liang Jia, L.X. Qin, Q.Z. Dong, Isolation and characterization of exosomes for cancer research, *Journal of Hematology & Oncology* 13:1 (2020) 1–24. https://doi.org/10.1186/S13045-020-00987-Y.
[48] S. Sonbhadra, Mehak, L.M. Pandey, Biogenesis, isolation, and detection of exosomes and their potential in therapeutics and diagnostics, *Biosensors* 13 (2023) 802. https://doi.org/10.3390/BIOS13080802.
[49] M.L. Heinemann, M. Ilmer, L.P. Silva, D.H. Hawke, A. Recio, M.A. Vorontsova, E. Alt, J. Vykoukal, Benchtop isolation and characterization of functional exosomes by sequential filtration, *Journal of Chromatography A* 1371 (2014) 125–135. https://doi.org/10.1016/J.CHROMA.2014.10.026.

[50] A. Gámez-Valero, M. Monguió-Tortajada, L. Carreras-Planella, M. Franquesa, K. Beyer, F.E. Borràs, Size-exclusion chromatography-based isolation minimally alters extracellular vesicles' characteristics compared to precipitating agents, *Scientific Reports* 6:1 (2016) 1–9. https://doi.org/10.1038/srep33641.
[51] K.E. Petersen, E. Manangon, J.L. Hood, S.A. Wickline, D.P. Fernandez, W.P. Johnson, B.K. Gale, A review of exosome separation techniques and characterization of B16-F10 mouse melanoma exosomes with AF4-UV-MALS-DLS-TEM, *Analytical and Bioanalytical Chemistry* 406 (2014) 7855–7866. https://doi.org/10.1007/S00216-014-8040-0/METRICS.
[52] P. Amrollahi, M. Rodrigues, C.J. Lyon, A. Goel, H. Han, T.Y. Hu, Ultra-sensitive automated profiling of EpCAM expression on tumor-derived extracellular vesicles, *Frontiers in Genetics* 10 (2019). https://doi.org/10.3389/FGENE.2019.01273.
[53] G.K. Patel, M.A. Khan, H. Zubair, S.K. Srivastava, M. Khushman, S. Singh, A.P. Singh, Comparative analysis of exosome isolation methods using culture supernatant for optimum yield, purity and downstream applications, *Scientific Reports* 9 (2019) 5335. https://doi.org/10.1038/S41598-019-41800-2.
[54] C. Chen, J. Skog, C.H. Hsu, R.T. Lessard, L. Balaj, T. Wurdinger, B.S. Carter, X.O. Breakefield, M. Toner, D. Irimia, Microfluidic isolation and transcriptome analysis of serum microvesicles, *Lab on a Chip* 10 (2010) 505–511. https://doi.org/10.1039/B916199F.
[55] S. Lin, Z. Yu, D. Chen, Z. Wang, J. Miao, Q. Li, D. Zhang, J. Song, D. Cui, Progress in microfluidics-based exosome separation and detection technologies for diagnostic applications, *Small* 16 (2020). https://doi.org/10.1002/SMLL.201903916.
[56] J. Skog, T. Würdinger, S. van Rijn, D.H. Meijer, L. Gainche, W.T. Curry, B.S. Carter, A.M. Krichevsky, X.O. Breakefield, Glioblastoma microvesicles transport RNA and proteins that promote tumour growth and provide diagnostic biomarkers, *Nature Cell Biology* 10 (2008) 1470–1476. https://doi.org/10.1038/NCB1800;KWRD=LIFE+SCIENCES.
[57] D.F. Quail, J.A. Joyce, Microenvironmental regulation of tumor progression and metastasis, *Nature Medicine* 19 (2013) 1423–1437. https://doi.org/10.1038/nm.3394.
[58] F.M. Barros, F. Carneiro, J.C. Machado, S.A. Melo, Exosomes and immune response in cancer: Friends or foes? *Frontiers in Immunology* 9 (2018). https://doi.org/10.3389/fimmu.2018.00730.
[59] A. Rai, D.W. Greening, M. Chen, R. Xu, H. Ji, R.J. Simpson, Exosomes derived from human primary and metastatic colorectal cancer cells contribute to functional heterogeneity of activated fibroblasts by reprogramming their proteome, *Proteomics* 19 (2019). https://doi.org/10.1002/pmic.201800148.
[60] S. Sergazy, R. Seydahmetova, A. Gulyayev, Z. Shulgau, M. Aljofan, The role of exosomes in cancer progression and therapy, *Biology (Basel)* 14 (2025) 27. https://doi.org/10.3390/biology14010027.
[61] T.L. Whiteside, Tumor-derived exosomes and their role in cancer progression, in: 2016: pp. 103–141. https://doi.org/10.1016/bs.acc.2015.12.005.
[62] M.H. Yang, J.H. Lee, J.-H. Ko, S.H. Jung, G. Sethi, K.S. Ahn, Brassinin represses invasive potential of lung carcinoma cells through deactivation of PI3K/Akt/mTOR signaling cascade, *Molecules* 24 (2019) 1584. https://doi.org/10.3390/molecules24081584.
[63] J.-H. Ko, D. Nam, J.-Y. Um, S. Jung, G. Sethi, K. Ahn, Bergamottin suppresses metastasis of lung cancer cells through abrogation of diverse oncogenic signaling cascades and epithelial-to-mesenchymal transition, *Molecules* 23 (2018) 1601. https://doi.org/10.3390/molecules23071601.
[64] X. Dai, K.S. Ahn, L.Z. Wang, C. Kim, A. Deivasigamni, F. Arfuso, J.-Y. Um, A.P. Kumar, Y.-C. Chang, D. Kumar, G.C. Kundu, J. Magae, B.C. Goh, K.M. Hui, G. Sethi, Ascochlorin enhances the sensitivity of doxorubicin leading to the reversal of epithelial-to-mesenchymal transition in hepatocellular carcinoma, *Molecular Cancer Therapeutics* 15 (2016) 2966–2976. https://doi.org/10.1158/1535-7163.MCT-16-0391.
[65] M.S. Ong, S. Deng, C.E. Halim, W. Cai, T.Z. Tan, R.Y.-J. Huang, G. Sethi, S.C. Hooi, A.P. Kumar, C.T. Yap, Cytoskeletal proteins in cancer and intracellular stress: A therapeutic perspective, *Cancers (Basel)* 12 (2020) 238. https://doi.org/10.3390/cancers12010238.
[66] J.E. Park, H. Sen Tan, A. Datta, R.C. Lai, H. Zhang, W. Meng, S.K. Lim, S.K. Sze, Hypoxic tumor cell modulates its microenvironment to enhance angiogenic and metastatic potential by secretion of proteins and exosomes, *Molecular & Cellular Proteomics* 9 (2010) 1085–1099. https://doi.org/10.1074/mcp.M900381-MCP200.
[67] H.W. King, M.Z. Michael, J.M. Gleadle, Hypoxic enhancement of exosome release by breast cancer cells, *BMC Cancer* 12 (2012) 421. https://doi.org/10.1186/1471-2407-12-421.
[68] T. Umezu, H. Tadokoro, K. Azuma, S. Yoshizawa, K. Ohyashiki, J.H. Ohyashiki, Exosomal miR-135b shed from hypoxic multiple myeloma cells enhances angiogenesis by targeting factor-inhibiting HIF-1, *Blood* 124 (2014) 3748–3757. https://doi.org/10.1182/blood-2014-05-576116.
[69] H. Wang, L. Wang, X. Zhou, X. Luo, K. Liu, E. Jiang, Y. Chen, Z. Shao, Z. Shang, OSCC exosomes regulate miR-210-3p targeting EFNA3 to promote oral cancer angiogenesis through the PI3K/AKT pathway, *Biomed Research International* 2020 (2020). https://doi.org/10.1155/2020/2125656.
[70] C. Yang, S.-H. Kim, N.R. Bianco, P.D. Robbins, Tumor-derived exosomes confer antigen-specific immunosuppression in a murine delayed-type hypersensitivity model, *PLoS One* 6 (2011) e22517. https://doi.org/10.1371/journal.pone.0022517.
[71] H.-G. Zhang, W.E. Grizzle, Exosomes: A novel pathway of local and distant intercellular communication that facilitates the growth and metastasis of neoplastic lesions, *The American Journal of Pathology* 184 (2014) 28–41. https://doi.org/10.1016/j.ajpath.2013.09.027.
[72] R. Gaballa, H.E.A. Ali, M.O. Mahmoud, J.S. Rhim, H.I. Ali, H.F. Salem, M. Saleem, M.A. Kandeil, S. Ambs, Z.Y. Abd Elmageed, Exosomes-mediated transfer of Itga2 promotes migration and invasion of prostate cancer cells by inducing epithelial-mesenchymal transition, *Cancers (Basel)* 12 (2020) 2300. https://doi.org/10.3390/cancers12082300.
[73] T.L. Whiteside, Exosomes in cancer: Another mechanism of tumor-induced immune suppression, in: 2017: pp. 81–89. https://doi.org/10.1007/978-3-319-67577-0_6.
[74] T.L. Whiteside, Exosomes carrying immunoinhibitory proteins and their role in cancer, *Clinical and Experimental Immunology* 189 (2017) 259–267. https://doi.org/10.1111/cei.12974.

[75] E.U. Wieckowski, C. Visus, M. Szajnik, M.J. Szczepanski, W.J. Storkus, T.L. Whiteside, Tumor-derived microvesicles promote regulatory T cell expansion and induce apoptosis in tumor-reactive activated CD8+ T lymphocytes, *The Journal of Immunology* 183 (2009) 3720–3730. https://doi.org/10.4049/jimmunol.0900970.

[76] C.L. Montes, A.I. Chapoval, J. Nelson, V. Orhue, X. Zhang, D.H. Schulze, S.E. Strome, B.R. Gastman, Tumor-induced senescent T cells with suppressor function: A potential form of tumor immune evasion, *Cancer Research* 68 (2008) 870–879. https://doi.org/10.1158/0008-5472.CAN-07-2282.

[77] Y. Zhang, L.W. Pfannenstiel, E. Bolesta, C.L. Montes, X. Zhang, A.I. Chapoval, R.B. Gartenhaus, S.E. Strome, B.R. Gastman, Interleukin-7 inhibits tumor-induced CD27–CD28– suppressor T cells: Implications for cancer immunotherapy, *Clinical Cancer Research* 17 (2011) 4975–4986. https://doi.org/10.1158/1078-0432.CCR-10-3328.

[78] M. Morikawa, R. Derynck, K. Miyazono, TGF-β and the TGF-β family: Context-dependent roles in cell and tissue physiology, *Cold Spring Harbor Perspectives in Biology* 8 (2016) a021873. https://doi.org/10.1101/cshperspect.a021873.

[79] J.-C. Neel, L. Humbert, J.-J. Lebrun, The dual role of TGFβ in human cancer: From tumor suppression to cancer metastasis, *ISRN Molecular Biology* 2012 (2012) 1–28. https://doi.org/10.5402/2012/381428.

[80] A. Thakur, D.C. Parra, P. Motallebnejad, M. Brocchi, H.J. Chen, Exosomes: Small vesicles with big roles in cancer, vaccine development, and therapeutics, *Bioactive Materials* 10 (2022) 281–294. https://doi.org/10.1016/J.BIOACTMAT.2021.08.029.

[81] N. Dilsiz, Exosomes as new generation vehicles for drug delivery systems, *Journal of Drug Delivery Science and Technology* 95 (2024) 105562. https://doi.org/10.1016/J.JDDST.2024.105562.

[82] H. Fu, Y. Chen, Q. Fu, Q. Lv, J. Zhang, Y. Yang, P. Tan, X. Wang, Y. Yang, Z. Wu, From conventional to cutting-edge: Exosomes revolutionizing nano-drug delivery systems, *Chemical Engineering Journal* 500 (2024) 156685. https://doi.org/10.1016/J.CEJ.2024.156685.

[83] L. Nie, J. Ma, Y. Yu, Y. Tao, Z. Song, J. Li, Exosomes as carriers to stimulate an anti-cancer immune response in immunotherapy and as predictive markers (2024). https://doi.org/10.1016/j.bcp.2024.116699.

[84] L. Cunha Silva, F. Branco, J. Cunha, C. Vitorino, C. Gomes, M.A. Carrascal, A. Falcão, B. Miguel Neves, M. Teresa Cruz, The potential of exosomes as a new therapeutic strategy for glioblastoma, *European Journal of Pharmaceutics and Biopharmaceutics* 203 (2024) 114460. https://doi.org/10.1016/J.EJPB.2024.114460.

[85] P.M. Pordanjani, A. Bolhassani, M.H. Pouriayevali, A. Milani, F. Rezaei, Engineered dendritic cells-derived exosomes harboring HIV-1 Nefmut-Tat fusion protein and heat shock protein 70: A promising HIV-1 safe vaccine candidate, *International Journal of Biological Macromolecules* 270 (2024) 132236. https://doi.org/10.1016/J.IJBIOMAC.2024.132236.

[86] F. Ghorbaninezhad, H. Alemohammad, B. Najafzadeh, J. Masoumi, M.A. Shadbad, M. Shahpouri, H. Saeedi, O. Rahbarfarzam, B. Baradaran, Dendritic cell-derived exosomes: A new horizon in personalized cancer immunotherapy? *Cancer Letters* 562 (2023) 216168. https://doi.org/10.1016/J.CANLET.2023.216168.

[87] J. Li, J. Li, Y. Peng, Y. Du, Z. Yang, X. Qi, Dendritic cell derived exosomes loaded neoantigens for personalized cancer immunotherapies, *Journal of Controlled Release* 353 (2023) 423–433. https://doi.org/10.1016/j.jconrel.2022.11.053.

[88] W. Zhou, X. Chen, Y. Zhou, S. Shi, C. Liang, X. Yu, H. Chen, Q. Guo, Y. Zhang, P. Liu, C. Li, Y. Chu, Y. Luo, Y. Wang, Z. Zhou, Z. Zhao, Q. Chen, T. Sun, C. Jiang, Exosomes derived from immunogenically dying tumor cells as a versatile tool for vaccination against pancreatic cancer, *Biomaterials* 280 (2022) 121306. https://doi.org/10.1016/j.biomaterials.2021.121306.

[89] D. Zhang, Y. Jiang, M. Wang, J. Zhao, J. Wan, Z. Li, D. Huang, J. Yu, J. Li, J. Liu, F. Huang, S. Hao, A novel costimulatory molecule gene-modified leukemia cell-derived exosome enhances the anti-leukemia efficacy of DC vaccine in mouse models, *Vaccine* 42 (2024) 126097. https://doi.org/10.1016/j.vaccine.2024.06.064.

[90] S. Madhan, R. Dhar, A. Devi, Plant-derived exosomes: A green approach for cancer drug delivery, *Journal of Materials Chemistry B* 12 (2024) 2236–2252. https://doi.org/10.1039/D3TB02752J.

[91] K. Kalele, S. Nyahatkar, D. Mirgh, R. Muthuswamy, M.D. Adhikari, K. Anand, Exosomes: A cutting-edge theranostics tool for oral cancer, *ACS Applied Bio Materials* 7 (2024) 1400–1415. https://doi.org/10.1021/acsabm.3c01243.

[92] C.A. Fitts, N. Ji, Y. Li, C. Tan, Exploiting exosomes in cancer liquid biopsies and drug delivery, *Advanced Healthcare Materials* 8 (2019) 1801268. https://doi.org/10.1002/ADHM.201801268.

[93] I. Kimiz-Gebologlu, S.S. Oncel, Exosomes: Large-scale production, isolation, drug loading efficiency, and biodistribution and uptake, *Journal of Controlled Release* 347 (2022) 533–543. https://doi.org/10.1016/j.jconrel.2022.05.027.

[94] G. Tuluwengjiang, I. Rasulova, S. Ahmed, B.A. Kiasari, I. Sârbu, C.I. Ciongradi, T.M. Omar, F. Hussain, M.J. Jawad, R.Y. Castillo-Acobo, T. Hani, N. Lakshmaiya, S.S.C. Samaniego, Dendritic cell-derived exosomes (Dex): Underlying the role of exosomes derived from diverse DC subtypes in cancer pathogenesis, *Pathology – Research and Practice* 254 (2024) 155097. https://doi.org/10.1016/J.PRP.2024.155097.

[95] G. Sareen, M. Mohan, A. Mannan, K. Dua, T.G. Singh, A new era of cancer immunotherapy: Vaccines and miRNAs, *Cancer Immunology, Immunotherapy* 74:5 (2025) 1–28. https://doi.org/10.1007/S00262-025-04011-5.

[96] M. Negahdaripour, B. Vakili, N. Nezafat, Exosome-based vaccines and their position in next generation vaccines, *International Immunopharmacology* 113 (2022) 109265. https://doi.org/10.1016/J.INTIMP.2022.109265.

[97] G. Zhao, Y. Wang, S. Xing, Y. Jiang, J. Ding, Y. Cai, P. Ma, H. Miao, Y. Fang, N. Jiang, D. Cui, Y. Yu, Q. Tang, S. Wang, N. Li, Exosome-based anticancer vaccines: From bench to bedside, *Cancer Letters* 595 (2024) 216989. https://doi.org/10.1016/J.CANLET.2024.216989.

[98] C. Wang, X. Huang, Y. Wu, J. Wang, F. Li, G. Guo, Tumor cell-associated exosomes robustly elicit anti-tumor immune responses through modulating dendritic cell vaccines in lung tumor, *International Journal of Biological Sciences* 2020 (2020) 633–643. https://doi.org/10.7150/ijbs.38414.

[99] L. Zhang, D. Yu, Exosomes in cancer development, metastasis, and immunity, *Biochimica et Biophysica Acta (BBA) –*

Reviews on Cancer 1871 (2019) 455–468. https://doi.org/10.1016/J.BBCAN.2019.04.004.
[100] Y.S. Lee, S.H. Kim, J.A. Cho, C.W. Kim, Introduction of the CIITA gene into tumor cells produces exosomes with enhanced anti-tumor effects, *Experimental & Molecular Medicine* 43 (2011) 281–290. https://doi.org/10.3858/emm.2011.43.5.029.
[101] Y. Xie, O. Bai, H. Zhang, J. Yuan, S. Zong, R. Chibbar, K. Slattery, M. Qureshi, Y. Wei, Y. Deng, J. Xiang, Membrane-bound HSP70-engineered myeloma cell-derived exosomes stimulate more efficient CD8+ CTL- and NK-mediated antitumour immunity than exosomes released from heat-shocked tumour cells expressing cytoplasmic HSP70, *Journal of Cellular and Molecular Medicine* 14 (2010) 2655–2666. https://doi.org/10.1111/J.1582-4934.2009.00851.X.
[102] A. Del Prete, V. Salvi, A. Soriani, M. Laffranchi, F. Sozio, D. Bosisio, S. Sozzani, Dendritic cell subsets in cancer immunity and tumor antigen sensing, *Cellular & Molecular Immunology* 20:5 (2023) 432–447. https://doi.org/10.1038/s41423-023-00990-6.
[103] H. Zhu, K. Wang, Z. Wang, D. Wang, X. Yin, Y. Liu, F. Yu, W. Zhao, An efficient and safe MUC1-dendritic cell-derived exosome conjugate vaccine elicits potent cellular and humoral immunity and tumor inhibition in vivo, *Acta Biomaterialia* 138 (2022) 491–504. https://doi.org/10.1016/j.actbio.2021.10.041.
[104] S. Viaud, M. Terme, C. Flament, J. Taieb, F. André, S. Novault, B. Escudier, C. Robert, S. Caillat-Zucman, T. Tursz, L. Zitvogel, N. Chaput, Dendritic cell-derived exosomes promote natural killer cell activation and proliferation: A role for NKG2D ligands and IL-15Rα, *PLoS One* 4 (2009) e4942. https://doi.org/10.1371/JOURNAL.PONE.0004942.
[105] J. Cui, X. Wang, J. Li, A. Zhu, Y. Du, W. Zeng, Y. Guo, L. Di, R. Wang, Immune exosomes loading self-assembled nanomicelles traverse the blood-brain barrier for chemo-immunotherapy against glioblastoma, *ACS Nano* 17 (2023) 1464–1484. https://doi.org/10.1021/ACSNANO.2C10219/ASSET/IMAGES/LARGE/NN2C10219_0008.JPEG.
[106] N. Behranvand, F. Nasri, R. Zolfaghari Emameh, P. Khani, A. Hosseini, J. Garssen, R. Falak, Chemotherapy: A double-edged sword in cancer treatment, *Cancer Immunology, Immunotherapy* 71:3 (2021) 507–526. https://doi.org/10.1007/S00262-021-03013-3.
[107] M.F.S. Lindenbergh, R. Wubbolts, E.G.F. Borg, E.M. van 'T Veld, M. Boes, W. Stoorvogel, Dendritic cells release exosomes together with phagocytosed pathogen; potential implications for the role of exosomes in antigen presentation, *Journal of Extracellar Vesicles* 9 (2020) 1798606. https://doi.org/10.1080/20013078.2020.1798606.
[108] L. Xiao, U. Erb, K. Zhao, T. Hackert, M. Zöller, Efficacy of vaccination with tumor-exosome loaded dendritic cells combined with cytotoxic drug treatment in pancreatic cancer, *Oncoimmunology* 6 (2017). https://doi.org/10.1080/2162402X.2017.1319044.
[109] A. Hedaoo, P. Khairnar, G. Vambhurkar, D.A. Srinivasarao, M. Negi, A. Shinde, S. Srivastava, Unveiling the potential of inorganic nanoarchitecture-mediated photothermal therapy: An illustration on melanoma, *European Polymer Journal* 216 (2024) 113282. https://doi.org/10.1016/J.EURPOLYMJ.2024.113282.
[110] Y. Cai, T. Chai, W. Nguyen, J. Liu, E. Xiao, X. Ran, Y. Ran, D. Du, W. Chen, X. Chen, Phototherapy in cancer treatment: Strategies and challenges, *Signal Transduction and Targeted Therapy* 10:1 (2025) 1–46. https://doi.org/10.1038/s41392-025-02140-y.
[111] Y. Gao, J. Liu, M. Wu, Y. Zhang, M. Wang, Q. Lyu, W. Zhang, Y. Zhou, Y.C. Cheuk, X. Wang, Y. Liu, W. Wang, W. Tu, Photosensitive hybrid γδ-T exosomes for targeted cancer photoimmunotherapy, *ACS Nano* 19 (2025) 4251–4268. https://doi.org/10.1021/ACSNANO.4C11024/ASSET/IMAGES/LARGE/NN4C11024_0008.JPEG.
[112] M. Wang, J. Rao, M. Wang, X. Li, K. Liu, M.F. Naylor, R.E. Nordquist, W.R. Chen, F. Zhou, Cancer photo-immunotherapy: From bench to bedside, *Theranostics* 11 (2021) 2218–2231. https://doi.org/10.7150/THNO.53056.
[113] Q. Liu, T. Fan, Y. Zheng, S.L. Yang, Z. Yu, Y. Duo, Y. Zhang, D. Adah, L. Shi, Z. Sun, D. Wang, J. Xie, H. Wu, Z. Wu, C. Ge, L. Qiao, C. Wei, L. Huang, Q. Yan, Q. Yang, S. Bao, L.P. Liu, H. Zhang, Immunogenic exosome-encapsulated black phosphorus nanoparticles as an effective anticancer photo-nanovaccine, *Nanoscale* 12 (2020) 19939–19952. https://doi.org/10.1039/D0NR05953F.
[114] S. Meng, A.G. Whitt, B.F. Stamp, J.W. Eaton, C. Li, K. Yaddanapudi, Exosome-based cancer vaccine for prevention of lung cancer, *Stem Cell Investigation* 10 (2023). https://doi.org/10.21037/SCI-2022-030/COIF.
[115] J.M. Diamond, C. Vanpouille-Box, S. Spada, N.P. Rudqvist, J.R. Chapman, B.M. Ueberheide, K.A. Pilones, Y. Sarfraz, S.C. Formenti, S. Demaria, Exosomes shuttle TREX1-sensitive IFN-stimulatory dsDNA from irradiated cancer cells to DCs, *Cancer Immunology Research* 6 (2018) 910–920. https://doi.org/10.1158/2326-6066.CIR-17-0581/470561/AM/EXOSOMES-SHUTTLE-TREX1-SENSITIVE-IFN-STIMULATORY.
[116] W.S. Kim, D. Choi, J.M. Park, H.Y. Song, H.S. Seo, D.E. Lee, E.B. Byun, Comparison of exosomes derived from non- and gamma-irradiated melanoma cancer cells as a potential antigenic and immunogenic source for dendritic cell-based immunotherapeutic vaccine, *Vaccines* 8 (2020) 699. https://doi.org/10.3390/VACCINES8040699.
[117] J. Taieb, N. Chaput, N. Schartz, S. Roux, S. Novault, C. Ménard, F. Ghiringhelli, M. Terme, A.F. Carpentier, G. Darrasse-Jèse, F. Lemonnier, L. Zitvogel, Chemoimmunotherapy of tumors: Cyclophosphamide synergizes with exosome-based vaccines, *The Journal of Immunology* 176 (2006) 2722–2729. https://doi.org/10.4049/JIMMUNOL.176.5.2722.
[118] M.K. Khan, K.K. Jella, T. Nasti, Z. Li, D.H. Lawson, R. Ahmed, W. Dynan, Identification of radiation induced exosomes as a potential in situ vaccine for melanoma growth delay, *Journal of Clinical Oncology* 36 (2018) e21535–e21535. https://doi.org/10.1200/JCO.2018.36.15_SUPPL.E21535.
[119] Y. Jang, H. Kim, S. Yoon, H. Lee, J. Hwang, J. Jung, J.H. Chang, J. Choi, H. Kim, Exosome-based photoacoustic imaging guided photodynamic and immunotherapy for the treatment of pancreatic cancer, *Journal of Controlled Release* 330 (2021) 293–304. https://doi.org/10.1016/J.JCONREL.2020.12.039.
[120] W. Lin, Y. Xu, X. Chen, J. Liu, Y. Weng, Q. Zhuang, F. Lin, Z. Huang, S. Wu, J. Ding, L. Chen, X. Qiu, L. Zhang, J. Wu, D. Lin, S. Qiu, Radiation-induced small extracellular vesicles as "carriages" promote tumor antigen release and trigger antitumor immunity, *Theranostics* 10 (2020) 4871–4884. https://doi.org/10.7150/THNO.43539.
[121] S. Shi, Q. Rao, C. Zhang, X. Zhang, Y. Qin, Z. Niu, Dendritic cells pulsed with exosomes in combination with PD-1 antibody increase the efficacy of sorafenib in hepatocellular

carcinoma model, *Translational Oncology* 11 (2018) 250–258. https://doi.org/10.1016/J.TRANON.2018.01.001.

[122] B. Besse, M. Charrier, V. Lapierre, E. Dansin, O. Lantz, D. Planchard, T. Le Chevalier, A. Livartoski, F. Barlesi, A. Laplanche, S. Ploix, N. Vimond, I. Peguillet, C. Théry, L. Lacroix, I. Zoernig, K. Dhodapkar, M. Dhodapkar, S. Viaud, J.C. Soria, K.S. Reiners, E. Pogge von Strandmann, F. Vély, S. Rusakiewicz, A. Eggermont, J.M. Pitt, L. Zitvogel, N. Chaput, Dendritic cell-derived exosomes as maintenance immunotherapy after first line chemotherapy in NSCLC, *Oncoimmunology* 5 (2016). https://doi.org/10.1080/2162402X.2015.1071008.

[123] Vaccines, Blood & Biologics | FDA, (n.d.). https://www.fda.gov/vaccines-blood-biologics (accessed April 2, 2025).

[124] Advanced therapy medicinal products: Overview | European Medicines Agency (EMA), (n.d.). https://www.ema.europa.eu/en/human-regulatory-overview/advanced-therapy-medicinal-products-overview (accessed April 2, 2025).

[125] J. Rezaie, M. Feghhi, T. Etemadi, A review on exosomes application in clinical trials: Perspective, questions, and challenges, *Cell Communication and Signaling* 20 (2022) 1–13. https://doi.org/10.1186/S12964-022-00959-4/TABLES/7.

[126] Z. Lu, B. Zuo, R. Jing, X. Gao, Q. Rao, Z. Liu, H. Qi, H. Guo, H.F. Yin, Dendritic cell-derived exosomes elicit tumor regression in autochthonous hepatocellular carcinoma mouse models, *Journal of Hepatology* 67 (2017) 739–748. https://doi.org/10.1016/J.JHEP.2017.05.019.

[127] P. Santos, F. Almeida, Exosome-based vaccines: History, current state, and clinical trials, *Frontiers in Immunology* 12 (2021) 711565. https://doi.org/10.3389/FIMMU.2021.711565/PDF.

[128] L. Debbi, S. Guo, D. Safina, S. Levenberg, Boosting extracellular vesicle secretion, *Biotechnology Advances* 59 (2022) 107983. https://doi.org/10.1016/J.BIOTECHADV.2022.107983.

9 Exosome-Based Therapeutic Insights in Metabolic Disorders
Exosome Therapies

Harish Bhardwaj[1] *and Rajendra Kumar Jangde*[2]
[1]Faculty of Pharmacy, Shri Rawatpura Sarkar University, Raipur, Chhattisgarh, India
[2]University Institute of Pharmacy, Pt. Ravishankar Shukla University, Raipur, Chhattisgarh, India

9.1 INTRODUCTION

Metabolic syndrome (MetS) is a multifaceted disorder defined by insulin resistance, central obesity, dyslipidemia, hypertension, hyperglycemia, type 2 diabetes, heart disease, stroke, and chronic renal disease, and many cancers are made more likely by this. Epidemics of metabolic disorders in the medical community have primarily been ascribed to genetic predisposition and alterations in nutrition, physical activity, and age. Nonetheless, substantial data currently indicate that several environmental factors may play a role in the swift rise of obesity, type 2 diabetes, and other components of metabolic syndrome observed in the last 30 years. An environmental factor garnering attention is a category of compounds that can disrupt hormonal activity, including metabolic hormones. These substances, referred to as endocrine-disrupting chemicals (EDCs), are present in numerous consumer items, and human exposure to them is widespread. Particularly concerning is the evidence that exposure to EDCs during crucial stages of adipocyte differentiation and organ development, including the pancreas, liver, brain, and others, can lead to impacts that emerge later in life, often as clinical diseases [1]. Metabolic disorders are becoming more common across genetically diverse populations worldwide, creating an escalating cost burden on public health systems in both emerging nations in Africa and Southeast Asia, as well as in industrialized countries such as Japan and Australia. Metabolic diseases are conditions characterized by aberrant metabolism, which refers to the biochemical processes within cells that convert food into energy and other essential substances [2]. At present, the discussion encompasses alterations in the prevalence of diabetes mellitus, obesity, nonalcoholic fatty liver disease, and their related hyperlipidemia, alongside the roles of genetics and the endocrine system in these metabolic pathologies. The worldwide obesity epidemic impacts infants, adolescents, and adults. For the first time globally, the population of obese and overweight individuals surpasses that of underweight individuals. Diabetes mellitus may arise from a decline in function and/or a reduction in pancreatic tissue mass. The liver is the principal organ engaged in lipid metabolism. Nonalcoholic fatty liver disease (NAFLD), marked by excessive lipid buildup in hepatocytes (steatosis), is considered by some to be the hepatic manifestation of metabolic syndrome (MetS) and obesity. MetS is ultimately characterized as a constellation of the most perilous risk factors for myocardial infarction, encompassing diabetes, prediabetes, abdominal obesity, hyperlipidemia, and hypertension [3]. The molecular pathways underlying disease have prompted researchers to explore genomic, transcriptomic, metabolomic, environmental, nutritional, and microbial variables that may contribute to the onset of muscular dystrophy (MD). Molecular methodologies encompass genetic screening for the early identification and diagnosis of genes associated with MDs in neonates, as well as the establishment of protocols for genetic screening of specific MDs. The impact of epigenetics on gene regulation, transcriptome function (including RNA transcript variations in MD), and the involvement of regulatory RNA all have been investigated in recent studies [4]. Analyzing global trends has shown an increase in research on this noncommunicable disease (NCD), diabetes mellitus (DM), the most common metabolic ailment recognized globally. Over the next 30 years, the number of people diagnosed with diabetes is expected to increase steadily from the current 463 million. This calls for a range of interventions, such as changes in lifestyle, control of gut microflora, stem cell transplantation, and genome editing using CRISPR-based technologies. Evidence for causative changes and genes in approximately 250 genomic regions has been linked to type 2 diabetes predisposition in genome-wide association studies [5].

Extracellular vesicles (EVs), including exosomes, are secreted by all live cells and encompass a variety of bioactive substances, including nucleic acids, proteins, lipids, and metabolites. Increasing evidence of EV-related capabilities has demonstrated that these small vesicles can facilitate

DOI: 10.1201/9781003623632-9

specialized intercellular communication [6]. Exosomes are spherical bilayer vesicles produced by mammalian cells via a series of regulatory mechanisms. These entities are typically found in cell culture supernatants, plasma, saliva, urine, amniotic fluid, malignant ascites, and various other biological fluids, containing a high concentration of biologically active substances, including proteins, lipids, nucleic acids, and metabolites. Exosomes serve as crucial mediators of cell-to-cell communication by transporting bioactive substances, which significantly influence the progression of various diseases, including multiple sclerosis, cardiovascular diseases, and neurological disorders [7]. Recent studies have thoroughly investigated exosomes due to their critical biochemical properties, positioning them as essential tools for disease diagnosis and drug delivery, owing to their biocompatibility, stability, and safety. The lack of efficient diagnostic techniques and treatment approaches has made the disease a significant public health concern, negatively impacting human health and overall quality of life [8]. Exosomes, as effective agents of intercellular communication, have attracted considerable interest due to their distinctive cargo-carrying capacities, enabling them to influence various physiological and pathological processes.

A comprehensive study has clarified the synthesis, secretion, and functions of exosomes. These vesicles are released by cells in multiple states, performing either protective or detrimental biological tasks [9]. The importance of extracellular vesicles in normal and pathological interorgan communication has made them a focal point for scientific inquiry. Crucial for cellular signaling and recognition processes are exosomes, which are extracellular vesicles composed of proteins and phospholipids and measuring 30–200 nm in size. They make up the cargo molecules, with microRNAs (miRNAs) playing a key role. The uptake process determines whether exosomes influence target tissues via cargo molecule transfer or interactions between target cell plasma membrane receptors and exosomal membrane proteins. Exosomes play a crucial role in the endocrine process of chemical translocation across tissues, as they can be captured and ingested by distant tissues through fusion with the plasma membrane. Type 2 diabetes mellitus, obesity, cardiovascular disease, and other metabolic illnesses highlight the growing importance of understanding the processes of exosome-mediated signaling. Particularly noteworthy are the relationships between miRNA-containing exosomes and metabolically active organs, such as skeletal muscle, adipose tissue, the small intestine, the liver, and pancreatic beta cells. Crucial for metabolic regulation, pancreatic beta cells primarily respond to glucose concentrations by secreting insulin. Insulin controls metabolic processes by binding to specific organ receptors upon secretion; these organs include skeletal muscle and the liver [10].

Nonetheless, exomes that target beta cells enhance putative pathways analogous to insulin signaling; evidence suggests the existence of more extensive systems that may also involve other metabolically active tissues. Appropriate nutritional therapy, centered on energy equilibrium and the incorporation of bioactive chemicals that enhance physiological function, may alter the exosomal secretion profile in metabolically active tissues. These methods may serve as a foundation for creating novel therapeutics by modifying metabolic regulatory proteins or enhancing controlled release systems based on their biological responses [11]. Immune cells send out exosomes that control immune reactions by sending signaling molecules to specific cells. Depending on the situation, these exchanges can either make immune responses stronger or weaker [12, 13]. Exosomes play a pivotal role in neuronal signaling because they change the way synapses work and how neurons send messages. They carry signaling molecules, growth factors, and synaptic proteins, which can be directed to certain parts of the brain. Neuronal exosome-mediated communication is studied to learn more about learning, memory, and neurodegenerative illnesses. Exosomes derived from stem cells have been shown to aid in tissue healing and reduce inflammation. Understanding the signals caused by the molecules carried by exosomes opens the door to new ways to promote cell regeneration and healing without the need for direct transplantation. Exosomes have become crucial for cells to communicate with each other and hold considerable promise as biomarkers for various diseases and conditions. Their unique cargo makes them very useful for diagnostics and personalized treatment because it shows how the cells that produced them are functioning physiologically [14]. Studying exosomes also helps us understand fundamental biological processes, which enables us to understand how genes function, how diseases progress, how the immune system operates, how neurons communicate with each other, and how to generate new cells [15]. With the advancement of research methodologies, exosomes hold great promise for reshaping our understanding of cellular communication and the development of novel therapeutic approaches [15].

9.2 BIOGENESIS AND CHARACTERISTICS OF EXOSOMES

The generation of exosomes is a meticulously regulated process that involves four primary stages: initiation, endocytosis, development of multivesicular bodies (MVBs), and the secretion of exosomes. The cell membrane invaginates to create early endosomes, which contain accumulating intraluminal vesicles (ILV) within their lumen. Subsequently, the maturation of early endosomes results in the creation of multivesicular bodies (MVBs). Typically, MVBs are eliminated through fusion with lysosomes; however, certain MVBs merge with the plasma membrane to release exosomes into the extracellular environment. At present, the most comprehensive mechanisms elucidating the formation of ILVs and MVBs are the ESCRT-independent and ESCRT-dependent routes. During endosomal maturation, the late endosomal membrane invaginates to form ILVs and MVBs. MVBs can fuse with lysosomes or plasma membranes to exocytose ILVs as exosomes. The endosomal sorting complex needed for transport (ESCRT) produces primary exosomes. Multivesicular bodies (MVBs) are formed, cargo is sorted, MVBs are transported, and they fuse with the plasma membrane during exosome biogenesis. ESCRT is typically required to form MVBs. The main component of MVB/ILV synthesis is ESCRT, a multiprotein complex that has been extensively studied [14]. Class E vacuolar protein sorting (VPS) proteins are involved in transporting and sorting proteins to the vacuole/lysosome for degradation. They play key roles in vesicle formation, cargo recognition, and membrane trafficking. To produce ESCRT-0, ESCRT-I, ESCRT-II, and ESCRT-III multimeric protein core complexes. These complexes interact with auxiliary proteins, including TSG101, ALIX, and VPS4. ESCRT-0 interacts with phosphatidylinositol-3-phosphate to trap ubiquitinated protein cargo in the endosomal limiting membrane and its clathrin-coated microdomain, which faces the cytoplasm. ESCRT-I produces a protein complex that captures ubiquitinated cargo with TSG101 and ESCRT-II. The ESCRT-III components form spiral oligomers that restrict endosomal inward budding and cargo distribution on the limiting membrane. ESCRT-III subunits attaching to ESCRT-II cause successive polymerization, membrane deformation, and fission of ILVs and MVBs. VPS4-facilitated ATP hydrolysis disassembles the ESCRT complex, causing subunit turnover. Additional pathways that use ESCRT recruit a specific complex via auxiliary components. ESCRT-III and VPS4 recruitment by ALIX and HD-PTP boosts ILV development. The removal of ESCRT-related subunits may still permit ILV formation, suggesting alternative regulatory mechanisms. To promote MVB budding and ILV formation, the nSMase2-ceramide-dependent pathway self-assembles raft-like structures. Exosomes contain biomarkers from the conserved transmembrane protein family CD9, CD63, CD81, and CD82. Tetraspanin-enriched microdomains may regulate ILV formation alongside other proteins. Routes that use or do not use ESCRT can create distinct ILV populations in the same MVB. MVB formation and freight sorting often occur together [16]. Hence, we discuss cargo sorting to exosomes later, which is a complex, controlled, and selective process. Multivesicular bodies (MVBs) merge with the plasma membrane and release exosomes without lysosomal degradation after cargo sorting. Rab7 and ISG15 proteins fuse MVBs with lysosomes to inhibit exosome release [17]. Secreting exosomes are carried to the plasma membrane by actin and microtubules. The fusion of MVBs with the plasma membrane is assisted by Ras-associated binding protein (RAB) GTPases and soluble *N*-ethylmaleimide-sensitive factor activating protein receptor (SNARE). Research suggests that exosomes may have a half-life of 30 minutes or less, which is necessary for biogenesis and absorption. Phagocytosis or direct interaction with phagocytic cells degrades exosome cargo [18]. Other recipient cells mainly accept exosomes via endocytosis. After entering the endosome, exosomes must exit to prevent cargo destruction and transfer functioning cargo into cells. When they combine with the plasma membrane, certain exosomes release their cargo into the cytoplasm. Functional transport of cargo from exosomes affects gene expression and cellular signaling networks. Abbreviations include intraluminal vesicles (ILVs), MHC-IIs, mRNA, miRNA, multivesicular bodies (MVBs), Ras-associated binding proteins (RABs), and sEVs [19]. Figure 9.1 depicts exosome production, cargo, and uptake.

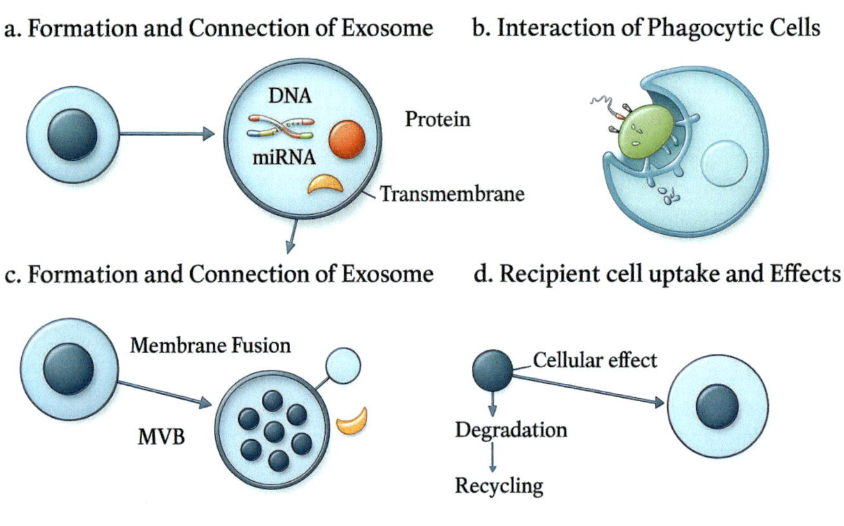

FIGURE 9.1 Mechanisms of exosome synthesis, cargo, and uptake. Ectosomes and exosomes (sEVs) have different biogenic routes. Exosomes form via the endocytic route, while ectosomes directly bud from the plasma membrane. Exosome biogenesis produces late endosomes that bud inward to generate ILVs. MVBs arise after maturation and can merge with lysosomes or release their ILVs as exosomes. Biogenesis selectively sorts proteins, miRNAs, mRNAs, and DNA in exosomes. The lipid bilayer membrane of exosomes comprises proteins and lipids.

Exosomes are nanosized extracellular vesicles, commonly termed "cargo," composed of bilipid layers. Exosomes encapsulate numerous substances, including proteins, nucleic acids, lipids, and metabolites. The composition of exosomal cargo is significantly diverse, contingent upon the source of cell types, the cellular state, and the milieu. The cholesterol, sphingolipids, ceramides, glycolipids, and glycerophospholipid chains, key components of the cellular membrane environment, are prevalent in the outer membrane of exosomes. The lipids of exosome membranes primarily function to stabilize bioactive components, enhance their bioavailability, and inhibit their degradation. Research indicates that exosomal lipids participate in lipid transport in vivo [20]. Exosomes produced by many cell types exhibit biological enzymes, integrins, cytoskeletal proteins, heat-shock proteins, three transmembrane proteins (CD9, CD63, and CD81), and major histocompatibility complex (MHC) I and II molecules on their surfaces. Exosomes predominantly comprise proteins, including cytokines, with miRNA, mRNA, DNA, and lipids. Among the 286 recent investigations on exosomes, over 41,860 distinct proteins, 7,540 RNAs, and 1,160 lipid molecules have been identified as originating from exosomes produced by various cell types [8]. Figure 9.2 illustrates the biological features of exosomes. Exosomes can convey information to recipient cells through three distinct mechanisms. The protein on the exosomal membrane interacts with the protein on the receptor cell membrane, subsequently initiating the intracellular signaling cascade; the exosomal membrane fuses with the recipient cell membrane, releasing its contents into the recipient cell; targeting cells directly phagocytose exosomes and internalize them into their components [21].

9.3 ROLES OF ADIPOSE EXOSOMAL MICRORNA IN METABOLIC DISEASE

microRNAs (miRNAs) are classified within the small noncoding RNA (sncRNA) category, measuring 19–25 nucleotides in length, and they inhibit target gene expression through posttranscriptional regulation. Significantly, a single miRNA can engage with many mRNAs and regulate gene expression. Conversely, lncRNA and circRNA can act as miRNA sponges, thereby inhibiting miRNA activity. According to new research, exosomes derived from adipose tissue contain numerous miRNAs that facilitate intercellular communication between cells and affect the expression of genes involved in metabolism [22].

9.3.1 Insulin Resistance

Insulin resistance is characterized by reduced blood glucose-lowering effects due to impaired insulin sensitivity: a feature commonly observed in individuals with type 2 diabetes mellitus (T2DM). Obesity-induced insulin resistance significantly contributes to various metabolic disorders. Extensive evidence indicates that adipose exosomal miRNA might have a role in how insulin-sensitive organs develop insulin resistance, including the liver, skeletal muscle, and cardiac muscle [23]. Exosomal miRNA derived from adipose tissue contributes to the progression of insulin resistance via multiple molecular pathways. Downregulation of IRS-1 expression is a crucial signaling channel for adipose tissue-derived exosomal miRNA, resulting in insulin resistance. MiR-222 in adipocytes has been identified as a possible modulator of estrogen receptor-α expression in estrogen-induced insulin resistance. Moreover, serum exosomes containing miR-222 produced from gonadal white adipose tissue (gWAT) have been shown to induce insulin resistance in the liver and skeletal muscle of obese mice subjected to a high-fat diet (HFD) by directly inhibiting insulin receptor substrate 1 (IRS1) expression. A study showed that exosomes produced from adipose tissue, enriched with miR-141-3p, diminish insulin sensitivity and glucose absorption in hepatocytes of mice by upregulating phosphatase and tensin homologue deleted on chromosome 10 (PTEN), which inhibits AKT phosphorylation.

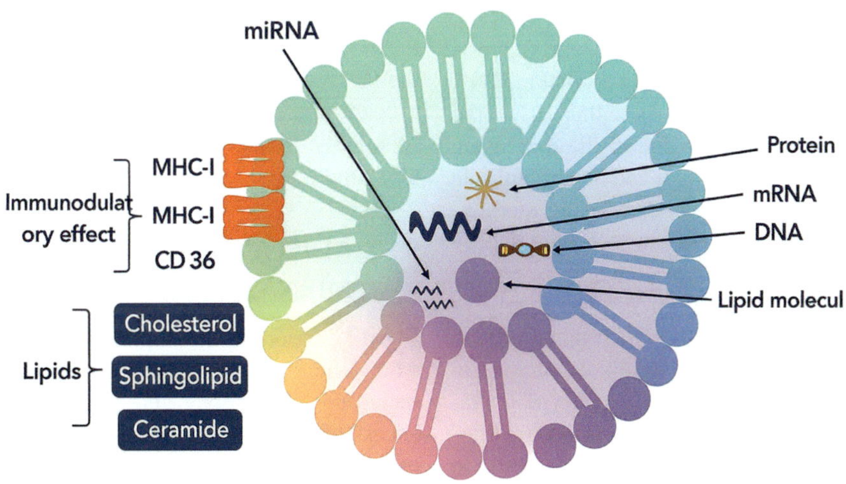

FIGURE 9.2 Characteristics of exosomes.

MiR-409-3p, originating from brown adipose tissue (BAT) endothelial cells (ECs), has been recognized as an antiangiogenic microRNA in both humans and animals. Research indicates that miR-409-3p influences the ZEB1-MAP 4K3-PLGF signaling pathway to regulate insulin sensitivity through endothelial-adipose signaling [24].

9.3.2 Diabetes Mellitus and Its Complications

Diabetes is characterized by chronic hyperglycemia resulting from impaired insulin action or secretion, including type 1 diabetes (T1D), type 2 diabetes (T2D), and gestational diabetes mellitus (GDM). Type 1 diabetes (T1D) is characterized by inadequate insulin production resulting from the death of β-cells in the pancreatic islets, with risk factors that encompass both genetic and environmental influences. Insulin resistance and decreased insulin secretion characterize T2D, with increasing data indicating that adipose-derived exosomal microRNAs have several functions in diabetes mellitus and its consequences, particularly T2D. Preclinical studies investigating the potential utility of extracellular microvesicles (EMVs) as biomarkers for type II diabetes revealed the following: db/db mice exhibit markedly elevated levels of circulating total and platelet-derived EMVs compared to db/+ mice; male db/db mice subjected to a two-week treatment with an antidiabetic agent show no change in platelet-derived EMVs, but total and tissue-derived EMVs are much lower, leading to weight gain, lowered blood glucose levels, increased plasma adiponectin, and decreased plasma insulin. Circulating EMVs display elevated levels and changed signatures in healthy individuals compared to those with type II diabetes or metabolic syndrome. Streptozotocin-induced mice show substantially elevated levels of total, platelet-derived, and tissue-derived EMVs prior to the clinical onset of diabetes [25, 26].

9.3.3 Obesity

Obesity is a significant risk factor that contributes to the global disease burden, with its primary adverse outcomes including type 2 diabetes mellitus, cardiovascular disease, asthma, and some malignancies. Recent evidence suggests a significant correlation between adipose exosomal miRNA and obesity. Research into the processes that drive the growth of adipose tissue is ongoing due to the increasing prevalence of obesity in modern culture and the crucial role of this tissue in maintaining energy balance and metabolism. Obesity is thought to be caused by a chronic energy imbalance between the energy we consume and the energy we expend. This imbalance builds up surplus energy in the form of triacylglycerol in fat cells, which then undergo hypertrophy and hyperplasia. The substantial increase in adipose tissue mass in people with obesity and type II diabetes is attributed, as is commonly known, to both adipocyte hypertrophy and hyperplasia. Inadequate exploration of the underlying molecular mechanisms has limited our understanding. The paracrine transmission of information regarding the lipogenic versus lipolytic states between large and small adipocytes inside adipose tissue depots via adipokines or extracellular microvesicles may mediate these processes [27].

9.3.4 Other Metabolic Diseases

Owing to its widespread global incidence, diagnostic problems, and absence of viable therapies, NAFLD has emerged as a significant global issue. NAFLD is a key contributor to liver cirrhosis and liver cancer, especially those not linked to alcohol consumption. Recent discoveries indicate that miRNA originating from adipose tissue presents a compelling avenue for understanding the pathophysiology of liver disorders. In 2017, Qu et al. engineered adipose-derived stem cells (ADSCs) to overexpress miR-181-5p and administered them to a mouse model of liver fibrosis. In their research, the study found that exosomes originating from ADSCs, which overexpress miR-181-5p, reduced levels of Stat3, Bcl-2, α-SMA, collagen I, fibronectin, and vimentin in the liver, while also activating autophagy in mouse hepatic stellate (HST-T6) cells, thereby alleviating liver fibrosis. Furthermore, Baranova et al. indicated that adipose tissue can release miR-122 to impede the progression of fibrosis through various mechanisms, including the targeting of YY1 or Wnt, thereby significantly mitigating NAFLD [24].

9.4 THE FUNCTION OF EXOSOMAL MICRORNA IN TYPE 2 DIABETES

The mechanisms by which miRNAs are loaded into exosomes remain incompletely understood. However, the fact that miRNA distribution is not uniform among tissues and exosomes implies that this loading process is not random. Their exosomal miRNA patterns frequently indicate the biological state of cells. Moreover, the intercellular transfer of miRNAs through exosomes seems to modulate mRNA translation and posttranscriptional regulation in recipient cells. microRNAs identified in bodily fluids (blood, urine, saliva, tears, and breast milk) work as optimal biomarkers for diseases, including type 2 diabetes and its related endothelial dysfunction. Good biomarkers exhibit several significant properties. The considerable attributes encompass specificity, sensitivity, stability, noninvasiveness, robustness, and reproducibility [28].

9.5 ADIPOCYTE-DERIVED EXOSOMAL MIRNAS AND T2DM

The likelihood of developing T2DM is greatly enhanced when one is obese. Changes in adipocyte morphologies, including hypertrophy and hyperplasia brought on by an excess of calories consumed, are hallmarks of obesity. At the same time, these fat cells are where most of the microRNAs that are released into the exosome originate from. They both stay in adipose tissue and send miRNA-enriched exosomes to faraway insulin-targeting regions. Through autocrine signaling, miRNAs secreted by adipocytes affect both adipogenesis and adipocyte differentiation. One possible mechanism by which miR-122 overexpression in adipocyte-derived exosomes promotes adipogenesis is by its targeting of the vitamin D3 receptor (VDR) gene, a negative regulator of sterol production. Adipose tissue dysfunction is a primary contributor to diabetes associated with obesity. The condition is characterized by dysregulated adipokine production, which exacerbates insulin resistance in key metabolic tissues, including the liver, skeletal muscles, and adipose tissue. The inter-organ communication between adipose tissue and other organs, along with the intra-organ interaction between adipocytes and macrophages within adipose tissue, formerly mediated by hormones, has recently been shown to be regulated by exosomes generated from adipose tissue. Exosomes are nanoscale membrane-bound vesicles released by donor cells that alter intercellular communication by transferring nucleic acids and proteins to target cells [29].

9.6 EXOSOMAL MIRNAS PRODUCED BY IMMUNE CELLS AND TYPE 2 DIABETES

The generation of pro-inflammatory cytokines by immune cells plays a crucial role in the etiology of type 2 diabetes by encouraging insulin resistance and low-grade tissue inflammation. To improve insulin action and glucose uptake in adipocytes, macrophages – the primary immune cells in adipose tissue – release exosomes that are high in microRNAs. Reduced glucose uptake in adipocytes and decreased expression of the glucose transporter 4 (GLUT4) were caused by the inhibition of PPAR-γ by exosomal miR-155 generated from adipose tissue macrophages (ADMs). Following this, studies have shown that miR-155-enriched exosomes can travel to distant locations, such as the pancreas, where β-cells absorb them and thereby suppress insulin release in response to glucose. In the initial stages of diabetes, elevated blood glucose levels can result in endothelial dysfunction and microvascular rarefaction [30]. Inadequate myocardial angiogenesis is the primary symptom of ischemic cardiovascular disease induced by diabetes. Considering that diabetic hearts have inadequate angiogenesis, it is essential to evaluate whether exosomes produced from diabetic cardiomyocytes have any potential in regulating angiogenesis. Wang et al. have demonstrated that in the context of T2D, markedly elevated fluxes of exosomes containing miR-320-3p have been seen in cardiac cell cultures [31].

9.7 EXOSOMAL MIRNAS PRODUCED FROM SKELETAL MUSCLE CELLS AND TYPE 2 DIABETES

Skeletal muscle, being one of the largest tissues, plays a significant role in the production of exosomes. Research on exosomes derived from skeletal muscle cells is limited, with only a few studies published on this topic. The process of β-cell hyperplasia may be better understood after exosomes were brought to the pancreas and stimulated β-cell proliferation by reducing the expression of the protein patched homolog 1 (Ptch1) gene. The repair and regeneration of skeletal muscle are controlled by numerous factors, with exosomes playing a crucial role in these processes. In the following discussion, we will explore the complex role of exosomes in skeletal muscle biology. This investigation will encompass three primary domains: the role of exosomes in the proliferation, differentiation, and migration of skeletal muscle cells; their function in immunoregulatory processes within skeletal muscles; and their significance in the induction of angiogenesis in skeletal muscle tissue. These features, combined, highlight the vital relevance of exosomes in skeletal muscle repair and regeneration [32].

9.8 EXOSOMAL MIRNAS PRODUCED FROM PANCREATIC β-CELLS AND TYPE 2 DIABETES

Pancreatic β-cells not only secrete insulin but also act as glucose sensors, which help regulate the production of both insulin and glucagon. To maintain normal blood sugar levels, β-cells increase their proliferation during the initial phase of hyperglycemia. Endoplasmic reticulum (ER) stress reduces their proliferation potential as illness progresses. Accelerating the development of type 2 diabetes is the gradual loss of β-cells, leading to a decrease in insulin production. Insulin resistance in individuals with type 2 diabetes is characterized by reduced glucose transport and metabolism in adipose tissue and skeletal muscle, as well as diminished suppression of hepatic glucose output [23]. A study identified a cross talk pathway between β-cells and macrophages that is mediated by the miRNA-29–TNF-receptor-associated factor 3 (TRAF3) axis. MiR-29 facilitates the recruitment and activation of monocytes and macrophages, resulting in inflammation and disrupted glucose homeostasis through miR-29 exosomes in a TRAF3-dependent manner. The findings indicate that β-cells can influence systemic inflammatory tone and glucose metabolism via miR-29 in response to nutrient overload [33].

9.9 THE ROLE OF EXOSOMAL MICRORNAS IN TYPE 2 DIABETES RESEARCH

Multiple clinical investigations have demonstrated that patients with type 2 diabetes and healthy controls exhibit distinct miRNA expression in exosomes, corroborated by experimental results from cell culture and preclinical studies. The results showed that, compared to individuals with standard glucose tolerance, the levels of exosomal miR-375-3p in the serum of individuals with early-stage type 2 diabetes were 2.8 times higher. The majority of miR-375-3p was found to be released by β-cells in response to harmful substances and cytokines, according to later in vitro model research [34]. Exosomal miR-210, originating from adipose tissue macrophages in mice subjected to a high-fat diet, suppresses the expression of NADH dehydrogenase (ubiquinone) 1α subcomplex 4 (NDUFA4) and thus reduces glucose absorption in adipocytes. NDUFA4 constitutes one of the subunits of complex IV within the mitochondrial respiratory chain. Consequently, they proposed that the downregulation of NDUFA4 expression by miR-210 may induce mitochondrial dysfunction, thereby affecting insulin signaling in the context of obesity [35]. Adipose tissue, as a significant regulatory organ of human health, is implicated in the regulation of several metabolic illnesses, particularly in the context of insulin resistance and T2DM [36, 37].

9.10 THE ROLE, FORMATION, AND ABSORPTION OF EXTRACELLULAR VESICLES

Originally, extracellular vesicles (EVs) were waste transporters that removed excess material from cells. They are generally found in bacteria, eukaryotes, and humans and serve as an alternative avenue for intercellular communication in both physiological and pathological situations. Exosome intermediates are multivesicular entities produced by the endosomal limiting membrane budding inward. Combining multivesicular bodies with plasma membranes releases exosomes. Multivesicular bodies can merge with lysosomes to disintegrate or recycle to the trans-Golgi network. Blood, saliva, urine, cerebrospinal fluid (CSF), semen, and other extracellular fluids contain exosomes. In addition to endosomal sorting complexes, exosomes contain key proteins involved in their biogenesis, including accessory proteins like Alix, HSC70, HSP90β, and TSG101. Integrins, immunomodulatory proteins, and tetraspanins (CD63, CD9, and CD81) are exosomes' main transmembrane proteins [38]. They are no longer exosome markers and may be unreliable. EV exosomes may potentially influence the immune response by presenting antigens and contributing to the development of neurological, cardiovascular, and cancerous diseases. Plasma membrane outward budding creates microvesicles (microparticles) without internal membranes. They have more phosphorylated and glycosylated proteins than exosomes and may contain tetraspanins, integrins, HSPs, cytoskeletal proteins, and the microvesicle marker proteins Annexin A1 and ARF6. Microvesicles, like exosomes, communicate between cells and contribute to cancer and cardiometabolic illnesses. Microvesicles aid inflammation, clotting, and endothelial dysfunction. Cancer cells release oncosomes (1–10 μm) and exosomes, which promote angiogenesis, invasiveness, immune suppression, pre- and pro-metastatic niches, and treatment resistance [39].

9.11 THE ROLE OF ADIPOSE TISSUE–DERIVED EXTRACELLULAR VESICLES IN INSULIN RESISTANCE

Adipose tissue–derived extracellular vesicles (AT-EVs) may exhibit either positive or negative adiponectin expression and originate from adipocytes, macrophages, endothelial cells, or stem cells. EVs containing adiponectin are believed to originate from adipocytes, whereas those lacking adiponectin are derived from other adipose tissue cells. Extracellular vesicles in adipocytes encompass a range of cytokines and adipokines, including IL-6, MIF, MCP-1, TNF-α, adiponectin, resistin, RBP4, and various microRNAs [40]. Extracellular vesicles derived from adipocytes exhibit differences between individuals with obesity and those who are lean, as well as between subcutaneous and visceral adipose tissues. Visceral adipose tissue–derived extracellular vesicles contain higher levels of MIF-1, MCP-1, and IL-6 compared to those derived from subcutaneous adipose tissue [41].

9.12 THERAPEUTIC USAGE OF EXTRACELLULAR VESICLES

Mesenchymal stem cell–derived extracellular vesicles (MSC-EVs) have the potential to address obesity-related metabolic disorders, in contrast to adipocyte-derived EVs, which are detrimental to conditions such as NAFLD/MASLD and PCOS. Following MSC-EV therapy for liver fibrosis, there was a reduction in TGF-β and Smad2 levels. In vitro treatment of hepatic stellate cells with extracellular vesicles derived from hematopoietic stem cells containing miR-92a-3p results in reduced synthesis of αSMA, collagen Iα1, fibronectin, and TIMP-1, alongside diminished chemotaxis and proliferation [42]. MSC-EVs have the potential to enhance fertility in a mouse model of polycystic ovarian syndrome. In vitro exosomes derived from MSCs decreased the expression of CYP17A1, CYP11A1, and DENND1a genes in H293R cells, thereby inhibiting androgen synthesis [41].

9.13 VESICLES IN THE EXTRACELLULAR MEMBRANE AS BIOMARKERS: OVERALL FACTORS

The following characteristics suggest that extracellular vesicles (EMVs) may be a promising area for exploring new biomarkers. Distinct intrinsic features, such as phosphatidylserine concentration, size, and sedimentation behavior, allow for the identification and isolation of EMVs. Table 9.1 depicts the key miRNAs involved in metabolic regulation and their functions. Expression of cell-lineage markers, total chemical composition, luminal and surface content patterns, and EMV specificity are all features of EMVs. Since EMVs are released early during stimulatory or microenvironmental changes and in the pathogenic cascade of diseases, their signatures are greatly affected by these factors. On the other hand, EMVs are noninvasive and can be detected in various bodily fluids. They are also translatable since their release is not limited to a single cell type or species. Lastly, EMVs serve as vectors, transporting and protecting biological messages; these messages typically consist of mRNAs, microRNAs, proteins, and phospholipids [27].

TABLE 9.1
Key microRNAs Involved in Metabolic Regulation and Their Functions

miRNA	Functions	Disease	References
MiR33	Hemostasis	Atherosclerosis, nonalcoholic fatty liver disease	[43]
miR-103 and miR-107	Target caveolin-1 and insulin receptor substrate-1 (IRS-1) to aggressively control insulin sensitivity. Affect lipid and glucose metabolism	Diabetic mellitus, obesity, and metabolic syndrome	[44]
miR-34a	Controls lipid metabolism genes like SIRT1. Enhances hepatic lipid buildup and apoptosis	Hepatocellular carcinoma	[45]
miR-335	Adipogenesis regulation	Metabolic syndrome and obesity	[46]
let-7	Adipogenesis, inflammation, angiogenesis	Tumor suppressor, oncogene	[47]
miR-17	Differentiation, resistance to insulin, adipogenesis	Oncology, Alzheimer's disease, and polycystic kidney disease	[48]
miR-21	Apoptosis, adipogenesis, inflammation	Malignancies, cardiovascular ailments, autoimmune conditions, and skeletal abnormalities	[49]
miR-27b	Adipocyte dysfunction, adipogenesis,	Cardiovascular diseases and cancer	[50]
miR-34a, miR-92, miR-130a	Adipogenesis	Tumor suppression, autoimmune disease	[51]
miR-125b	Adipocyte dysfunction, adipogenesis, inflammation	Cancer	[52]
miR-132	Insulin resistance, inflammation	Age-related dementia (AD) and tauopathies	[53]
miR-143	Adipogenesis, proliferation, and adipocyte dysfunction	Diabetes, cancer	[54]
miR-221	Adipogenesis, angiogenesis, inflammation, and adipocyte dysfunction	Liver, pancreatic	[55]
miR-222	Adipogenesis, angiogenesis, inflammation, and adipocyte dysfunction	Cancer and cardiovascular diseases	[56]

9.14 MESENCHYMAL STEM CELL THERAPY

In treating metabolic diseases like diabetes mellitus, obesity, metabolic syndrome, nonalcoholic fatty liver disease (NAFLD), and some inherited metabolic illnesses, mesenchymal stem cell (MSC) therapy has shown promise. Derived from bone marrow, adipose tissue, umbilical cord, and other tissues, MSCs are multipotent stromal cells able to differentiate into several mesenchymal lineages. Their special qualities of self-renewal, immunomodulating ability, anti-inflammatory action, and paracrine factor secretion make them relatively fit for use in regenerative medicine [57]. MSCs have demonstrated the capacity to protect and repair pancreatic β-cells, boost insulin production, and mitigate autoimmune damage in type 1 diabetes mellitus. In type 2 diabetes, MSCs downregulate inflammatory cytokines and increase insulin receptor signaling, hence improving insulin sensitivity. Specifically, adipose-derived MSCs have been shown to mitigate metabolic dysregulation and inflammation associated with obesity by modulating macrophage activity and cytokine levels. MSC treatment reduces hepatic inflammation, fibrosis, and steatosis in NAFLD through immunomodulation and antifibrotic characteristics. MSC-derived exosomes are a cell-free therapeutic method with similar regenerative and anti-inflammatory effects without the risks of live cell transplantation [58]. MSCs can be utilized as enzyme delivery vehicles in Gaucher's disease or as a supportive therapy to enhance enzyme replacement. Despite these promising results, long-term safety data, cell engraftment inefficiencies, tumorigenicity, and allogeneic immunological rejection remain key barriers to clinical translation. However, advances in MSC engineering, gene editing, and personalized therapy make MSC-based metabolic illness interventions more practical and successful [59].

9.15 CLINICAL POTENTIAL OF EXOSOMES

The process of intercellular communication relies on exosomes, which are vesicles located outside of cells. The following bodily fluids contain blood, saliva, amniotic, cerebrospinal fluid, ascites, and lymph. Biomarkers or fingerprints for the donor cell can be found in the contents of exosomes, which mimic the origin and physiological activity of cells. These contents include unique microRNAs, lipids, and proteins. Therefore, exosomes can be used as biomarkers to detect pathological diseases due to their condition-specific payloads, which accurately reflect cellular activity. As disease biomarkers, exosomes may provide a decreased benefit.

9.16 REGULATORY NONCODING RNAS

9.16.1 Diabetic Retinopathy

It is a well-documented and common consequence in patients with diabetes, primarily classified as a microvascular illness. It is the primary cause of vision impairment, especially among adults aged 45–75 years, as reported by the *Lancet Diabetes & Endocrinology* journal. As global diabetes prevalence escalates, so does the occurrence of diabetic retinopathy, highlighting the necessity for early diagnosis and effective treatment to avert vision impairment. Certain studies categorize diabetic retinopathy (DR) as a microvascular condition, highlighting its possible involvement in retinal neurodegeneration [60].

9.16.2 Three Types of Regulatory Noncoding RNAs

The transcript length of long noncoding RNAs (lncRNAs) is more than 200 nucleotides. The fact that they are associated with numerous human malignancies makes them useful diagnostic tools among other indicators, even if they do not encode proteins. Different ocular organs show distinct expression patterns of long noncoding RNAs. Long noncoding RNAs (lncRNAs) act as decoys by attaching to microRNAs and transcription factors, trapping them and preventing them from carrying out their usual functions. By interacting with transcription factors, the long noncoding RNA, HOX transcript antisense intergenic RNA (HOTAIR), regulates cell proliferation and differentiation by reducing the expression of specific genes [61].

9.16.3 miRNA

Noncoding RNAs that regulate gene expression posttranscriptionally are known as microRNAs (miRNAs). These molecules are highly conserved and typically range in length from 19 to 25 nucleotides. Essential regulators of gene expression, microRNAs (miRNAs) enable post-transcriptional silencing by binding to complementary mRNA targets. In the nucleus, RNA polymerase II transcribes miRNAs to produce primary miRNAs (pri-miRNAs), which are usually longer than 1,000 nucleotides. Thereafter, the RNase III enzyme Drosha, along with the DGCR8 complex, converts the pri-miRNA into a precursor miRNA (pre-miRNA) of roughly 65 nucleotides in length. The exportin-5 protein then transports the pre-miRNAs to the cytoplasm, where the Dicer enzyme converts them into double-stranded RNA molecules that are approximately 21 nucleotides long. This process typically involves the destruction of the miRNA's passenger strand and the association of its guide strand with an Argonaute protein, such as AGO2, to create the RNA-induced silencing complex (RISC). In the RISC, the "seed" region (nucleotides 2–7) of the target mRNA is used by the miRNA to bind to the 3′-untranslated region (UTR) [60].

9.16.4 circRNA

Noncoding RNAs known as circRNAs are generated from linear precursor mRNA using a noncanonical splicing mechanism. Their lengths range from a few hundred to several thousand nucleotides. EcircRNAs and EIciRNAs are two types of circular RNAs that their structural features can distinguish. Blood samples can be noninvasively tested for circRNAs, which show distinct tissue-specific expression patterns. More stable than linear RNAs, circRNAs are characterized by a covalently closed circular shape that gives substantial resistance to degradation. Retinal tissue exhibits a common and highly conserved pattern of circRNA expression. Consistent with this, aberrant expression of circRNAs appears to characterize retinal diseases [62].

9.17 APPLICATION OF EXOSOMES

Exosomes, due to their remarkable features, play a pivotal role in disease diagnosis and treatment. Initially, exosomes can be retrieved from bodily fluids, including saliva, blood, and urine, using a technique known as liquid biopsy. This offers a noninvasive alternative to traditional biopsies [63]. This enables the analysis of exosome composition using mass spectrometry and other methods, providing extensive information about molecular carriers both intracellularly and extracellularly, thereby facilitating a thorough diagnosis across various stages of disease progression. A total of 120 plasma exosome samples obtained from patients were utilized to identify specific biomarkers for 16 distinct tumor types by proteomic analysis. The findings indicate that they achieved 95% sensitivity and 90% specificity in classifying diverse types of cancer. The evidence, as mentioned earlier, collectively suggests that exosomes possess significant promise for the diagnosis and treatment of illnesses, including those situated in inaccessible deep areas [8].

9.17.1 Exosomes in Personalized Medicine for Metabolic Disorders

Exosomes are increasingly recognized as valuable tools in personalized treatment, particularly concerning metabolic diseases such as diabetes mellitus, obesity, metabolic syndrome, and nonalcoholic fatty liver disease (NAFLD). Personalized medicine aims to align a patient's environmental, biochemical, and genetic traits with appropriate treatment strategies. Exosomes, which are naturally occurring nanovesicles derived from cells, function as both biomarkers and therapeutic delivery vehicles, offering a highly individualized approach to diagnosis, monitoring, and treatment. The application of exosomes as diagnostic indicators represents a significant advancement in personalized medicine. miRNAs, mRNAs, and proteins, along with other exosomal components, reflect the health or disease state of the originating cell. Exosomal miRNAs, such as miR-122, miR-375, and miR-21, serve as markers for insulin resistance, β-cell dysfunction, and liver steatosis in metabolic disorders. The early and minimally invasive identification of metabolic imbalances enables timely care tailored to the individual's risk profile [64]. Exosomes can be engineered to deliver specific medications, gene-editing tools (such as CRISPR/Cas9), or therapeutic RNAs to targeted tissues in a patient-specific manner. Exosomes derived from a patient's mesenchymal stem cells (MSCs) can be enriched with miRNAs that diminish hepatic lipogenesis, inflammation, or compounds that enhance insulin sensitivity. Autologous exosomes exhibit natural biocompatibility, thereby reducing the risk of immunological rejection and improving the safety of long-term treatment. This method enables patients with type 2 diabetes or NAFLD to achieve customized regulation of metabolic pathways according to their genetic and molecular profiles. Exosomal profiling facilitates the monitoring of disease progression and therapeutic response, enabling timely adjustments to treatment protocols. A decrease in circulating exosomal miR-122 levels in NAFLD patients after treatment may indicate improved liver function. Assessing variations in exosome load from adipose tissue could aid in evaluating inflammatory conditions or the efficacy of pharmaceutical and lifestyle interventions in obesity and metabolic syndrome. Despite their potential, outstanding challenges remain in scalable exosome manufacturing, cargo-loading efficiency, and regulatory approval. Continuous advancements in omic-based profiling, liquid biopsy technologies, and exosome bioengineering position exosomes as a fundamental component of precision treatment for metabolic diseases. They offer the potential for targeted and efficient treatment, as well as tailored therapeutic plans based on each patient's unique molecular profile [65, 66].

9.18 CONCLUSION

Extracellular vesicles may serve as drug delivery mechanisms in several scenarios. The ability of exosomes to transport various synthetic and biological substances has opened up new possibilities for metabolic disorders and cellular therapy. The effective application of exosomes in treating metabolic disorders hinges on three key factors: the development of cost-efficient large-scale production techniques, advanced isolation and characterization technologies to assess batch-to-batch consistency and biological effects, and universally applicable drug-loading methods. The use of exosomes as a drug delivery carrier offers several benefits, including the avoidance of drug accumulation or absorption in the liver and the bypass of first-pass metabolic processes before reaching their intended targets. The intrinsic capacity to transport intercellular nucleic acids and therapeutic compounds across complex barriers, such as the blood–brain barrier, combined with the long-term safety of well-characterized exosomes, makes them very useful. The growing availability of innovative analytical tools is anticipated to provide deeper insights into the distinct properties of exosomes, potentially accelerating the development of next-generation synthetic systems for metabolic disorder therapies.

ACKNOWLEDGMENTS

The author thanks the Faculty of Pharmacy, Shri Rawatpura Sarkar University, and University Institute of Pharmacy, Pt Ravishankar Shukla University, for providing the necessary facilities.

AI DISCLOSURE STATEMENT

During the preparation of this work the authors used Grammarly, QuillBot, and ChatGPT in order to enhance grammatical structure and narrative coherence. After using this tool/service, the authors reviewed and edited the content as needed and take full responsibility for the content of the publication.

CONFLICT OF INTEREST

The author declares no financial or other conflict of interest.

FUNDING

No funding.

REFERENCES

[1] Heindel JJ, Blumberg B, Cave M, Machtinger R, Mantovani A, Mendez MA, et al. Metabolism disrupting chemicals and metabolic disorders. *Reprod Toxicol* 2017;68:3–33. https://doi.org/10.1016/j.reprotox.2016.10.001.

[2] Boddu SK, Giannini C, Marcovecchio ML. Metabolic disorders in young people around the world. *Diabetologia* 2025. https://doi.org/10.1007/s00125-025-06450-2.

[3] Clemente-Suárez VJ, Martín-Rodríguez A, Redondo-Flórez L, López-Mora C, Yáñez-Sepúlveda R, Tornero-Aguilera JF. New insights and potential therapeutic interventions in metabolic diseases. *Int J Mol Sci* 2023;24. https://doi.org/10.3390/ijms241310672.

[4] Mostosi D, Molinaro M, Saccone S, Torrente Y, Villa C, Farini A. Exploring the gut microbiota–muscle axis in duchenne muscular dystrophy. *Int J Mol Sci* 2024;25. https://doi.org/10.3390/ijms25115589.

[5] Rodrigues KF, Yong WTL, Bhuiyan MSA, Siddiquee S, Shah MD, Venmathi Maran BA. Current understanding on the genetic basis of key metabolic disorders: A review. *Biology (Basel)* 2022;11. https://doi.org/10.3390/biology11091308.

[6] Yokoi A, Ochiya T. Exosomes and extracellular vesicles: Rethinking the essential values in cancer biology. *Semin Cancer Biol* 2021;74:79–91. https://doi.org/10.1016/j.semcancer.2021.03.032.

[7] Ganesh BH, Padinjarathil H, Rajendran RL, Ramani P, Gangadaran P, Ahn BC. The role of extracellular vesicles in aging and age-related disorders. *Antioxidants* 2025;14. https://doi.org/10.3390/antiox14020177.

[8] Wang N, Li J, Hu Z, Ngowi EE, Yan B, Qiao A. Exosomes: New insights into the pathogenesis of metabolic syndrome. *Biology (Basel)* 2023;12. https://doi.org/10.3390/biology12121480.

[9] Zhang S, Yang Y, Lv X, Liu W, Zhu S, Wang Y, et al. Unraveling the intricate roles of exosomes in cardiovascular diseases: A comprehensive review of physiological significance and pathological implications. *Int J Mol Sci* 2023;24. https://doi.org/10.3390/ijms242115677.

[10] Mizgier ML, Casas M, Contreras-Ferrat A, Llanos P, Galgani JE. Potential role of skeletal muscle glucose metabolism on the regulation of insulin secretion. *Obes Rev* 2014;15:587–97. https://doi.org/10.1111/obr.12166.

[11] Chimal-Vega B, Maldonado-Arvizu JE, Hernández Avalos AD, Díaz-Villanueva JF, Avila-Barrientos LP, García González VG. Inter-tissue communication mechanisms via exosomes and their implications in metabolic diseases: Opportunities for pharmacological regulation. *Future Pharmacol* 2025;5:11. https://doi.org/10.3390/futurepharmacol5010011.

[12] Smalheiser NR. Exosomal transfer of proteins and RNAs at synapses in the nervous system. *Biol Direct* 2007;2. https://doi.org/10.1186/1745-6150-2-35.

[13] Meerson A, Traurig M, Ossowski V, Fleming JM, Mullins M, Baier LJ. Human adipose microRNA-221 is upregulated in obesity and affects fat metabolism downstream of leptin and TNF-α. *Diabetologia* 2013;56:1971–9. https://doi.org/10.1007/s00125-013-2950-9.

[14] Yang B, Chen Y, Shi J. Exosome biochemistry and advanced nanotechnology for next-generation theranostic platforms. *Adv Mater* 2019;31. https://doi.org/10.1002/adma.201802896.

[15] Umeche IE, Olaniyan MF. Exosomes: Emerging biomarkers unveiling cellular mysteries – a narrative review. *J BioX Res* 2023;6:104–15. https://doi.org/10.1097/JBR.0000000000000146.

[16] Jangam TC, Desai SA, Patel VP, Pagare NB, Raut ND. Exosomes as therapeutic and diagnostic tools: Advances, challenges, and future directions. *Cell Biochem Biophys* 2025. https://doi.org/10.1007/s12013-025-01730-5.

[17] Kenific CM, Zhang H, Lyden D. An exosome pathway without an ESCRT. *Cell Res* 2021;31:105–6. https://doi.org/10.1038/s41422-020-00418-0.

[18] Xu M, Ji J, Jin D, Wu Y, Wu T, Lin R, et al. The biogenesis and secretion of exosomes and multivesicular bodies (MVBs): Intercellular shuttles and implications in human diseases. *Genes Dis* 2023;10:1894–907. https://doi.org/10.1016/j.gendis.2022.03.021.

[19] Cunha Rocha K, Ying W, Olefsky JM. Exosome-mediated impact on systemic metabolism. *Annu Rev Physiol*. Downloaded from WwwAnnualreviewsOrg Guest 2025;01:33. https://doi.org/10.1146/annurev-physiol-042222.

[20] Yeh HC, Gupta K, Lu Y-H, Srinivasan A, Delila L, Tran N, et al. Platelet extracellular vesicles as natural delivery vehicles for mitochondrial dysfunction therapy? n.d. https://doi.org/10.1021/acsbiomaterials.5c00473.

[21] Yang E, Wang X, Gong Z, Yu M, Wu H, Zhang D. Exosome-mediated metabolic reprogramming: The emerging role in tumor microenvironment remodeling and its influence on cancer progression. *Signal Transduct Target Ther* 2020;5. https://doi.org/10.1038/s41392-020-00359-5.

[22] Sassi A, Bakhtiar MT Bin, Khattak MMAK, Haron NB, Kaderi MA Bin, Rostam MA Bin, et al. Biological roles of selected microRNAs in glucose metabolism as a candidate biomarker for diabetes mellitus. *Mol Nutr Food Res* 2025. https://doi.org/10.1002/mnfr.70077.

[23] Accili D, Deng Z, Liu Q. Insulin resistance in type 2 diabetes mellitus. *Nat Rev Endocrinol* 2025. https://doi.org/10.1038/s41574-025-01114-y.

[24] Wu YL, Lin ZJ, Li CC, Lin X, Shan SK, Guo B, et al. Adipose exosomal noncoding RNAs: Roles and mechanisms in metabolic diseases. *Obes Rev* 2024;25. https://doi.org/10.1111/obr.13740.

[25] Pradhan D, Sahu PK, Purohit S, Ranajit SK, Acharya B, Sangam S, et al. Therapeutic interventions for diabetes mellitus-associated complications. *Curr Diabetes Rev* 2025;21. https://doi.org/10.2174/0115733998291870240408043837.

[26] Wang Y, Chen LM, Liu ML. Microvesicles and diabetic complications-novel mediators, potential biomarkers and therapeutic targets. *Acta Pharmacol Sin* 2014;35:433–43. https://doi.org/10.1038/aps.2013.188.

[27] Müller G. Microvesicles/exosomes as potential novel biomarkers of metabolic diseases. *Diabet Metab Syndr Obes* 2012;5:247–82. https://doi.org/10.2147/dmso.s32923.

[28] Fluitt MB, Mohit N, Gambhir KK, Nunlee-Bland G. To the future: The role of exosome-derived microRNAs as markers, mediators, and therapies for endothelial dysfunction in type 2 diabetes mellitus. *J Diabetes Res* 2022;2022. https://doi.org/10.1155/2022/5126968.

[29] Zhang B, Yang Y, Xiang L, Zhao Z, Ye R. Adipose-derived exosomes: A novel adipokine in obesity-associated diabetes. *J Cell Physiol* 2019;234:16692–702. https://doi.org/10.1002/jcp.28354.

[30] Ghoshal K. Exosomal RNA in type 2 diabetes mellitus and other metabolic diseases. *Exosomal RNA*, Elsevier; 2024, p. 135–65. https://doi.org/10.1016/B978-0-443-14008-2.00009-7.

[31] Beuzelin D, Kaeffer B. Exosomes and miRNA-loaded biomimetic nanovehicles, a focus on their potentials preventing type-2 diabetes linked to metabolic syndrome. *Front Immunol* 2018;9. https://doi.org/10.3389/fimmu.2018.02711.

[32] Aas V, Øvstebø R, Brusletto BS, Aspelin T, Trøseid AMS, Qureshi S, et al. Distinct microRNA and protein profiles of extracellular vesicles secreted from myotubes from morbidly obese donors with type 2 diabetes in response to electrical pulse stimulation. *Front Physiol* 2023;14. https://doi.org/10.3389/fphys.2023.1143966.

[33] Zheng W, Ji X, Yin QQ, Wu C, Xu C, Pan H, et al. Exosomes as emerging regulators of immune responses in type 2 diabetes mellitus. *J Diabetes Res* 2024;2024. https://doi.org/10.1155/2024/3759339.

[34] Sou YL, Chilian WM, Ratnam W, Zain SM, Syed Abdul Kadir SZ, Pan Y, et al. Exosomal miRNAs and isomiRs: Potential biomarkers for type 2 diabetes mellitus. *Precis Clin Med* 2024;7. https://doi.org/10.1093/pcmedi/pbae021.

[35] Tian F, Tang P, Sun Z, Zhang R, Zhu D, He J, et al. miR-210 in exosomes derived from macrophages under high glucose promotes mouse diabetic obesity pathogenesis by suppressing NDUFA4 expression. *J Diabetes Res* 2020;2020:1–12. https://doi.org/10.1155/2020/6894684.

[36] Lu T, Zheng Y, Chen X, Lin Z, Liu C, Yuan C. The role of exosome derived miRNAs in inter-cell crosstalk among insulin-related organs in type 2 diabetes mellitus. *J Physiol Biochem* 2024. https://doi.org/10.1007/s13105-024-01026-x.

[37] Wondmkun YT. Obesity, insulin resistance, and type 2 diabetes: Associations and therapeutic implications. *Diabet Metab Syndr Obes* 2020;13:3611–6. https://doi.org/10.2147/DMSO.S275898.

[38] Mallia A, Gianazza E, Zoanni B, Brioschi M, Barbieri SS, Banfi C. Proteomics of extracellular vesicles: Update on their composition, biological roles and potential use as diagnostic tools in atherosclerotic cardiovascular diseases. *Diagnostics* 2020;10. https://doi.org/10.3390/diagnostics10100843.

[39] Kumar MA, Baba SK, Sadida HQ, Marzooqi S Al, Jerobin J, Altemani FH, et al. Extracellular vesicles as tools and targets in therapy for diseases. *Signal Transduct Target Ther* 2024;9. https://doi.org/10.1038/s41392-024-01735-1.

[40] Kumar V, Kiran S, Kumar S, Singh UP. Extracellular vesicles in obesity and its associated inflammation. *Int Rev Immunol* 2022;41:30–44. https://doi.org/10.1080/08830185.2021.1964497.

[41] Mladenović D, Vesković M, Šutulović N, Hrnčić D, Stanojlović O, Radić L, et al. Adipose-derived extracellular vesicles – a novel cross-talk mechanism in insulin resistance, non-alcoholic fatty liver disease, and polycystic ovary syndrome. *Endocrine* 2024;85:18–34. https://doi.org/10.1007/s12020-024-03702-w.

[42] Povero D, Pinatel EM, Leszczynska A, Goyal NP, Nishio T, Kim J, et al. Human induced pluripotent stem cell–derived extracellular vesicles reduce hepatic stellate cell activation and liver fibrosis. *JCI Insight* 2019;4. https://doi.org/10.1172/jci.insight.125652.

[43] Morishita A, Oura K, Tadokoro T, Fujita K, Tani J, Kobara H, et al. MicroRNAs and nonalcoholic steatohepatitis: A review. *Int J Mol Sci* 2023;24. https://doi.org/10.3390/ijms241914482.

[44] Vivacqua A, Marco P De, Belfiore A, Maggiolini M. Recent advances on the role of microRNAs in both insulin resistance and cancer. *Curr Pharm Des* 2017;23. https://doi.org/10.2174/1381612823666170622105123.

[45] Wang L, Sun M, Cao Y, Ma L, Shen Y, Velikanova AA, et al. miR-34a regulates lipid metabolism by targeting SIRT1 in non-alcoholic fatty liver disease with iron overload. *Arch Biochem Biophys* 2020;695:108642. https://doi.org/10.1016/j.abb.2020.108642.

[46] Nakanishi N, Nakagawa Y, Tokushige N, Aoki N, Matsuzaka T, Ishii K, et al. The up-regulation of microRNA-335 is associated with lipid metabolism in liver and white adipose tissue of genetically obese mice. *Biochem Biophys Res Commun* 2009;385:492–6. https://doi.org/10.1016/j.bbrc.2009.05.058.

[47] Zhu Y, Zhang J, Hu X, Wang Z, Wu S, Yi Y. Extracellular vesicles derived from human adipose-derived stem cells promote the exogenous angiogenesis of fat grafts via the let-7/AGO1/VEGF signalling pathway. *Sci Rep* 2020;10. https://doi.org/10.1038/s41598-020-62140-6.

[48] Tian L, Song Z, Shao W, Du WW, Zhao LR, Zeng K, et al. Curcumin represses mouse 3T3-L1 cell adipogenic differentiation via inhibiting miR-17-5p and stimulating the Wnt signalling pathway effector Tcf7l2. *Cell Death Dis* 2017;8. https://doi.org/10.1038/cddis.2016.455.

[49] Xu D, Yuan L, Che M, Lu D, Liu W, Meng F, et al. Molecular mechanism of Gan-song Yin inhibiting the proliferation of renal tubular epithelial cells by regulating miR-21-5p in adipocyte exosomes. *J Ethnopharmacol* 2024;321. https://doi.org/10.1016/j.jep.2023.117530.

[50] Lin Q, Gao Z, Alarcon RM, Ye J, Yun Z. A role of miR-27 in the regulation of adipogenesis. *FEBS J* 2009;276:2348–58. https://doi.org/10.1111/j.1742-4658.2009.06967.x.

[51] Yang S, Yang G, Wang X, Li L, Li Y, Xiang J, et al. MicroRNA-92b in the skeletal muscle regulates exercise capacity via modulation of glucose metabolism. *J Cachexia Sarcopenia Muscle* 2023;14:2925–38. https://doi.org/10.1002/jcsm.13377.

[52] Li T, Cao H, Zhuang J, Wan J, Guan M, Yu B, et al. Identification of miR-130a, miR-27b and miR-210 as serum biomarkers for atherosclerosis obliterans. *Clin Chim Acta* 2011;412:66–70. https://doi.org/10.1016/j.cca.2010.09.029.

[53] Hanin G, Yayon N, Tzur Y, Haviv R, Bennett ER, Udi S, et al. miRNA-132 induces hepatic steatosis and hyperlipidaemia by synergistic multitarget suppression. *Gut* 2018;67:1124–34. https://doi.org/10.1136/gutjnl-2016-312869.

[54] Xiao Q, Yang S, Yang Y, Ni H, Li Z, Wang C, et al. LncRNA A2ml2 inhibits fatty liver hemorrhage syndrome progression and function as ceRNA to target LPL by sponging miR-143-5p. *Poult Sci* 2025;104. https://doi.org/10.1016/j.psj.2025.105003.

[55] Tang Y, Yang LJ, Liu H, Song YJ, Yang QQ, Liu Y, et al. Exosomal miR-27b-3p secreted by visceral adipocytes contributes to endothelial inflammation and atherogenesis. *Cell Rep* 2023;42. https://doi.org/10.1016/j.celrep.2022.111948.

[56] Bibiloni P, Pomar CA, Palou A, Sánchez J, Serra F. miR-222 exerts negative regulation on insulin signaling pathway in 3T3-L1 adipocytes. *BioFactors* 2023;49:365–78. https://doi.org/10.1002/biof.1914.

[57] Alvites R, Branquinho M, Sousa AC, Lopes B, Sousa P, Maurício AC. Mesenchymal stem/stromal cells and their paracrine activity – immunomodulation mechanisms and how to influence the therapeutic potential. *Pharmaceutics* 2022;14. https://doi.org/10.3390/pharmaceutics14020381.

[58] Ma H, Siu WS, Leung PC. The potential of MSC-based cell-free therapy in wound healing – a thorough literature review. *Int J Mol Sci* 2023;24. https://doi.org/10.3390/ijms24119356.

[59] Trounson A, McDonald C. Stem cell therapies in clinical trials: Progress and challenges. *Cell Stem Cell* 2015;17:11–22. https://doi.org/10.1016/j.stem.2015.06.007.

[60] Ribeiro AL, Dallagiovanna B. The role of long non-coding RNAs in human endoderm differentiation. *Noncoding RNA* 2025;11. https://doi.org/10.3390/ncrna11020029.

[61] Pan S, Chen Y, Yan J, Li F, Chen X, Xu X, et al. The emerging roles and mechanisms of exosomal non-coding RNAs in the mutual regulation between adipose tissue and other related tissues in obesity and metabolic diseases. *Front Endocrinol (Lausanne)* 2022;13. https://doi.org/10.3389/fendo.2022.975334.

[62] Zuo Z, Li N, Zhang Q, Liu Q, Qin H, Yao K. The role of non-coding RNAs in diabetic retinopathy: Mechanistic insights and therapeutic potential. *Mol Neurobiol* 2025. https://doi.org/10.1007/s12035-025-04863-z.

[63] Hossam Abdelmonem B, Kamal LT, Wardy LW, Ragheb M, Hanna MM, Elsharkawy M, et al. Non-coding RNAs: Emerging biomarkers and therapeutic targets in cancer and inflammatory diseases. *Front Oncol* 2025;15. https://doi.org/10.3389/fonc.2025.1534862.

[64] Thomou T, Mori MA, Dreyfuss JM, Konishi M, Sakaguchi M, Wolfrum C, et al. Adipose-derived circulating miRNAs regulate gene expression in other tissues. *Nature* 2017;542:450–5. https://doi.org/10.1038/nature21365.

[65] Masyuk AI, Masyuk TV, Larusso NF. Exosomes in the pathogenesis, diagnostics and therapeutics of liver diseases. *J Hepatol* 2013;59(3):621–5. https://doi.org/10.1016/j.jhep.2013.03.028.

[66] Pang JL, Shao H, Xu XG, Lin ZW, Chen XY, Chen JY, et al. Targeted drug delivery of engineered mesenchymal stem/stromal-cell-derived exosomes in cardiovascular disease: Recent trends and future perspectives. *Front Bioeng Biotechnol* 2024;12. https://doi.org/10.3389/fbioe.2024.1363742.

10 Exosome-Mediated Immune Regulation in Autoimmune and Inflammatory Diseases

Smriti Dewangan[1], *Anamika Singh*[2], *Gregory Smith*[3], *and Jaya Patel*[4]
[1]Amity Institute of Pharmacy, Amity University Chhattisgarh, Raipur, Chhattisgarh, India
[2]Indukala Ipcowala College of Pharmacy, CVM University, Anand, Gujarat
[3]Department of Chemistry, University of Cape Town, South Africa
[4]Department of Pharmacognosy, Parul Institute of Pharmacy and Research, Parul University, Vadodara, Gujarat

10.1 INTRODUCTION

10.1.1 Overview of Autoimmune and Inflammatory Diseases

Autoimmune diseases occur when the immune system fails to recognize the body's cells as "self" and instead attacks them as foreign invaders. This misguided immune response can affect nearly any organ or system in the body. Common autoimmune diseases include rheumatoid arthritis (RA), systemic lupus erythematosus (SLE), type 1 diabetes, multiple sclerosis (MS), and autoimmune thyroid diseases like Hashimoto's thyroiditis and Graves' disease. The exact cause of autoimmune diseases remains unclear, but it is believed to involve a combination of genetic predisposition and environmental triggers such as infections, toxins, or stress. Women are disproportionately affected by autoimmune diseases, comprising approximately 75% of all cases. Symptoms vary depending on the specific condition and affected organs, but common features include fatigue, joint pain, skin rashes, and fever (1). Diagnosis typically involves a combination of clinical evaluation, laboratory testing (including autoantibody panels), and imaging studies. Treatment focuses on controlling the immune response and managing symptoms. Immunosuppressive medications, corticosteroids, and disease-modifying antirheumatic drugs (DMARDs) are commonly used, with newer biologic therapies offering more targeted approaches. Inflammatory diseases encompass autoimmune and nonautoimmune conditions marked by chronic or recurrent inflammation. Some inflammatory disorders, such as inflammatory bowel disease (IBD) – including Crohn's disease and ulcerative colitis – are believed to involve both immune dysregulation and environmental factors. Others, like gout or asthma, may not be autoimmune but involve excessive or inappropriate inflammatory responses. Inflammation, while part of the body's defense system, becomes problematic when it persists without a clear threat. Chronic inflammation can damage tissues over time and has been linked to a wide range of conditions, including cardiovascular disease, certain cancers, neurodegenerative diseases, and metabolic disorders like type 2 diabetes (2).

10.1.2 Role of Immune Dysregulation

Immune dysregulation refers to an abnormal or impaired functioning of the immune system, where the balance between immune activation and suppression is disrupted. This imbalance can result in either overactive immune responses, leading to autoimmunity and chronic inflammation, or underactive responses, leading to immunodeficiency and increased susceptibility to infections and cancer. The immune system is normally tightly regulated to distinguish between self and non-self, and to maintain a balance between attacking harmful agents and preserving healthy tissues. When this regulation is disturbed, the immune response becomes inappropriate or harmful (3).

10.1.3 Emergence of Exosomes in Immunological Research

Exosomes are small extracellular vesicles (30–150 nm in diameter) released by nearly all cell types, including immune cells. Originally thought to be cellular waste products, exosomes are now recognized as important mediators of intercellular communication, especially in immunology. They carry proteins, lipids, and nucleic acids (such as mRNA and microRNA), reflecting the physiological state of the cells they originate from.

In immunological research, exosomes have gained significant attention due to their roles in modulating immune responses. Immune cells such as dendritic cells (DCs), T cells, B cells, and macrophages release exosomes that can either stimulate or suppress immune activity depending on their content and target cells. For example, DC-derived

 DOI: 10.1201/9781003623632-10

exosomes can present antigens to T cells, promoting an immune response, while regulatory T-cell-derived exosomes may suppress immune activity, maintaining immune tolerance (4). Exosomes also play a role in autoimmune diseases, cancer immunity, and infectious diseases. In autoimmune diseases like systemic lupus erythematosus and rheumatoid arthritis, exosomes may carry autoantigens or inflammatory molecules that exacerbate immune dysregulation. In cancer, tumor-derived exosomes can modulate the immune system to create an immunosuppressive environment that helps tumors evade immune surveillance. Conversely, exosomes are being explored as cancer vaccines by delivering tumor antigens to activate antitumor immunity. Technological advances in exosome isolation and characterization have expanded their research potential. Exosomes are also being studied as biomarkers due to their stability in bodily fluids and as therapeutic vehicles for targeted drug or gene delivery (5).

10.2 BIOGENESIS AND COMPOSITION OF EXOSOMES

10.2.1 Mechanism of Exosome Formation

Exosomes are formed through a multistep process involving the endosomal pathway shown in Figure 10.1.

- *Endocytosis:* The cell membrane invaginates to form early endosomes.
- *Maturation:* Early endosomes mature into late endosomes or multivesicular bodies (MVBs), which contain intraluminal vesicles (ILVs).
- *ILV formation:* ILVs are formed by inward budding of the endosomal membrane, packaging proteins, lipids, and RNA into vesicles inside the MVB.
- *Exosome release:* MVBs fuse with the plasma membrane, releasing ILVs into the extracellular space as exosomes (6).

10.2.2 Molecular Cargo: Proteins, RNAs, Lipids

Exosomes are extracellular vesicles that carry a wide range of molecular cargo from their cell of origin. This cargo is selectively packaged during exosome formation and plays crucial roles in intercellular communication, immune modulation, and disease progression. The three major categories of exosome cargo are proteins, RNAs, and lipids. The other is a surface protein. Each class contributes to the biological function and diagnostic potential of exosomes presented in Figure 10.1. Exosomal proteins are involved in vesicle formation, trafficking, immune modulation, and signal transduction. These proteins make exosomes important players in immune response, cancer metastasis, and inflammation. Exosomes carry various RNA molecules, which can be delivered to recipient cells to alter gene expression. These RNA cargos are protected from degradation due to the lipid bilayer of exosomes. Exosomal RNAs are considered promising biomarkers for diseases such as cancer, neurodegenerative disorders, and autoimmune conditions. Exosomal lipids can serve as ligands for receptors on recipient cells and play roles in inflammation, apoptosis, and cellular recognition. Exosomes possess specific surface proteins that help in their identification and characterization (7). These markers are used in characterization methods such as Western blotting, flow cytometry, and nanoparticle tracking analysis. Different types of proteins, RNAs, and lipids are mentioned in Table 10.1. Exosome isolation is critical for research and clinical applications.

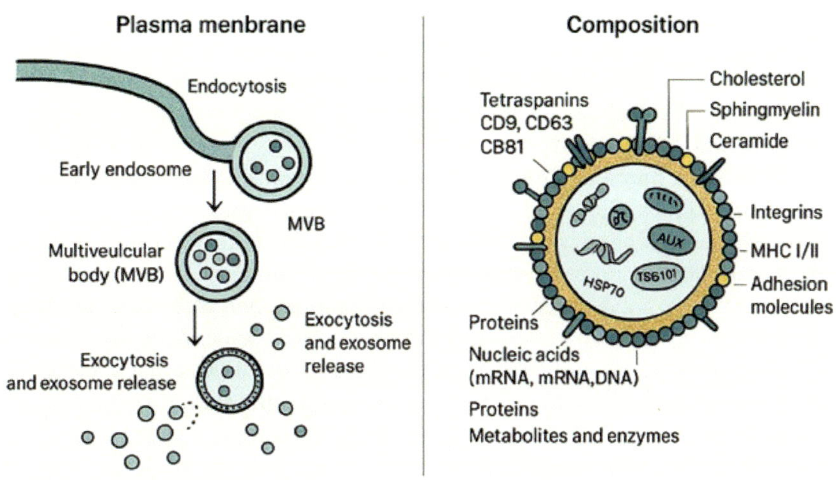

FIGURE 10.1 Biogenesis and composition of exosomes.

TABLE 10.1
Composition and Isolation Technique of Exosomes

Protein	RNAs	Lipids	Surface Markers	Isolation	References
Tetraspanins	microRNAs (miRNAs)	Cholesterol and sphingomyelin	Tetraspanins	Ultracentrifugation	(5)
ESCRT proteins	mRNAs	Ceramides	ESCRT-related proteins	Size-exclusion chromatography (SEC)	(6)
Heat shock proteins (HSP70, HSP90)	Long noncoding RNAs (lncRNAs)	Phosphatidylserine	Cell-type-specific markers	Precipitation-based methods	(5)
Cytoskeletal proteins	Circular RNAs (circRNAs)	Glycosphingolipids	Other markers like heat shock proteins (HSP70, HSP90), integrins, and MHC molecules	Immunoaffinity capture	(7)
Antigen-presenting molecules	–	–	–	Microfluidic devices	(6)

10.3 IMMUNOMODULATORY FUNCTIONS OF EXOSOMES

10.3.1 Exosome Interactions with Innate Immune Cells

Exosomes are essential in controlling innate immune responses because they permit communication between cells. Innate immune cells, including macrophages, dendritic cells (DCs), neutrophils, and natural killer cells (NK cells), can secrete or receive exosomes, which regulate inflammation, pathogen defense, and modulate immune cells (8). Macrophages can take up danger signal-laden exosomes (DAMPs or PAMPs) and stimulate cytokine production and activation. On the contrary, tumor-associated exosomes could inhibit macrophage activity or mobilize them toward the tumor-facilitating (M2) phenotype. Dendritic cells secrete MHC I and MHC II molecule-containing exosomes, targeting the antigen presentation to T cells, and thereby establishing a relationship between the innate and adaptive immunity (9). Neutrophils release exosomes containing enzymes and inflammatory mediators that can modulate other immune cells or contribute to tissue damage during excessive inflammation. Natural killer (NK) cells can be activated or suppressed by exosomes depending on their origin. For example, exosomes from infected or stressed cells may enhance NK cytotoxicity, while tumor-derived exosomes may inhibit NK cell function. Overall, exosome-based communication fine-tunes the innate immune response, leading to host defense or, in a disease context, contributes to immune evasion (Figure 10.2).

10.3.2 Effects on Adaptive Immune Responses

Exosomes have a major function of controlling adaptive immune response through interaction with T and B cells. These nanoscale vesicles can take antigens, MHC molecules, co-stimulatory signals, and regulatory RNA or proteins, to enable them to regulate immune activation and suppression. The exosomes of antigen-presenting cells, including dendritic cells (DCs), are abundant in MHC class I and II molecules, in addition to co-stimulatory molecules such as CD80 and CD86. Such exosomes are capable of directly loading antigens for $CD4^+$ helper T cells or $CD8^+$ cytotoxic T cells, thereby triggering a specific immune reaction. Such a function is especially important for anti-tumor immunity and vaccine development (10). Specific exosomes, including exosomes derived from regulatory T cells (Tregs), mesenchymal stem cells, or tumors, may inhibit adaptive immunity. They could contain immunosuppressive cytokines (e.g., TGF-β, IL-10) or checkpoint molecules such as PD-L1 that also block T-cell proliferation and function. These exosomes are responsible for immune tolerance, which is essential to preventing autoimmunity, but cancer cells are also able to exploit this for evasion of immune detection (11, 12).

10.3.3 Cross Talk between Immune Cells via Exosomes

Exosomes are critically immunomodulatory as they provide cross talk between immune cells to coordinate and regulate immune responses. Such miniature extracellular vesicles (30–150 nm) serve as shuttles for bioactive components – proteins, lipids, and nucleic acids – with the result that immune cells can act on each other across space.

B cells can take up exosomes that contain antigens and immune-regulating RNAs, which control their activation, proliferation, and secretion of antibodies. T follicular helper cell exosomes have the potential to facilitate B-cell differentiation in germinal centers. Exosomes taken from various immune cells may have consequences on macrophage polarization. For instance, T-cell-derived exosomes can reprogram macrophages toward a pro-inflammatory (M1) or anti-inflammatory (M2) phenotype by the secreted signaling molecules. Immune cells can also

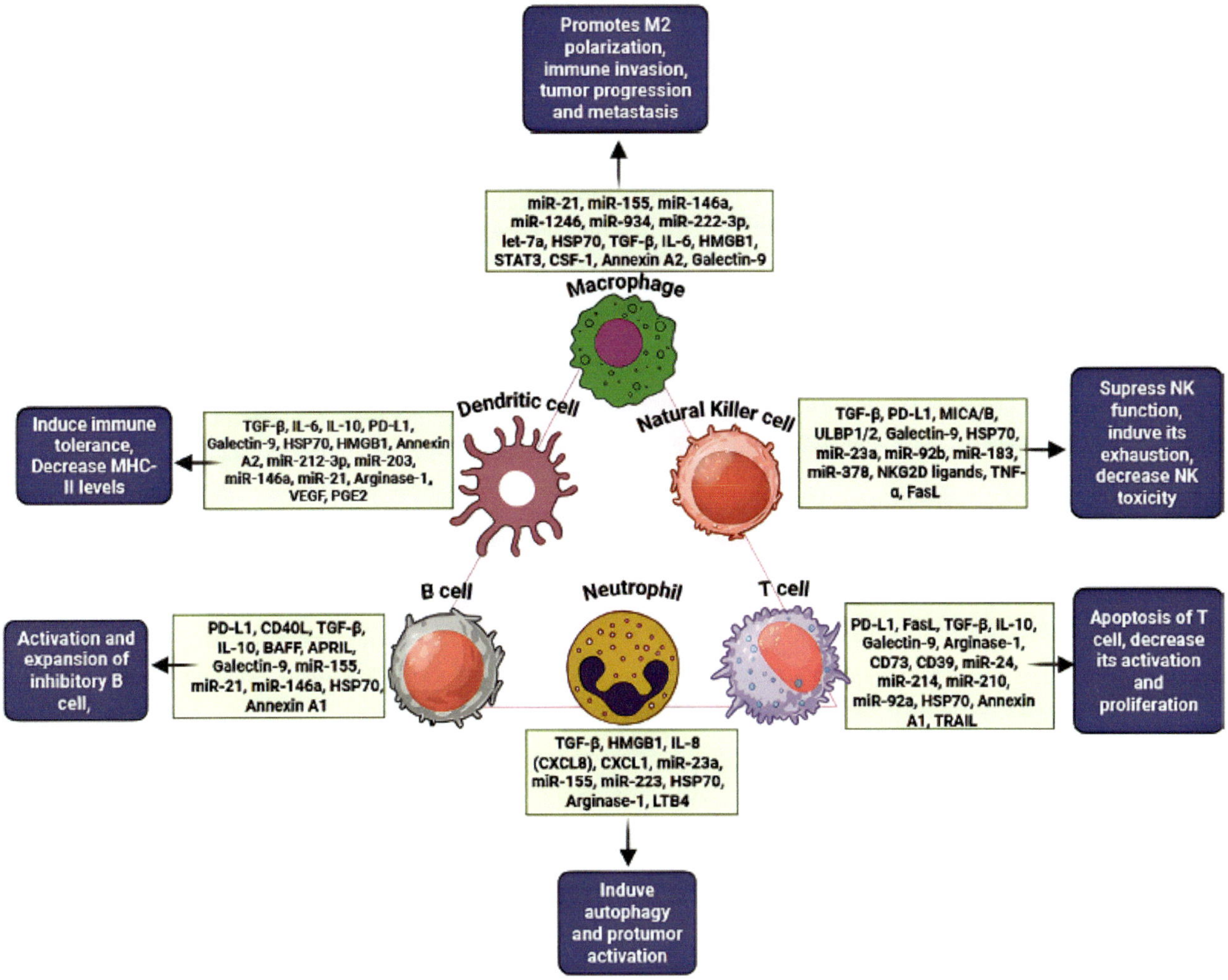

FIGURE 10.2 Functional molecules in tumor-derived exosomes for immunomodulation and exosome-mediated regulation of immune cells.

employ exosomes for negative feedback to preclude an unwarranted increase in immune activation. For instance, apoptotic-cell-derived exosomes might deliver suppressive signals to further inhibit activation of the immune system (13).

10.4 EXOSOMES IN AUTOIMMUNE DISEASES

10.4.1 Rheumatoid Arthritis

Rheumatoid arthritis is a chronic systemic autoimmune disease characterized mostly by symptoms of the joints. It involves chronic inflammation of the synovial membrane that destroys the joint, causing pain and disability. The disease is caused by an aberrant response of the immune system, innately predisposed and adaptively programmed, accompanied by numerous inflammatory mediators. Over the past decade, it has been discovered that exosomes are a major player in the pathogenesis, development, and treatment of RA. Below are described the functions of exosomes in RA pathogenesis.

Exosomes produced by T cells, B cells, and macrophages in patients of rheumatoid arthritis (RA) carry the pro-inflammatory cytokines (e.g., TNF-α, IL-1β, IL-6) along with autoantigens (e.g., citrullinated peptides). These exosomes can increase inflammation by activating other immune cells and stimulating further production of inflammatory mediators (14). In RA, synovial fibroblasts are hyperactive and invasive. Their exosomes contain matrix metalloproteinases (MMPs) and other enzymes that help to break down cartilage and cause joint damage. Several exosomes in RA synovial fluid frequently include citrullinated proteins, which are critical in the production of anti-citrullinated protein antibodies (ACPAs), which characterize RA.

10.4.2 Systemic Lupus Erythematosus

Exosomes in SLE are small extracellular vesicles (30–150 nm in diameter) produced by immune and nonimmune cells. These exosomes contain autoantigens, pro-inflammatory molecules, and regulatory RNAs, leading to the immune dysregulation, autoantibodies production, and tissue damage occurring in the lupus patients. The role of exosomes in SLE pathogenesis is explained below. Nuclear components of exosomes in SLE, such as DNA, histones, and RNA, are identified as autoantigens. These may induce the formation of antinuclear antibodies (ANAs), a characteristic feature of SLE. Exosome RNA and DNA can stimulate plasmacytoid dendritic cells (pDCs) through Toll-like receptors (TLR7 and TLR9), and type I interferons, key elements of lupus pathology, are produced. Lupus exosomes can stimulate B cells and T cells, which spur the autoimmune response. Some exosomal microRNAs (e.g., miR-146a, miR-155) change in SLE and may be used as biomarkers of SLE activity or flare prediction (15).

10.4.3 Multiple Sclerosis

Exosomes in MS are small extracellular vesicle particles (30–150 nm) released by a myriad of cells, such as immune cells, neurons, and others. These vesicles transport proteins and lipids, as well as nucleic acids (such as microRNAs), and enable intercellular signaling that can boost neuroinflammation or promote neuroprotection and repair depending on their source and cargoes. The role of exosomes in MS pathogenesis is as follows (16):

- Immune modulation and neuroinflammation
- Myelin damage
- Neuron and glial cell communication

10.4.4 Type 1 Diabetes

In type 1 diabetes, the exosomes are nonreceptor-mediated extracellular vesicles secreted by pancreatic beta cells, immune cells, and other tissues, which participate in intercellular communication as well as immune modulation. Below are explained the roles of exosomes in the pathogenesis of type 1 diabetes (17). Beta-cell-derived exosomes carry autoantigens (e.g., insulin, GAD65, IA-2) that can be taken up by antigen-presenting cells (APCs) and presented to T cells, triggering autoimmune responses. Exosomes can promote activation of autoreactive $CD4^+$ and $CD8^+$ T cells, leading to beta-cell destruction. Inflammatory exosomes derived from immune cells may promote exacerbation of the islet inflammation (insulitis) (18, 19).

10.5 EXOSOMES IN CHRONIC INFLAMMATORY DISORDERS

10.5.1 Inflammatory Bowel Disease

In the case of inflammatory bowel disease (IBD), comprising Crohn's disease and ulcerative colitis, exosomes perform quite an important role in controlling gut inflammation, epithelial barrier function, as well as immune responses. Their actions may be pathogenic and protective, depending on the source and cargo. Exosomes are being researched toward the establishment of biomarkers for IBD diagnosis, as well as therapeutic applications for exosomes in the targeted delivery of anti-inflammatory agents. Below, the pathogenic role and the protective role of exosomes are given (20):

Pathogenic roles: Immune cell–derived exosomes, like pro-inflammatory exosomes from activated immune cells (e.g., macrophages, T cells, TNF-α, IL-6) that worsen the intestinal inflammation, or miRNA that enhances the inflammatory state. Epithelial cell exosomes interfere with immune responses and the integrity of barriers, amplifying IBD pathology. Microbiota-derived exosomes could activate inflammation by PRRs (e.g., TLRs) (21).

Protective roles: Mesenchymal stem cell (MSC)-derived exosomes possess anti-inflammatory activities, helping damage reparation and regulation of immune responses through the transfer of miRNAs (i.e., miR-146a) that suppress NF-κB signaling. Exosomes of epithelial cells could play a regulatory role in immune tolerance and preservation of mucosal integrity. Probiotic-derived vesicles inhibit inflammation and increase the epithelial barrier function. Table 10.2 briefly shows the role of exosomes in chronic inflammatory disorders.

10.5.2 Psoriasis

Pathogenic roles: Keratinocyte-derived exosomes loaded with inflammatory mediators (e.g., IL-17, IL-23, TNF-α) and specific microRNAs (like miR-203, miR-21) that activate dendritic cells and T cells, driving chronic inflammation. Exosomes from activated T cells and dendritic cells can amplify the immune response by promoting Th17 polarization and cytokine release, key events in psoriasis pathogenesis. Exosomes act as carriers for inflammatory signals between immune cells and skin cells, perpetuating the inflammatory loop characteristic of psoriatic lesions (22).

10.5.3 Chronic Obstructive Pulmonary Disease (COPD)

Exosomes play a role in promoting COPD and have diagnostic and therapeutic potential. Exosomes in COPD are both causes and possible targets of therapy and diagnosis. Explanation of the pathogenic and protective roles of the exosomes is presented below:

Pathogenic roles: In COPD patients, exosomes from airway epithelial cells and immune cells transport pro-inflammatory cytokines, proteases, and molecules that

TABLE 10.2
Exosomes in Chronic Inflammatory Disorders

Autoimmune Disease	Source of Exosomes	Role/Mechanism	Therapeutic/Diagnostic Potential	References
Rheumatoid arthritis (RA)	Synovial fibroblasts, dendritic cells, and MSCs	Modulate immune response; exosomes carry TNF-α, IL-6, miR-150, citrullinated peptides	Diagnostic biomarker potential; MSC-derived exosomes reduce joint inflammation	(19)
Systemic lupus erythematosus (SLE)	B cells, dendritic cells, and MSCs	Exosomes contain autoantigens (e.g., dsDNA), pro-inflammatory cytokines	Diagnostic markers (e.g., exosomal miR-21); therapeutic modulation via MSC exosomes	(20)
Multiple sclerosis (MS)	T cells, oligodendrocytes, and microglia	Transfer of inflammatory signals, demyelination factors, or regulatory miRNAs	MSC-derived exosomes may promote remyelination and reduce neuroinflammation	(21)
Type 1 diabetes mellitus (T1DM)	Pancreatic β-cells, immune cells	Carry autoantigens (e.g., insulin, GAD65) and trigger an immune response	Early diagnosis, potential for immune modulation via tolerogenic exosomes	(21)
Inflammatory bowel disease (IBD)	Intestinal epithelial cells, macrophages	Mediate gut inflammation via cytokine and miRNA delivery	MSC exosomes suppress inflammation, promote epithelial repair	(21)
Psoriasis	Keratinocytes, T cells	Contain pro-inflammatory cytokines and miRNAs contributing to skin inflammation	Potential target for modulating immune response and reducing skin lesions	(22)

cause oxidative stress that destroy lung tissue and stimulate chronic inflammation. Smoking stimulates exosome release that is loaded with harmful miRNAs (e.g., miR-210, miR-21) that will damage tissue repair, leading to fibrosis and emphysema. Exosomes may affect macrophage and neutrophil activity, creating chronic inflammation and tissue destruction (23).

Protective roles: Mesenchymal stem cell (MSC) exosomes have anti-inflammatory and regenerative potential in preclinical models and could decrease lung injury and improve repair. Exosome miRNA signatures (such as miR-223, miR-21) are being investigated in research as noninvasive biomarkers for COPD detection and monitoring in the early stage.

10.5.4 Atherosclerosis and Cardiovascular Inflammation

Atherosclerosis is a chronic inflammatory disorder involving the accumulation of lipid-rich plaques in the walls of the arteries, leading to cardiovascular events such as heart attacks and strokes. Today, exosomes, small extracellular vesicles (up to 150 nm), are known as important facilitators in the development and proliferation of atherosclerosis and cardiovascular inflammation. They can prove to be of pro-atherogenic or protective nature depending on their origin and molecular cargo. Pathogenic roles of exosomes in the pathogenesis of the disease are as follows (24);

Pathogenic roles: Under stress (e.g., oxidized LDL, high glucose, or turbulent blood flow), the endothelial cells send exosomes loaded with high levels of miR-155, miR-92a, and adhesion molecules (ICAM-1, VCAM-1). These exosomes induce monocyte adhesion, inflammation, and endothelial dysfunction – early steps of atherogenesis. Macrophage-derived and foam cell–derived exosomes contain pro-inflammatory cytokines (for instance, TNF-α, IL-1β) and miRNAs that stimulate the neighboring immune cells and vascular smooth muscle cells (VSMCs). Platelet-derived exosomes have tissue factor, prothrombotic molecules, and mediators of inflammation. They promote vascular inflammation and enhance the risk of plaque rupture and thrombosis (25).

Protective roles of stem-derived exosomes: Stem-derived exosomes are abundant in anti-inflammatory miRNAs (miR-126, miR-21, and miR-146a) and growth factors. They stimulate endothelial repair, angiogenesis, and resolution of inflammation; decrease plaque size; and stabilize lesions in animal models. Healthy endothelial exosomes preserve vascular homeostasis to transport protective signals to VSMCs and the immune cells. They inhibit oxidative stress and maintain NO bioavailability.

10.6 EXOSOME-MEDIATED MOLECULAR MECHANISMS

10.6.1 miRNAs and Cytokine Signaling

Exosomes are nanosized extracellular vesicles (30–150 nm) released by nearly all cell types and are responsible for crucially important intercellular communication. Their most significant cargo consists of microRNAs (miRNAs) and cytokines, both are highly regulators of cytokine signaling pathways. The two coordinately assemble a highly complex network of regulation that is responsible for immunity, inflammation, and the progression of disease (7).

10.6.1.1 Exosome miRNAs: Regulators of Cytokine Signaling

miRNAs refer to the small noncoding RNAs that regulate gene expression at posttranscriptional level by binding to complementary sites on target mRNA and exhibiting translational suppression or degradation. Encapsulated into exosomes, miRNAs are not subject to enzymatic degradation and transported to distant recipient cells, where they have functional effects. Exosomal miRNA has a central role in regulating cytokine signaling by upregulating or downregulating cytokine expression and their pathways. Different miRNAs are miR-21, which is frequently increased in inflammatory diseases and cancers. miR-21 fosters cytokine signaling by targeting negative regulators of the NF-κB pathway, resulting in the upregulation of IL-6, IL-8, and TNF-α. miR-146a plays a negative feedback role. It inhibits IRAK1 and TRAF6, which are upstream NF-κB activators, thus inhibiting the expression of inflammatory cytokines. It is commonly present in MSCs and immune cells' exosomes in the resolution of inflammation. miR-155: recognized for being pro-inflammatory, expression of miR-155 increases the Th1 and Th17 immune response by targeting suppressors of cytokine signaling (e.g., SOCS1), thus enhancements of IFN-γ and IL-17 production (26).

10.6.1.2 Exosomal Cytokines and Receptors

Besides miRNAs, exosomes can contain cytokines (e.g., IL-1β, TNF-α, IL-6) and cytokine receptors (e.g., TNFR, IL-6R). These components can directly interact with the target cells and regulate their immune or inflammatory reactions. Cytokine receptors on exosomes can also function as decoys, which bind to unbound cytokine in the extracellular area and thus regulate in terms of bioactivity, a regulation mechanism during chronic inflammation or autoimmune dysfunction. Exosomes secreted by the immune cells are usually loaded with pro-inflammatory cytokines that can potentiate the immune reactions locally or transfer to general systemic inflammation. T regulatory (Treg) cell-derived or MSC-derived exosomes may communicate with cells bearing TGF-β or IL-10, which helps to suppress immune activation and induce immune tolerance (27).

10.6.1.3 Dynamic Interplay: Cytokines Influence miRNA Loading

Cytokine environment can also control the exosomal miRNAs content. For example, exposure to IL-1β or TNF-α can increase the levels of miR-155 or miR-21 in exosomes of cells of the immune system. The activation of TGF-β can promote the increase in miR-21 in exosomes derived from fibroblasts, thus participating in tissue remodeling or fibrosis.

Exosome miRNAs and cytokine cargoes are used for studying environmental risk factors, which can be measured. Biomarkers are used for the identification of disease and prognosis (e.g., cancer, cardiovascular disease, autoimmune disease). Therapeutic agents or targets, such as exosomes, are designed to transport anti-inflammatory miRNAs or decoy receptors to the inflamed tissues (28).

10.6.2 Antigen Presentation and Immune Cell Activation

It is an important process in the immune response, whereby the antigen-presenting cells, (APCs) like dendritic cells, macrophages, and B cells, process and present antigens to the surface of their cells, using major histocompatibility complex (MHC) molecules. MHC class I presents antigens to CD8+ cytotoxic T cells, while MHC class II presents to CD4+ helper T cells. On the recognition of an antigen-MHC complex via the T-cell receptor (TCR), T cells get activated. This activation needs co-stimulatory signals (e.g., CD80/CD86 binding to CD28 on T cells) in addition to antigen recognition. Activated T cells then proliferate and differentiate into effector cells – cytotoxic T cells to kill infected or cancer cells, or helper T cells to secrete cytokines to facilitate the immune response. The exosomes shed by APCs are also able to transfer MHC–antigen complexes and co-stimulatory molecules to the effect of distant antigen presentation as well as immune activation complexification. This process is vital for beginning adaptive immunity, clearing infections, and establishing immunological memory. Dysregulation may result in deficiencies of the immune system or autoimmune diseases (14).

10.6.3 Modulation of Treg/Th17 Balance

10.6.3.1 Modulation of Treg/Th17 Balance by Exosomes

Exosomes have a strong impact on the balance of regulatory T cells (Tregs) and T helper 17 (Th17) cells; it is one of the main tasks of immune regulation. The balance of Treg/Th17 in itself determines whether the immune system can induce tolerance or inflammation, and the disruption is associated with autoimmune diseases, cancer, and chronic inflammation.

10.6.3.2 Exosomal Influence on Treg/Th17 Differentiation

Exosomes originated from mesenchymal stem cells (MSCs) and transmit anti-inflammatory miRNAs (e.g., miR-146a, miR-21, miR-155) and TGF-β, leading to Treg amplification and neutralization of Th17 maturation. These exosomes suppress the STAT3/RORγt axis, one of the major pathways in Th17 cells development, and increase FoxP3 expression, critical in Treg function. When produced by activated dendritic cells or Th17 cells, exosomes can contain IL-6, IL-23, and miR-326 that will promote the Th17 polarization and inhibit Treg activity. On the other hand, exosomes secreted from Treg cells may present immunosuppressive molecules such as IL-10, TGF-β, and CTLA-4, preventing the activation of effector T cells as well as maintaining immune tolerance (29).

TABLE 10.3
Therapeutic Potential of Exosomes

Therapeutic Area	Mechanism/Role of Exosomes	Examples/Applications	References
Rheumatoid arthritis (RA)	Inhibit pro-inflammatory cytokines (e.g., TNF-α, IL-1β); modulate T cell and macrophage activity	MSC-derived exosomes reduce joint inflammation and synovial hyperplasia in RA models	(29)
Inflammatory bowel disease (IBD)	Regulate intestinal immune homeostasis; inhibit NF-κB signaling and Th17 differentiation	Exosomes from human umbilical cord MSCs alleviate colitis symptoms in IBD animal models	(30)
Systemic lupus erythematosus (SLE)	Deliver anti-inflammatory miRNAs; modulate B- and T-lymphocyte responses	Exosomes enriched with miR-146a reduce renal inflammation in lupus nephritis models	(31)
Multiple sclerosis (MS)	Suppress autoreactive T cells and microglial activation; promote regulatory T cells (Tregs)	Neural stem cell–derived exosomes mitigate CNS inflammation and demyelination in MS mice	(29)
Psoriasis	Modulate dendritic cell maturation and keratinocyte proliferation	Adipose-derived MSC exosomes reduce epidermal thickening and cytokine levels in psoriasis	(30)
Type 1 diabetes mellitus (T1DM)	Induce immune tolerance; protect pancreatic islets from immune-mediated destruction	Exosomes from tolerogenic dendritic cells prevent β-cell autoimmunity in T1DM models	(30)
Autoimmune uveitis	Inhibit Th1 and Th17 responses; reduce ocular inflammation	MSC-derived exosomes show therapeutic effect in experimental autoimmune uveitis	(30)
Systemic sclerosis (SSc)	Modulate fibroblast activity and immune cell infiltration	Exosomes downregulate fibrotic markers and inflammatory mediators in SSc skin models	(31)

10.7 THERAPEUTIC POTENTIAL OF EXOSOMES

10.7.1 Exosome-Based Drug Delivery Systems

These systems are becoming an attractive perspective for targeting and effective delivery of therapeutic agents mentioned in Table 10.3. Exosomes are natural nanocarriers (30–150 nm) with the ability to encapsulate various bioactive molecules, including proteins, nucleic acids (siRNA, mRNA), and small molecules. Compared to traditional drug delivery systems, this system has the following advantages:

- *Biocompatibility:* Exosomes are obtained from natural cells and hence are naturally biocompatible and not prone to cause immune reactions.
- *Targeted delivery:* By decorating the surface of exosomes with ligands, antibodies, or peptides, their targets may be adapted to distinct tissues or cells, thus enhancing the drug efficacy and diminishing side effects.
- *Protection of cargo:* The exosomes preserve sensitive cargo from the degradation of enzymes in the bloodstream, increasing the stability and half-life of drugs.
- *Gene therapy:* Exosomes are used as vehicles not only to deliver RNA-based therapies (siRNA, mRNA) for gene silencing or protein replacement, but also to initiate any kind of therapy through purification.
- *Cancer treatment:* Exosome-based systems can deliver chemotherapeutics and/or immune-modulatory agents directly to tumors, wherein other forms of toxicity are avoided.
- *Regenerative medicine:* Exosomes made of stem cells are being employed to support the repair of tissues and reduce the level of inflammation in diseases such as cardiovascular disease or neurodegeneration (30, 31).

10.7.2 Engineered Exosomes for Immune Modulation

The deliberate engineering of exosomes toward immune modulation is emerging as a new approach to the manipulation of immune processes in many diseases, such as autoimmune disease, cancer, and chronic inflammation (31). The key strategies and features of engineered exosomes have been summarized in Table 10.4.

10.8 CHALLENGES AND STRATEGIES IN CLINICAL TRANSLATION

Exosomes possess a high potential for use as drug transporters and agents of control of the body's defense mechanisms, but clinical uses have been handicapped by several challenges, which need rectification for their general use (35, 36).

Challenges: Efficient bulk isolation and purification of the high-quality exosomes continues to be a major barrier. Such techniques as ultracentrifugation, size-exclusion chromatography, and immunoaffinity capture can be slow and can cause contamination, therefore compromising

TABLE 10.4
Key Features of Engineered Exosomes

Engineering Strategy	Mechanism/Function	Target Immune Cells/Pathways	Application in Autoimmune/ Inflammatory Disorders	References
Surface modification with targeting ligands	Display of ligands/antibodies/peptides enables specific binding to immune cells (e.g., T cells, DCs)	CD4+ T cells, dendritic cells	Targeted delivery to autoreactive T cells in MS or RA for immune modulation	(32)
Surface engineering with MHC/co-stimulatory molecules	Present antigens or checkpoint molecules to induce tolerance or immunity	Antigen-specific T cells	Induction of antigen-specific tolerance in type 1 diabetes or SLE	(32)
Cargo loading of anti-inflammatory cytokines	Encapsulation of cytokines such as TGF-β, IL-10 promotes immune suppression	Regulatory T cells (Tregs), macrophages	Promotes immune tolerance in colitis, RA, psoriasis, and autoimmune uveitis	(33)
Loading with immune checkpoint inhibitors	Blockade of inhibitory receptors to enhance anti-inflammatory signaling or immune suppression	PD-1/PD-L1, CTLA-4 pathways	Rebalancing immune signaling in lupus or chronic inflammation	(33)
MSC-derived exosomes (natural immunosuppressors)	Naturally enriched in tolerogenic miRNAs and proteins that reduce inflammation	Tregs, macrophages, effector T cells	Broad application in RA, SLE, MS, IBD, and transplant rejection	(34)
Engineering exosomes with homing peptides	Peptides that guide exosomes to inflamed tissues or immune organs	Inflamed tissue microenvironment	Precise delivery in psoriatic lesions, inflamed synovium, and CNS inflammation	(34)

purity and function. There are no standardized and robust techniques for the characterization of exosomes (size, cargo, surface markers). Diversity of exosome populations between various sources may result in incongruous results in preclinical and clinical settings (37, 38). Rather, achieving a targeted delivery to a particular tissue or cell remains a challenge. Although exosomes possess inherent homing qualities, the specificity toward the tissues and the rapid clearance rate of these engineered exosomes by the mononuclear phagocyte system (MPS) might not be so efficient (39). Exosome production on a large scale is difficult, especially for clinical-grade exosomes. This results in a high cost of production; thus, it may curtail accessibility and applicability in clinical settings. The regulatory environment, as far as the exosome-based therapies are concerned, is not clear. Chapter 11 proves that safety, purity, and efficacy in clinical trials demand a full regulatory set of guidelines enacted (40).

10.8.1 Strategies for Overcoming Challenges

- *Improved isolation techniques:* Advancements in the automation and high throughput of isolated methods and microfluidic platforms can make the process faster and higher in purity.
- *Surface engineering:* Bringing in the surface modification utilizing targeting ligands or peptides is capable of increasing the targeting as well as the bio-distribution, thus improving the curative efficacy (41).
- *Standardization:* Writing standard operating procedures (SOPs) for exosome isolation and quality control will help in terms of reproducibility and reliability.
- *Clinical trials and regulatory framework:* Joint work with regulatory bodies may be used to develop categorical clinical guidelines and to support clinical trial approval for exosome-based therapies (42, 43).

10.9 EXOSOMES AS DIAGNOSTIC AND PROGNOSTIC BIOMARKERS

10.9.1 Circulating Exosomes in Body Fluids

Exosome is a small extracellular vesicle (30–150 nm) produced by different cell types in the body fluid such as blood, urine, saliva, cerebrospinal fluid (CSF), and pleural fluid. They play such important roles in intercellular communication, immunoregulation, and disease course, and their presence in body fluids helps establish them as excellent biomarkers for diagnosing and monitoring diseases and eventually for treatment purposes (44).

10.9.1.1 Biological Functions of Circulating Exosomes

Exosomes that circulate in the body are important regulators of the immune system because they can facilitate communication between cells. Most notably, dendritic cell (DCs)-derived exosomes serve as antigen-presenting vesicles that can prime immune responses by presenting peptide–MHC complexes to CD4+ and CD8+ T cells, thus stimulating adaptive immunity. More so, exosomes released by activated immune cells, including T lymphocytes and macrophages, may contain pro-inflammatory cytokines, chemokines, and other immunomodulatory molecules. These exosomal cargoes are able to change the behavior of the adjacent immune cells, resulting in potentiation of inflammatory responses, regulation of immune cell stimulation, and attraction of other immune effectors to inflamed sites (45, 46). Therefore, exosomes are crucial players in the development of the innate and adaptive immune responses by conveying immunological messages.

10.9.1.2 Disease Progression

Exosomes also play significant roles in the development and pathophysiology of different diseases. Within the oncological setting, tumor-derived exosomes have been frequently loaded with oncogenic proteins (e.g., EGFRvIII, HER2) and regulatory miRNAs (e.g., miR-21, miR-155), which can stimulate tumor growth, angiogenesis, metastasis, and immune evasion through modulation of the tumor microenvironment. These exosomes also have the capability to mirror the molecular portrait of the parent tumor, which can provide information about the stage and subtype of the tumor as well as its possible drug resistance (47). In cardiovascular diseases, endothelial cell-derived and activated platelet-derived exosomes carry pro-inflammatory cytokines and pro-thrombotic factors, including ICAM-1 and tissue factor, which promote vascular inflammation, endothelial dysfunction, and the formation of atherosclerotic plaques. Likewise, in neurodegenerative disorders, cerebrospinal fluid (CSF) exosomes have been demonstrated to take pathological proteins, including amyloid-beta peptides and hyperphosphorylated tau, involved in Alzheimer's disease. These exosomal markers can mirror the development of neuropathology, but can also become the early injured parties in diagnosis and prognosis (47).

10.9.1.3 Clinical Applications

Exosomes have the already acknowledged property of recapitulating the molecular signature of their originating cells, which makes them extremely promising as noninvasive diagnostic agents. Biomolecules like exosomal miRNAs and proteins in liquid body fluids, including blood and urine, can be accurate biomarkers in the early diagnosis of cancers, cardiovascular diseases, and neurodegenerative disorders. Besides their diagnostic opportunities, circulating exosomes may be exploited as natural drug delivery mechanisms, and they may carry therapeutic cargoes, delivering them right to the target tissues. This is not only targeted delivery, which increases the efficacy of treatment, but also reduces the systemic toxicity, making it a safer and more precise mode of delivery (48–50).

10.9.2 Disease-Specific Exosome Signatures

Exosomes are nanosized vesicles released by virtually all cell types in several body fluids. They have a high load of proteins, lipids, and RNAs (including miRNAs, lncRNAs, and mRNAs) that carry information regarding the physiological or pathological condition of their cell of origin. Disease-specific exosome signatures are molecular patterns within exosomes that characterize certain diseases, which are strong diagnostic, prognostic, and therapeutic monitoring tools (24).

Cancer-associated signatures: Tumor-derived exosomes (TDEs) are frequently loaded *with* oncogenic proteins (e.g., EGFR, KRAS) and miRNAs (e.g., miR-21, miR-155), which promote the growth and spread of the tumor. For example:

- *Lung cancer:* Elevated exosome miR-21 and EGFR mutations.
- *Breast cancer:* HER2 and exosome miR–373.
- *Pancreatic cancer:* Exosome GPC1 protein is a very specific marker. Such signatures allow differentiating one type of the disease and its subtypes and can be used for early diagnostics and detecting the efficiency of treatment (8).

Neurodegenerative diseases: In diseases such as Alzheimer's and Parkinson's, disease-specific proteins such as amyloid-β, tau, or α-synuclein carried in cerebrospinal fluid (CSF) or blood exosomes may be utilized for early diagnosis and the mechanism of disease.

Cardiovascular and metabolic disorders: In cardiovascular diseases, exosomes have miRNAs (such as miR-1, miR-133, miR-208) that are responsible for cardiac injury and repair. In diabetes, exosome signatures as well as altered levels of miR-29 or miR-375 may reflect insulin resistance and β-cell stress (40, 8).

Autoimmune and inflammatory diseases: Examples of diseases where exosome content is altered include rheumatoid arthritis, lupus, and the inflammatory bowel disease; dysregulation of pro-inflammatory cytokines and miRNAs can regulate the immune system (miR-146a, miR-155).

10.9.3 Early Detection and Disease Monitoring

Exosomes are developing as an important tool that can be used to detect illnesses at an early stage and to monitor changes in disease in terms of progression. These little vesicles that are secreted into body fluids by cells contain molecular signatures (proteins, miRNAs, lipids, and RNAs) that are indicative of their cell of origin condition (52, 53).

Early detection: In cancer, tumor-derived exosomes (TDEs) contain oncogenic proteins (e.g., EGFR, HER2) and miRNAs (e.g., miR-21, miR-155), which can be diagnosed by tests in blood, urine, or saliva for noninvasive screening. In neurodegenerative diseases, exosomes in cerebrospinal fluid (CSF) or blood could have proteins (tau proteins and amyloid-β), which provide indications on the early stage of Alzheimer's disease or Parkinson's disease. In cardiovascular diseases, exosome miRNAs (e.g., miR-1, miR-133a) can point to a myocardial injury or vascular damage, indicating early warning symptoms about cardiovascular conditions (54).

Disease monitoring – cancer: Serial assessment of exosome content can monitor progression of a tumor/response to therapy based on the alterations in exosome miRNA profiles or protein markers (e.g., GPC1 in pancreatic cancer).

Autoimmune diseases: Exosome levels of pro-inflammatory cytokines and immune-related miRNAs (e.g., miR146a) in conditions such as rheumatoid arthritis or lupus can serve to track the activity of the disease as well as response to treatment (55).

10.10 LIMITATIONS AND FUTURE PERSPECTIVES

10.10.1 Standardization Issues in Exosome Research

Exosome research has vast applications in terms of diagnostics, therapeutics, and drug delivery. There are, however, some problems of standardization of exosome isolation, characterization, and utilization, which have effects on reproducibility and clinical translation.

Isolation methods: Exosome isolation is a very important initial step of exosome-based research and clinical applications. However, there is no magic formula for the isolation of an exosome, which results in a difference in purity and yield. Conventional methods like ultracentrifugation, size-exclusion chromatography, or immunoaffinity capture all have advantages and disadvantages:

- *Characterization:* There is no standard known protocol for the characterization of exosomes. Even though some of the procedures, such as nanoparticle tracking analysis (NTA), dynamic light scattering (DLS), electron microscopy (EM), and Western blotting, are quite frequently used, each of them possesses limitations concerning resolution, accuracy, or reproducibility (56).
- *Exosome heterogeneity*: Exosomes are not a homogeneous population but a deeply heterogeneous population in terms of size/content and origin. This heterogeneity will make it very challenging to develop consistent markers or benchmarks for defining the populations of exosomes, which will result in a variability in the results, particularly where clinical trials are involved.
- *Lack of standardized protocols in clinical applications:* As exosomes get translated from laboratory applications to clinical applications (such as biomarkers or therapeutic agents), there is an urgent need to develop uniform protocols for acquiring and conserving exosomes, as their quality and functions may depend on it. Quality control is used to provide quality and dependable exosome-based products for clinical use (57).

10.10.2 Regulatory and Ethical Considerations

Regulatory Considerations

Exosome-based therapies and diagnostics have a great potential for application; however, several regulatory and ethical issues emerge and which need to be overcome to achieve clinical translation:

- *Lack of standardized guidelines:* Governmental regulatory bodies such as the FDA and EMA do not have clear regulations for exosome-based products. Exosome-based therapies enter a gray zone between biologics, cell therapies, and nanomedicines, and, therefore, confuse the paths of their regulation (58).
- *Manufacturing and quality control:* For the manufacture of exosomes, there is no generalized standardized manufacturing process for their isolation, purification, and storage. Patient safety and effectiveness can only be achieved if consistency, purity, and quality are enforced in exosome products.
- *Safety and toxicity:* Over time, the exosome-based therapeutics' safety is not well-established. Exosomes might be transporting pathogenic factors or harmful cargo. Risk evaluation tools for immunogenicity, off-target, and unintended interaction will be required by regulatory agencies in clinical applications.
- *Clinical trial protocols:* Specific requirements are needed to develop clinical trials with the help of exosomes, in particular, in terms of dosing and administering routes, and the process of monitoring side effects (59, 60).

Ethical Considerations

- *Source of exosomes:* Exosomes are isolated from various sources such as human tissues, stem cells, and tumors, among other sources. The ethical issues emerge in relation to the consent to use biological materials from the patients or donors, specifically for tumor-derived exosomes in cancer treatments.
- *Personalized medicine:* Personalized medicine obtained through exosome-based methods may cause ethical dilemmas, especially in aspects such as privacy or the usage of genetic data. Processing sensitive health information and getting informed consent are essential.
- *Equity and access:* Exosome-based treatments may prove costly, which may be a concern for accessibility and equality in healthcare. One ethical issue, in particular, is to guarantee that these advanced therapies are accessible to various populations (38).

10.10.3 Emerging Technologies: AI, Nanotech, and Omics

Exosomes (small extracellular vesicles 30–150 nm) are attracting considerable interest in biomedical science due to their ability to provide delivery, diagnosis, and therapy. Recent technologies, especially artificial intelligence (AI), nanotechnology, and omics, are playing prominent roles in promoting developmental research of exosomes and their clinical uses (61).

Artificial intelligence in exosome research: AI and ML are changing how exosomes are isolated, characterized, and used in healthcare. These technologies allow the analysis of complex omics data, thus hastening the separation of disease-specific exosome signatures and enhancing the effectiveness of exosome-based diagnostics and therapeutics. Major applications of AI:

- *Data integration:* AI can utilize multi-omics data (i.e., genomics, proteomics, and metabolomics) of exosome research to detect biomarkers for early disease detection, particularly in cancer, cardiovascular, and neurodegenerative disorders (62).
- *Predictive models:* Machine learning models themselves can be used to predict exosomes' behavior, including their biodistribution within the body, specificity to a target, and their efficacy as carriers for drugs (63).
- *Exosome characterization:* BI algorithms can analyze massive datasets from the microscopy images, DLS, and the NTA to characterize exosome size, shape, and markers on the surface with greater speed and precision than the conventional methods.

Nanotechnology in exosome research: Drug delivery function and potential of exosomes are improved through the vital role of nanotechnology. Nanomaterials will be applied to alter the properties of exosomes and load cargo, thereby achieving the possibility of targeted exosomes at diseased tissues. Key applications of nanotechnology:

- *Exosome engineering:* Nanotechnology allows one to make the surfaces of exosomes modifiable by adding target ligands (e.g., antibodies, peptides) to make targeting exosome-mediated delivery more selective (i.e., to cancer cells and inflamed regions).
- *Drug loading:* Nanomaterials, for example, lipids, polymeric nanoparticles, or gold nanoparticles, are being employed to improve the load of exosomes. These can enclose the chemotherapeutic agents or gene therapies, or RNA drugs to increase the therapeutic efficacy (64).
- *Enhanced stability:* Nanoparticles can also be employed to enhance exosome stability and prolong their shelf life by protecting against degradation and enhancing their times of circulation in the body (65).

10.10.4 Omics Technologies and Exosome Profiling

Omics technologies (genomics, proteomics, transcriptomics, and metabolomics) are inherent in the study of the molecular makeup of exosomes as well as their functions, in diverse diseases (66). Exosomes are cargo carriers that reflect the molecular condition of their source cell, for diagnostic and monitoring purposes. Key applications of omics:

- *Exosome RNA Profiling:* The content of miRNA and mRNA in exosomes is important for revealing disease mechanisms. For example, exosome miRNAs are being used to detect cancer-related miRNAs so that early tumor diagnosis can be made through liquid biopsy (blood, urine). High-throughput RNA sequencing enables the identification of exosome miRNA signatures for diseases (67).
- *Proteomics:* The proteome of exosomes can inform about the molecular pathways associated with diseases such as cancer, neurodegenerative disorders, and cardiovascular diseases. State-of-the-art approaches such as mass spectrometry allow the discovery of markers of exosome proteins useful in disease monitoring or target for therapy (68).
- *Metabolomics:* Exosomes carry metabolites that provide information concerning the parent cell's metabolic state. Either in tumor diseases or inflammatory diseases, it is possible to obtain information on processes that underlie the disease based on a metabolomic profiling of exosomes, thus facilitating the unearthing of new biomarkers (69).

Integration of AI, nanotechnology, and omics: The combination of AI, nanotechnology, and omics brings to precision medicine powerful tools. By integrating AI-based omics data analysis, on the one hand, and nanotechnological advances in exosome engineering, on the other hand, researchers can create personalized therapies and diagnostics, which are extremely targeted to particular patients. Synergistic applications:

- *Personalized medicine:* AI models can study immense datasets of exosome profiles to elucidate the way particular patients will react to treatments, pointing the way to individualized delivery of drugs with engineered exosomes (69, 70)
- *Targeted therapy:* Exosomes can be loaded with drugs or with genetic material using

nanotechnology, while the design of those therapies can be optimized with AI algorithms relying on molecular data from the omics analysis that is also personalized.
- *Early diagnosis and monitoring:* AI and omics technologies facilitate the ability of round-the-clock monitoring of exosome changes from patients' samples to detect diseases early and monitor therapeutic responses in real time (70).

10.11 CONCLUSION

Exosomes are small extracellular entities that have immense promise in diagnostics, therapeutics, and drug delivery. They represent the molecular state of the cell of origin, making them useful for early-stage disease detection and monitoring. Exosome-based applications are especially promising in cancer, neurodegenerative diseases, and cardiovascular diseases, which can be used as biomarkers or for the delivery of targeted therapies. Emerging technologies, such as artificial intelligence (AI), nanotechnology, and omics, are improving exosome studies. AI is enhancing exosome identification, optimizing drug delivery, and marking disease markers. Nanotechnology is applied to engineer exosomes, including loading, targeting specificity, and stability. While omics technologies are assisting the comprehensive profiling of exosomal content, thus for the identification of biomarkers and where it comes to understanding mechanisms of the diseases, a set of limitations persists in standardization, regulation, and scalability to clinical environments. Problems of isolation, characterization, and long-term safety of exosomes are to be solved to feasibly maintain reproducibility and efficacy. Despite these, exosomes are an exciting drug tool for precision medicine and are on the cusp of transforming the diagnostics, personalized treatments, and delivery models of drugs in the coming years.

ACKNOWLEDGMENT

All the authors want to credit their affiliated institutes.

AI DISCLOSURE STATEMENT

During the preparation of this work the authors used Grammarly, QuillBot, and ChatGPT in order to enhance grammatical structure and narrative coherence. After using this tool/service, the authors reviewed and edited the content as needed and take full responsibility for the content of the publication.

CONFLICT OF INTEREST

None.

REFERENCES

[1] Anel A, Gallego-Lleyda A, de Miguel D, Naval J, Martínez-Lostao L. Role of exosomes in the regulation of t-cell mediated immune responses and in autoimmune disease. *Cells*. 2019;8(2). https://doi.org/10.3390/cells8020154

[2] Tan L, Wu H, Liu Y, Zhao M, Li D, Lu Q. Recent advances of exosomes in immune modulation and autoimmune diseases. *Autoimmunity*. 2016;49(6):357–65.

[3] Ocansey DKW, Zhang L, Wang Y, Yan Y, Qian H, Zhang X, et al. Exosome-mediated effects and applications in inflammatory bowel disease. *Biol Rev*. 2020;95(5):1287–307. https://doi.org/10.1111/brv.12608

[4] Zhang B, Yeo RWY, Lai RC, Sim EWK, Chin KC, Lim SK. Mesenchymal stromal cell exosome–enhanced regulatory T-cell production through an antigen-presenting cell–mediated pathway. *Cytotherapy*. 2018;20(5):687–96. https://doi.org/10.1016/j.jcyt.2018.02.372

[5] Ji Y, Mi L, Zhao M, He X, Hu Y, Gao Y, et al. Innovative diagnosis and therapeutic modalities: Engineered exosomes in autoimmune disease. *Int J Nanomed*. 2024;19:3943–56. https://doi.org/10.2147/ijn.s452184

[6] Console L, Scalise M, Indiveri C. Exosomes in inflammation and role as biomarkers. *Clin Chim Acta*. 2019;488:165–71. https://doi.org/10.1016/j.cca.2018.11.009

[7] Kugeratski FG, Kalluri R. Exosomes as mediators of immune regulation and immunotherapy in cancer. *FEBS J*. 2021;288(1):10–35. https://doi.org/10.1111/febs.15558

[8] Tian Y, Cheng C, Wei Y, Yang F, Li G. The role of exosomes in inflammatory diseases and tumor-related inflammation. *Cells*. 2002;11(6):1005. https://doi.org/10.3390/cells11061005

[9] Tran TH, Mattheolabakis G, Aldawsari H, Amiji M. Exosomes as nanocarriers for immunotherapy of cancer and inflammatory diseases. *Clin Immunol*. 2015;160(1):46–58. https://doi.org/10.1016/j.clim.2015.03.021

[10] Li Q, Wang H, Peng H, Huyan T, Cacalano NA. Exosomes: Versatile nano mediators of immune regulation. *Cancers (Basel)*. 2019;11(10). https://doi.org/10.3390/cancers11101557

[11] Wang Y, Ma H, Zhang X, Xiao X, Yang Z. The increasing diagnostic role of exosomes in inflammatory diseases to leverage the therapeutic biomarkers. *J Inflamm Res*. 2024;17:5005–24. https://doi.org/10.2147/jir.s475102

[12] Kowal J, Tkach M, Théry C. Biogenesis and secretion of exosomes. *Curr Opin Cell Biol*. 2014;29(1):116–25. [accessed 23 Aug 2025] Available from: https://www.sciencedirect.com/science/article/abs/pii/S095506741400057X?via%3Dihub

[13] Burkova EE, Sedykh SE, Nevinsky GA. Human placenta exosomes: Biogenesis, isolation, composition, and prospects for use in diagnostics. *Int J Mol Sci*. 2021;22(4):1–20. https://doi.org/10.3390/ijms22042158

[14] Heydari R, Koohi F, Rasouli M, Rezaei K, Abbasgholinejad E, Bekeschus S, et al. Exosomes as rheumatoid arthritis diagnostic biomarkers and therapeutic agents. *Vaccines (Basel)*. 2023;11(3):1–13. https://doi.org/10.3390/vaccines11030687

[15] Jella KK, Nasti TH, Li Z, Malla SR, Buchwald ZS, Khan MK. Exosomes, their biogenesis and role in inter-cellular communication, tumor microenvironment and cancer immunotherapy. *Vaccines (Basel)*. 2018;6(4:69):1–15. https://doi.org/10.3390/vaccines6040069

[16] Sonbhadra S, Mehak, Pandey LM. Biogenesis, isolation, and detection of exosomes and their potential in therapeutics

and diagnostics. *Biosensors (Basel)*. 2023;13(8). https://doi.org/10.3390/bios13080802
[17] Lau NCH, Yam JWP. From exosome biogenesis to absorption: Key takeaways for cancer research. *Cancers (Basel)*. 2023;15(7). https://doi.org/10.3390/cancers15071992
[18] Dyball LE, Smales CM. Exosomes: Biogenesis, targeting, characterization and their potential as 'Plug & Play' vaccine platforms. *Biotechnol J*. 2022;17(11). https://doi.org/10.1002/biot.202100646
[19] Elsner C, Ergün S, Wagner N. Biogenesis and release of endothelial extracellular vesicles: Morphological aspects. *Ann Anat Anat Anz*. 2023;245:152006. [accessed 23 Aug 2025] Available from: https://www.sciencedirect.com/science/article/abs/pii/S0940960222001212
[20] Xie S, Zhang Q, Jiang L. Current knowledge on exosome biogenesis, cargo-sorting mechanism and therapeutic implications. *Membranes*. MDPI; 2022;12. https://doi.org/10.3390/membranes12050498
[21] Beninson LA, Fleshner M. Exosomes: An emerging factor in stress-induced immunomodulation. *Semin Immunol*. 2014;26(5):394–401. https://doi.org/10.1016/j.smim.2013.12.001
[22] Wang M, Zhang B. The Immunomodulation potential of exosomes in tumor microenvironment. *J Immunol Res*. 2021;2021(1):3710372. https://doi.org/10.1155/2021/3710372
[23] Khan FH, Reza MJ, Shao YF, Perwez A, Zahra H, Dowlati A, et al. Role of exosomes in lung cancer: A comprehensive insight from immunomodulation to theragnostic applications. *Biochim Biophys Acta: RevCancer*. 2022;1877(5):188776.
[24] Alshahrani MY, Jasim SA, Altalbawy FMA, Bansal P, Kaur H, Al-Hamdani MM, et al. A comprehensive insight into the immunomodulatory role of MSCs-derived exosomes (MSC-Exos) through modulating pattern-recognition receptors (PRRs). *Cell Biochem Funct*. 2024;42(4):e4029. [accessed 23 Aug 2025] Available from: https://doi.org/10.1002/cbf.4029
[25] Fan L, Guan P, Xiao C, Wen H, Wang Q, Liu C, et al. Exosome-functionalized polyetheretherketone-based implant with immunomodulatory property for enhancing osseointegration. *Bioact Mater*. 2021;6(9):2754–66. https://doi.org/10.1016/j.bioactmat.2021.02.005
[26] Hong Y, Lee J, Vu TH, Lee S, Lillehoj HS, Hong YH. Immunomodulatory effects of poly(I:C)-stimulated exosomes derived from chicken macrophages. *Poult Sci*. 2021;100(8). https://doi.org/10.1016/j.psj.2021.101247
[27] Lai RC, Yeo RWY, Lim SK. Mesenchymal stem cell exosomes. *Semin Cell Dev Biol*. 2015;40:82–8. https://doi.org/10.1016/j.semcdb.2015.03.001
[28] Xin L, Lin X, Zhou F, Li C, Wang X, Yu H, et al. A scaffold laden with mesenchymal stem cell-derived exosomes for promoting endometrium regeneration and fertility restoration through macrophage immunomodulation. *Acta Biomater*. 2020;113:252–66. https://doi.org/10.1016/j.actbio.2020.06.029
[29] Chang TH, Wu CS, Chiou SH, Chang CH, Liao HJ. Adipose-derived stem cell exosomes as a novel anti-inflammatory agent and the current therapeutic targets for rheumatoid arthritis. *Biomedicines*. 2022;10(7). https://doi.org/10.3390/biomedicines10071725
[30] Alzahrani FA, Saadeldin IM, Ahmad A, Kumar D, Azhar EI, Siddiqui AJ, et al. The potential use of mesenchymal stem cells and their derived exosomes as immunomodulatory agents for COVID-19 patients. *Stem Cells Int*. 2020;2020. https://doi.org/10.1155/2020/8835986
[31] Favaro RR, Murrieta-Coxca JM, Gutiérrez-Samudio RN, Morales-Prieto DM, Markert UR. Immunomodulatory properties of extracellular vesicles in the dialogue between placental and immune cells. *Am J Reprod Immunol*. 2021;85(2). https://doi.org/10.1111/aji.13383
[32] Wang Y, Lin Q, Zhang H, Wang S, Cui J, Hu Y, et al. M2 macrophage-derived exosomes promote diabetic fracture healing by acting as an immunomodulator. *Bioact Mater*. 2023;28:273–83. https://doi.org/10.1016/j.bioactmat.2023.05.018
[33] Lee BC, Kang I, Yu KR. Therapeutic features and updated clinical trials of mesenchymal stem cell (Msc)-derived exosomes. *J Clin Med*. 2021;10(4):1–18. https://doi.org/10.3390/jcm10040711
[34] Chen H, Han Z, Ma Y, Meng Q. Advances in macrophage-derived exosomes as immunomodulators in disease progression and therapy. *Int Immunopharmacol*. 2024;142:113248. https://doi.org/10.1016/J.INTIMP.2024.113248
[35] Miao C, Wang X, Zhou W, Huang J. The emerging roles of exosomes in autoimmune diseases, with special emphasis on microRNAs in exosomes. *Pharmacol Res*. 2021;169:105680. https://doi.org/10.1016/j.phrs.2021.105680
[36] Fang Y, Ni J, Wang YS, Zhao Y, Jiang LQ, Chen C, et al. Exosomes as biomarkers and therapeutic delivery for autoimmune diseases: Opportunities and challenges. *Autoimmun Rev*. 2023;22(3):103260. https://doi.org/10.1016/j.autrev.2022.103260
[37] Duan L, Lin W, Zhang Y, Jin L, Xiao J, Wang H, et al. Exosomes in autoimmune diseases: A review of mechanisms and diagnostic applications. *Clin Rev Allergy Immunol*. 2025;68(1):1–19. https://doi.org/10.1007/S12016-024-09013-2
[38] Mirzaei R, Zamani F, Hajibaba M, Rasouli-Saravani A, Noroozbeygi M, Gorgani M, et al. The pathogenic, therapeutic and diagnostic role of exosomal microRNA in the autoimmune diseases. *J Neuroimmunol*. 2021;358:577640. https://doi.org/10.1016/j.jneuroim.2021.577640
[39] Riazifar M, Mohammadi MR, Pone EJ, Yeri A, Lasser C, Segaliny AI, et al. Stem cell-derived exosomes as nanotherapeutics for autoimmune and neurodegenerative disorders. *ACS Nano*. 2019;13(6):6670–88. https://doi.org/10.1021/acsnano.9b01004
[40] Suh JH, Joo HS, Hong EB, Lee HJ, Lee JM. Therapeutic application of exosomes in inflammatory diseases. *Int J Mol Sci*. 2021;22(3):1–22. https://doi.org/10.3390/ijms22031144
[41] Hejrati A, Hasani B, Esmaili M, Bashash D, Tavakolinia N, Zafari P. Role of exosome in autoimmunity, with a particular emphasis on rheumatoid arthritis. *Int J Rheum Dis*. 2021;24(2):159–69. https://doi.org/10.1111/1756-185X.14021
[42] Saadh MJ, Allela OQB, Al-Hussainy AF, Baldaniya L, Rekha MM, Nathiya D, et al. Exosomal non-coding RNAs: Gatekeepers of inflammation in autoimmune disease. *J Inflamm*. 2025;22(1). https://doi.org/10.1186/s12950-025-00443-z
[43] Kapsogeorgou EK, Abu-Helu RF, Moutsopoulos HM, Manoussakis MN. Salivary gland epithelial cell exosomes: A source of autoantigenic ribonucleoproteins. *Arthritis Rheum*. 2005;52(5):1517–21. https://doi.org/10.1002/art.21005
[44] Tian Y, Zhang T, Li J, Tao Y. Advances in development of exosomes for ophthalmic therapeutics. *Adv Drug Deliv*

Rev. 2023;199:114899. https://doi.org/10.1016/j.addr.2023.114899

[45] Gu J, Jin T, Li Z, Chen H, Xia H, Xu X, et al. Exosomes expressing neuronal autoantigens induced immune response in antibody-positive autoimmune encephalitis. *Mol Immunol.* 2021;131:164–70. https://doi.org/10.1016/j.molimm.2020.12.034

[46] Arabpour M, Saghazadeh A, Rezaei N. Anti-inflammatory and M2 macrophage polarization-promoting effect of mesenchymal stem cell-derived exosomes. *Int Immunopharmacol.* 2021;97. https://doi.org/10.1016/j.intimp.2021.107823

[47] Lin SQ, Wang K, Pan XH, Ruan GP. Mechanisms of stem cells and their secreted exosomes in the treatment of autoimmune diseases. *Curr Stem Cell Res Ther.* 2024;19(11):1415–28. https://doi.org/10.2174/011574888X271344231129053003/CITE/REFWORKS

[48] Bu N, Wu HQ, Zhang GL, Zhan SQ, Zhang R, Fan QY, et al. Immature dendritic cell exosomes suppress experimental autoimmune myasthenia gravis. *J Neuroimmunol.* 2015;285:71–5. https://doi.org/10.1016/j.jneuroim.2015.04.009

[49] Fei Y, Liu Q, Peng N, Yang G, Shen Z, Hong P, et al. Exosomes as crucial players in pathogenesis of systemic lupus erythematosus. *J Immunol Res.* 2022;2022(1):8286498. https://doi.org/10.1155/2022/8286498

[50] Li Q, Park J, Kim JS, Le QV, Lee J, Oh YK. Anti-inflammatory macrophage-derived exosomes modified with self-antigen peptides for treatment of experimental autoimmune encephalomyelitis. *Adv Sci.* 2025;12(13):2415265. https://doi.org/10.1002/advs.202415265

[51] Wang WM, Wu C, Jin HZ. Exosomes in chronic inflammatory skin diseases and skin tumors. *Exp Dermatol.* 2019;28(3):213–8. https://doi.org/10.1111/exd.13857

[52] Harrell CR, Jovicic N, Djonov V, Arsenijevic N, Volarevic V. Mesenchymal stem cell-derived exosomes and other extracellular vesicles as new remedies in the therapy of inflammatory diseases. *mdpi.com.* 2019;8(12). https://doi.org/10.3390/cells8121605

[53] Selmaj I, Mycko MP, Raine CS, Selmaj KW. The role of exosomes in CNS inflammation and their involvement in multiple sclerosis. *J Neuroimmunol.* 2017;306:1–10. https://doi.org/10.1016/j.jneuroim.2017.02.002

[54] Han J, Zhang Y, Ge P, Dakal TC, Wen H, Tang S, et al. Exosome-derived CIRP: An amplifier of inflammatory diseases. *Front Immunol.* 2023;14:1066721. https://doi.org/10.3389/FIMMU.2023.1066721/XML

[55] Zhang C, Li D, Wang Z, Wang D, Lu Y, Zhang K, et al. M1 type macrophage targeted anti-inflammatory exosomes derived from BMSC for the treatment of acute and chronic inflammatory diseases. *Mater Des.* 2024;240:112844. https://doi.org/10.1016/j.matdes.2024.112844

[56] Cheng Z, Yang L, Chu H. The role of gut microbiota, exosomes, and their interaction in the pathogenesis of ALD. *J Adv Res.* 2025;72:353–67. https://doi.org/10.1016/j.jare.2024.07.002

[57] Bai L, Yu L, Ran M, Zhong X, Sun M, Xu M, et al. Harnessing the potential of exosomes in therapeutic interventions for brain disorders. *Int J Mol Sci.* 2025;26(6):2491. https://doi.org/10.3390/IJMS26062491

[58] Bian Y, Li W, Jiang X, Yin F, Yin L, Zhang Y, et al. Garlic-derived exosomes carrying miR-396e shapes microphage metabolic reprograming to mitigate the inflammatory response in obese adipose tissue. *J Nutr Biochem.* 2023;113:109249. https://doi.org/10.1016/j.jnubtio.2022.109249

[59] Gangadaran P, Madhyastha H, Madhyastha R, Rajendran RL, Nakajima Y, Watanabe N, et al. The emerging role of exosomes in innate immunity, diagnosis and therapy. *Front Immunol.* 2023;13:1085057. https://doi.org/10.3389/FIMMU.2022.1085057/XML

[60] Tang X, He Y, Liu J, Xu J, Peng Q. Exosomes: The endogenous nanomaterials packed with potential for diagnosis and treatment of neurologic disorders. *Colloids Surf B Biointerfaces.* 2024;239:113938. https://doi.org/10.1016/J.COLSURFB.2024.113938

[61] Chen LY, Kao TW, Chen CC, Niaz N, Lee HL, Chen YH, et al. Frontier review of the molecular mechanisms and current approaches of stem cell-derived exosomes. *Cells.* 2023;12(7):1018. https://doi.org/10.3390/CELLS12071018

[62] Seo N, Akiyoshi K, Shiku H. Exosome-mediated regulation of tumor immunology. *Cancer Sci.* 2018;109(10):2998–3004. https://doi.org/10.1111/cas.13735

[63] Liu G, Liu J, Li S, Zhang Y, He R. Exosome-mediated chemoresistance in cancers: Mechanisms, therapeutic implications, and future Directions. *Biomolecules.* 2025;15(5):685. https://doi.org/10.3390/Biom15050685

[64] Khan MI, Alsayed RKME, Choudhry H, Ahmad A, Mione MC, Peña C, et al. Exosome-mediated response to cancer therapy: Modulation of epigenetic machinery. *Int J Mol Sci.* 2022;23(11):6222. https://doi.org/10.3390/IJMS23116222

[65] Luo L, Zhang X, Rousuli Y, Aini A. Exosome-mediated transfer of miR-3613-5p enhances doxorubicin resistance by suppression of PTEN expression in breast cancer cells. *J Oncol.* 2022;2022(1):9494910. https://doi.org/10.1155/2022/9494910

[66] Cui X, Lai W, Zhao Y, Chen C. The exosome-mediated cascade reactions for the transfer and inflammatory responses of fine atmospheric particulate matter in macrophages. *Environ Sci Technol.* 2023;57(21):7891–901. https://doi.org/10.1021/acs.est.3c01436

[67] Soares Martins T, Trindade D, Vaz M, Campelo I, Almeida M, Trigo G, et al. Diagnostic and therapeutic potential of exosomes in Alzheimer's disease. *J Neurochem.* 2021;156(2):162–81. https://doi.org/10.1111/jnc.15112

[68] Xunian Z, Kalluri R. Biology and therapeutic potential of mesenchymal stem cell-derived exosomes. *Cancer Sci.* 2020;111(9):3100–10. https://doi.org/10.1111/cas.14563

[69] Li T, Li X, Han G, Liang M, Yang Z, Zhang C, et al. The therapeutic potential and clinical significance of exosomes as carriers of drug delivery system. *Pharmaceutics.* 2022;15(1):21. https://doi.org/10.3390/pharmaceutics15010021

[70] Li JY, Li QQ, Sheng R. The role and therapeutic potential of exosomes in ischemic stroke. *Neurochem Int.* 2021;151:105194. https://doi.org/10.1016/j.neuint.2021.105194

11 Exploring Exosome-Based Therapies in Respiratory Diseases

Santosh Kumar Prajapti[1], Aarti Prajapati[2], Naimish Nanda[1], Prince Kumar Jha[1], Dinesh Kumar[3], and Ashish Kumar Agrawal[3]
[1]MATS School of Pharmacy, MATS University, Raipur, Chhattisgarh, India
[2]Department of Pharmacy, School of Health Sciences, Quantum University, Roorkee, Uttarakhand, India
[3]Department of Pharmaceutical Engineering & Technology, Indian Institute of Technology (Banaras Hindu University), Varanasi, Uttar Pradesh, India

11.1 INTRODUCTION

Exosome-based therapies have emerged as a promising avenue for the treatment of chronic respiratory diseases, having the potential to facilitate cell-to-cell intercellular communication, leading to the delivery of therapeutic agents directly to target cells [1]. These nanosized vesicles, derived from various cells with the affinity to cross the biological barrier and modulating the immune response to promote tissue repair, make them an ideal candidate for the advanced treatments in chronic conditions such as asthma, chronic obstructive pulmonary disease (COPD), and pulmonary fibrosis [2]. Exosome-based treatments have been demonstrated in recent research to improve the effectiveness of drug delivery, along with minimizing the side effects, opening the door to more successful treatment of these crippling respiratory diseases [3].

Exosomes were first recognized in the mid-1980s as small vesicles to eliminate certain membrane-bound proteins and obsolete cellular components by participating in various biological and pathological processes [4, 5]. The discovery of genetic exchange between cells carrying functional nucleic acids, such as mitochondrial DNA, mRNA, and miRNA, by microvesicles was a breakthrough in exosome research [6]. The mast cell–derived exosomes containing over 1,200 mRNAs were extensively studied in 2007 and was found that they can be easily translated into proteins after being transferred to recipient cells; this has been confirmed by the presence of mRNAs and miRNAs in exosomes originating from the cells [7]. Generally, exosomes contain several components essential for intercellular communications, such as integrins, tetraspanins, adhesion molecules, membrane-trafficking, membrane transport, and fusion proteins. Tetraspanins like CD9, CD63, and CD81, along with heat shock protein (Hsp70), are universally present in exosomes and serve as key exosomal markers [8]. These exosomal components are protected through lipid bilayer encapsulation and enhance their stability in the physiological state of originating cells [9]. Exosomes are widely distributed in body fluids, including urine, breast milk, plasma, saliva, seminal fluid, amniotic fluid, cerebrospinal fluid (CSF), and synovial fluid. This ubiquitous presence has drawn significant attention in potential medical applications. The properties such as stability, low immunogenicity, and biocompatibility make them promising vehicles to deliver small RNAs, signaling molecules, or therapeutic drugs [10]. Additionally, it can also exert autocrine, paracrine, and endocrine effects on other cells, which opens avenues for their therapeutic application [11]. Interestingly, the mesenchymal stem cells (MSCs)-derived exosome and other ancestor cells mimic the properties and functions of their parent cells, exhibiting angiogenic and regenerative effects [12]. This chapter explores the roles of exosomes derived from various cells in lung disease, focusing on their influence on lung function and their involvement in respiratory diseases, and discloses the therapeutic potential of exosomes in chronic respiratory disorders.

11.2 EXOSOMES IN PULMONARY MICROENVIRONMENTAL REGULATION

Respiratory diseases are one of the leading causes of mortality worldwide. Exposure to several injurious substances, such as air pollution, smoking, toxic agents, and infectious pathogens, significantly contributes to damaging airway structures and leads to the development of conditions such as asthma, COPD, tuberculosis, and histoplasmosis.

The defense mechanisms of the lungs involve a range of innate immunity, including physical barriers, protective enzymes, the mucociliary apparatus, and immunoglobulin A (IgA). In addition, dendritic cells (DCs) and alveolar macrophages (AM) induce an immune response by arresting the microorganism within the microenvironment rich in immunoglobulin G (IgG), fibronectin, and other immune components. However, chronic inflammation can damage epithelial cells, triggering further immune responses along with altering epithelial cell morphology. The epithelial–mesenchymal transition (EMT) is one such change that aids in airway remodeling and is connected to respiratory

DOI: 10.1201/9781003623632-11

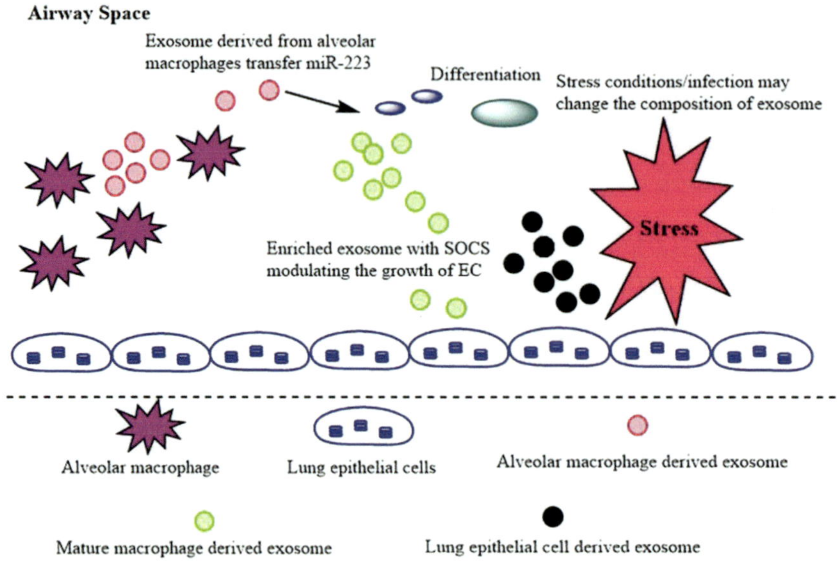

FIGURE 11.1 Exosomes in stressed airway physiology. Various respiratory cell types within the airway are capable of releasing exosomes, particularly under conditions of stress. Factors such as infections or oxidative stress not only increase exosome production but also alter their molecular makeup. When macrophages are damaged, the exosomes they release often carry pro-inflammatory cytokines, which can influence the body's immune response. Exosomes from alveolar macrophages play key roles in maintaining cellular equilibrium and promoting cell differentiation through the transfer of microRNA-223 (miR-223). They also contribute to immune regulation by delivering suppressor of cytokine signaling (SOCS) proteins to lung epithelial cells. Moreover, exosomes secreted by lung endothelial cells (ECs) under stress conditions can enter the circulation and transport altered biomolecules, further impacting respiratory physiology.

conditions such as lung cancer, asthma, and COPD. The cell-to-cell communication among endothelial, epithelial, mesenchymal, and immune cells is crucial for maintaining pulmonary homeostasis and mediating pathological responses. In recent years, exosomes have garnered significant attention due to their critical roles in intercellular and paracrine signaling.

Admyer et al. investigated the presence of exosomes, which contain major histocompatibility complex (MHC) and bronchoalveolar lavage (BAL), a co-stimulatory molecule in fluid from healthy individuals [13]. These exosomes carry functional proteins, RNAs, and microRNAs (miRNAs) capable of influencing gene expression and cell function in recipient cells. For instance, AM-derived exosomes promote monocyte differentiation by delivering miR-223 [14]. Moreover, exosomes secreted by epithelial cells contain membrane-bound mucins, which affect the local mucociliary defense mechanisms within the lungs' innate immune system [15].

Exosomes are also crucial for regulating airway inflammation by transporting signaling molecules such as macrophage-derived exosomes, which deliver suppressor of cytokine signaling 1 (SOCS1), and downregulate the STAT signaling pathway in AECs to reduce the inflammation [16]. Interestingly, epithelial-cell-derived exosomes have shown antiviral properties by neutralizing the human influenza virus through the secretion of mucins and sialic acid content [17, 18] (Figure 11.1). In sarcoidosis, exosomes can aggravate the symptoms by increasing the production of cytokines such as interferon-γ (IFN-γ), interleukin-13 (IL-13), and chemokine CXCL-8 in the respiratory tract [19]. Furthermore, infections can increase the secretion of exosomes and increase the load of HSP70, which activate immune responses [20]. In addition, pathological conditions may also lead to specific changes in exosomal miRNA profiles, which reflect the cellular stress responses [21–23]. In hypoxic pulmonary hypertension, exosomes derived from mesenchymal stromal cells (MSCs) exhibit protective effects by modulating inflammatory responses via paracrine signaling [24]. Nearly all cell types in the respiratory system, including epithelial cells, fibroblasts, endothelial cells, and immune cells, release exosomes into the extracellular space, contributing to physiological regulation and the maintenance of pulmonary homeostasis.

11.3 THERAPEUTIC POTENTIAL OF EXOSOMES IN RESPIRATORY DISEASES

Exosomes have emerged as a novel and promising tool for both therapeutic and diagnostic applications in the management of respiratory diseases and are utilized in multiple ways (Table 11.1). Notably, exosomes derived from mesenchymal/stromal cells (MSCs) possess regenerative properties that support lung tissue repair and wound healing, making them valuable in mitigating airway remodeling

TABLE 11.1
Therapeutic Potential of Exosomes Derived from Different Cells in Respiratory Disease

Source	System/Model	Therapeutic Effect	References
Alveolar epithelial cells–derived exosome	In vivo SD rats; in vitro alveolar epithelial cell line RLE-6TN	Activate AMs, causing pulmonary inflammation; a potential diagnostic biomarker for ALI	[40]
Alveolar macrophage-derived exosomes	In vivo Balb/c mice; in vitro rat alveolar macrophage line NR8383	BALF-exosomes as modulators of inflammatory response and cell communication during ALI	[41]
BMSCs-Exos	In vivo SD rats	Effects on phosgene-induced ALI by regulating inflammation, decreasing MMP-9, and increasing SP-C	[42]
BMSCs-Exos	In vivo adult male C57BL/6 mice; in vitro murine alveolar macrophage line MH-S	Inhibited endotoxin-induced glycolysis and alleviated LPS-induced inflammation and lung pathological damage	[43]
ADSC-Exos	In vivo C57BL/6 mice	Inhibited IL-27 production in macrophages, alleviated sepsis-induced ALI	[44]

associated with chronic pulmonary diseases [25]. In addition, exosomes circulating in the body often carry key molecules involved in respiratory disorders. These can be easily targeted by modulating the processes of exosome release/uptake [26].

Another innovative approach involves using exosomes as delivery systems for therapeutic agents, as they can transport drugs directly to specific cellular targets while minimizing degradation and reducing the risk of triggering an adverse immune response. The exosomes carrying tumor-associated antigens for immunotherapeutic interventions in advanced non-small-cell lung cancer have also been explored. Moreover, exosomes have potential as a carrier for mucosal vaccines, contributing to the development of exosome-based immunotherapy [27, 28].

MSCs-derived exosomes have shown promising results in treating various lung conditions due to their anti-inflammatory, immunomodulatory, and regenerative properties. MSCs are recognized for their diverse therapeutic capabilities, and their application in asthma treatment has garnered significant interest. For instance, intranasal administration of MSC-derived exosomes (MSC-Exo) has been promising in mitigating allergic asthma. Intranasal delivery of MSC-Exo significantly reduced the airway inflammation in BALB/c mice sensitized and challenged by ovalbumin (OVA). This effect is attributed to the expansion of IL-10-producing interstitial macrophages from the spleen, which contributes to protection against allergic asthma in mice [29]. The exosome derived from adipose-derived stem cells (ADSCs), particularly those enriched with mmu_circ_0001359, contributes to reduced airway remodeling in OVA-induced asthma in mice. This effect is achieved by encouraging M2-like macrophage polarization and reprogramming the pulmonary microenvironment [30].

Additionally, MSC-derived exosomes induce tolerogenic DCs via low expression of surface co-stimulatory molecules, and the treatment of DCs with exosomes leads to suppressing lymphocyte activity by enhancing IL-10 and TGF-β release and reducing IL-6 production [31]. Furthermore, the use of small extracellular vesicles (sEVs) from MSCs in allergic disease therapy proposed by Fang et al. highlights the role of miR-146a-5p carried by MSC-sEVs in reducing group 2 innate lymphoid cell-mediated airway inflammation [32].

The effectiveness of MSC-Exo has been investigated by Dong et al. in severe steroid-resistant asthma (SSRA). Exosomes derived from human umbilical cord MSCs (hUCMSCs) alleviate the airway hyperresponsiveness, histopathological changes, and inflammation in SSRA mice. This is achieved through the promotion of M2 macrophage polarization via inhibition of the TRAF1 pathway [33].

COPD also presents a therapeutic challenge due to the limited efficacy of current treatments. To address this, MSCs and MSC-Exo have been studied in models of cigarette smoke (CS)-induced lung inflammation. These investigations conferred protection against acute CS-induced pulmonary injury when MSCs are administered via the intraperitoneal route by targeting mitochondrial dysfunction [34]. Moreover, artificial nanovesicles derived from adipose-derived stem cells (ASCs) express both exosomal and stem cell markers as regenerative potential in an elastase-induced emphysema model, in a lower dose than ASCs or their naturally secreted exosomes [35].

The therapeutic applications of MSCs have also extended to idiopathic pulmonary fibrosis (IPF). The purified exosomes from human bone marrow MSCs are administered intravenously to prevent pathological features of bleomycin-induced pulmonary fibrosis. This effect is supposed to be mediated through the modulation of an anti-inflammatory, immunomodulatory monocyte phenotype [36]. Another innovative approach for IPF involves inhalation therapy using lung spheroid cell-derived secretome (LSC-Sec) and exosomes (LSC-Exo). These treatments are shown to be

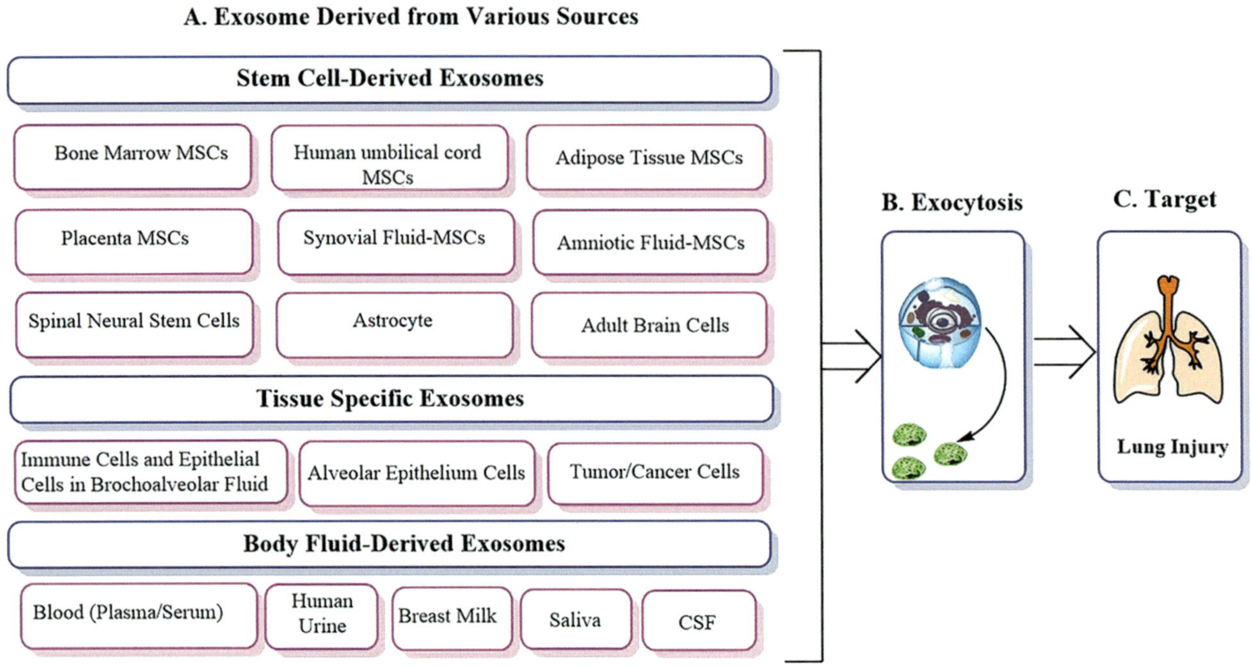

FIGURE 11.2 (a) Exosomes derived from different cells the therapeutic efficacy. (b) Exosome biogenesis extracted from donor cells. (c) Delivery of exosomes to the lungs for their therapeutic efficacy.

effective in the models of bleomycin- and silica-induced fibrosis, restoring alveolar architecture and decreasing collagen deposition and myofibroblast activity [37].

microRNA-based (miR-based) therapies are also being explored for targeted interventions in IPF. According to Yao et al., M2 macrophage-derived silencing miR-328 exosomes have the ability to inhibit fibroblast proliferation and attenuate fibrosis in a rat model, possibly via upregulation of FAM13A, a gene involved in COPD pathogenesis [38]. Similarly, exosomal miR-22 has been shown to regulate fibroblast-to-myofibroblast differentiation, thereby reducing fibrotic lesions in pulmonary tissue [39].

Exosomes from macrophages, neutrophils, and epithelial cells are found to be active in bronchoalveolar lavage fluid (BALF) after acute lung injury (ALI). Early after injury, exosomes in BALF mainly carry pro-inflammatory cytokines, which activate neutrophils to produce more cytokines, including IL-10. Moreover, exosomes derived from alveolar epithelial cells (AECs) also contributed to inflammation by activating AMs, especially in sepsis-related ALI [40].

In a rat model, Liu et al. showed that GW4869, an exosome inhibitor, reduced lung damage. Exosomes from LPS-treated AECs (LPS-Exos) activated AMs, increasing lung permeability and inflammation. This was linked to miR-92a-3p in the exosomes, which activated AMs through the NF-κB pathway and reduced PTEN expression. Blocking miR-92a-3p reduced these effects, suggesting it as a possible treatment target for ALI [41].

MSCs-derived exosomes improved lung function in phosgene-induced ALI rats by lowering inflammation and lung fluid buildup. They decreased TNF-α, IL-1β, and IL-6 while increasing IL-10 and surfactant protein C (SP-C), and reducing MMP-9 levels [42].

Bone marrow MSC–derived exosomes also helped in sepsis-related lung damage by lowering glycolysis in macrophages. They reduced M1 (pro-inflammatory) macrophages and increased M2 (anti-inflammatory) ones by suppressing HIF-1α and related glycolysis proteins. In LPS-induced ARDS mice, these exosomes reduced both inflammation and lung injury [43]

Exosomes from adipose-derived MSCs (ADMSCs) were taken up by macrophages and reduced IL-27 production. In live animal models, deleting IL-27 improved outcomes after ALI. ADMSC exosomes lowered lung inflammation, macrophage buildup, and vascular leakage, improving survival. However, adding IL-27 reversed these benefits, showing that ADMSC exosomes protect against sepsis-induced ALI by suppressing IL-27 [44]. Despite these encouraging results, further research is required to establish safe and clinically effective exosome-based therapies for chronic respiratory disorders. Figure 11.2 depicts the various sources of exosomes utilized for drug delivery and targeting of lung disorders.

11.4 EXOSOMES AND RESPIRATORY DISORDERS

11.4.1 Asthma

In asthma, a diverse array of immune cells from both the innate and adaptive systems contribute to the initiation of an inflammatory response within the lung epithelium. Bronchial tissues and airway mucus in individuals with asthma commonly exhibit a high presence of lymphocytes, plasma cells, eosinophils, and neutrophils. Exosomes play a significant role in the pathological processes of asthma, given their biological and structural characteristics.

Inflammation, the crucial pathological feature of asthma, is influenced by exosomes through their ability to regulate immune cell behavior, including recruitment, activation, and differentiation [45].

Exosomes bearing the mucin-1 marker indicative of BEC origin have been detected in BAL fluid from asthmatic patients. These vesicles can induce the expression of inflammatory mediators like CXCL-8 and leukotriene C4 (LTC4) in recipient epithelial cells [18]. Additionally, IL-13 has been shown to boost exosome secretion from airway epithelial cells, which subsequently stimulates the proliferation of undifferentiated lung macrophages [46]. Notably, in murine asthma models, pharmacological inhibition of exosome generation using GW4869 was associated with a reduction in asthmatic symptoms [26].

Eosinophils from individuals with asthma produce and secrete more exosomes than those from healthy controls, which increases the production of pro-inflammatory molecules, including chemokines, reactive oxygen species (ROS), and nitric oxide (NO) in recipient eosinophils along with the expression of intercellular adhesion molecule-1 (ICAM-1) and integrin β2 [47].

Interestingly, exosomes also have the potential to suppress allergic responses. BAL-derived exosomes from allergen-tolerized mice have demonstrated the ability to prevent allergic sensitization when administered intranasally. This finding supports the possibility of exosome-based vaccines for treating asthma and related allergic disorders. Furthermore, CD8+ T cell-derived exosomes decorated with antigen-specific antibodies and containing microRNA-150 have been shown to induce antigen-specific tolerance in mice upon uptake by T cells [48].

Mast cell (MC)-derived exosomes also play a crucial role by binding to and lowering free IgE levels in the serum, thereby reducing MC activation and its downstream effects [49]. Exosomes from fibroblasts of asthmatic patients containing transforming growth factor-beta 2 (TGF-β2) promote epithelial cell proliferation. However, these exosomes show reduced TGF-β2 levels compared to those from healthy individuals [50].

Asthmatic exosomes also show altered microRNA (miRNA) content. The expression of certain dysregulated miRNAs has been associated with clinical indicators such as eosinophil count and forced expiratory volume (FEV1), implicating their role in pathways that maintain airway integrity. In severe asthma patients, differentially expressed exosomal miRNAs have been linked to the TGF-β signaling pathway, ErbB signaling, focal adhesion, and IL-13-associated mechanisms [51].

11.4.2 Lung Cancer

Lung cancer remains one of the most prevalent forms of cancer globally. Despite advancements in diagnostic tools and therapeutic strategies, it continues to be the leading cause of cancer-related mortality worldwide. Exosomes, due to their critical role in intercellular communication, can significantly influence the physiological behavior of recipient cells, potentially contributing to pathological outcomes, including cancer [52]. Tumor cells actively secrete a specific exosome referred to as tumor-derived exosomes (TEXs). TEXs serve as crucial mediators of cellular communication within the tumor microenvironment, facilitating cancer cell adaptation, survival, and proliferation by modifying surrounding cellular and molecular landscapes [53, 54].

These vesicles carry a variety of tumor-promoting messages, including immune modulators, oncoproteins, and microRNAs, such as K-RAS, MET, and oncogenic miRNAs. When delivered to neighboring or distant cells, these factors can reprogram cellular functions, driving tumor initiation and progression [55].

One of the pivotal roles of TEXs is in establishing pre-metastatic niches and guiding metastatic cells to future colonization sites, a process regulated by the integrins expressed on the exosome surfaces. However, TEXs can also have paradoxical effects by engaging natural killer (NK) cell receptors like NKG2D, potentially triggering antitumor immune responses [56, 57].

In lung cancer specifically, TEXs contribute to a pro-metastatic inflammatory environment. For instance, their microRNA cargo miR-21 and miR-29a interact with toll-like receptors (TLRs) in immune cells, activating NF-κB signaling, thereby promoting tumor growth and metastasis [58].

11.4.2.1 Impact of Exosomes on Immune Cell Function in Lung Cancer

The presence of TEXs in the murine model of lung cancer has been linked to a decrease in NK cell populations within lung tissues. NK cells typically exert cytotoxic effects through NKG2D, a receptor involved in recognizing and eliminating tumor cells. TEXs impair NK cell function by delivering TGF-β1, which downregulates NKG2D expression. Additionally, TEXs contribute to the loss of NKG2D signaling by causing the shedding of its ligands on tumor cells, leading to receptor internalization and inactivation [59–61].

Lung cancer-derived TEXs also carry miR-23a, which interacts with CD107a, potentially protecting NK cells from damage due to cytotoxic granules, while still dampening their immune response. Moreover, TEXs may inhibit other aspects of NK cell activity, such as IL-2 signaling, perforin release, cyclin D3 production, and Janus kinase 3 (Jak3) activation [62, 63].

A significant proportion of exosomes isolated from lung cancer biopsies can induce the development of tolerogenic DCs and regulatory T cells, largely due to their epidermal growth factor receptor (EGFR) content. These changes can inhibit the activation of CD8+ T cells that are specific to tumor antigens. Additionally, miR-203a found in TEXs reduces TLR4 expression on DCs, subsequently decreasing the production of cytokines like TNF-α and IL-12, which are critical for initiating immune responses [64].

The maturation of DCs is further impaired by TGF-β1 transported by TEXs, promoting the expansion of myeloid-derived suppressor cells (MDSCs). These immature cells possess a strong immunosuppressive capacity and secrete cytokines such as TGF-β and prostaglandin E2. TEXs also guide CD14+ monocytes to differentiate into MDSCs rather than functional DCs. Furthermore, TEXs bearing HSP72 activate STAT3 signaling and enhance IL-6 production, reinforcing the suppressive activity of MDSCs [65].

11.4.3 Tuberculosis

Tuberculosis (TB) is a contagious yet preventable and treatable disease caused by the bacterium *Mycobacterium tuberculosis* (M.tb). Proteomic investigations of exosomes derived from the serum of TB patients have identified numerous M.tb proteins linked to bacterial virulence, oxidative and nitrogen stress responses, as well as adhesion mechanisms [66]. Notably, these exosomal proteins, which vary across different *Mycobacterium* species, include lipoproteins such as TLR2 agonists that have demonstrated immunomodulatory capabilities in mouse models. Furthermore, elevated levels of heat shock proteins (HSP90, HSP60, and HSP70) have been found in the exosomes of individuals with latent TB [67].

M.tb-derived exosomes are also rich in lipids as well as phospholipids such as phosphatidylglycerol (PG), phosphatidylcholine (PE), phosphatidylinositol (PI), and cardiolipin. Additionally, acylated phosphatidylinositol dimannosides (PIM2 and PIM6) and polyacylated trehalose are profusely present in these structures [68].

Moreover, exosomal microRNAs (miRNAs) from M. tb modulate the metabolic processes such as metabolism, cellular signaling, and bacterial invasion. The miRNA expression patterns in serum exosomes are specifically associated with TB (e.g., hsa-miR-1246, hsa-miR-2110, hsa-miR-370–3P, hsa-miR-28–3p, and hsa-miR-193b-5p). miRNAs also show promising results as diagnostic biomarkers. Some studies revealed that miR-484, miR-425, and miR-96 are significantly upregulated in TB patient serum and correlated with disease severity [68, 69].

Moreover, macrophages infected by M.tb release exosomes with unique RNA profiles, including specific miRNAs and mRNAs, which influence the host immune response. These exosomes also carry mycobacterial transcripts; for example, the presence of the 19-kDa lipoarabinomannan protein has been shown to modulate immune signaling by affecting cytokine production, such as TNF-α which is essential for eliminating M.tb. This result suggests that M.tb uses exosomes as a strategic mechanism to deliver its components into host cells, thereby enhancing its survival [70].

11.4.4 Histoplasmosis

Histoplasmosis is a respiratory illness caused by inhalation of spores from the fungus *Histoplasma capsulatum*. Common symptoms include fever, chills, headache, muscle pain, dry cough, chest pain, and general fatigue.

From *H. capsulatum* G186AR and G217B strains, the RNA content of extracellular vesicles (EVs) is studied, and 124 distinct miRNAs are identified present in the EVs of both strains. This RNA cargo displayed substantial variation among these two strains, suggesting the formation of strain-specific transcripts responsible for unique biological behaviors of each fungal strain [71].

Furthermore, the murine model treating *H. capsulatum* cells with monoclonal antibodies (mAbs) altered the infection outcome. The binding of antibodies to fungal cells influenced the composition of their EV cargo, affecting the fungus's interaction with host immune defenses. These findings revealed that the antibody-mediated opsonization altered the quantity and type of molecules packaged into fungal EVs in a dose-dependent manner, ultimately influencing immune response mechanisms [72].

Comprehensive proteomic and lipidomic analyses of EVs derived from *H. capsulatum* culture supernatants have revealed a wide range of bioactive molecules such as phospholipids, chaperones, and nuclear-associated proteins. These components are involved in virulence, cell signaling, and metabolic regulation [73].

11.4.5 COPD

COPD is marked by gradually worsening airflow restriction, often accompanied by infiltration of inflammatory cells, thickening of the alveolar walls, and eventual destruction of lung tissue, including the parenchyma and epithelial layers [74]. A defining feature of COPD is persistent inflammation, and exosomes are increasingly recognized for their significant involvement in the disease's progression [75].

In emphysema, endothelial cell (EC) injury within the lung parenchyma plays a central role. In stable COPD patients, exacerbation phases are associated with elevated levels of endothelial-derived exosomes [76]. These vesicles are closely linked with the extent of lung tissue destruction and airflow limitation. They often contain molecules such as vascular endothelial cadherin, platelet-endothelial cell adhesion molecules, and E-selectin, highlighting their relevance in COPD pathogenesis [77]. Injured epithelial cells secrete inflammatory mediators, including TNF-α, IL-1α, GM-CSF, TGF-β, and CXCL-8, which exert their effects through both autocrine and paracrine mechanisms. Among these, TGF-β plays a crucial role by promoting myofibroblast differentiation, leading to fibrosis and airway remodeling [78]. These effects are partially mediated by exosomes. Continued exposure leads to enriched exosomes with CCN1, a key extracellular matrix protein involved in tissue repair and remodeling [79].

Furthermore, a low population of COPD cases is associated with Alpha-1 Antitrypsin (AAT1) deficiency. Exosomes facilitate the transport of AAT1 from endothelial cells to alveolar epithelium and air spaces, where it is rapidly absorbed by epithelial cells, offering a protective mechanism against CS-related injury [77].

Beyond their pathological role, exosomes also show promise as biomarkers for tracking COPD progression and therapeutic response. The extent of endothelial cell damage in COPD patients may be inferred from circulating exosome levels. Research suggests that plasma exosomal miRNA concentrations are linked to skeletal muscle damage and could serve as biomarkers for muscle dysfunction in COPD. Moreover, variations in five specific exosomal miRNAs found in the BAL fluid of COPD patients have been correlated with disease severity. These miRNAs regulate ribosomal S6 kinase (S6K), a component of the TORC1 signaling pathway and a key regulator of muscle wasting, underscoring their potential as noninvasive biomarkers for diagnosing and monitoring COPD [80].

11.4.6 Sarcoidosis

Pulmonary sarcoidosis is a systemic inflammatory disorder that primarily targets the lungs. It is defined by the presence of non-caseating granulomas (NCGs) and interferon-producing T cells within the affected organs, which leads to persistent inflammation and tissue damage, particularly in lung tissue. Although the exact cause of sarcoidosis remains unclear, current molecular insights into its pathophysiology suggest that exosomes play a key role in initiating and sustaining the inflammatory response associated with this disease [81].

Studies have shown that the BAL fluid of sarcoidosis patients contains a significantly higher concentration of exosomes compared to that of healthy individuals. Exosomes isolated from this fluid have been found to enhance IFN-γ and IL-13 expression in peripheral blood mononuclear cells (PBMCs), as well as stimulate CXCL-8 production in epithelial cells. These findings underscore their potential role in promoting inflammation [82].

While the evidence points to a significant involvement of exosomes in the development and advancement of sarcoidosis-related inflammation, further detailed and practical research is necessary to fully explore their therapeutic potential in treating this complex disease.

11.5 CONCLUSION

In summary, exosomes are of significant promise as a new therapeutic modality in the management of lung injury and inflammation. Their distinct properties, like high biocompatibility, low immunogenicity, toxicity, and capability to cross cellular barriers, make them ideal for targeted drug delivery. These vesicles can be engineered and designed to have specific biological activities and targeting capabilities, and have been effective in delivering proteins, nucleic acids, small molecules, and nanoparticles to the site of inflammation. Of interest, exosomes from MSCs and inflammatory cells are highly attracted to inflamed tissue and thus represent efficient vehicles for immune modulation and delivery of therapeutic agents directly to inflammatory cells. Even with developments in the comprehension of interactions among exosomes, the lung microenvironment, and inflammatory processes, several significant points remain uncertain. Especially, improved characterization and phenotyping of associated conditions are urgently required. In addition, extracellular vesicles could potentially be used as markers for disrupted cellular behavior in chronic lung disease. Therefore, exosomes hold promise to direct personalized treatment approaches and advance knowledge of disease pathophysiology, which in turn will lead to better patient outcomes.

ABBREVIATIONS

ADSC(s)	adipose-derived stem cells
ARDS	acute respiratory distress syndrome
BMSC(s)	bone marrow stem cells
CCN1	cellular communication network factor 1 (CYR61)
CS	cigarette smoke
EGFR	epidermal growth factor receptor
FAM13A	family with sequence similarity 13 member A
FGF2	fibroblast growth factor 2
Hsp70	heat shock protein 70
HSP72	heat shock protein 72
Jak3	Janus kinase 3
LPS	lipopolysaccharide
LSC-Sec/Exo	lung spheroid cell secretome/exosome
M2	M2 macrophage (alternatively activated)
MDSC(s)	myeloid-derived suppressor cells
MMP-9	matrix metalloproteinase-9
NK	natural killer (cell)
NKG2D	natural killer group 2 member D
OVA	ovalbumin
PBMC(s)	peripheral blood mononuclear cells
PE/PG/PI	phosphatidylethanolamine/ phosphatidylglycerol/phosphatidylinositol
PIM2/6	phosphatidylinositol dimannosides (2/6 acyl chains)
SOCS	suppressor of cytokine signaling
sEVs	small extracellular vesicles
SP-C	surfactant protein-C
STAT	signal transducer and activator of transcription
TEX(s)	tumor-derived exosomes
TLR(s)	Toll-like receptors
TORC1	target of rapamycin complex 1

ACKNOWLEDGMENTS

I extend my heartfelt thanks to the management of MATS University for their valuable support and encouragement throughout the development of this chapter. I am especially thankful to the Hon'ble Chancellor, Vice Chancellor, whose guidance and commitment to fostering scholarly activities have been a continuous source of inspiration.

AI DISCLOSURE STATEMENT

During the preparation of this work the authors used Grammarly, QuillBot, and ChatGPT in order to enhance grammatical structure and narrative coherence. After using this tool/service, the authors reviewed and edited the content as needed and take full responsibility for the content of the publication.

CONFLICT OF INTEREST

There are no conflicts to declare.

REFERENCES

[1] Alipoor SD, Mortaz E, Garssen J, Movassaghi M, Mirsaeidi M, Adcock IM, Exosomes and exosomal MIRNA in respiratory diseases, *Mediators Inflamm*, **2016** (2016) 1–11. https://doi.org/10.1155/2016/5628404

[2] Hough KP, Chanda D, Duncan SR, Thannickal VJ, Deshane JS, Exosomes in immunoregulation of chronic lung diseases, *Allergy*, **72** (2017) 534. https://onlinelibrary.wiley.com/doi/10.1111/all.13086

[3] Lucchetti D, Santini G, Perelli L, Ricciardi-Tenore C, Colella F, Mores N, Macis G, Bush A, Sgambato, A, Montuschi P, Detection and characterisation of extracellular vesicles in exhaled breath condensate and sputum of COPD and severe asthma patients, *Eur Respir J*, **58** (2021) 2003024. https://doi.org/10.1183/13993003.03024-2020

[4] Pan BT, Johnstone RM, Fate of the transferrin receptor during maturation of sheep reticulocytes in vitro: Selective externalization of the receptor, *Cell*, **33** (1983) 967. https://doi.org/10.1016/0092-8674(83)90040-5

[5] Trams EG, Lauter CJ, Salem JN, Heine U, Exfoliation of membrane ecto-enzymes in the form of micro-vesicles, *BBA Biomembranes*, **645** (1981) 63. https://doi.org/10.1016/0005-2736(81)90512-5

[6] Valadi H, Ekström K, Bossios A, Sjöstrand M, Lee JJ, Lötvall JO, Exosome-mediated transfer of mRNAs and microRNAs is a novel mechanism of genetic exchange between cells, *Nat Cell Biol*, **9** (2007) 654. https://doi.org/10.1038/ncb1596

[7] Eissa NT, The exosome in lung diseases: Message in a bottle, *J Allergy Clin Immunol*, **131** (2013) 904. https://doi.org/10.1016/j.jaci.2013.01.021

[8] Tickner JA, Urquhart AJ, Stephenson S, Richard DJ, Kenneth O' Byrne KJ, Functions and therapeutic roles of exosomes in cancer, *Front Oncol*, **4** (2014) 127. https://doi.org/10.3389/fonc.2014.00127

[9] Cheng L, Sharples RA, Scicluna BJ, Hill AF, Exosomes provide a protective and enriched source of miRNA for biomarker profiling compared to intracellular and cell-free blood, *J Extracell Vesicles*, **3** (2014) 1–14. https://doi.org/10.3402/jev.v3.23743

[10] Samanta S, Rajasingh S, Drosos N, Zhou Z, Dawn B, Rajasingh J, Exosomes: New molecular targets of diseases, *Acta Pharmacol Sin*, **39** (2017) 501. https://doi.org/10.1038/aps.2017.162

[11] Andaloussi SE, Mäger I, Breakefield XO, Wood MJA, Extracellular vesicles: Biology and emerging therapeutic opportunities, *Nat Rev Drug Discov*, **12** (2013) 347. https://doi.org/10.1038/nrd3978

[12] Jo H, Brito S, Kwak BM, Park S, Lee M, Bin B, Applications of mesenchymal stem cells in skin regeneration and rejuvenation, *Int J Mol Sci*, **22** (2021) 2410. https://doi.org/10.3390/ijms22052410

[13] Admyre C, Grunewald J, Thyberg J, Gripenbäck S, Tornling G, Eklund A, Scheynius A, Gabrielsson S, Exosomes with major histocompatibility complex class II and co-stimulatory molecules are present in human BAL fluid, *Eur Respir J*, **22** (2003) 578. https://doi.org/10.1183/09031936.03.00041703

[14] Ismail N, Wang Y, Dakhlallah D, Moldovan L, Agarwal K, Batte K, Shah P, Wisler J, Eubank, TD, Tridandapani S, Paulaitis, ME, Piper MG, Marsh CB, Macrophage microvesicles induce macrophage differentiation and miR-223 transfer, *Blood*, **121** (2012) 984. https://doi.org/10.1182/blood-2011-08-374793

[15] Rose MC, Voynow JA, Respiratory tract Mucin genes and mucin glycoproteins in health and disease, *Physiol Rev*, **86** (2005) 245. https://doi.org/10.1152/physrev.00010.2005

[16] Fujita Y, Kosaka N, Araya J, Kuwano K, Ochiya T, Extracellular vesicles in lung microenvironment and pathogenesis, *Trends Mol Med*, **21** (2015) 533. https://doi.org/10.1016/j.molmed.2015.07.004

[17] Kesimer M, Scull M, Brighton B, DeMaria G, Burns K, O'Neal W, Pickles RJ, Sheehan JK, Characterization of exosome-like vesicles released from human tracheobronchial ciliated epithelium: A possible role in innate defense, *FASEB J*, **23** (2009) 1858. https://doi.org/10.1096/fj.08-119131

[18] Kulshreshtha A, Ahmad T, Agrawal, Ghosh B, Proinflammatory role of epithelial cell–derived exosomes in allergic airway inflammation, *J Allergy Clin Immunol*, **131** (2013) 1194. https://doi.org/10.1016/j.jaci.2012.12.1565

[19] Qazi KR, Paredes PT, Dahlberg B, Grunewald J, Eklund A, Gabrielsson S, Proinflammatory exosomes in bronchoalveolar lavage fluid of patients with sarcoidosis, *Thorax*, **65** (2010) 1016. https://doi.org/10.1136/thx.2009.132027

[20] Cordazzo C, Petrini S, Neri T, Lombardi S, Carmazzi Y, Pedrinelli R, Paggiaro P, Celi A, Rapid shedding of proinflammatory microparticles by human mononuclear cells exposed to cigarette smoke is dependent on Ca^{2+} mobilization, *Inflamm Res*, **63** (2014) 539. https://doi.org/10.1007/s00011-014-0723-7

[21] Alipoor SD, Mortaz E, Garssen J, Movassaghi M, Mirsaeidi M, Adcock IM, miR-1224 Expression is increased in human macrophages after infection with Bacillus Calmette-Guerin (BCG), *Iran J Allergy Asthma Immunol*, **17** (2018) 250.

[22] Alipoor SD, Mortaz E, Garssen J, Movassaghi M, Mirsaeidi M, Adcock I, Bovis Bacillus Calmette-Guerin (BCG) infection induces exosomal miRNA release by human macrophages, *J Transl Med*, **15** (2017) 105. https://doi.org/10.1186/s12967-017-1203-6

[23] Levänen B, Bhakta NR, Torregrosa Paredes P, Barbeau R, Hiltbrunner S, Pollack JL, Sköld CM, Svartengren M, Grunewald J, Gabrielsson S, Larsson BM, Woodruff PG, Erle DJ, Eklund A, Altered microRNA profiles in bronchoalveolar lavage fluid exosomes in asthmatic patients, *J Allergy Clin Immunol*, **131** (2013) 894. https://doi.org/10.1016/j.jaci.2012.11.039

[24] Lee C, Mitsialis SA, Aslam M, Vitali SH, Vergadi E, Konstantinou G, Sdrimas K, Kourembanas S, Exosomes mediate the cytoprotective action of mesenchymal stromal cells on hypoxia-induced pulmonary hypertension, *Circulation*, **126** (2012) 2601. https://doi.org/10.1161/CIRCULATIONAHA.112.114173

[25] Lener T, Gimona M, Aigner L, Börger V, Buzas E, Camussi G, Chaput N, Chatterjee D, Court FA, del Portillo HA, O'Driscoll L, Fais S, Falcon-Perez JM, Felderhoff-Mueser U, Fraile L, Gho YS, Görgens A, Gupta RC, Hendrix A, Hermann DM, Hill AF, Hochberg F, Horn PA, de Kleijn D, Kordelas L, Kramer BW, Krämer-Albers EM, Laner-Plamberger S, Laitinen S, Leonardi T, Lorenowicz MJ,

Lim SK, Lötvall J, Maguire CA, Marcilla A, Nazarenko I, Ochiya T, Patel T, Pedersen S, Pocsfalvi G, Pluchino S, Quesenberry P, Reischl IG, Rivera FJ, Sanzenbacher R, Schallmoser K, Slaper-Cortenbach I, Strunk D, Tonn T, Vader P, van Balkom BWM, Wauben M, El Andaloussi S, Théry C, Rohde E, Giebel B, Applying extracellular vesicles based therapeutics in clinical trials – an ISEV position paper, *J Extracell Vesicles*, **4** (2015) 30087. https://doi.org/10.3402/jev.v4.30087

[26] Mortaz E, Alipoor SD, Adcock IM, Mumby S, Exosomes in severe asthma: Update in their roles and potential in therapy, *Biomed Res Int*, **2018** (2018) 2862187. https://doi.org/10.1155/2018/2862187

[27] Morse MA, Garst J, Osada, T, Khan S, Hobeika A, Clay TM, Valente N, Shreeniwas R, Sutton, MA, Delcayre A, Hsu DH, Le Pecq, J B, Lyerly HK, A phase I study of dexosome immunotherapy in patients with advanced non-small cell lung cancer, *J Transl Med*, **3** (2005) 9. https://doi.org/10.1186/1479-5876-3-9

[28] Hough KP, Trevor JL, Strenkowski, JG, Wang Y, Chacko BK, Tousif S, Chanda D, Steele C, Antony VB, Dokland T, Ouyang X, Zhang J, Duncan SR, Thannickal VJ, Darley-Usmar V, Deshane JS, Exosomes in immunoregulation of chronic lung diseases, *Allergy*, **72** (2017) 534. https://doi.org/10.1111/all.13056

[29] Ren J, Liu X, Ma H, Zhao L, Chen Y, Intranasal delivery of MSC-derived exosomes attenuates allergic asthma via expanding IL-10 producing lung interstitial macrophages in mice, *Int Immunopharmacol*, **91** (2021) 107288. https://doi.org/10.1016/j.intimp.2020.107288

[30] Shang Y, Song X, Wang X, Ge L, Wang N, Exosomes from mmu_circ_0001359-modified ADSCs attenuate airway remodeling by enhancing FoxO1 signaling-mediated M2-like macrophage activation, *Mol Ther Nucleic Acids*, **19** (2020) 951. https://doi.org/10.1016/j.omtn.2019.11.018

[31] Shahir M, Hashemi SM, Asadirad A, Varahram M, Kazempour-Dizaji M, Folkerts G, Garssen J, Adcock I, Mortaz E, Effect of mesenchymal stem cell-derived exosomes on the induction of mouse tolerogenic dendritic cells, *J Cell Physiol*, **235** (2020) 7043. https://doi.org/10.1002/jcp.29601

[32] Fang SB, Zhang HY, Wang C, He BX, Liu XQ, Meng XC, Peng YQ, Xu ZB, Fan XL, Wu ZJ, Chen D, Zheng L, Zheng S G, Fu QL, Small extracellular vesicles derived from human mesenchymal stromal cells prevent group 2 innate lymphoid cell-dominant allergic airway inflammation through delivery of miR-146a-5p, *J Extracell Vesicles*, **9** (2020) 1723260. https://doi.org/10.1080/20013078.2020.1723260

[33] Dong B, Wang C, Zhang J, Zhang J, Gu Y, Guo X, Zuo X, Pan H, Hsu ACY, Wang G, Wang F, Exosomes from human umbilical cord mesenchymal stem cells attenuate the inflammation of severe steroid-resistant asthma by reshaping macrophage polarization, *Stem Cell Res Ther*, **12** (2021) 204. https://doi.org/10.1186/s13287-021-02272-2

[34] Maremanda KP, Sundar IK, Rahman I, Protective role of mesenchymal stem cells and mesenchymal stem cell-derived exosomes in cigarette smoke-induced mitochondrial dysfunction in mice, *Toxicol Appl Pharmacol*, **385** (2019) 114788. https://doi.org/10.1016/j.taap.2019.114788

[35] Kim YS, Kim JY, Cho R, Shin DM, Lee SW, Oh YM, Adipose stem cell-derived nanovesicles inhibit emphysema primarily via an FGF2-dependent pathway, *Exp Mol Med*, **49** (2017) e284. https://doi.org/10.1038/emm.2017.150

[36] Mansouri N, Willis GR, Fernandez-Gonzalez A, Reis M, Nassiri S, Mitsialis SA, Kourembanas S, Mesenchymal stromal cell exosomes prevent and revert experimental pulmonary fibrosis through modulation of monocyte phenotypes, *JCI Insight*, **4** (2019) e128060. https://doi.org/10.1172/jci.insight.128060

[37] Dinh PUC, Paudel D, Brochu H, Popowski KD, Gracieux MC, Cores J, Huang K, Hensley MT, Harrell E, Vandergriff AC, George AK, Barrio RT, Hu S, Allen TA, Blackburn K, Caranasos TG, Peng X, Schnabel LV, Adler KB, Lobo LJ, Goshe MB, Cheng K, Inhalation of lung spheroid cell secretome and exosomes promotes lung repair in pulmonary fibrosis, *Nat Commun*, **11** (2020) 1064. https://doi.org/10.1038/s41467-020-14344-7

[38] Yao MY, Zhang WH, Ma WT, Liu QH, Xing LH, Zhao GF, microRNA-328 in exosomes derived from M2 macrophages exerts a promotive effect on the progression of pulmonary fibrosis via FAM13A in a rat model, *Exp Mol Med*, **51** (2019) 1–16. https://doi.org/10.1038/s12276-018-0176-1

[39] Kuse N, Kamio K, Azuma A, Matsuda K, Inomata M, Usuki J, Morinaga A, Tanaka T, Kashiwada T, Atsumi K, Hayashi H, Saito Y, Seike M, Gemma A, Exosome-derived microRNA-22 ameliorates pulmonary fibrosis by regulating fibroblast-to-myofibroblast differentiation in vitro and in vivo, *J Nippon Med Sch*, **87** (2020) 118. https://doi.org/10.1272/jnms.JNMS.2020_87-300

[40] Ye C, Li H, Bao M, Zhuo R, Jiang G, Wang W, Alveolar macrophage-derived exosomes modulate severity and outcome of acute lung injury, *Aging*, **12** (2020) 6120. https://doi.org/10.18632/aging.103019

[41] Liu F, Peng W, Chen J, Xu Z, Jiang R, Shao Q, Zhao N, Qian K, Exosomes derived from alveolar epithelial cells promote alveolar macrophage activation mediated by miR-92a-3p in sepsis-induced acute lung injury, *Front Cell Infect Microbiol*, **11** (2021) 646546. https://doi.org/10.3389/fcimb.2021.646546

[42] Xu N, Shao Y, Ye K, Qu Y, Memet O, He D, Shao G, Shao F, Gao Y, Mesenchymal stem cell-derived exosomes attenuate phosgene-induced acute lung injury in rats, *Inhal Toxicol*, **31** (2019) 52. https://doi.org/10.1080/08958378.2019.1580509

[43] Deng H, Wu L, Liu M, Zhu L, Chen Y, Zhou H, Hu X, Wang S, Zhang J, Jin H, Zhang Z, Xie L, Bone marrow mesenchymal stem cell-derived exosomes attenuate LPS-induced ARDS by modulating macrophage polarization through inhibiting glycolysis in macrophages, *Shock*, **54** (2020) 828. https://doi.org/10.1097/SHK.0000000000001519

[44] Wang X, Liu D, Zhang X, Yang L, Xia Z, Zhang Q, Exosomes from adipose-derived mesenchymal stem cells alleviate sepsis-induced lung injury in mice by inhibiting the secretion of IL-27 in macrophages, *Cell Death Discov*, **8** (2022) 18. https://doi.org/10.1038/s41420-021-00780-6

[45] Rossios C, Pavlidis S, Hoda U, Kuo CH, Wiegman C, Russell KE, Sun K, Loza MJ, Baribaud, F, Durham AL, Adcock IM, Chung KF, Impaired innate immune gene profiling in airway smooth muscle cells from chronic cough patients, *Biosci Rep*, **37** (2017) BSR20171090. https://doi.org/10.1042/BSR20171090

[46] Cazzoli R, Buttitta F, Di Nicola M, Malatesta S, Marchetti A, Rom WN, Pass HI, microRNAs derived from circulating exosomes as noninvasive biomarkers for screening and diagnosing lung cancer, *J Thorac Oncol*, **8** (2013) 1156. https://doi.org/10.1097/JTO.0b013e318299ac32

[47] Mazzeo C, Calzada D, Alonso R, Merino JM, Sastre B, Echániz P, Izquierdo M, del Pozo V, Exosome secretion by eosinophils: A possible role in asthma pathogenesis, *J Allergy Clin Immunol*, **135** (2015) 1603. https://doi.org/10.1016/j.jaci.2014.10.019

[48] Rebane A, Akdis CA, MicroRNAs in allergy and asthma, *Curr Allergy Asthma Rep*, **14** (2014) 424. https://doi.org/10.1007/s11882-014-0424-8
[49] Xie G, Dong H, Liang Y, Ham JD, Rizwan R, Chen J, Mast cell exosomes can suppress allergic reactions by binding to IgE, *J Allergy Clin Immunol*, **141** (2018) 788. https://doi.org/10.1016/j.jaci.2017.03.040
[50] Haj-Salem I, Fakhfakh R, Daghfous J, Gagliardo R, Hamzaoui, K, Hamzaoui A, Fibroblast-derived exosomes promote epithelial cell proliferation through TGF-β2 signalling pathway in severe asthma, *Allergy*, **73** (2018) 178. https://doi.org/10.1111/all.13230
[51] Suzuki M, Yamada T, Nakatani Y, Chikaishi Y, Mizuno K, Goto T, Tanaka F, Kuwano H, Altered circulating exosomal RNA profiles detected by next-generation sequencing in patients with severe asthma, *Eur Respir J*, **48** (2016) 3410. https://erj.ersjournals.com/content/48/suppl_60/PA3410
[52] Andre F, Escudier B, Angevin E, Tursz T, Zitvogel L, Exosomes for cancer immunotherapy, *Ann Oncol*, **15** (2004) 141. https://doi.org/10.1093/annonc/mdh910
[53] Muller L, Simms P, Hong CS, Nishimura MI, Jackson EK, Watkins SC, Whiteside TL, Tumor-derived exosomes regulate expression of immune function-related genes in human T cell subsets, *Sci Rep*, **6** (2016) 20254. https://doi.org/10.1038/srep20254
[54] Haratani K, Hayashi H, Tanaka T, Kaneda H, Hayashi T, Nakagawa K, Watanabe S, Kato R, Takegawa N, Takeda M, Iwasa T, Okamoto I, Nakagawa K, Tumor immune microenvironment and nivolumab efficacy in EGFR mutation-positive non-small cell lung cancer based on T790M status after disease progression during EGFR-TKI treatment, *Ann Oncol*, **28** (2017) 1532. https://doi.org/10.1093/annonc/mdx183
[55] Sun T, Yin Y, Li X, Wang L, Role of exosomal noncoding RNAs in lung carcinogenesis, *BioMed Res Int*, **2015** (2015) 125807. https://doi.org/10.1155/2015/125807
[56] Théry C, Ostrowski M, Segura E, Membrane vesicles as conveyors of immune responses, *Nat Rev Immunol*, **9** (2009) 581. https://doi.org/10.1038/nri2567
[57] Viaud S, Théry C, Ploix S, Tursz T, Lapierre V, Lantz O, Zitvogel L, Chaput N, Dendritic cell-derived exosomes promote natural killer cell activation and proliferation: A role for NKG2D ligands and IL-15Rα, *PLoS ONE*, **4** (2009) e4942. https://doi.org/10.1371/journal.pone.0004942
[58] Fabbri M, Paone A, Calore F, Galli R, Gaudio E, Santhanam R, Lovat F, Fadda P, Mao C, Nuovo GJ, Zanesi N, Crawford M, Ozer GH, Wernicke D, Alder H, Caligiuri MA, Nana-Sinkam P, Perrotti D, Croce CM, MicroRNAs bind to toll-like receptors to induce prometastatic inflammatory response, *PNAS*, **109** (2012) E2110. https://doi.org/10.1073/pnas.1209414109
[59] Malmberg KJ, Carlsten M, Björklund A, Sohlberg E, Bryceson YT, Ljunggren HG, Natural killer cell-mediated immunosurveillance of human cancer, *Semin Immunol*, **31** (2017) 20. https://doi.org/10.1016/j.smim.2017.08.002
[60] Berchem G, Noman MZ, Bosseler M, Paggetti J, Baconnais S, Le Cam E, Nanbakhsh A, Moussay E, Mami-Chouaib F, Janji B, Chouaib S, Hypoxic tumor-derived microvesicles negatively regulate NK cell function by a mechanism involving TGF-β and miR23a transfer, *Oncoimmunology*, **5** (2016) e1062968. https://doi.org/10.1080/2162402X.2015.1062968
[61] Clayton A, Mitchell JP, Court J, Linnane S, Mason MD, Tabi Z, Human tumor-derived exosomes down-modulate NKG2D expression, *J Immunol*, **180** (2008), 7249. https://doi.org/10.4049/jimmunol.180.11.7249
[62] Cohnen A, Chiang SCC, Stojanovic A, Schmidt H, Claus M, Saftig P, Watzl C, Surface CD107a/LAMP-1 protects natural killer cells from degranulation-associated damage, *Blood*, **122** (2013) 1411. https://doi.org/10.1182/blood-2012-10-460220
[63] Filipazzi P, Burdek M, Villa A, Rivoltini L, Huber V, Recent advances on the role of tumor exosomes in immunosuppression and disease progression, *Semin Cancer Biol*, **22** (2012) 342. https://doi.org/10.1016/j.semcancer.2012.02.010
[64] Huang SH, Li YL, Lin YC, Chou, CH, Hung MH, Liao WY, Chen HC, Chou HL, Epidermal growth factor receptor-containing exosomes induce tumor-specific regulatory T cells, *Cancer Investig*, **31** (2013) 330. https://doi.org/10.3109/07357907.2013.793366
[65] Chalmin F, Ladoire S, Mignot G, Vincent J, Bruchard M, Remy-Martin JP, Boireau W, Rouleau, A, Simon B, Lanneau D, De Thonel A, Multhoff G, Hamman A, Martin F, Chauffert B, Solary E, Bengrine A, Bernard C, Rébé C, Ghiringhelli F, Membrane-associated Hsp72 from tumor-derived exosomes mediates STAT3-dependent immunosuppressive function of mouse and human myeloid-derived suppressor cells, *J Clin Invest*, **120** (2010) 457. https://doi.org/10.1172/JCI40483
[66] Kruh-Garcia NA, Wolfe LM, Chaisson LH, Worodria WO, Nahid P, Schorey JS, Davis JL, Dobos KM, Detection of *Mycobacterium tuberculosis* peptides in the exosomes of patients with active and latent *M. tuberculosis* infection using MRM-MS, *PLoS ONE*, **9** (2014) e103811. https://doi.org/10.1371/journal.pone.0103811
[67] Prados-Rosales R, Baena A, Martinez LR, Luque-Garcia J, Kalscheuer R, Veeraraghavan U, Camara C, Nosanchuk JD, Besra GS, Chen B, Jimenez J, Glatman-Freedman A, Jacobs WR, Porcelli SA, Casadevall A, Mycobacteria release active membrane vesicles that modulate immune responses in a TLR2-dependent manner in mice, *J Clin Invest*, **121** (2011) 1471. https://doi.org/10.1172/JCI44261
[68] Alipoor SD, Mortaz E, Tabarsi P, Farnia P, Mirsaeidi M, Garssen J, Movassaghi M, Adcock IM, *Bovis* Bacillus Calmette–Guérin (BCG) infection induces exosomal miRNA release by human macrophages, *J Transl Med*, **15** (2017) 105. https://doi.org/10.1186/s12967-017-1199-2
[69] Lyu L, Zhang X, Li C, Yang T, Wang J, Pan L, Jia H, Li Z, Sun Q, Yue L, Chen F, Zhang Z, Small RNA profiles of serum exosomes derived from individuals with latent and active tuberculosis, *Front Microbiol*, **10** (2019) 1174. https://doi.org/10.3389/fmicb.2019.01174
[70] Bhatnagar S, Schorey JS, Exosomes released from infected macrophages contain *Mycobacterium avium* glycopeptidolipids and are proinflammatory, *J Biol Chem*, **282** (2007) 25779. https://doi.org/10.1074/jbc.M705356200
[71] Alves LR, Peres da Silva R, Sanchez DA, Zamith-Miranda D, Rodrigues ML, Goldenberg S, Puccia R, Nosanchuk JD, Extracellular vesicle-mediated RNA release in Histoplasma capsulatum, *mSphere*, **4** (2019) e00176-19. https://doi.org/10.1128/mSphere.00176-19
[72] Matos Baltazar L, Nakayasu ES, Sobreira TJP, Choi H, Casadevall A, Nimrichter L, Nosanchuk JD, Antibody binding alters the characteristics and contents of extracellular vesicles released by Histoplasma capsulatum, *mSphere*, **1** (2016) e00085-15. https://doi.org/10.1128/mSphere.00085-15
[73] Albuquerque PC, Nakayasu ES, Rodrigues ML, Frases S, Casadevall A, Zancope-Oliveira RM, Almeida IC, & Nosanchuk JD, Vesicular transport in *Histoplasma capsulatum*: An effective mechanism for trans-cell wall

transfer of proteins and lipids in ascomycetes, *Cell Microbiol*, **10** (2008) 1695. https://doi.org/10.1111/j.1462-5822.2008.01160.x
[74] Pearson M, Is the primary mechanism underlying COPD: Inflammation or ischaemia? COPD, *J Chronic Obstr Pulm Dis*, **10** (2013) 536. https://doi.org/10.3109/15412555.2013.804291
[75] Barnes PJ, Shapiro SD, Pauwels RA, Chronic obstructive pulmonary disease: Molecular and cellular mechanisms, *Eur Respir J*, **22** (2003) 672. https://doi.org/10.1183/09031936.03.00085703
[76] Kratzer A, Lutter N, Goke T, Witte K, Endothelial cell adhesion molecule CD146: Implications for its role in the pathogenesis of COPD, *J Pathol*, **230** (2013) 388. https://doi.org/10.1002/path.4117
[77] Takahashi T, Kubo H, The role of microparticles in chronic obstructive pulmonary disease, *Int J Chron Obstruct Pulmon Dis*, **9** (2014) 303. https://doi.org/10.2147/COPD.S52774
[78] Ardekani AM, Naeini MM, The role of microRNAs in human diseases. *Avicenna J Med Biotechnol*, **2** (2010) 161.
[79] Moon HG, Park S, Lee JH, Song JW, Kang SG, Park JH, Kubo H, CCN1 secretion and cleavage regulate the lung epithelial cell functions after cigarette smoke, *Am J Physiol Lung Cell Mol Physiol*, **307** (2014) L326. https://doi.org/10.1152/ajplung.00054.2014
[80] Donaldson AL, Goss A, Cocks SE, Dore MP, Increased skeletal muscle-specific microRNA in the blood of patients with COPD, *Thorax*, **68** (2013) 1140. https://doi.org/10.1136/thoraxjnl-2012-203129.
[81] Maertzdorf J, Weiner J, Ota M, Dheda K, Common patterns and disease-related signatures in tuberculosis and sarcoidosis, *PNAS*, **109** (2012) 7853. https://doi.org/10.1073/pnas.1202807109
[82] Lin J, Yang Y, Zhang M, Li G, Exosomes: Novel biomarkers for clinical diagnosis, *Sci World*, **2015** (2015) 657086. https://doi.org/10.1155/2015/657086

12 Exosome Applications in Skin Regeneration and Wound Healing

Hemanathan Elango[1], *Shyam Ganesh*[1], *Adithya Prabhakaran*[1], and *Abhijit Saha*[1]
[1]Department of Chemistry, College of Engineering and Technology, SRM Institute of Science and Technology, Kattankulathur, Tamil Nadu, India

12.1 INTRODUCTION

The most important part of the human body is our skin, acting as a barrier for external factors. However, when the skin is injured, whether by a cut, burn, surgery, or a chronic condition like diabetes, its healing process becomes complex and delicate [1–3]. When your body heals a wound, it follows a careful and well-organized process. Each step happens in a specific order and at the right time, like a schedule [4, 5]. This healing process generally divides into four overlapping, distinct phases: hemostasis (stopping the bleeding), inflammation (fighting infection), proliferation (rebuilding tissue), and remodeling (strengthening the tissue and restoring function) [6]. When the tissue repairs, our network of cells and signaling molecules works to repair the damage to the tissue [7]. Yet, in many patients, this natural process does not proceed efficiently. Factors such as infection, poor blood flow, or systemic disorders, including diabetes, can interfere with healing and eventually lead to chronic wounds that fail to close. Problems such as infections, poor blood circulation, or systemic conditions like diabetes can interfere with recovery, often resulting in chronic wounds that fail to heal. These long-lasting wounds are not only physically harmful but also emotionally draining, as patients must cope with persistent pain and reduced quality of life. Over the years, researchers have tested many strategies to improve healing, ranging from specialized dressings and topical antibiotics to more advanced options such as stem cell therapy. Stem cells are promising because of their ability to regenerate damaged tissues, but their clinical application remains difficult. Once inside the body, stem cells can be unpredictable, may not survive for long, and in some cases, trigger immune reactions or other safety concerns [8]. Because of these challenges, attention has shifted toward exosomes as a safer and more reliable alternative. Exosomes are tiny vesicles, usually between 30 nm and 150 nm in size, naturally released by a lot of cells, including MSCs [9, 10]. These tiny vesicles are essential to cell-to-cell communication and are often referred to as the body's messenger. The exosome is enclosed by a lipid bilayer and contains several biologically active molecules, such as proteins, lipids, RNAs (including miRNA and messenger RNA), DNA fragments, and signal components that can affect the way the recipient cell behaves and functions [11, 12]. Importantly, exosomes can form in a wide range of bodily fluids, such as urine, umbilical fluid, blood, breast milk, and saliva, making them useful and available tools for diagnosis and treatment [13, 14]. Furthermore, their unique ability to transport water-soluble and fat-soluble drugs together makes them promise the delivery of drugs by trucks [15, 16]. Exosomes, particularly those derived from MSCs, have shown remarkable potential to repair skin regeneration and function during skin renewal. These vesicles help regenerate the skin using several major techniques. They encourage the skin cells named fibroblasts to produce a crucial systemic protein called elastin and collagen, which enhances the skin's potency, firmness, and resistance. They also help other skin cells, such as keratinocytes, to grow and move to areas that need repair, which is important for the development of a good skin barrier. Exosomes facilitate the growth of new blood vessels, called angiogenesis, which brings oxygen and nutrients to the injured tissues. Exosomes can also reduce inflammation and protect the skin cells from damage caused by oxidative stress, which plays an important role in old age. Exosomes are being studied for their use in treatment, preferentially for scar healing, wrinkle reduction, sun damage repair, and general skin rejuvenation. In addition, exosomes contribute to injury healing by modulating the various stages of repair. They control the inflammatory phase by reducing excessive inflammation, support the proliferative phase by increasing cell migration and tissue formation, and support the maturation phase by promoting collagenation and tissue maturation. Exosomes, by this mechanism, contribute equally to the speed and quality of wound closure, making them an effective tool for regenerative medicine. Exosomes are immediately available loads in hydrogel, cream, and injectable carriers which can be directly applied for wound healing. Once used, these exosomes interact with nearby cells, forming and supporting the healing system [17, 18]. These approaches are nevertheless under the clinical microscope, inspiring findings in preclinical animal studies have shown faster wound closure, improved collagen restructuring, improved vascularization, and in some cases, complete restoration of the skin structure and function [19, 20]. A detailed discussion on the use of engineer exosomes and exosome-loaded scaffolds in skin renewal and wound healing is presented here. It explores the double nature of exosomes in skin renewal and wound healing, focusing on

DOI: 10.1201/9781003623632-12

their mechanism of action, curative objectives, and rising capability as a next-generation solution in dermatology and regenerative therapy.

12.2 EXOSOMES APPLICATION IN SKIN-REPAIRING AND WOUND-HEALING MECHANISMS

Exosomes are enriched in bioactive molecules that significantly contribute to skin regeneration and wound healing. These small vesicles carry proteins, including important growth factors like vascular endothelial growth factor (VEGF), transforming growth factor-beta (TGF-β), epidermal growth factor (EGF), and fibroblast growth factor (FGF) [21]. These proteins help form new blood vessels, encourage skin cells to grow, and speed up the repair process. Exosomes also include cytokines that reduce inflammation and attract immune cells to the wound area for faster healing. Another group of proteins called matrix metalloproteinases (MMPs) perform a crucial role in damaged tissue, supporting its repair and regeneration. Exosomes are also made up of lipids such as ceramides and phospholipids, which help protect the skin barrier and support healing [22]. One of the most powerful features of exosomes is their genetic material, especially small RNAs called microRNAs. These include miR-21, miR-126, and miR-146a, which control skin cell growth, reduce swelling, improve blood flow, and help form new tissue. In addition, mRNAs in exosomes can make new proteins that support skin repair. Special surface proteins like CD9, CD63, and CD81 help exosomes find and enter the right skin cells. All these components work together to heal wounds by reducing inflammation, encouraging new skin and blood vessel growth, and rebuilding the damaged area.

12.3 TECHNIQUES USED IN EXOSOME ISOLATION

Several exosome isolation methods have already been discussed in previous chapters (Figure 12.1). Among these, differential ultracentrifugation remains the most widely used standard, while techniques like size-exclusion chromatography (SEC), field-flow fractionation (FFF), and microfluidic filtration offer label-free and structurally gentle alternatives. Each technique offers distinct advantages depending on the intended application and the balance between yield and purity [23–25].

12.3.1 Exosome Isolation from Mesenchymal Stem Cells (MSC)

MSCs derived from bone marrow (BM) and adipose tissue (AD) are two of the most widely used sources for exosome production due to their accessibility, proliferative capacity, and robust paracrine activity [26–28]. Their isolation mechanism has been further examined for skin regeneration and wound healing.

12.3.1.1 Isolation of Adipose Mesenchymal Stem Cell (ADMSC) Exosome

ADMSCs are collected from fat tissue. These cells are a valuable source of exosomes because they are easily obtainable, grow well in the laboratory, and release exosomes that aid in healing. Exosomes derived from ADMSCs are first cultured in a growth medium. After the cell releases exosomes, a sufficient quantity of exosomes is released into the surrounding supernatant, and the medium is collected for subsequent use. It is then processed using several rounds of centrifugation at different speeds to remove cells and debris. Finally, the exosomes are collected by spinning the liquid at a very high speed (100,000*g*) for 70 minutes, and this step is repeated to ensure purity. The purified exosomes are then mixed in a buffer solution (PBS) and stored at –80°C [29].

12.3.2 Isolation of Bone Marrow–Derived Mesenchymal Stem Cells (BM-MSCs) Exosomes

Exosomes are isolated from BM-MSCs through a cold, stepwise centrifugation process designed to separate them

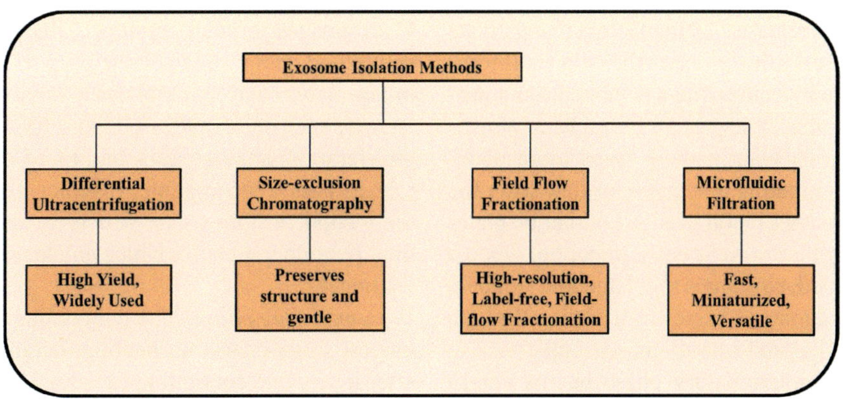

FIGURE 12.1 Isolation and separation methods for exosomes.

from cells and larger vesicles. The initial spins at 300*g*, 2,000*g*, and 10,000*g* (each for increasing durations) gradually removed unwanted cellular components. Once the sample was cleared of debris, ultracentrifugation of the mixture was performed at 100,000*g* at 4°C (70 minutes). This step yielded a pellet rich in exosomes. After a gentle PBS wash, the sample is centrifuged again at the same high speed to reduce microvesicle contamination. The final purified exosomes are stored at –80°C until further use [30].

12.4 THE MECHANISM OF SKIN REGENERATION AND WOUND HEALING

Wound healing is the normal process by which the body repairs damaged skin or tissue after an injury [31]. The ability of intact tissue to carry out protective and regulatory activities is made possible by the structural integrity of its three important layers: the epidermis, dermis, and subcutaneous fat [32, 33]. When the skin's natural protection is damaged by injury, infection, or other harmful causes, it can initiate tissue repair [34]. The healing process usually happens in four main stages: hemostasis, inflammation, proliferation, and remodeling. During each of these phases, various types of cells, MSCs, including ADMSCs and BM-MSCs, actively secrete exosomes [35]. These stem cell–derived exosomes carry important proteins, lipids, and microRNAs that play crucial roles in regulating the healing process [36, 37]. The process of tissue regeneration typically follows a structured timeline, though the exact duration can vary depending on the tissue type (e.g., skin, bone, liver), the extent of injury, and the organism's health. The sequential stages of tissue regeneration are depicted in the timeline shown in Figure 12.2a.

12.4.1 Stages of Skin Regeneration and the Wound Healing Process

Hemostasis: It is an immediate response of the body to injury, beginning within seconds after tissue damage (Figure 12.2b). This phase, which typically lasts about 5–10 minutes, aims to quickly stop bleeding by forming a temporary clot at the wound site [38]. When a blood vessel is damaged, platelets and circulating erythrocytes (red blood cells) accumulate at the injury site. Platelets adhere to the exposed vessel wall and release clotting factors that activate the coagulation cascade. In this process, the enzyme thrombin plays a crucial role by converting a soluble plasma protein fibrinogen into an insoluble protein fibrin, which forms a mesh-like network. This fibrin mesh traps more platelets and erythrocytes, creating a steady clot that seals the wound and prevents further blood loss. Additionally, platelets release molecules that prevent early infection. Once a sufficient clot has formed, clotting is regulated and stopped, and the repair of the damaged vessel wall begins through the activity of endothelial and smooth muscle cells.

Inflammation: The inflammatory phase begins within minutes after injury and typically lasts up to day 4. It plays a vital role in wound healing by preparing the wound site for tissue regeneration. This inflammation stage has a major purpose to clear the debris and other pathogens for tissue repair (Figure 12.2b). This phase starts when damaged cells give a signal known as damage-associated molecular patterns (DAMPs), and signals from any invading bacteria or viruses, called pathogen-associated molecular patterns (PAMPs), activate special receptors on immune cells [39]. This activation initiates the release of messenger proteins called cytokines and chemokines, which widen blood vessels and make them more permeable [40]. As a result, immune cells, mainly neutrophils, quickly move to the wound site to destroy germs and clean up debris. Soon after, another type of immune cell called a monocyte arrives and transforms into a macrophage, which continues cleaning the wound and releases substances that control inflammation and encourage tissue repair. Together, these immune cells coordinate to remove harmful substances, protect the wound from infection, and prepare it for the next stage of healing with careful control of the process that prevents unnecessary tissue damage.

Proliferation: The proliferative phase begins around day 3 after injury and may continue up to approximately day 24, depending on the wound's severity (Figure 12.2b). During this phase, the body actively rebuilds the damaged tissue through several tightly coordinated processes, including granulation tissue formation, fibroblast proliferation, angiogenesis, and re-epithelialization. Fibroblasts, which are specialized connective tissue cells, travel into the wound area and produce key ECM application such as collagen, elastin, and glycoproteins that form a supportive structural network for other cells to attach and grow. This newly formed ECM builds the granulation tissue, providing a scaffold that sustains the migration and differentiation of wound healing cells. Simultaneously, endothelial cells, which line the inner walls of blood vessels, trigger angiogenesis, the process of creating new capillaries to deliver nutrients and oxygen to the regenerating tissue. At the wound edges, keratinocytes (the primary cells of the skin's outer layer) become activated and undergo partial epithelial–mesenchymal transition (EMT), a transformation that boosts their mobility. These keratinocytes release MMPs, which are enzymes that break down and remodel the ECM, allowing cells to migrate efficiently and reconstruct the basement membrane, the thin layer that anchors the epidermis to underlying tissue. Macrophages, crucial immune cells, assist in remodeling new blood vessels and secrete growth factors that regulate inflammation and promote tissue organization. The growth factors, such as VEGF, FGF, and TGF-β, orchestrate cell proliferation, angiogenesis, and ECM production to ensure robust tissue regeneration. Together, these processes close the wound, restore tissue structure, and prepare it for the final remodeling phase, which enhances tissue strength and resilience.

Remodeling phase: The remodeling phase is the last and longest stage of wound healing, beginning around day 24 after injury and continuing for several months, and in some cases up to two years, depending on the extent of the wound and the individual's healing capacity (Figure 12.2b). The main purpose of this phase is to strengthen and reorganize

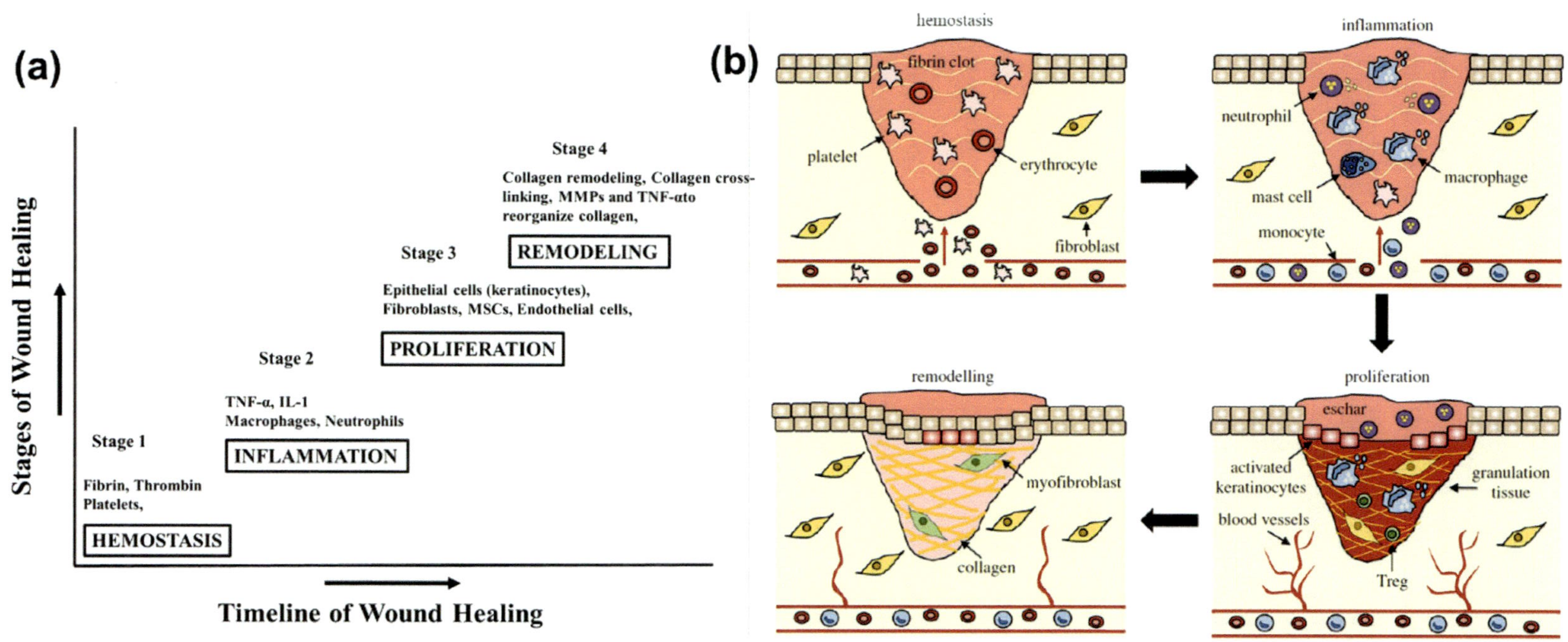

FIGURE 12.2 (a) Timeline of wound healing. (b) The four stages of wound healing. It starts with hemostasis, followed by inflammation, ensuring the debris removal and preventing infection. The third stage is the proliferation stage at which the keratinocytes migrate to close the gap of the wound. Then remodeling of the wound is happening for the wound contraction. *Source:* Figure 12.2b was adapted with permission from The Royal Society, *Open Biology*, an open-access journal, September 30, 2020 [38].

the newly repaired tissue. This involves replacing the temporary ECM, a network of proteins and structural molecules that provides support and scaffolding between cells and the granulation tissue. However, the repaired tissue rarely regains the full strength of the original skin [41]. During remodeling, collagen fibers, which give skin its strength and flexibility, are carefully realigned and cross-linked to increase the tissue's tensile strength and resistance to physical stress. However, even after full remodeling, the repaired tissue usually does not regain the exact strength of healthy, uninjured skin. In this process, type III collagen holds the wound together at first. It is weaker and forms quickly, so it is gradually changed by the stronger type I collagen. This type of collagen is a lot tougher, and it sticks around for a longer time. Exosomes play a big role in this change by guiding the ECM by regulating MMPs. ECM components and TIMPs, when they work together, collagen gets rearranged without breaking down too much or making too much scar tissue. While myofibroblasts are doing their thing, they also help seal up the cut. These cells pull the edges tight and squeeze, closing the gap and making the scar tougher. Meanwhile, as our skin layers are aging just like everything else, the top layer (epidermis) and the deeper skin (dermis) get tougher, and new blood vessels pop up and get settled, which helps bring back great blood flow. These things happen one at a time to get rid of extra scar tissue and restore the skin's normal strength and barrier function. That's when the healing's all done, completing the wound-healing process.

12.5 THE EXOSOME ROLE IN SKIN REPAIR AND WOUND HEALING

Exosomes have gained attention as a healing tool for fixing up skin and speeding up wound healing because they are super involved in the whole healing process [42]. Exosomes, especially those produced from mesenchymal stem cells (MSCs), actively engage in all four traditional phases of wound healing when given externally using topical gels, dressings, or injections. During the hemostasis phase, they do not actually make the clot, but they help make it stronger by getting the platelets to work better and helping with the early fix-up signals [43, 44]. During the inflammatory phase of wound healing, when a wound starts healing and gets inflamed, these tiny exosomes spit out little bits called microRNAs, like miR-21 and miR-146a, [45], which are key in managing how these tiny RNA things that cut down on the swelling cause signals that help prevent too much inflammation and harm to tissues [46]. Exosomes also play a role in controlling macrophages, those key immune cells that deal with inflammation and healing, by nudging them from a pro-inflammatory (M1) vibe to an anti-inflammatory (M2) state that supports tissue repair [47, 48]. This regulation of the immune system, or immunomodulation, is key to stopping long-term swelling and setting up a good balance that lets the wound heal just right [49, 50]. As the wound gets deeper into the healing stage, exosomes are like a boost for keratinocytes, the main skin cells that help patch up the epidermis [51]. They help cells move to the wound area, grow in number, and change into the right types of cells needed to rebuild the skin's surface. At the same time, they activate other cells called fibroblasts to make collagen and other materials that form new tissue, thereby helping the wound heal properly. They also drive angiogenesis, or the growth of new blood vessels, by delivering VEGF and angiogenesis-related microRNAs25 as miR-126 and miR-210, which are part of a larger family of noncoding RNAs, specifically microRNA, which enhance oxygen and nutrient delivery to regenerating tissues. In the final remodeling phase of wound healing, exosomes help organize and strengthen the repaired tissue. Exosomes regulate enzymes called matrix metalloproteinases (MMPs), which help remove damaged components of the extracellular matrix (ECM), the structural framework that supports surrounding cells [52].

By managing this process, exosomes help build stronger new tissue and help reduce the formation of scars [53]. Because of these effects, treatments with exosomes are looking good for better healing and reducing scar formation. Another big plus for exosomes is that they are already linked with living cell therapies, which is an advanced tool for regenerative medicine. Exosomes are more stable, less likely to cause an immune response, and way easier to work with compared to live cells. They do not cause the body to reject them, so they are safer to use over and over. Moreover, exosomes can be engineered, stored, and delivered with greater precision, enhancing their therapeutic effectiveness. Studies show that these tiny exosomes from human mesenchymal stem cells (MSCs) are rich in bioactive components [54, 55]. We are talking about a huge number here, like over 350,000 proteins, about 27,000 mRNAs, and nearly 10,000 miRNAs [56, 57]. Each of these molecules act as a controller, its controlling the cellular processes in wound healing, involved less inflammation, increasing cell proliferation and movement, and helping new tissue form.

12.6 ENGINEERED EXOSOME

Researchers have developed strategies to load exosomes with specific bioactive molecules, such as microRNAs, siRNAs, proteins, and small-molecule drugs to enhance therapeutic efficacy of exosomes. Exosomes laden with miR-155 are one well-researched example of how immune responses can be modulated, especially in cancer therapy, by encouraging immune cell activation [58]. Similarly, exosomes loaded with miR-210 have shown promise in regenerating tissues under low-oxygen conditions by promoting angiogenesis. Exosomes have also been used to deliver small interfering RNA (siRNA) molecules, such as those designed to silence harmful genes in cancer [59]. For example, siRNA targeting the Kirsten rat sarcoma viral oncogene homologue (KRAS) mutation, common in pancreatic cancer, can be delivered via exosomes directly to tumor cells, shutting down one of their key survival pathways [60]. Even traditional drugs have entered the picture. When paclitaxel, a chemotherapy drug, is encapsulated inside exosomes, it can more efficiently reach cancer cells while sparing healthy tissues [61]. Natural compounds like curcumin, known for their anti-inflammatory effects, have also been delivered using exosomes to improve their stability and effectiveness, especially in brain inflammation models. In neurological disorders, researchers have gone a step further, engineering exosomes to carry antioxidant proteins like catalase. These specialized vesicles help reduce oxidative stress in brain cells, offering hope for diseases like Parkinson's. Altogether, these advances show how exosomes are being transformed from passive messengers into smart, customizable couriers capable of delivering healing instructions exactly where they are needed, with precision and care.

12.7 APPLICATION OF EXOSOMES AND ENGINEERED EXOSOMES IN SKIN REGENERATION

Exosomes are derived from different types of stem cells from various studies that can significantly enhance wound healing by regulating important cell-signaling pathways. Ti et al. [62] reported that exosomes from human umbilical cord stem cells helped heal diabetic skin wounds in rats. This was achieved by changing macrophages (immune cells) into the M2 anti-inflammatory type, which reduces harmful inflammation. This change made blocked inflammation pathways (TLR4 and NF-κB) and activated repair pathways (STAT3 and AKT) modulated through exosomes. As a result, the diabetic skin wounds healed faster and with better tissue repair. Similarly, Dalirfardouei et al. found that stem cell exosomes improved chronic wound healing by boosting fibroblast movement and lowering inflammation [63]. Wang et al. showed that specially modified exosomes carry microRNAs that reduced inflammation and helped form new blood vessels in diabetic wounds [64]. He et al. established that exosomes from dental pulp stem cells supported skin repair by increasing collagen and controlling inflammation [65]. Li et al. reported that mesenchymal stem cell exosomes sped up wound closure by managing immune reactions and raising growth factor levels [66].

Further, Qiu et al. observed that exosomes from fat-derived stem cells helped deep skin wounds heal better by promoting new blood vessels and fibroblast growth [67]. Zhang et al. discovered that exosomes from gingival stem cells lowered scarring and encouraged healthy skin structure [68]. Choi and colleagues demonstrated that exosomes containing therapeutic agents could directly deliver healing molecules to wounded epidermis, thereby reducing inflammation [69]. Hu et al. emphasized that engineered exosomes improved tissue regeneration by stimulating blood vessel growth and regulating immune cells [70]. Kciuk et al. explained that exosomes can transport therapeutic microRNAs and proteins to control key repair pathways and reduce oxidative stress, which is crucial for proper wound healing [71]. Together, these findings highlight how exosome-based therapies are emerging as powerful, cell-free strategies to treat various skin injuries and chronic wounds more safely and effectively. These findings highlight how exosome-based therapies are emerging as powerful, cell-free strategies to treat various skin injuries and chronic wounds more safely and effectively, as detailed in Table 12.1.

12.8 APPLICATION OF ENGINEERED EXOSOMES AND EXOSOME-BASED SCAFFOLDS IN SKIN WOUND HEALING

Xiong et al. engineered exosomes from BMSCs by loading them with miR-542-3p using an electrophoresis technique for boosting the healing power of the exosomes when tested on skin wounds [72]. The miR-542-3p-rich exosomes helped skin cells grow faster, increased collagen production, and improved the new blood cell formation, all of which are dynamic for healthy tissue repair. As a result, the treated skin wounds in mice healed more quickly and with better quality tissue regeneration. Further, Wu et al. reported that exosomes from BMSCs when combined with Fe_3O_4 (iron oxide) nanoparticles were capable of treating bone injuries in a rat calvarial (skull) defect model [73]. The magnetic nanoparticles helped to guide the exosomes to reach the damaged bone site, improving their healing action. This smart combination boosted both new bone formation (osteogenesis) and the growth of new blood vessels (angiogenesis) in the skull defect area. These results highlight a promising approach for repairing cranial bone injuries using exosome–nanoparticle systems. Zhang et al. showed that exosomes from UCMSCs, when grown in a low-oxygen (hypoxic) condition and a tissue-like environment with blood cells, became more effective for healing chronic skin wounds that require new blood vessel formation [74]. These specially prepared exosomes helped blood vessel cells grow and survive better, which is important for healing hard-to-close skin wounds, especially in poor blood supply areas like diabetic ulcers. Additionally, Huang et al. prepared engineered exosomes to carry miR-31-5p by transducing HEK293 cells with a lentiviral vector encoding this microRNA [75]. These miR-31-rich exosomes were applied to chronic diabetic wound models, which typically suffer from impaired healing. The treatment improved fibrogenesis, enhanced angiogenesis, and stimulated re-epithelialization. This multifaceted regenerative effect highlights the ability of miR-31 to modulate multiple wound healing pathways, thereby offering a promising approach to overcome delayed healing associated with diabetes. Su et al. derived exosomes from monocytic cells and loaded them with immunomodulatory compounds to boost their regenerative efficacy [76]. These engineered exosomes were applied to HUVECs in vitro. The exosomes knowingly enhanced tube formation (a proxy for angiogenesis), as well as proliferation processes and cell migration involved in wound healing. The results suggest that immunomodulation of exosome-producing cells can shift the functional cargo of exosomes toward promoting vascular and skin cell regeneration. Born et al. conducted a study using BMSCs that were genetically modified to overexpress long noncoding RNA (lncRNA) derived from the Homeobox (HOX) gene transcript [77]. The resulting exosomes, enriched with HOX antisense RNA, were administered to db/db mice (a

TABLE 12.1
Therapies with Mesenchymal Stem Cell–Derived Exosomes for the Repair and Regeneration of the Skin

Exosome Source	Model	Mechanism of Action	Type of Application	References
Human umbilical cord	In vitro (human leukemia monocytic cell line) + LPS (lipopolysaccharide) + rats + STZ (streptozotocin)	Macrophages were converted to an M2 state by MSCs. The TLR4 (Toll-like receptor 4)/NF-κB (nuclear factor kappa-light-chain-enhancer of activated B cells)/STAT (signal transducer and activator of transcription)3/AKT (protein kinase B) regulatory signaling pathways were responsible for the regulation of macrophage polarization	Diabetic wound healing	[62]
Menstrual blood sample	C57BL/6 (a common laboratory mouse strain) + STZ	TLR4 and NF-κB were reduced to activate the M2 stage. STAT (signal transducer and activator of transcription) 3 and STAT6 were activated, which in turn caused the expression of M2 (macrophage) transcripts, including Arg-1 and Arg-2	Triggering wound healing mechanisms in diabetic ulcers	[63]
ADSCs	BALB/c (Bagg Albino Line B, subline c) mice + high fat diet + STZ + skin injury	The PI3K (phosphoinositide 3-kinase)/AKT pathway was activated, thereby promoting fibroblast proliferation and migration, expediting the healing of diabetic ulcers. The proliferation of fibroblasts was also stimulated by the increase in TGF (transforming growth factor) levels	Enhancement of diabetic wound repair	[64]
Bone marrow from the human jaw and the iliac crest	C57BL/6J female adult mice + skin excision	By secreting miR-223, exosomes facilitated anti-inflammatory responses, resulting in the M2 polarization of macrophages and the acceleration of wound healing. The expression of M2 factors, RELM- (resistin-like molecule) and Arginase 1 increased	Dermal wound healing	[65]
Human umbilical cord: hSFCs and hUCMSCs	SD (Sprague Dawley) rats + full thickness burn wounds	White blood cell infiltration and TLR4 pathway suppression play a critical role in modulating inflammation which were reduced by hUCMSC exosomes; it prevented the burns which releases inflammatory factors such as TNF and IL-1	Reducing inflammation that is the consequence of severe burns	[66]
C57BL/6J mice's bone marrow mesenchymal stem cells and human umbilical vein endothelial cells	C57BL/6J (a common laboratory mouse strain) + pentobarbitone sodium + dorsal wounds	The AKT-mediated VEGF (vascular endothelial growth factor) pathway was activated, resulting in the stimulation of angiogenesis by the increased expression of *p*-AKT and *p*-eNOS	Cutaneous wounds	[67]
Human umbilical cord blood stem cells	SD (Sprague Dawley) rats + excisional wound	The inhibition of TGF (transforming growth factor)-receptors by miR-21-5p and miR-125b-5p leads to a reduction in myofibroblast differentiation, thereby stopping the formation of extreme scar tissue. The late phases of the healing process result in the inhibition of collagen formation	Scar tissue	[68]
Primary human adipose stem cells	In vitro (HDFs) (human dermal fibroblasts) + UVB radiation (ultraviolet B light)	The recovery of damaged fibroblasts is facilitated by the proliferation of elastin, TGF-1, TIMP-1 (tissue inhibitor of metalloproteinases), and collagen types I, II, III, IV, and V, which is facilitated by a reduction of MMP production	Facilitating the regeneration of skin tissue that has been damaged by ultraviolet radiation	[69]
Mesenchymal stem cells that are derived from bone marrow	Dermal fibroblasts/mice + UVB radiation	The expressions of MMP1 (matrix metalloproteinase) and MMP9 that were induced by UV light were significantly reduced. The synthesis of type I procollagen was improved. MSCs generate growth factors, including VEGF, EGF (epidermal growth factor), and bFGF (basic fibroblast growth factor), that aid in the repair of injuries and angiogenesis	Decreasing the effects of UV radiation and degeneration, including inadequate collagen production	[70]
Commercialized milk and bovine colostrum	HDF (human dermal fibroblasts) + UV, HaCaTs (human adult keratinocytes) + UV	Col M-exo transport	Damaged skin by UV light	[71]

diabetic mouse model with a leptin receptor gene mutation) suffering from chronic skin wounds. These engineered exosomes significantly improved wound healing and angiogenesis, likely by influencing endothelial cell behavior and promoting new vessel formation. Yuan et al. showed that human adipose-derived mesenchymal stem cells (hAD-MSCs) were genetically modified to produce exosomes containing miR-29a, a microRNA known to prevent excessive scarring [78]. When these engineered exosomes were given to mice with thermal burn wounds, the exosomes helped to control scar formation. This approach provided a targeted way to heal burn wounds with less fibrotic scarring and better skin function. Jiang et al. described that MSCs were genetically engineered using lentiviral vectors to overexpress the anti-inflammatory gene TSG-6 (tumor necrosis factor-stimulated gene-6) [79]. The exosomes from TSG-6-overexpressing cells showed remarkable therapeutic potential in a mouse full-thickness wound model. They reduced local inflammation and significantly minimized scar formation. These findings highlight TSG-6 as a potent molecular target for engineering exosomes with anti-inflammatory functions in wound healing. Sung et al. found that when mesenchymal stem cells were treated with thrombin (a clotting factor) before collecting their extracellular vesicles (EVs), the EVs became more effective at healing skin wounds [80]. In lab tests, these thrombin-treated EVs helped blood vessel cells (HUVECs) to grow, move, and form new blood vessels. When applied to cutaneous (skin) wounds in animals, they sped up skin repair and improved tissue regeneration. This shows that using thrombin can boost the healing power of MSC-derived exosomes for treating skin injuries. The above methods and outcomes from these studies are summarized in Table 12.2. Das et al. demonstrated that exosomes isolated from the human fibrosarcoma cell line HT1080 were used as natural thrombolytic agents to break down blood clots [81]. These exosomes were encapsulated within a thermo-responsive polymer hydrogel to enhance their delivery and controlled release of exosomes. When applied to damaged blood vessels with thrombi, this approach achieved about 28% clot lysis within 10 hours and maintained 22% lysis over 72 hours. This strategy shows great promise for safe thrombus degradation and vascular endothelial healing. Further, Shi et al. reported a novel method to address delayed wound healing in diabetic patients [82]. They used a chitosan/silk fibroin hydrogel loaded with exosomes derived from gingival mesenchymal stem cells (GMSCs). When applied to skin defects in a diabetic rat model, this exosome–hydrogel combination accelerated healing, enhanced tissue regeneration, and promoted denser collagen deposition with better collagen remodeling. This demonstrates an effective, minimally invasive approach for treating chronic diabetic skin wounds. Chachques et al. come up with a new type of scaffold by integrating mesenchymal stem cell (MSC)-derived exosomes into a polycaprolactone (PCL) elastomeric scaffold [83]. These exosomes, immobilized on the scaffold's surface, created a highly regenerative and anti-inflammatory microenvironment. This system modulated macrophage behavior toward a healing (M2 phenotype) and showed strong potential for myocardial tissue renewal and advanced wound repairing applications. Geng et al. explored a multifunctional hydrogel system for treating chronic wounds in diabetes [84]. They combined bone marrow–derived mesenchymal stem cell exosomes (MSC-Exos) with a carboxyethyl chitosan–dialdehyde carboxymethyl cellulose hydrogel (CEC-DCMC hydrogel) to form a MSC-Exos@CEC-DCMC hydrogel. This hydrogel blocked bacteria from infecting the wound and could self-heal after damage. It released exosomes that shifted wound macrophages from a pro-inflammatory (M1) to a pro-healing (M2) state, encouraging new blood vessel formation (neovascularization) and improving chronic diabetic wound repair. Liu et al. presented a new exosome-based gel for enhanced wound healing. They loaded adipose-derived mesenchymal stem cell (AMSC) exosomes into a beta-chitin nanofiber hydrogel and tested it both in vitro on mouse fibroblast cells and in vivo in a rat skin wound model [85]. The exosome hydrogel improved endothelial cell migration and tube formation in vitro and accelerated wound closure, neovascularization, and collagen deposition in vivo. It also promoted the shift of macrophages from the pro-inflammatory (M1) to the pro-regenerative (M2) state, ensuring faster and more efficient skin wound healing. Table 12.2 summarizes the materials, mechanisms of action, and outcomes of these advanced exosome–scaffold systems.

12.8.1 Diabetic Wound Healing by miR-155 Inhibitor-Loaded Engineered Exosome

Diabetes is a growing global health challenge, affecting millions of people worldwide. One of the most serious problems faced by people with diabetes is slow or poor wound repairing, especially in the skin [86, 87]. These diabetic wounds can lead to severe complications, including infections and sometimes even amputations. Scientists have found that small molecules called microRNAs (miRNAs) play an important role in the wound-healing process [88, 89]. These miRNAs control important processes such as cell growth, movement, and the development of new blood vessels, which are essential for wound repair [90]. In diabetic wounds, some miRNAs behave abnormally, disrupting the healing process [91]. Among these, miR-155 stands out because it is found at higher levels in diabetic wounds and is linked to increased inflammation and poor tissue repair. So, controlling or blocking this miRNA could be a promising way to improve healing in diabetic patients. However, delivering molecules like miRNA inhibitors directly to the wound is tricky. They can be quickly broken down in the body and might not reach the damaged tissues effectively. Scientists have found that exosomes from MSCs, a type of stem cell involved in tissue repair, can deliver miRNA inhibitors safely and efficiently to the wound site. They explored exosomes loaded with a miR-155 inhibitor to treat diabetic wounds. First, in lab-grown skin cells exposed to high sugar levels (mimicking diabetes), the treatment helped restore the production of growth factors needed for healing and reduced harmful inflammation.

TABLE 12.2
Therapies Using Engineered Exosomes and Exosome-Based Scaffolds in Skin Wound Healing

Exosome Type	Model	Scaffold/Engineering Strategy	Functions	References
BMSC	Mouse model with a skin wound	Using electrophoresis to put miRNA-542-3p into exosomes	Encourage cellular growth, collagen deposition, neovascularization, and enhanced wound healing	[72]
BMSC exosomes	Rat with a calvarial defect	When BMSC exosomes were mixed with Fe_3O_4 nanoparticles	Improve angiogenesis and osteogenesis	[73]
UCMSC (umbilical cord mesenchymal stem cells) exosomes	In vitro study	Hypoxic three-dimensional coculture of UCMSCs and vascular cells	Promotes proliferation and inhibits apoptosis	[75]
Engineered miR-31 exosomes	Chronic diabetic wounds	HEK293 (Human embryonic kidney) cells with a miR-31-5p lentiviral carrier	Improves fibrogenesis, angiogenesis, and re-epithelization	[76]
Engineered exosomes that are derived from monocytic cells	HUVEC	Exosomes are treated with some immunomodulating compounds	Encourage the formation of HUVEC tubes and improve the growth and movement of skin cells	[77]
miR-29a-modified hADMSC	Thermal mouse model	HADSC (human adipose-derived stem cells) were transfected with miR-29a copies	It lesser additional scar formation	[78]
TSG-6 (tumor necrosis factor-stimulated gene-6) modified MSC (mesenchymal stem cells) exosomes	Full-thickness wound model for a mouse	Increased and decreased TSG-6 lentivirus infection in hBMSCs (human bone marrow-derived mesenchymal stem cells)	Minimalized scar formation by lowering inflammation and stopping the buildup of collagen	[79]
MSC-EV (mesenchymal stem cell–derived extracellular vesicles)	In vitro: HUVECs (human umbilical vein endothelial cells) In vivo: cutaneous wound healing	Thrombin preconditioning regimen	HUVECs' growth, motility, and tube formation were improved in vitro, and skin wound healing was improved in vivo	[80]
Fibrosarcoma cell line HT1080- (human fibrosarcoma cell line) Exosomes	In vitro study	A thermo-responsive polymer of poly(*N*-vinyl caprolactam) (PNVCL) for the encapsulation of exosomes	Facilitate the degradation of thrombus and the regeneration of the endothelial membrane	[81]
GMSCs-(gingiva-derived mesenchymal stem cells) exosomes	Diabetic rats	A chitosan/silk hydrogel sponge in combination with GMSC-derived exosomes	Encouraging the re-epithelialization, deposition, and remodeling of collagen, as well as promoting angiogenesis and neuronal ingrowth	[82]
hADMSCs (human adipose-derived mesenchymal stem cells) exosomes	In vitro study	Elastomeric scaffolds (polycaprolactone)	Enhancing the wound-healing properties and collagen type I and vitronectin of the MSC, and enhancing the M2 phenotype of the macrophages	[83]
BMSC (bone marrow–derived mesenchymal stem cells) exosomes	Chronic diabetic wound healing	MSC exosomes@CECDCMCHG is a carboxyethyl chitosan dialdehyde carboxymethyl cellulose (DCMC) hydrogel that has been infused with BMSC exosomes	Promoted neovascularization, accelerated wound healing, and adjusted the wound inflammation microenvironment in type 1 diabetic rodents	[84]
ADMSC (adipose-derived mesenchymal stem cells) exosomes	The rat's full-thickness epidermis injury model	β-Chitin nanofiber hydrogel stuffed with AMSC (amniotic mesenchymal stem cells) exosomes	Increased rate of wound closure	[85]

Then, when tested in diabetic mice, the treatment sped up wound closure, improved new blood vessel formation, and enhanced skin repair. The encouraging results from in vitro experiments, the wound-healing potential of miR-155-inhibitor-loaded MSC-derived exosomes was further assessed using a diabetic mouse model. After confirming the diabetic condition in mice, full-thickness wounds were created on their dorsal skin. These wounds were then treated with various formulations, including the exosome-loaded system. The experimental workflow is shown in Figure 12.3. In this study, the researchers used C57BL/6 mice, which are a common laboratory mouse strain (the '6' indicates the specific inbred subline) known for their consistent genetic background and making these mice a reliable model for studying diabetes and testing wound-healing therapies (Figure 12.3a). Diabetes was induced in these mice by injecting streptozotocin (50 mg/kg), a chemical that specifically damages the high blood sugar (hyperglycemia) was causing because insulin-producing beta cells in the pancreas. This step mimics diabetes in humans. After the injections, the mice entered a diabetes induction phase, lasting about ten days (from day 6 to day 15), during which their blood glucose was monitored regularly to ensure that hyperglycemia was successfully established. Once stable diabetes was confirmed, the researchers moved to the next critical step: creating wounds on the mice to mimic diabetic ulcers that are difficult to heal. Immediately after the wounds were made (around day 20), different groups of mice received various treatments. One group was treated with plain MSC-Exo, which are naturally known to support tissue repair. Another group received MSC-derived exosomes that were specially loaded with a miR-155 inhibitor (MSC Exo-miR-155 INH). These engineered exosomes were designed to deliver a molecule that blocks microRNA-155 (miR-155), a small RNA that usually promotes inflammation and slows down wound healing. The researchers aimed to accelerate tissue regeneration by specifically inhibiting this damaging microRNA at the wound site. To guarantee an efficient dosage without inflicting damage, the exosome treatment's concentration was maintained at 0.1 μg/μL: the mice's blood sugar levels and body

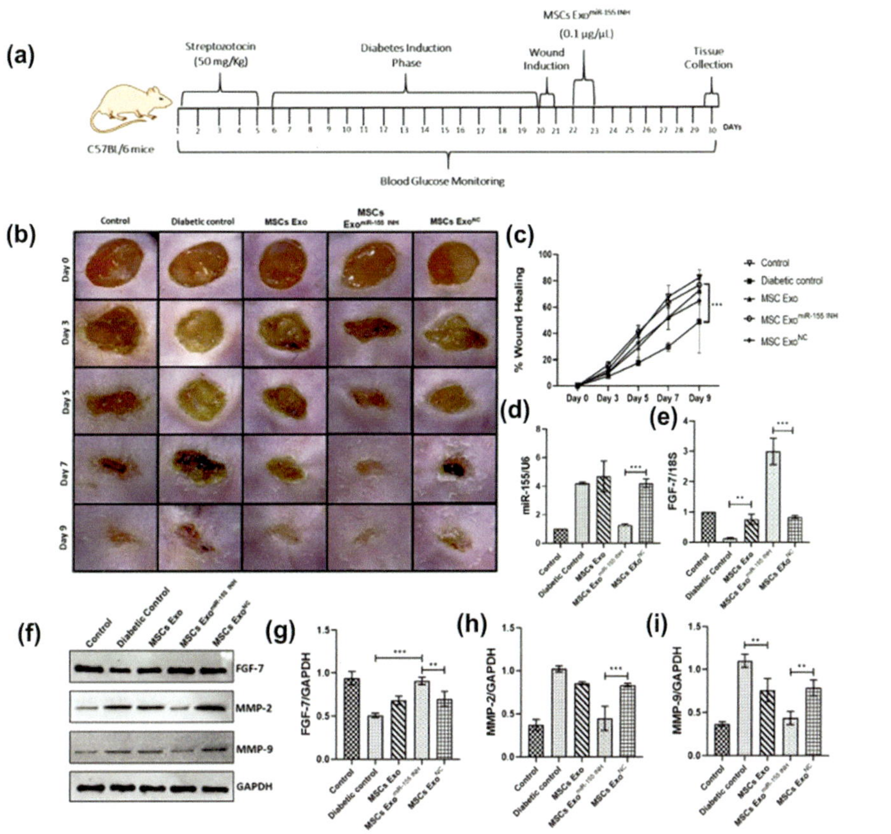

FIGURE 12.3 MSC-derived exosomes loaded with miR-155 inhibitor enhance wound healing in diabetic mice. (a) Overview of the experimental design, (b) Representative images showing wound-healing progression on days 0, 3, 5, 7, and 9 in diabetic mice treated with (i) MSC-derived exosomes, (ii) miR-155-inhibitor-loaded exosomes, and (iii) negative control (NC)-loaded exosomes. Untreated normal and diabetic mice were used as controls. (c) Quantitative analysis of wound closure (%) across different treatment groups. (d) miR-155 expression in wound tissues measured by qRT-PCR and normalized to U6. (e) mRNA levels of FGF-7 in the wound bed measured by qRT-PCR and normalized to 18S rRNA. (f) Western blot analysis showing protein expression of FGF-7, MMP-2, and MMP-9 in the different groups. (g–i) Densitometric quantification of Western blot bands for FGF-7 (g), MMP-2 (h), and MMP-9 (i). *Source:* Adapted with permission from the American Chemical Society, May 1, 2022 [89].

weight. Throughout this phase (from day 21 to day 30), we kept an eye on the mice's weight and blood sugar to see how they were doing overall and to check on the healing of their wound. On day 30, the mice were euthanized, and wound biopsies were collected for further analysis. This allowed the researchers to study the effects of the wound-healing treatment on a cellular level. Following the treatment timeline, the team was able to assess the biological impacts of exosome wound treatment and the visible wound-healing effect. As demonstrated in Figure 12.3b, the researchers monitored the different wound treatments and their effects on wound healing in diabetic mice. They compared five groups: diabetic mice treated with exosomes from normal mesenchymal stem cells (MSC-Exo), diabetic mice treated with exosomes from MSCs loaded with a microRNA-155 inhibitor (MSC Exo-miR-155 INH), diabetic mice treated with MSC exosomes loaded with a nonspecific negative control (MSC Exo-NC). Serial photographs taken on days 0, 3, 5, 7, and 9 showed that the wounds in the MSC Exo-miR-155 INH group healed up quicker than in the mice without treatment. It could significantly improve healing. To quantify these visual changes, Figure 12.3c shows the percentage of wound closure calculated for each time point. This confirmed that the inhibitor-treated group achieved a much higher rate of wound contraction than untreated diabetic mice, providing objective evidence that blocking miR-155 accelerates healing. To understand the underlying biological mechanism, the researchers measured the level of miR-155 itself in the wound tissues (Figure 12.3.d) and found it was significantly lower in the inhibitor-treated group, proving that the specially loaded exosomes effectively silenced this harmful RNA at the wound site. Next, they assessed fibroblast growth factor 7 (FGF-7), a protein essential for promoting keratinocyte growth and migration during skin repair. FGF-7 expression shown in Figure 12.3e was markedly higher in the inhibitor group compared to diabetic controls, indicating enhanced signals for new skin formation. To verify that this effect also translated to the protein level, the team performed Western blot analysis (Figure 12.3f), which clearly showed more FGF-7 protein and less MMP-2 and MMP-9 in the treated group. MMP-2 and MMP-9 are enzymes that degrade the ECM; while some matrix breakdown is needed for tissue remodeling, excessive MMP activity can hinder healing by destroying newly formed tissue. Quantitative analysis of these blots is shown in Figure 12.3g–i, which confirms significantly increased FGF-7 and decreased MMP-2 and MMP-9 levels in the MSC Exo-miR-155 INH group compared to diabetic controls. Together, these results demonstrate that engineering exosomes to deliver a miR-155 inhibitor directly to the wound site effectively silences a negative regulator, boosts pro-healing growth factors, and reduces excessive tissue breakdown, resulting in faster and more robust wound healing even in a diabetic environment.

12.8.2 Exosome-Loaded Hydrogel for Skin Wound Healing

One promising method being explored for wound healing involves embedding exosomes into hydrogel systems. Hydrogels, especially those derived from naturally occurring materials like chitosan, act as three-dimensional scaffolds that can hold and gradually release exosomes at the injury site. One popular material used in such hydrogels is gelatin methacrylate (GelMA), which is known for being safe, easy to use, and suitable for irregular wounds [88]. Researchers have designed smart wound dressings such as multifunctional hydrogels that not only preserve a moist healing location but also offer antimicrobial and antioxidant protection. Among these, the OD/GM/QCS hydrogel, composed of oxidized dextran (OD), gelatin methacrylate (GM), and quaternized chitosan (QCS), has shown promising therapeutic potential. Exosomes were isolated from adipose-derived mesenchymal stem cells (ADMSCs) and loaded into OD/GM/QCS hydrogel. The hydrogel exosome scaffold provides a biocompatible scaffold that promotes cell proliferation, supports neovascularization, modulates the local immune response, and offers inherent antimicrobial activity, collectively contributing to an enhanced and efficient healing process [92].

The hydrogel exhibited significant antibacterial activity against both *Escherichia coli* (Gram-negative) and *Staphylococcus aureus* (Gram-positive), two common wound-infecting pathogens. This was demonstrated by a marked reduction in bacterial colony formation compared to untreated controls (Figure 12.4a). Quantitative assessment confirmed that bacterial survival rates were drastically reduced to $2.9 \pm 0.6\%$ for *E. coli* and $3.5 \pm 1.2\%$ for *S. aureus*, a performance nearly equivalent to that of the antibiotic carbapenem (Figure 12.4b). SEM shows direct visual evidence of this bactericidal effect. Bacteria treated with the hydrogel showed distinct morphological damage, including membrane wrinkling and pore formation, signs of membrane disruption (Figure 12.4c). This effect is primarily attributed to quaternized chitosan (QCS), a positively charged derivative of chitosan. QCS electrostatically interacts with the negatively charged bacterial cell membranes, destabilizing their structure and leading to cell lysis. Furthermore, the hydrogel network includes Schiff base linkages, dynamic covalent bonds between aldehydes and amines that incorporate aromatic rings, which may enhance antimicrobial efficiency through additional membrane interaction or oxidative stress (Figure 12.4d). In vitro assays were conducted using two representative cell lines, NIH-3T3 fibroblasts and HUV endothelial cells (HUVECs), to check the cytocompatibility of the OD/GM/QCS hydrogel. The NIH-3T3 cell line is commonly used to assess fibroblast activity, which is crucial for extracellular matrix (ECM) remodeling and tissue regeneration during wound healing. HUVECs, on the other hand, are primary endothelial cells widely used to study angiogenesis, a vital process for restoring blood supply to damaged tissues. Cell viability and proliferation were confirmed by

CCK-8 assays and live/dead staining, which demonstrated that the hydrogel supported healthy cell growth without inducing cytotoxic effects (Figure 12.4e and f). The ability of the hydrogel to maintain viability in both fibroblasts and endothelial cells is important to support multiple stages of wound healing, including tissue formation and neovascularization. Moreover, in vivo biocompatibility was evaluated by subcutaneous implantation in rats. Histological analysis of surrounding tissues revealed minimal inflammatory response, indicating good tissue integration and biological safety (Figure 12.4g and h). Another critical feature of the OD/GM/QCS hydrogel is its antioxidant capacity, which is vital because hydroxyl radicals (•OH), reactive oxygen species (ROS), superoxide, and ABTS$\bullet^{+}$ radicals anions ($O_2\bullet^{-}$) are often elevated in chronic wounds and can hinder tissue regeneration. Antioxidant assays demonstrated that the hydrogel significantly quenched the absorbance peaks associated with these radicals, suggesting strong free radical scavenging ability (Figure 12.4i–k). Therefore, the OD/GM/QCS hydrogel offers a powerful combination of antibacterial, antioxidant, and biocompatible properties, making it a strong candidate for effective wound management.

Cell migration and angiogenesis are crucial steps in the wound repairing process, as they facilitate the employment of regenerative cells to the injury site and promote the creation of new blood vessels to source nutrients to spoiled tissue, thereby accelerating tissue regeneration and enhancing overall healing outcomes. Incorporating OD/GM/QCS

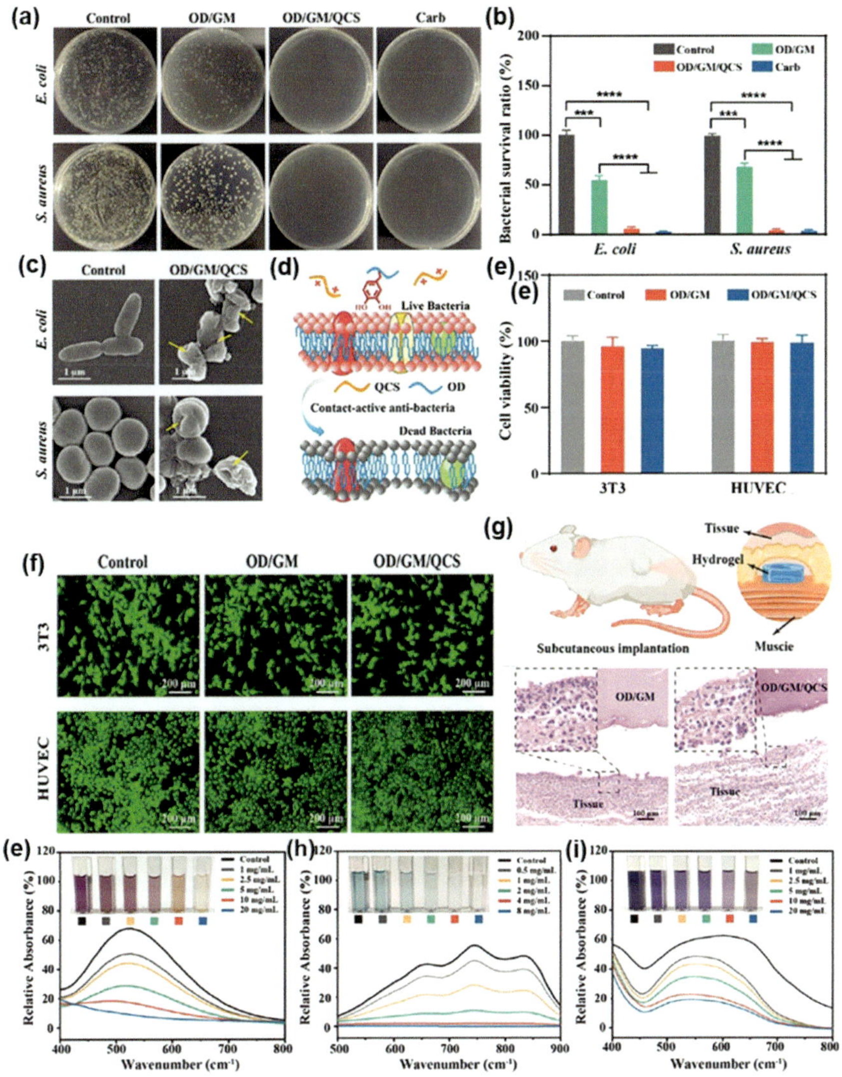

FIGURE 12.4 (a) Antibacterial efficacy, hydrogel samples were co-cultured with *E. coli* and *S. aureus*. (b) The results showed a marked reduction in bacterial colony formation, as seen in the photographic images and quantified survival rates. (c) Scanning electron microscopy (SEM) images. (d) The proposed antibacterial mechanism of the OD/GM/QCS hydrogel is illustrated. (e) Cell viability assay. (f) Live/dead staining images confirming the material's biocompatibility. (g) Pictorial representation of in vivo compatibility, the hydrogel was implanted subcutaneously in rats. (h) Staining of the surrounding tissue showed minimal inflammation, further supporting its safety. (i–k) The hydrogel exhibited significant antioxidant properties, demonstrated by its ability to scavenge hydroxyl radicals (•OH), ABTS radicals, and superoxide anions ($O_2\bullet^{-}$), indicating its potential to reduce oxidative stress in wound environments. *Source:* Adapted with permission from the American Chemical Society, May 1, 2024 [89].

hydrogels has been shown to provide a supportive microenvironment that not only sustains cell viability but also actively modulates these biological processes by releasing bioactive cues and maintaining a moist, protective wound bed. The transwell migration assay measured the ability of cells from different treatment groups, negative control (NC), hydrogel alone (Gel), exosomes alone (Exo), and exosome-loaded hydrogel composite (Exo/Gel) to migrate through a porous membrane over time. The migrated cells were stained blue using a dye such as crystal violet, which highlights the cells that successfully passed through the membrane and adhered to the lower surface, allowing for clear visualization of cell migration (Figure 12.5a). Results showed a time-dependent increase in traveled cells, with the Exo/Gel group exhibiting the highest migration at 24 and 48 hours. This qualitative data is supported quantitatively in Figure 12.5c, which presents a bar graph of the number of migrated cells counted per field. These findings indicate that the combined use of chitosan-based hydrogel and ADMSC-derived exosomes significantly enhances cell motility beyond the effect of either component alone. In parallel, the angiogenesis tube formation assay evaluated the ability of endothelial cells to form capillary-like tube structures, simulating new blood vessel formation. The tube networks formed by cells in each treatment group are shown in Figure 12.5b. The Exo and Exo/Gel groups developed more interconnected and complete vascular networks compared to the NC and Gel groups, indicating enhanced angiogenic potential. It provides quantitative analysis of these networks, typically including metrics such as total tube length, number of junctions, or branch points, confirming the superior pro-angiogenic activity in the groups treated with exosomes (Figure 12.5d). Overall, the Exo/Gel composite hydrogel demonstrated a synergistic enhancement of both cell migration and angiogenesis, highlighting its potential to promote efficient wound repair by combining a supportive scaffold with bioactive, pro-regenerative factors [29].

After observing promising results in cell-based experiments, the next logical step was to test whether the exosome-loaded hydrogel could promote effective healing in a living system. A full-thickness skin wound model was

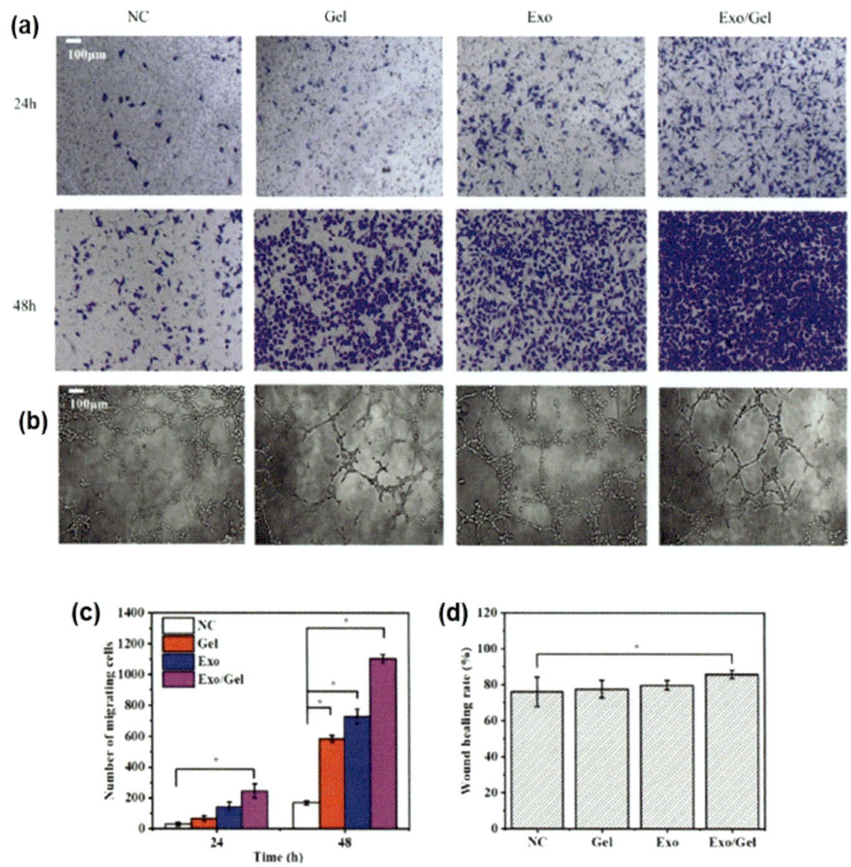

FIGURE 12.5 Impact of the composite hydrogel on HUVEC migration and angiogenesis. (a) Representative images from the transwell migration assay showing HUVECs after 24 and 48 hours under different treatment conditions: untreated control (NC), hydrogel alone (Gel, 200 μL), exosomes at a final concentration of 30 μg/mL (Exo), and hydrogel loaded with 40 μg of exosomes (Exo/Gel). Scale bars represent 200 μm. (b) Images illustrating tube formation by HUVECs following the same treatment groups, scale bars represent 200 μM. (c) Quantification of the migrated cell numbers derived from the trans well assay images shown in part (a). (d) Analysis of the number of lumen structures formed, based on the tube formation images in part (b). Statistical significance is indicated by $*p < 0.05$. *Source:* Adapted with permission from the American Chemical Society, February 1, 2024 [29].

established in rats, designed to mimic real-life injuries and evaluate how well the treatment performs in complex biological environments (Figure 12.6a). The study involved four groups: a control group without treatment (NC), a group treated with hydrogel alone, a group treated with exosomes alone, and a group receiving the exosome-loaded hydrogel (Exo/Gel). Throughout the 14-day observation period, the animals were regularly monitored not only for wound healing but also for general health. Body weight was recorded as an indicator of overall well-being, and results showed normal weight gain in all groups (Figure 12.6b), confirming that none of the treatments caused systemic stress or adverse effects. Visual tracking of wound recovery (Figure 12.6c) revealed progressive healing in every group. By day 14, nearly complete wound closure was achieved across the board, with healing percentages reaching close to 99% (Figure 12.6d). However, differences began to emerge when the quality of the regenerated tissue was examined more closely. New hair growth became a distinguishing feature. The Exo/Gel-treated wounds showed significantly more hair regrowth, about 85% of the wound area, compared to only 76% in the untreated NC group (Figure 12.6e). This difference suggested that the exosome-loaded hydrogel not only accelerated healing but also contributed to better-quality skin regeneration, including recovery of hair follicles. Another noteworthy observation was the presence of yellowish exudate around the wounds in the NC group, a sign of lingering inflammation. In contrast, none of the treated groups showed such fluid accumulation. This absence of inflammatory residue in the Exo/Gel group supports the anti-inflammatory action of MSC-derived exosomes, indicating that the treatment may also help establish a cleaner, more balanced environment for tissue repair. These in vivo findings highlight the dual advantage of the

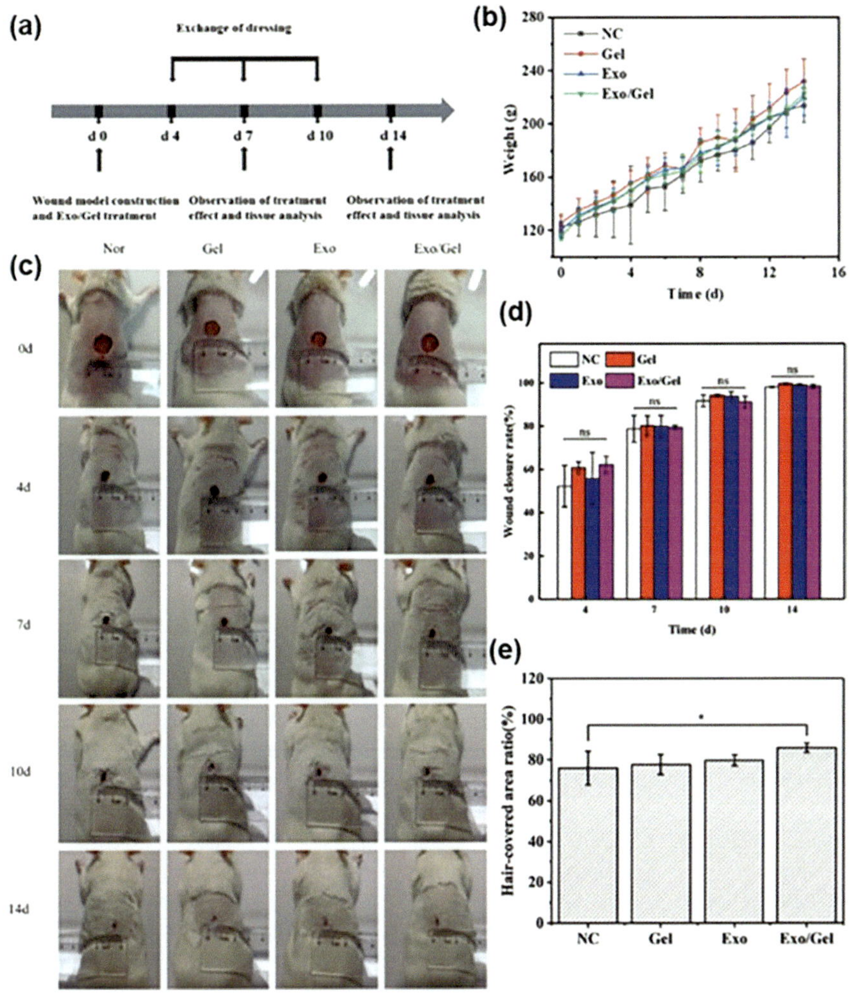

FIGURE 12.6 Evaluation of wound healing efficacy using the ADMSC-exosome-loaded hydrogel. (a) Schematic overview illustrating the creation of the full-thickness skin wound model in rats and the corresponding treatment groups. (b) Changes in body weight of rats were monitored over the 14-day healing period to assess overall health and treatment safety. (c) Representative images showing the progression of wound closure at various time points: day 0, 4, 7, 10, and 14. (d) Quantitative analysis of wound closure rates based on the images in panel c. (e) Measurement of the hair regrowth area over the healed wound on day 14, demonstrating improved tissue regeneration in treated groups (*p <0.05). *Source:* Adapted with permission from the American Chemical Society, February 1, 2024. [29].

Exo/Gel composite: it promotes faster wound closure while also enhancing tissue quality and reducing inflammation, key factors in achieving complete and functional healing.

12.9 APPLICATION OF ENGINEERED EXOSOME-BASED SCAFFOLDS IN SKIN WOUND HEALING

Researchers have also been developing innovative engineered scaffolds that can carry exosomes to wounds in a more controlled and effective way. Sun et al. developed an intelligent nanoplatform by integrating exosomes into a responsive hydrogel system to attain precise and specific precise release [93]. This design ensured that the exosomes remained active longer, effectively promoting angiogenesis and tissue regeneration while also providing an antibacterial barrier, making it highly suitable for chronic wound care. Qian et al. created a water-repelling/water-loving membrane dressing with a top hydrophobic layer and a bottom hydrophilic layer. It is efficiently loaded into mesenchymal stem cell–derived exosomes (MSC exosomes) [94]. This smart dressing really helps wounds heal better by keeping up the supply of healing exosomes, getting fibroblasts to move in, and cutting down on swelling. Wang et al. showed a natural healing gel that is loaded with exosomes from adipose-derived mesenchymal stem cells (ADMSCs) [95]. This hydrogel speeds up wound healing, helps new tissue grow, and keeps the wound safe from germs, which is super important for those stubborn wounds that just won't heal. Li et al. pushed the field forward by making a new kind of gel that mixes hydroxyapatite (HAP) and chitosan (CS) to deliver synovium mesenchymal stem cell–derived exosomes (SMSCs–exosomes) to carry stem cell–made exosomes loaded with microRNA-126 (miR-126) [96]. This bioactive scaffold boosted blood vessel growth, improved skin structure, and helped heal wounds faster, making it a great tool for fixing long-lasting injuries. These designs, materials used, and therapeutic implications of these systems are summarized in Table 12.3.

12.10 LIMITATIONS AND ADVANTAGES OF EXOSOMES AND EXOSOME-BASED SCAFFOLDS IN SKIN AND WOUND HEALING

Exosomes have an important role in skin regeneration and wound repair due to their capacity to promote angiogenesis, modulate inflammation, and migration. However, despite their promise, the application of unmodified or naturally derived exosomes alone is associated with several critical limitations that hinder their clinical effectiveness.

- *Weak therapeutic effect when used alone:* Exosomes do not always have strong enough biological activity unless modified or combined with other agents.
- *Poor targeting:* They do not naturally go to the exact place in the body where healing is needed.
- *Short lifespan:* They degrade quickly or are cleared by the immune system.
- *Difficult to isolate and purify:* Making clinical-grade exosomes is hard and not standardized.

TABLE 12.3
Therapies Using Engineering Exosome-Loaded Scaffolds in Skin Wound Healing

Exosome Type	Model	Engineering Strategy	Function	References
hADSCs-exosome (human adipose-derived stem cells)	A diabetic wound festers model.	Engineered exosomes (E-Exos) collected from TNF-α-treated MSCs under hypoxia are immobilized using reductive 2D COFs as nanocarriers	Promoting angiogenesis, removing bacterial infection, and reducing oxidative injury and tissue inflammation	[93]
HUMSCs (human umbilical mesenchymal stem cells)–exosome	Mouse skin defect model infected with *P. aeruginosa*	An asymmetric wettable dressing, CTS-SF/SA/Ag-Exo (chitosan/silk fibroin/sodium alginate), has been developed using exosomes and silver nanoparticles	Wound healing was improved by the CTS-SF/SA/Ag-Exo dressing through increased collagen accumulation, angiogenesis, and neural repair	[94]
ADMSC exosomes (adipose-derived mesenchymal stem cell)	A diabetic wound that is chronically full-thickness and does not resolve	Engineering bioactive natural healing exosomes that are antibacterial hydrogel	The wound site experiences enlarged wound closure rates, re-epithelialization, rapid angiogenesis, and collagen deposition	[95]
Synovium MSC-126 Exos (synovium-derived mesenchymal stem cells overexpressing miR-126)	Chronic wound (diabetic)	HAP-CS-SMSCs-126-Exos (hydroxyapatite/chitosan composite hydrogel loaded with SMSCs-126-derived exosomes	Accelerate angiogenesis, expedite collagen maturation, and promote wound surface re-epithelialization	[96]

- *Low stability and fast clearance:* They don't stay long at the wound site, so multiple doses are needed.
- *Clinical translation is limited:* No universal protocols, concerns over safety, and lack of regulation make it hard to bring exosome therapies into hospitals.

The above limitations of exosomes could be overcome by the use of engineered exosomes and exosome-loaded scaffolds. The external therapeutic molecules enhance the wound healing potential of engineered exosomes. In addition, the engineered exosome could be targeted to the wound site. When exosomes are loaded in several scaffolds, like hydrogel, their life span is increased due to the slow release of exosomes from the hydrogel. Engineered exosomes loaded in a scaffold have more advantages. Therapeutic molecules, along with exosomes, are released in the wound site slowly from the engineered exosomes loaded in a scaffold. Those materials can be used as new-generation dressing materials for skin regeneration and wound healing.

12.11 FUTURE PROSPECTS

Exosomes have emerged as powerful cell-free therapeutic tools with remarkable potential to enhance tissue repair by promoting cell proliferation, modulating inflammation, and stimulating tissue regeneration. Exosome-based therapies in clinical translation will benefit from deeper insights into their mechanisms of action and improved techniques for large-scale isolation and purification. The use of exosomes derived from MSCs, such as BMSCs and ADSCs, is expected to increase due to their proven safety and regenerative effects. Building upon the advantages of natural exosomes, engineered exosomes are designed or modified to enhance therapeutic efficiency by loading them with specific molecules, improving targeting ability, or combining multiple functional modifications. Scaffold-based exosome delivery systems have attracted growing attention. By incorporating exosomes into biocompatible scaffolds or hydrogels, researchers aim to attain the constant and controlled release of exosomes straight at the wound site, providing a favorable microenvironment for cell migration, angiogenesis, and tissue remodeling. This combination strategy holds great promise for overcoming the speed clearance and short half-life of free exosomes, enhancing their retention and therapeutic efficacy. Moving forward, continued research is needed to optimize scaffold materials, loading techniques, and release kinetics, and to understand how scaffold–exosome interactions influence wound-healing processes. Overall, the integration of exosomes, engineered exosomes, and smart scaffold systems represents a powerful future direction for developing advanced, targeted, and effective therapies for clinical wound repair.

12.12 CONCLUSION

There are many steps in the biological process of skin wound healing. Chronic wounds are still a major health issue that affects many individuals and makes their lives worse. They often have inadequate blood flow and continuous inflammation. Traditional therapies can be expensive, take a long time, and don't always work well, especially for wounds that won't heal. This shows that we need to find better ways to treat people. Recently, therapies that involve the therapeutic potential of MSCs have been widely recognized in the field of regenerative medicine. since these cells can change into numerous types of cells. But some limitations make it hard to employ them more widely, including immunological rejection, stability, mobility, and the chance of uncontrolled cell proliferation (Jiang et al., 2002). Studies now indicate that an important portion of the therapeutic efficacy of MSCs is derived from the signals they emit, particularly via nanoscale entities known as extracellular vesicles (EVs), including exosomes. Exosomes from MSCs have garnered interest due to their capacity to transport significant molecules, such as proteins and RNA, to other cells, hence replicating many of the therapeutic effects of MSCs themselves (Zhang et al., 2014). Research on animals suggests that using MSC exosomes might be an option than using MSCs. This is because they seem to cause reactions from the immune system and are less likely to lead to tumors or other harmful effects. However, there are hurdles to overcome before these treatments can be used in hospitals. One of the issues is that there is no established method for isolating, purifying, and identifying MSC exosomes, which is a crucial step in making them safe and effective for patients. The quality control process, which involves keeping an eye on things like consistency, potency, safety, and stability, is still a work in progress. This makes it tough to produce things on a large scale. The current classification systems, such as the International Society for Cellular Therapy (ISCT) criteria and the Minimal Information for Studies of Extracellular Vesicles (MISEV) guidelines, are insufficient to clearly define or separate MSC–exosomes from other exosomal types. Functional assays to validate the biological activity of these exosomes are also lacking. Storage and transport conditions further complicate clinical application. Research suggests that freeze-drying or lyophilization might be a way to keep MSC exosomes stable when they're stored at room temperature. However, more work is needed to make sure this method is reliable and always produces the desired results. Recent advances in bioengineering and nanotechnology have opened up avenues for enhancing MSC exosomes. One approach is to modify the parent MSCs or directly alter the exosomes themselves, enabling them to effectively locate and repair damaged cells. EVs have shown promise than nanoparticles in delivering RNA and proteins to their intended targets. Nonetheless, there is still much to be discovered about how these altered MSC exosomes function, and further research is needed to refine their design, ensuring both their effectiveness and protection. In conclusion, exosomes derived from MSCs present a promising cell-free therapy for improving wound healing and skin regeneration. This chapter discusses how exosomes, including engineered and scaffolded ones, can support skin repair and speed up the healing process. It covers methods for isolating and characterizing these exosomes, with an emphasis on those from MSCs. Natural exosomes help with healing, but they do have some limits. By modifying them and combining them with scaffolds or hydrogels, we can improve their ability to target specific areas and control their release. For example, exosomes loaded with miR-155 inhibitors and hydrogels containing exosomes show great promise, especially for treating chronic wounds or wounds caused by diabetes. Natural exosomes help with healing, but they do have some limits. Overall, exosome-based strategies offer a promising future for advanced wound care therapies.

ABBREVIATIONS

ABTS	2,2′-azino-bis(3-ethylbenzothiazoline-6-sulfonic acid)
AKT	protein kinase B
BALB/c	Bagg Albino Line B, subline c
C57BL/6	common laboratory mouse strain,
C57BL/6J	common laboratory mouse strain (Jackson subline)
CCK-8	cell counting Kit-8
CD9	cluster of differentiation 9
CD63	cluster of differentiation 63
CD81	cluster of differentiation 81
CEC-DCMC	carboxyethyl chitosan–dialdehyde carboxymethyl cellulose
COFs	covalent organic frameworks
Col M-exo	colostrum/milk exosomes
db/db	diabetic mouse strain with leptin receptor mutation
E-Exos	engineered exosomes
eNOS	endothelial nitric oxide synthase
Exo/Gel	exosome-loaded hydrogel
HaCaTs	human adult keratinocytes cell line
HAP-CS	hydroxyapatite/chitosan
HDFs	human dermal fibroblasts
HEK293	human embryonic kidney 293 cells
HOX	homeobox
hSFCs	human skin fibroblast cells
HT1080	human fibrosarcoma cell line 1080
lncRNA	long noncoding RNA
LPS	lipopolysaccharide
mRNAs	messenger RNAs
NC	negative control
OD/GM/QCS	oxidized dextran/gelatin/quaternized chitosan
p-AKT	phosphorylated AKT
PBS	phosphate-buffered saline
p-eNOS	phosphorylated endothelial nitric oxide synthase
PCL	polycaprolactone
PNVCL	poly(*N*-vinyl caprolactam)
RELM-α	resistin-like molecule alpha
ROS	reactive oxygen species
SMSCs–126-Exos	synovium-derived mesenchymal stem cell exosomes overexpressing miR-126

12.14 ACKNOWLEDGMENTS

H.E., S.G., and A.P. acknowledge the Department of Chemistry of SRM Institute of Science and Technology for providing the chemicals, fellowship, and lab facility. A.S. acknowledges the Science Engineering Research Board (SERB) for providing start-up research funding (grant No.: SRG/2023/000554) as well as the Department of Chemistry of the SRM Institute of Science and Technology for providing the instruments and lab facilities.

AI DISCLOSURE STATEMENT

During the preparation of this work, the authors used Grammarly, QuillBot, and ChatGPT in order to enhance grammatical structure and narrative coherence. After using this tool/service, the authors reviewed and edited the content as needed and take full responsibility for the content of the publication.

CONFLICT OF INTEREST

None.

REFERENCES

[1] Rodrigues, M.; Kosaric, N.; Bonham, C. A.; Gurtner, G. C. Wound Healing: A Cellular Perspective. *Physiol Rev*, **2019**, *99*, 665–706.

[2] Tottoli, E. M.; Dorati, R.; Genta, I.; Chiesa, E.; Pisani, S.; Conti, B. Skin Wound Healing Process and New Emerging Technologies for Skin Wound Care and Regeneration. *Pharmaceutics*, **2020**, *12* (8), 735.

[3] Guo, S.; DiPietro, L. A. Factors Affecting Wound Healing. *J Dent Res*, **2010**, *89*, 219–229.

[4] Lopes, B.; Sousa, P.; Alvites, R.; Branquinho, M.; Sousa, A.; Mendonça, C.; . . . Maurício, A. C. The Application of Mesenchymal Stem Cells in Wound Repair and Regeneration. *Appl Sci*, **2021**, *11* (7), 3000.

[5] Kirwan, H. O. L. L. I. E.; Pignataro, R. O. S. E. The Skin and Wound Healing. *Pathol Interv Musculoskelet Rehabil*, **2015**, *25* (8), 125–129.

[6] Yang, G.; Waheed, S.; Wang, C.; Shekh, M.; Li, Z.; Wu, J. Exosomes and Their Bioengineering Strategies in the Cutaneous Wound Healing and Related Complications: Current Knowledge and Future Perspectives. *Int J Biol Sci*, **2023**, *19* (5), 1430.

[7] Peña, O. A.; Martin, P. Cellular and Molecular Mechanisms of Skin Wound Healing. *Nat Rev Mol Cell Biol*, **2024**, *25* (8), 599–616.

[8] Mazini, L.; Rochette, L.; Admou, B.; Amal, S.; Malka, G. Hopes and Limits of Adipose-Derived Stem Cells (ADSCs) and Mesenchymal Stem Cells (MSCs) in Wound Healing. *Int J Mol Sci*, **2020**, *21* (4), 1306.

[9] Lv, Q.; Deng, J.; Chen, Y.; Wang, Y.; Liu, B.; Liu, J. Engineered Human Adipose Stem-Cell-Derived Exosomes Loaded with MiR-21-5p to Promote Diabetic Cutaneous Wound Healing. *Mol Pharm*, **2020**, *17* (5), 1723–1733.

[10] Hettich, B. F.; Ben-Yehuda Greenwald, M.; Werner, S.; Leroux, J. Exosomes for Wound Healing: Purification Optimization and Identification of Bioactive Components. *Adv Sci*, **2020**, *7* (23), 2002596.

[11] Ding, J.-Y.; Chen, M.-J.; Wu, L.-F.; Shu, G.-F.; Fang, S.-J.; Li, Z.-Y.; Chu, X.-R.; Li, X.-K.; Wang, Z.-G.; Ji, J.-S. Mesenchymal Stem Cell-Derived Extracellular Vesicles in Skin Wound Healing: Roles, Opportunities and Challenges. *Mil Med Res*, **2023**, *10* (1), 36.

[12] Zhang, W.; Wang, T.; Xue, Y.; Zhan, B.; Lai, Z.; Huang, W.; Peng, X.; Zhou, Y. Research Progress of Extracellular Vesicles and Exosomes Derived from Mesenchymal Stem Cells in the Treatment of Oxidative Stress-Related Diseases. *Front Immunol*, **2023**, *14*, 1238789.

[13] Michela, B. Liquid Biopsy: A Family of Possible Diagnostic Tools. *Diagnostics*, **2021**, *11* (8), 1391.

[14] Dayon, L.; Cominetti, O.; Affolter, M. Proteomics of Human Biological Fluids for Biomarker Discoveries: Technical Advances and Recent Applications. *Expert Rev Proteomics*, **2022**, *19* (2), 131–151.

[15] Karami, M. H.; Abdouss, M. Advancements in Utilizing Exosomes for Cancer Therapy Through Drug Delivery Systems: Recent Developments. *Nanomed J*, **2025**, *12* (2).

[16] Liu, W.-S.; Liu, Y.; Gao, J.; Zheng, H.; Lu, Z.-M.; Li, M. Biomembrane-Based Nanostructure-and Microstructure-Loaded Hydrogels for Promoting Chronic Wound Healing. *Int J Nanomedicine*, **2023**, 385–411.

[17] Safari, B.; Aghazadeh, M.; Davaran, S.; Roshangar, L. Exosome-Loaded Hydrogels: A New Cell-Free Therapeutic Approach for Skin Regeneration. *Eur J Pharm Biopharm*, **2022**, *171*, 50–59.

[18] Golchin, A.; Shams, F.; Basiri, A.; Ranjbarvan, P.; Kiani, S.; Sarkhosh-Inanlou, R.; Ardeshirylajimi, A.; Gholizadeh-Ghaleh Aziz, S.; Sadigh, S.; Rasmi, Y. Combination Therapy of Stem Cell-Derived Exosomes and Biomaterials in the Wound Healing. *Stem Cell Rev Rep*, **2022**, *18* (6), 1892–1911.

[19] Freedman, B. R.; Hwang, C.; Talbot, S.; Hibler, B.; Matoori, S.; Mooney, D. J. Breakthrough Treatments for Accelerated Wound Healing. *Sci Adv*, **2023**, *9* (20), eade7007.

[20] Huang, Y.-Z.; Gou, M.; Da, L.-C.; Zhang, W.-Q.; Xie, H.-Q. Mesenchymal Stem Cells for Chronic Wound Healing: Current Status of Preclinical and Clinical Studies. *Tissue Eng Part B Rev*, **2020**, *26* (6), 555–570.

[21] Sreeraj, H.; AnuKiruthika, R.; Tamilselvi, K. S.; Subha, D. Exosomes for Skin Treatment: Therapeutic and Cosmetic Applications. *Nano TransMed*, **2024**, *3*, 100048. https://doi.org/10.1016/j.ntm.2024.100048.

[22] Sun, T.; Li, M.; Liu, Q.; Yu, A.; Cheng, K.; Ma, J.; Murphy, S.; McNutt, P. M.; Zhang, Y. Insights into Optimizing Exosome Therapies for Acute Skin Wound Healing and Other Tissue Repair. *Front Med*, **2024**, *18* (2), 258–284. https://doi.org/10.1007/s11684-023-1031-9.

[23] Dilsiz, N. A Comprehensive Review on Recent Advances in Exosome Isolation and Characterization: Toward Clinical Applications. *Transl Oncol*, **2024**, *50*, 102121.

[24] Sidhom, K.; Obi, P. O.; Saleem, A. A Review of Exosomal Isolation Methods: Is Size Exclusion Chromatography the Best Option? *Int J Mol Sci*, **2020**, *21* (18), 6466.

[25] Yang, D.; Zhang, W.; Zhang, H.; Zhang, F.; Chen, L.; Ma, L.; Larcher, L. M.; Chen, S.; Liu, N.; Zhao, Q. Progress, Opportunity, and Perspective on Exosome Isolation-Efforts for Efficient Exosome-Based Theranostics. *Theranostics*, **2020**, *10* (8), 3684.

[26] Maqsood, M.; Kang, M.; Wu, X.; Chen, J.; Teng, L.; Qiu, L. Adult Mesenchymal Stem Cells and Their Exosomes: Sources, Characteristics, and Application in Regenerative Medicine. *Life Sci*, **2020**, *256*, 118002.

[27] Hassanzadeh, A.; Rahman, H. S.; Markov, A.; Endjun, J. J.; Zekiy, A. O.; Chartrand, M. S.; Beheshtkhoo, N.; Kouhbanani, M. A. J.; Marofi, F.; Nikoo, M. Mesenchymal Stem/Stromal Cell-Derived Exosomes in Regenerative Medicine and Cancer; Overview of Development, Challenges, and Opportunities. *Stem Cell Res Ther*, **2021**, *12* (1), 297.

[28] Mohammed, M. A.; Hay, N. H. A.; Mohammed, M. T.; Mahmoud, H. S.; Ahmed, M. Y.; Abdelmenem, A.; Abdelrahim, D. S. The Effect of Adipose-Derived Mesenchymal Stem Cells against High Fructose Diet Induced Liver Dysfunction and Dysbiosis. *Naunyn Schmiedebergs Arch Pharmacol*, **2024**, 1–13.

[29] Wu, D.; Tao, S.; Zhu, L.; Zhao, C.; Xu, N. Chitosan Hydrogel Dressing Loaded with Adipose Mesenchymal Stem Cell-Derived Exosomes Promotes Skin Full-Thickness Wound Repair. *ACS Appl Bio Mater*, **2024**, *7* (2), 1125–1134.

[30] Gondaliya, P.; Sayyed, A. A.; Bhat, P.; Mali, M.; Arya, N.; Khairnar, A.; Kalia, K. Mesenchymal Stem Cell-Derived Exosomes Loaded with MiR-155 Inhibitor Ameliorate Diabetic Wound Healing. *Mol Pharm*, **2022**, *19* (5), 1294–1308.

[31] Yang, F.; Bai, X.; Dai, X.; Li, Y. The Biological Processes During Wound Healing. *Regenerative Med*, **2021**, *16* (4), 373–390.

[32] Liu, Y.; Wang, H.; Wang, J. Exosomes as a Novel Pathway for Regulating Development and Diseases of the Skin. *Biomed Rep*, **2018**, *8* (3), 207–214.

[33] Choudhary, V.; Choudhary, M.; Bollag, W. B. Exploring Skin Wound Healing Models and the Impact of Natural Lipids on the Healing Process. *Int J Mol Sci*, **2024**, *25* (7), 3790.

[34] Mohd Zaid, N. A.; Sekar, M.; Bonam, S. R.; Gan, S. H.; Lum, P. T.; Begum, M. Y.; Mat Rani, N. N. I.; Vaijanathappa, J.; Wu, Y. S.; Subramaniyan, V. Promising Natural Products in New Drug Design, Development, and Therapy for Skin Disorders: An Overview of Scientific Evidence and Understanding Their Mechanism of Action. *Drug Des Devel Ther*, **2023**, 23–66.

[35] Tang, Y.; Zhou, Y.; Li, H.-J. Advances in Mesenchymal Stem Cell Exosomes: A Review. *Stem Cell Res Ther*, **2021**, *12* (1), 71.

[36] Asgarpour, K.; Shojaei, Z.; Amiri, F.; Ai, J.; Mahjoubin-Tehran, M.; Ghasemi, F.; ArefNezhad, R.; Hamblin, M. R.; Mirzaei, H. Exosomal MicroRNAs Derived from Mesenchymal Stem Cells: Cell-to-Cell Messages. *Cell Commun Signal*, **2020**, *18*, 1–16.

[37] Weiliang, Z.; Lili, G. Research Advances in the Application of Adipose-Derived Stem Cells Derived Exosomes in Cutaneous Wound Healing. *Ann Dermatol*, **2021**, *33* (4), 309.

[38] Wilkinson, H. N.; Hardman, M. J. Wound Healing: Cellular Mechanisms and Pathological Outcomes. *Open Biol*, **2020**, *10* (9), 200223.

[39] Harris, F. M.; Mou, Z. Damage-Associated Molecular Patterns and Systemic Signaling. *Phytopathology*, **2024**, *114* (2), 308–327.
[40] Mahaling, B.; Low, S. W. Y.; Beck, M.; Kumar, D.; Ahmed, S.; Connor, T. B.; Ahmad, B.; Chaurasia, S. S. Damage-Associated Molecular Patterns (DAMPs) in Retinal Disorders. *Int J Mol Sci*, **2022**, *23* (5), 2591.
[41] Singh, D.; Rai, V.; Agrawal, D. K. Regulation of Collagen I and Collagen III in Tissue Injury and Regeneration. *Cardiol Cardiovasc Med*, **2023**, *7* (1), 5.
[42] Ha, D. H.; Kim, H.; Lee, J.; Kwon, H. H.; Park, G.-H.; Yang, S. H.; Jung, J. Y.; Choi, H.; Lee, J. H.; Sung, S. Mesenchymal Stem/Stromal Cell-Derived Exosomes for Immunomodulatory Therapeutics and Skin Regeneration. *Cells*, **2020**, *9* (5), 1157.
[43] Zhou, C.; Zhang, B.; Yang, Y.; Jiang, Q.; Li, T.; Gong, J.; Tang, H.; Zhang, Q. Stem Cell-Derived Exosomes: Emerging Therapeutic Opportunities for Wound Healing. *Stem Cell Res Ther*, **2023**, *14* (1), 107.
[44] Bian, D.; Wu, Y.; Song, G.; Azizi, R.; Zamani, A. The Application of Mesenchymal Stromal Cells (MSCs) and Their Derivative Exosome in Skin Wound Healing: A Comprehensive Review. *Stem Cell Res Ther*, **2022**, *13* (1), 24.
[45] Jiang, Y.; Xu, X.; Xiao, L.; Wang, L.; Qiang, S. The Role of MicroRNA in the Inflammatory Response of Wound Healing. *Front Immunol*, **2022**, *13*, 852419.
[46] Das, K.; Rao, L. V. M. The Role of MicroRNAs in Inflammation. *Int J Mol Sci*, **2022**, *23* (24), 15479.
[47] Arabpour, M.; Saghazadeh, A.; Rezaei, N. Anti-Inflammatory and M2 Macrophage Polarization-Promoting Effect of Mesenchymal Stem Cell-Derived Exosomes. *Int Immunopharmacol*, **2021**, *97*, 107823.
[48] Gharavi, A. T.; Hanjani, N. A.; Movahed, E.; Doroudian, M. The Role of Macrophage Subtypes and Exosomes in Immunomodulation. *Cell Mol Biol Lett*, **2022**, *27* (1), 83.
[49] Zhao, W.; Zhang, H.; Liu, R.; Cui, R. Advances in Immunomodulatory Mechanisms of Mesenchymal Stem Cells-Derived Exosome on Immune Cells in Scar Formation. *Int J Nanomed*, **2023**, 3643–3662.
[50] Muire, P. J.; Thompson, M. A.; Christy, R. J.; Natesan, S. Advances in Immunomodulation and Immune Engineering Approaches to Improve Healing of Extremity Wounds. *Int J Mol Sci*, **2022**, *23* (8), 4074.
[51] An, Y.; Lin, S.; Tan, X.; Zhu, S.; Nie, F.; Zhen, Y.; Gu, L.; Zhang, C.; Wang, B.; Wei, W. Exosomes from Adipose-Derived Stem Cells and Application to Skin Wound Healing. *Cell Prolif*, **2021**, *54* (3), e12993.
[52] Liew, F. F.; Chew, B. C.; Ooi, D. J. Wound Healing Properties of Exosomes – a Review and Modelling of Combinatorial Analysis Strategies. *Curr Mol Med*, **2022**, *22* (2), 165–191.
[53] Kandhwal, M.; Behl, T.; Singh, S.; Sharma, N.; Arora, S.; Bhatia, S.; Al-Harrasi, A.; Sachdeva, M.; Bungau, S. Role of Matrix Metalloproteinase in Wound Healing. *Am J Transl Res*, **2022**, *14* (7), 4391.
[54] Sun, Y.; Liu, G.; Zhang, K.; Cao, Q.; Liu, T.; Li, J. Mesenchymal Stem Cells-Derived Exosomes for Drug Delivery. *Stem Cell Res Ther*, **2021**, *12*, 1–15.
[55] Janockova, J.; Slovinska, L.; Harvanova, D.; Spakova, T.; Rosocha, J. New Therapeutic Approaches of Mesenchymal Stem Cells-Derived Exosomes. *J Biomed Sci*, **2021**, *28*, 1–26.
[56] Wang, Y.; Wu, X. New Perspectives and Prospects of MicroRNA Delivery in Diabetic Wound Healing. *Mol Pharmacol*, **2024**, *106* (2), 84–91.
[57] Mansour, R. M.; Mageed, S. S. A.; Awad, F. A.; Sadek, M. M.; Adel, S. A.; Ashraf, A.; Alam-Eldein, K. M.; Ahmed, N. E.; Abdelaziz, R. Y.; Tolba, E. F. MiRNAs and Their Multifaceted Role in Cutaneous Wound Healing. *Funct Integr Genomics*, **2025**, *25* (1), 1–23.
[58] Sun, S.; Wang, Z.; Yao, F.; Sun, K.; Li, Z.; Sun, S.; Li, C. Breast Cancer Cell-Derived Exosome-Delivered MicroRNA-155 Targets UBQLN1 in Adipocytes and Facilitates Cancer Cachexia-Related Fat Loss. *Hum Mol Genet*, **2023**, *32* (13), 2219–2228.
[59] Zhang, Y.; Liu, Q.; Zhang, X.; Huang, H.; Tang, S.; Chai, Y.; Xu, Z.; Li, M.; Chen, X.; Liu, J. Recent Advances in Exosome-Mediated Nucleic Acid Delivery for Cancer Therapy. *J Nanobiotechnol*, **2022**, *20* (1), 279.
[60] Khusnutdinov, R. R. Could Drug Delivery Systems Be Enhancers for Therapy Against KRAS-Driven Cancers? *Rev Adv Chem*, **2025**, 1–32.
[61] Wang, K.; Ye, H.; Zhang, X.; Wang, X.; Yang, B.; Luo, C.; Zhao, Z.; Zhao, J.; Lu, Q.; Zhang, H. An Exosome-like Programmable-Bioactivating Paclitaxel Prodrug Nanoplatform for Enhanced Breast Cancer Metastasis Inhibition. *Biomaterials*, **2020**, *257*, 120224.
[62] Ti, D.; Hao, H.; Tong, C.; Liu, J.; Dong, L.; Zheng, J.; Zhao, Y.; Liu, H.; Fu, X.; Han, W. LPS-Preconditioned Mesenchymal Stromal Cells Modify Macrophage Polarization for Resolution of Chronic Inflammation via Exosome-Shuttled Let-7b. *J Transl Med*, **2015**, *13*, 1–14.
[63] Dalirfardouei, R.; Jamialahmadi, K.; Jafarian, A. H.; Mahdipour, E. Promising Effects of Exosomes Isolated from Menstrual Blood-Derived Mesenchymal Stem Cell on Wound-Healing Process in Diabetic Mouse Model. *J Tissue Eng Regen Med*, **2019**, *13* (4), 555–568.
[64] Wang, J.; Wu, H.; Peng, Y.; Zhao, Y.; Qin, Y.; Zhang, Y.; Xiao, Z. Hypoxia Adipose Stem Cell-Derived Exosomes Promote High-Quality Healing of Diabetic Wound Involves Activation of PI3K/Akt Pathways. *J Nanobiotechnol*, **2021**, *19*, 1–13.
[65] He, X.; Dong, Z.; Cao, Y.; Wang, H.; Liu, S.; Liao, L.; Jin, Y.; Yuan, L.; Li, B. MSC-Derived Exosome Promotes M2 Polarization and Enhances Cutaneous Wound Healing. *Stem Cells Int*, **2019**, *2019* (1), 7132708.
[66] Li, X.; Liu, L.; Yang, J.; Yu, Y.; Chai, J.; Wang, L.; Ma, L.; Yin, H. Exosome Derived from Human Umbilical Cord Mesenchymal Stem Cell Mediates MiR-181c Attenuating Burn-Induced Excessive Inflammation. *EBioMed*, **2016**, *8*, 72–82.
[67] Qiu, X.; Liu, J.; Zheng, C.; Su, Y.; Bao, L.; Zhu, B.; Liu, S.; Wang, L.; Wang, X.; Wang, Y. Exosomes Released from Educated Mesenchymal Stem Cells Accelerate Cutaneous Wound Healing via Promoting Angiogenesis. *Cell Prolif*, **2020**, *53* (8), e12830.
[68] Zhang, Y.; Pan, Y.; Liu, Y.; Li, X.; Tang, L.; Duan, M.; Li, J.; Zhang, G. Exosomes Derived from Human Umbilical Cord Blood Mesenchymal Stem Cells Stimulate Regenerative Wound Healing via Transforming Growth Factor-β Receptor Inhibition. *Stem Cell Res Ther*, **2021**, *12*, 1–14.
[69] Choi, J. S.; Cho, W. L.; Choi, Y. J.; Kim, J. D.; Park, H.-A.; Kim, S. Y.; Park, J. H.; Jo, D.-G.; Cho, Y. W. Functional Recovery in Photo-Damaged Human Dermal Fibroblasts by Human Adipose-Derived Stem Cell Extracellular Vesicles. *J Extracell Vesicles*, **2019**, *8* (1), 1565885.
[70] Hu, S.; Li, Z.; Cores, J.; Huang, K.; Su, T.; Dinh, P.-U.; Cheng, K. Needle-Free Injection of Exosomes Derived from Human Dermal Fibroblast Spheroids Ameliorates Skin Photoaging. *ACS Nano*, **2019**, *13* (10), 11273–11282.

[71] Kciuk, M.; Marciniak, B.; Mojzych, M.; Kontek, R. Focus on UV-Induced DNA Damage and Repair – Disease Relevance and Protective Strategies. *Int J Mol Sci*, **2020**, *21* (19), 7264.

[72] Xiong, Q.-H.; Zhao, L.; Wan, G.-Q.; Hu, Y.-G.; Li, X.-L. Engineered BMSCs-Derived Exosomal MiR-542–3p Promotes Cutaneous Wound Healing. *Endocr Metab Immune Disord Drug Targets*, **2023**, *23* (3), 336–346. https://doi.org/10.2174/1871530322666220523151713.

[73] Xu, P.; Xin, Y.; Zhang, Z.; Zou, X.; Xue, K.; Zhang, H.; Zhang, W.; Liu, K. Extracellular Vesicles from Adipose-Derived Stem Cells Ameliorate Ultraviolet B-Induced Skin Photoaging by Attenuating Reactive Oxygen Species Production and Inflammation. *Stem Cell Res Ther*, **2020**, *11* (1), 264. https://doi.org/10.1186/s13287-020-01777-6.

[74] Zhang, B.; Yin, Y.; Lai, R. C.; Tan, S. S.; Choo, A. B. H.; Lim, S. K. Mesenchymal Stem Cells Secrete Immunologically Active Exosomes. *Stem Cells Dev*, **2014**, *23* (11), 1233–1244.

[75] Huang, J.; Yu, M.; Yin, W.; Liang, B.; Li, A.; Li, J.; Li, X.; Zhao, S.; Liu, F. Development of a Novel RNAi Therapy: Engineered MiR-31 Exosomes Promoted the Healing of Diabetic Wounds. *Bioact Mater*, **2021**, *6* (9), 2841–2853.

[76] Su, Y.; Sharma, N. S.; John, J. V; Ganguli-Indra, G.; Indra, A. K.; Gombart, A. F.; Xie, J. Engineered Exosomes Containing Cathelicidin/LL-37 Exhibit Multiple Biological Functions. *Adv Healthc Mater*, **2022**, *11* (20), 2200849.

[77] Born, L. J.; Chang, K.; Shoureshi, P.; Lay, F.; Bengali, S.; Hsu, A. T. W.; Abadchi, S. N.; Harmon, J. W.; Jay, S. M. HOTAIR-loaded Mesenchymal Stem/Stromal Cell Extracellular Vesicles Enhance Angiogenesis and Wound Healing. *Adv Healthc Mater*, **2022**, *11* (5), 2002070.

[78] Yuan, R.; Dai, X.; Li, Y.; Li, C.; Liu, L. Exosomes from MiR-29a-Modified Adipose-Derived Mesenchymal Stem Cells Reduce Excessive Scar Formation by Inhibiting TGF-B2/Smad3 Signaling. *Mol Med Rep*, **2021**, *24* (5), 758.

[79] Jiang, L.; Zhang, Y.; Liu, T.; Wang, X.; Wang, H.; Song, H.; Wang, W. Exosomes Derived from TSG-6 Modified Mesenchymal Stromal Cells Attenuate Scar Formation During Wound Healing. *Biochimie*, **2020**, *177*, 40–49.

[80] Sung, D. K.; Chang, Y. S.; Sung, S. I.; Ahn, S. Y.; Park, W. S. Thrombin Preconditioning of Extracellular Vesicles Derived from Mesenchymal Stem Cells Accelerates Cutaneous Wound Healing by Boosting Their Biogenesis and Enriching Cargo Content. *J Clin Med*, **2019**, *8* (4), 533.

[81] Das, A.; Nikhil, A.; Kumar, A. Preparation of Thermo-Responsive Polymer Encapsulated Exosomes and Its Role as a Therapeutic Agent for Blood Clot Lysis. *Colloids Surf B Biointerfaces*, **2022**, *216*, 112580.

[82] Shi, Q.; Qian, Z.; Liu, D.; Sun, J.; Wang, X.; Liu, H.; Xu, J.; Guo, X. GMSC-Derived Exosomes Combined with a Chitosan/Silk Hydrogel Sponge Accelerates Wound Healing in a Diabetic Rat Skin Defect Model. *Front Physiol*, **2017**, *8*, 904.

[83] Chachques, J. C.; Gardin, C.; Lila, N.; Ferroni, L.; Migonney, V.; Falentin-Daudre, C.; Zanotti, F.; Trentini, M.; Brunello, G.; Rocca, T. Elastomeric Cardiowrap Scaffolds Functionalized with Mesenchymal Stem Cells-Derived Exosomes Induce a Positive Modulation in the Inflammatory and Wound Healing Response of Mesenchymal Stem Cell and Macrophage. *Biomedicines*, **2021**, *9* (7), 824.

[84] Geng, X.; Qi, Y.; Liu, X.; Shi, Y.; Li, H.; Zhao, L. A Multifunctional Antibacterial and Self-Healing Hydrogel Laden with Bone Marrow Mesenchymal Stem Cell-Derived Exosomes for Accelerating Diabetic Wound Healing. *Biomater Adv*, **2022**, *133*, 112613. https://doi.org/10.1016/j.msec.2021.112613.

[85] Liu, Y.; Liu, Y.; Zhao, Y.; Wu, M.; Mao, S.; Cong, P.; Zou, R.; Hou, M.; Jin, H.; Bao, Y. Application of Adipose Mesenchymal Stem Cell-Derived Exosomes-Loaded β-Chitin Nanofiber Hydrogel for Wound Healing. *Folia Histochem Cytobiol*, **2022**, *60* (2), 167–178.

[86] Okonkwo, U. A.; DiPietro, L. A. Diabetes and Wound Angiogenesis. *Int J Mol Sci*, **2017**, *18* (7), 1419.

[87] Anuradha, U.; Mehra, N. K.; Khatri, D. K. Understanding Molecular Mechanisms and MiRNA-Based Targets in Diabetes Foot Ulcers. *Mol Biol Rep*, **2024**, *51* (1), 82.

[88] Lv, Y.; Li, L.; Zhang, J.; Li, J.; Cai, F.; Huang, Y.; Li, X.; Zheng, Y.; Shi, X.; Yang, J. Visible-Light Cross-Linkable Multifunctional Hydrogels Loaded with Exosomes Facilitate Full-Thickness Skin Defect Wound Healing through Participating in the Entire Healing Process. *ACS Appl Mater Interfaces*, **2024**, *16* (20), 25923–25937. https://doi.org/10.1021/acsami.4c05512.

[89] Li, D.; Wu, N. Mechanism and Application of Exosomes in the Wound Healing Process in Diabetes Mellitus. *Diabetes Res Clin Pract*, **2022**, *187*, 109882.

[90] Ozdemir, D.; Feinberg, M. W. MicroRNAs in Diabetic Wound Healing: Pathophysiology and Therapeutic Opportunities. *Trends Cardiovasc Med*, **2019**, *29* (3), 131–137.

[91] Xiao, S.; Zhao, T.; Wang, J.; Wang, C.; Du, J.; Ying, L.; Lin, J.; Zhang, C.; Hu, W.; Wang, L. Gelatin Methacrylate (GelMA)-Based Hydrogels for Cell Transplantation: An Effective Strategy for Tissue Engineering. *Stem Cell Rev Rep*, **2019**, *15*, 664–679.

[92] Lv, Y.; Li, L.; Zhang, J.; Li, J.; Cai, F.; Huang, Y.; Li, X.; Zheng, Y.; Shi, X.; Yang, J. Visible-Light Cross-Linkable Multifunctional Hydrogels Loaded with Exosomes Facilitate Full-Thickness Skin Defect Wound Healing through Participating in the Entire Healing Process. *ACS Appl Mater Interfaces*, 2024, *16* (20), 25923–25937.

[93] Sun, B.; Wu, F.; Wang, X.; Song, Q.; Ye, Z.; Mohammadniaei, M.; Zhang, M.; Chu, X.; Xi, S.; Zhou, N. An Optimally Designed Engineering Exosome–Reductive COF Integrated Nanoagent for Synergistically Enhanced Diabetic Fester Wound Healing. Small, 2022, 18 (26), 2200895.

[94] Qian, Z.; Bai, Y.; Zhou, J.; Li, L.; Na, J.; Fan, Y.; Guo, X.; Liu, H. A Moisturizing Chitosan-Silk Fibroin Dressing with Silver Nanoparticles-Adsorbed Exosomes for Repairing Infected Wounds. *J Mater Chem B*, **2020**, *8* (32), 7197–7212.

[95] Wang, C.; Wang, M.; Xu, T.; Zhang, X.; Lin, C.; Gao, W.; Xu, H.; Lei, B.; Mao, C. Engineering Bioactive Self-Healing Antibacterial Exosomes Hydrogel for Promoting Chronic Diabetic Wound Healing and Complete Skin Regeneration: Erratum. *Theranostics*, **2021**, *11* (20), 10174.

[96] Li, M.; Ke, Q.-F.; Tao, S.-C.; Guo, S.-C.; Rui, B.-Y.; Guo, Y.-P. Fabrication of Hydroxyapatite/Chitosan Composite Hydrogels Loaded with Exosomes Derived from MiR-126-3p Overexpressed Synovial Mesenchymal Stem Cells for Diabetic Chronic Wound Healing. *J Mater Chem B*, **2016**, *4* (42), 6830–6841.

13 Exosome-Mediated Immunomodulation and Therapeutic Opportunities

S. Princely E. Gnanakani[1], *R. Vijayalakshmi*[2], *N. Saleem Basha*[3], *J. John Kirubakaran*[4], *and S. Arun Prasad*[5]

[1]Department of Pharmaceutics, Parul Institute of Pharmacy, Parul University, Vadodara, Gujarat, India

[2]Department of Pharmaceutical Analysis, GIET School of Pharmacy, Rajahmundry, Andhra Pradesh, India

[3]Department of Pharmaceutics, Faculty of Pharmacy, Dr MGR Educational and Research Institute, Chennai, Tamil Nadu, India

[4]Department of Pharmacy Practice, Parul Institute of Pharmacy and Research, Parul University, Vadodara, Gujarat, India

[5]Department of Pharmaceutical Sciences, Mercer University Drive, Atlanta, GA

13.1 INTRODUCTION

13.1.1 Overview of Exosomes

Endosomes initiate the creation of exosomes, which are small extracellular vesicles (EVs) characterized by a lipid bilayer, ranging in size from 30 nm to 150 nm. The majority of cell subtypes in the extracellular milieu, including blood, are responsible for secretion due to their ability to traverse distant tissues [1, 2]. Lipid bilayer-delimited, nonreplicating particles, which are discharged from the cell, have been labeled as "EVs." This decision was based on the most recent changes to the International Society for Extracellular Vesicles' (ISEV) recommendations to standardize investigation approaches, including limited data [3]. In cellular communication, these macromolecules operate as mediators [4, 5]. Endosomes are generated through endocytosis, which creates exosomes; both endocytosis and the scission of the invaginated plasma membrane are crucial to this process. These nascent vesicles progress from early endosomes to late endosomes with the aid of the Golgi complex [6, 7]. Intraluminal vesicles (ILVs) are produced in the lumen of the endosome that has been produced as a result of multiple membrane invasions [8]. The endosomal sorting complexes required for transport (ESCRT) are responsible for regulating the process of endosomal membrane invagination that occurs within the lumen [9].

ILVs are able to sort endosomal proteins and cell contents with the assistance of these proteins, which are responsible for the creation of mature endosomes and are often referred to as multivesicular bodies (MVBs) and following their merger with the cell membrane, these MVBs then release their internal vesicles, which are referred to as exosomes, into the extracellular space [10, 11]. Different classes of exosomes can be identified from one another according to their dimension, RNA, exosomal markers, and related proteins that are unique to them [12, 13].

13.1.2 Role in Cellular Communication

Exosomes are known to entail a range of bioactives, presenting cytokine receptors, proteins, RNAs, and lipids, among others [14]. They constitute an enormous part of the communication that takes place between cells, organs, and the body as a whole is through the utilization of signaling pathways that are either paracrine or endocrine in nature [4]. Exosomes are the cells that are responsible for mediating biological functions within the microenvironment of the skeletal system [15]. Exosomes can move around the body and direct their attention to particular cells because of their ability to interact with receptors and ligands during their movement, and additionally, they are able to enter target cells through the process of endocytosis, where they deliver their cargo and initiate signaling pathways that are associated with the target cells [16, 17]. Exosomes foster intercommunication among bone-derived cells and diverse cell groups, allowing for interaction between the two. This condition initiates a range of intracellular and intercellular signaling pathways that intricately coordinate and modulate the skeletal microenvironment, thereby influencing its structure, function, and homeostasis [18].

13.1.3 Historical Context

Since their isolation in the early 1980s, exosomes have been the subject of a significant amount of analysis and

DOI: 10.1201/9781003623632-13

investigation. Initially, researchers discovered that exosomes were EVs that were involved in the maturation of reticulocytes [19]. On the basis of subsequent research, it has been demonstrated that these originate from MVBs and are necessary for intercellular communication [20]. By the middle of the 1990s, it was clear that they could change elements of the immunological framework, especially the presentation of T-cell antigen. The landmark discovery in 2007 that exosomes contain mRNA and microRNA, which was made publicly accessible, significantly transformed the understanding of intercellular genetic transfer. Studies on the tumor expansion and invasion of metastases have shown that researchers are clearly important in the microenvironment of tumors [21].

As of 2013, it was established that exosomes have been identified as potentially useful biomarkers that could be utilized to detect noninvasive conditions such as cancer [22]. Conversely, exosomes have been identified as potentially useful biomarkers for the investigation of noninvasive diseases, such as cancer. The standardization initiatives launched by the ISEV in 2014 generated a stringent investigation framework, and a notable advancement in exosome-based interventions in medicine transpired in 2020 with the introduction of exosome-based therapeutic applications [23]. These applications encompassed trials that were approved by the US Food and Drug Administration (FDA) [24]. Collectively, these milestones accentuate the evolutionary promise of exosomes in diagnostics, therapeutics, and understanding disease mechanisms, laying the groundwork for future breakthroughs in personalized medicine.

13.2 EXOSOME COMPOSITION AND BIOGENESIS

13.2.1 Biogenic Process Pathways

Exosomes originate from the endosomal system, where MVBs are produced by the transition of early endosomes into late endosomes, which occurs through a number of different pathways. Besides endosomal membrane inward curvature, ILVs are produced in the lumen of the organelle where this process takes place [25, 26]. The contents of multivesicular bodies, also known as MVBs, are degraded in the majority of cells by fusing with lysosomes. ILVs can be liberated into the extracellular space when organelles that contain MVBs undergo direct fusion with the plasma membrane. Complete understanding of the coexistence of diverse specific populations of MVBs inside cells and the pathways they traverse, such as merging with lysosomes for degradation or fusing to the plasma membrane during exocytosis, remains insufficient [27]. The ESCRT framework operates as a key regulatory part in the formation of ILVs and MVBs, which is the most extensively studied process involving this system. This pathway has been investigated to determine how it works. There is evidence that this pathway has been a key factor in the emergence of both types of cells. An extensive amount of research has been done on this particular pathway. The ESCRT framework is formed by around 30 proteins organized into 4 complexes, designated ESCRT-0 to ESCRT-3, along with related proteins such as vacuolar protein sorting–associated protein 4 (VPS4), vesicle trafficking 1 (VTA1), and apoptosis-linked gene 2-interacting protein X (ALIX). This system has been extensively conserved from yeast to mammals [28]. A recently hypothesized method indicates that MVB biogenesis may transpire autonomously of the ESCRT-dependent pathway, as proposed by several investigators. Research suggests ILVs periodically emerge within MVBs, implying the possibility of mechanisms that operate independently of ESCRT [29].

13.2.2 Molecular Entities

The components that make up exosomes comprise multiple macromolecules, namely, nucleic acids, as well as copious lipid and protein structures. Exosomes are characterized by the presence of high concentrations of microRNAs (miRNAs), primarily pre-miRNAs that are inactive until they are transformed into mature miRNAs [30]. Exosomes are derived from endosomes and encompass proteins, namely, annexins, R_{as}-related protein, flotillin, and GTPases. These fusion and membrane transport proteins are implicated in the process of exosome formation. In addition, they contain proteins that are allied with lipid microdomains, like integrins and tetraspanins, besides proteins that are involved in the biogenesis of MVBs, like ALIX and tumor susceptibility gene (TSG101). Other proteins that are frequently discovered are also associated with metabolism (for example, glyceraldehyde-3-phosphate dehydrogenase [GADPH]) and the cytoskeleton (for example, tubulin, myosin, and actin) [31], tissue-specific proteins (for example, major histocompatibility complex class I [MHC II]) on exosomes radiated by B-lymphocytes or dendritic cells (DCs), heat shock proteins/cognate (HSC70, HSC90), or proteins that are specific to cancer cell lines (for example, breast cancer human epidermal growth factor receptor 2 [HER2], ovarian cancer cluster of differentiation 24 (CD24), and glioma epidermal growth factor receptor [EGFR]) [32].

13.2.3 Cargo Transport Mechanisms

The endosomal pathway, crucial for cellular trafficking and sorting, initiates exosome production when extracellular substances enter early endosomes, which mature into multivesicular bodies, releasing exosomes for intercellular communication [33]. MVBs originate from early endosomes, embarking on a complex journey guided by specialized maturation proteins that orchestrate their development through intricate cellular pathways, ensuring precise sorting and transport within the cell. MVBs are liable for the fabrication of ILVs utilizing a mechanism that involves the invagination of the endosomal membrane. Exosome biogenesis is distinguished by the encapsulation of cytoplasmic components within ILVs, in contrast to other types of biogenesis. MVBs can experience degradation via lysosomal fusion or release exosomes by merging with the plasma membrane. MVB trafficking and fusion both are governed by molecular mechanisms that are subject to stringent regulatory requirements. This controlled release makes it possible for exosomes to shuttle functional biomolecules, which in turn improve cellular communication and have an effect on the surrounding microenvironment [34]. It is essential for the correct incorporation of cargo into ILVs for exosomes to fulfill their many different functions [35]. Furthermore, ESCRT-independent pathways utilizing heat shock proteins, lipids, and tetraspanins facilitate cargo sorting.

13.3 IMMUNOMODULATORY FUNCTIONS OF EXOSOMES

Exosomes, once assumed to be microparticles from neoplastic cells, are now known to be secreted by most cells as membrane-bound EVs [36, 37]. Formed via early sorting endosomes from endocytic vesicles [38, 39], their production, release, and uptake are influenced by factors like radiation, oxidative stress, pH shifts, and intracellular calcium levels [40, 41]. Exosomes play diverse roles, including modulating the extracellular matrix, immunity, and intercellular signaling, with implications for cancer and neurological diseases, though many functions remain unclear [42, 43]. Notably, neuronal exosomes affect synaptic transmission and protein expression [44], which can activate intracellular signaling cascades and alter the immune system [45] as depicted in Figure 13.1.

13.3.1 Interaction with Innate Immunity: Innate Immune Signaling and Regulation

When viruses infect host cells, the innate immune system detects foreign nucleic acids through DNA sensors like Toll-like receptor 9 (TLR9), absent in melanoma (AIM2), and cyclic GMP–AMP synthase (cGAS), and RNA sensors, including TLR3, TLR7, TLR8, melanoma differentiation-associated protein 5 (MDA5), and retinoic acid–inducible gene I (RIG-I) [46, 47]. This detection activates signaling pathways involving the TANK-binding kinase 1-IκB kinase epsilon (TBK1-IKKε) and IKK complex, which trigger transcription factors interferon regulatory factor 3 (IRF3) and nuclear factor-kappa B (NF-κB), ultimately inducing type I interferons and pro-inflammatory cytokines to establish an antiviral state [48].

13.3.1.1 EV-Mediated Innate Immune Response Activation and Suppression

Exosomes carrying double-stranded DNA (dsDNA) or viral RNA can stimulate DCs through the IRF3-dependent interferon signaling and cGAS/STING (stimulator of interferon genes) pathway, enhancing antiviral immune responses [49]. T cell-derived EVs further modulate DC function to promote antigen-specific innate immunity [50]. Agents like topotecan and radiotherapy amplify exosomal DNA release, activating DCs and priming antitumor T-cell responses in vitro/in vivo through the cGAS-STING axis [51, 52]. Additionally, hepatitis C virus (HCV)-infected cells package viral RNA into exosomes that stimulate plasmacytoid DCs (pDCs) to produce type I interferons, even when intracellular sensing pathways are blocked, revealing a novel exosome-mediated immune surveillance mechanism [53]. Tumor-derived exosomes (TEXs) suppress host immunity by altering cytokine production, promoting apoptosis, and impairing both innate and adaptive immune responses [54]. They induce monocyte differentiation into myeloid-derived suppressor cells (MDSCs), inhibit dendritic cell maturation, and enhance the survival of tumor-associated macrophages (pre-TAMs) via transfer of EGFR and HER-2. TEXs also modulate programmed death-ligand 1 (PD-L1) and EGFR expression through ALIX regulation, where ALIX depletion increases surface PD-L1 and EGFR signaling while reducing exosomal PD-L1, weakening immune defense in melanoma [55]. Chronic lymphocytic leukemia (CLL)-derived exosomes carrying Y RNA upregulate monocyte PD-L1 and cytokines, aiding tumor progression, while EGFR+ gastric cancer exosomes trigger hepatocyte growth factor (HGF) release for liver metastasis [56]. Additionally, TEXs impair macrophage antimicrobial responses via mitogen-activated protein kinase kinase kinase (2MEKK2)-mediated interferon regulatory factor 3 (IRF3) phosphorylation [57].

13.3.1.2 Exosomes and Macrophages

Macrophage-originated exosomes (Mφ-Exos) are key mediators in immune modulation, inflammation, tumor progression, and drug delivery. These nanosized vesicles transport cytokines, RNAs, and proteins that modify the behavior of target cells. Exosomes stemming from M2-polarized macrophages activate the phosphatidylinositol 3-kinase/protein kinase B pathway (PI3K/AKT) signaling network in recipient cells, leading to anti-inflammatory responses and promoting a sustained immunosuppressive environment [58]. Exosomes released by tumor-associated macrophages in the tumor microenvironment (TAM-Exos) enhance cancer cell proliferation, invasion, angiogenesis, and metastasis by altering local signaling pathways and promoting stromal remodeling [59, 60]. Additionally, macrophage-derived exosomes are capable of traversing the BBB, offering a potential platform for CNS drug delivery [61]. Their biocompatibility and targeting capacity make them promising natural nanocarriers for therapeutic applications [62].

13.3.1.3 Exosomes and Dendritic Cells

DCs, initially characterized by Steinman et al. in 1973, function as antigen-presenting cells derived from bone marrow, crucial in bridging innate and adaptive immunity [63]. Originating from myeloid or plasmacytoid lineages, DCs mature under cytokines like granulocyte-macrophage colony-stimulating factor (GM-CSF), interleukin-4 (IL-4), and FMS-like tyrosine kinase 3 ligand (FLT3L) [64, 65]. Upon activation, they release exosomes (DC-Exos) – nanosized vesicles (30–150 nm) enriched with MHC I/II, CD86, and intercellular adhesion molecule 1 (ICAM-1) – that function as cell-free antigen-presenting cell (APCs), capable of activating naive T cells and inducing cytotoxic T-lymphocyte (CTL) responses [66, 67]. DC-Exos have shown promise in cancer immunotherapy by delivering tumor antigens and stimulating antitumor immunity [68]. Depending on their origin, DC-Exos can also mediate immune tolerance, promoting Tregs and suppressing effector T cells, making them potential tools in autoimmune and transplant settings [69]. In infections, DC-Exos enhance antigen-specific responses by presenting microbial antigens but can also be hijacked by viruses like HIV and EBV (Epstein-Barr Virus) for immune evasion [70, 71].

13.3.1.4 Exosomes and Natural Killer (NK) Cells

NK cells, once known primarily for their cytotoxic activity against tumor cells, are currently acknowledged for their cytokine-producing and governance roles, interacting with monocytes, DCs, macrophages, and T lymphocytes [72]. NK cells use a balance of activating and inhibitory receptors to recognize abnormal cells, with tissue-resident NK (trNK) cells found in organs like the lungs, liver, and uterus, comprising 10–15% of peripheral lymphocytes [73]. Exosomes (30–150 nm) produced by many cell types travel through bodily fluids and can alter immune responses [74]. Cancer-derived exosomes support tumor growth by enhancing angiogenesis, promoting metastasis, and enabling immune system evasion [75]. These exosomes are internalized by key immune players, notably NK cells, and can modulate their activity. Dendritic cell-originated exosomes enhance NK cell initiation and antitumor responses, including tumor necrosis factor alpha (TNF-α) and IFN-γ (Interferon gamma) release [76]. Genetically modified exosomes carrying 4–1BB ligand (4–1BBL), IL-15, and IL-18 enhance NK cytotoxicity [77], while hypoxic NK cell-stemmed exosomes impede breast and ovarian cancer cell expansion and invasion [78]. Despite ongoing research, the full impact of tumor-originated exosomes on immune modulation remains to be fully understood [79, 80].

13.3.2 Role in Adaptive Immunity: T-Cell Activation, B-Cell Regulation, Regulatory T-Cell Induction

13.3.2.1 Exosomes and T Cells

Exosomes (30–150 nm nanoparticles) intervene in cellular cross talk by trafficking proteins, lipids, miRNAs, and mRNAs, playing key roles in antigen presentation, T-cell activation, B-cell regulation, and Treg induction, crucial for adaptive immune responses and tolerance [13, 81], and are presented in Table 13.1.

13.3.2.2 Exosomes in T-Cell Activation

T cells necessitate three signals for activation, prominently featuring antigen presentation through MHC molecules, co-stimulatory signals (CD80/CD86 → CD28), and cytokine stimulation for differentiation. Exosomes from DCs and B cells carry MHC-I and MHC-II molecules to present antigens to $CD8^+$ cytotoxic and $CD4^+$ helper T cells and transfer peptide–MHC complexes to other APCs, enhancing immune responses [82, 83]. Exosomes from mature DCs contain CD80/CD86 for CD28-mediated T cell co-stimulation and cytokines like IFN-γ, IL-2, and IL-12, which accelerate the expansion of T cells and their differentiation into effector and memory phenotypes [84].

13.3.2.3 Exosomes in B-Cell Regulation

Antigen-presenting cell exosomes enhance T-cell receptor signaling by stabilizing the immunological synapse and influence $CD4^+$ T-cell differentiation through cytokines like IL-12 (Th1), IL-4 (Th2), IL-6, IL-21, transforming growth factor beta (TGF-β), and TGF-β (Th17) with IL-10 (Tregs). Exosomal microRNAs such as miR-155, miR-146a, and miR-181 regulate B-cell survival, activation, and isotype switching, with miR-155 promoting germinal center formation and immunoglobulin G (IgG) switching, while miR-146a restrains overactivation [85, 86].

B-cell-originated exosomes enhance immunity by transferring antibody–antigen complexes to DCs, while plasma cell-derived exosomes carry IgG and IgA to neutralize pathogens independently. Exosome-induced Tregs help suppress autoimmunity and control excessive immune responses [87, 88]. TGF-β and IL-10-rich exosomes from MSCs and tolerogenic dendritic cells (Tol-DCs) promote Treg formation by activating Forkhead box P3 (FoxP3) in naive T cells. Treg-derived exosomes suppress effector T-cell activity through cytotoxic T-lymphocyte-associated protein 4 (CTLA-4), IL-35, and miRNAs like miR-21 and miR-146a, which inhibit inflammation and T-cell activation [89, 90].

13.3.2.4 Exosomes in Immune Tolerance and Autoimmunity Regulation

Exosome therapy with TGF-β and IL-10-loaded exosomes has shown potential in suppressing autoreactive T cells in autoimmune diseases like lupus and rheumatoid arthritis (RA). In transplantation, donor-derived tolerogenic dendritic cell exosomes have been explored to trigger regulatory T cell maturation and enhance graft tolerance [91, 92]. Exosomes strongly regulate T-cell, B-cell, and Treg activation. They transport immune-modulatory chemicals, making them useful for cancer immunotherapy, vaccine development, and autoimmune diseases. Researchers are developing immune response boosting or inhibition techniques for particular diseases using exosomes [88, 93, 94].

13.3.3 Immune Tolerance and Autoimmunity

Small exosomes (30–150 nm) convey intercellular information. Immunological regulation is affected by their transport of proteins, lipids, and nucleic acids. Exosomes regulate pregnancy, transplantation, and autoimmune disease tolerance.

TABLE 13.1
Overview of Core Functional Attributes of Key Immunomodulatory Agents

Immunomodulators	Core Functions
Interaction with innate immunity	Exosomes control innate immunological signaling that activate or repress macrophages, DCs, and NK cells [49, 110]
EV-mediated innate immune response	Pathogen-associated molecular patterns (PAMPs)/damage-associated molecular patterns (DAMPs) in exosomes activate TLR or decrease pro-inflammatory responses [37, 52]
Macrophages	Accelerate M1 or M2 polarization using exosomal miRNA and cytokines [58]
DCs	Impact antigen presentation and dendritic cell maturation [63]
NK cells	Modulate NK cell cytotoxicity by exosomal ligands or cytokines [78, 79]
Role in adaptive immunity	Stimulate T- and B-cell interactions, tolerance, and antigen-specific responses [81]
T cells	Present antigens through MHC I/II, influence T-cell activation or exhaustion [82, 83]
T-cell activation	Operate as antigen-presenting cell-like entities to activate naive or memory T cells [84]
B-cell regulation	Modulate antibody synthesis and isotype switching [85, 86]
Immune tolerance and autoimmunity	Communicate tolerogenic signals (TGF-β, IL-10) that promote Tregs and diminish autoimmune [91–93]
Pregnancy and immune tolerance	pEXOs decrease immune activation to preserve maternal–fetal tolerance [99, 100]
Transplantation tolerance	Promote graft tolerance through Treg induction and donor antigen administration [102–104]
Autoimmune disease modulation	Mitigate inflammation and autoimmunity in multiple sclerosis (MS), RA, and SLE models [107, 108]

13.3.3.1 Exosomes in Pregnancy and Immune Tolerance

Immunoregulatory adaptation to the semi-allogeneic fetus is essential during pregnancy, and placental exosomes (pEXOs) assist this process by engaging with maternal immune components [95–97]. pEXOs trigger regulatory T cells (Tregs), which attenuate immune responses and enhance tolerance, aided by exosomal TGF-β and FasL (Fas ligand) that promote Treg development and limit maternal immune activation [95, 98]. These exosomes also attenuate inflammatory responses by repressing cytokines such as TNF-α (tumor necrosis factor alpha) and IL-6 via interactions with monocytes and DCs and contain miRNAs such as miR-146a that modulate immune signaling and inflammation [99]. Trophoblast-derived exosomes, enriched with immunosuppressive human leukocyte antigen G (HLA-G), inhibit maternal NK cell activation, preventing fetal rejection [100]. Impaired exosome function has been linked to pregnancy complications like hypertension and recurrent loss, highlighting their therapeutic potential in promoting maternal–fetal tolerance.

13.3.3.2 Exosomes in Transplantation Tolerance

Organ transplantation requires immunological tolerance to prevent graft rejection. Donor-stemmed exosomes carrying donor-specific antigens are internalized by recipient APCs, enabling "semi-direct presentation" that can activate graft tolerance pathways [101–103]. These exosomes, expressing donor MHC molecules, stimulate regulatory T cells (Tregs), thereby reducing alloimmune responses and enhancing graft survival. Additionally, exosomes derived from MSCs aid in suppressing inflammation and promoting tissue repair in transplant settings [104]. MSC-derived exosomes contain IL-10 and TGF-β, enhancing Treg proliferation and suppressing effector T cells to improve graft survival [13]. These exosomes also deliver miRNAs that mitigate pro-inflammatory pathways like NF-κB; namely, exosomal miR-155 reduces graft-versus-host disease (GVHD) by modulating immune responses [105]. Exosome therapy may replace immunosuppressants in transplant recipients, and donor-derived exosomes may improve long-term graft tolerance.

13.3.3.3 Exosomes in Autoimmune Disease Modulation

Pathophysiology and therapy of autoimmune diseases involve exosomes. They transfer autoantigens and inflammatory mediators between cells, advancing illness. Drug-like anti-inflammatory exosomes have been examined. Engineered exosomes are being studied for directing immunosuppressive medicines to autoimmune disease pathways.

In autoimmune diseases, immunological tolerance fails, causing an aberrant immune response to self-antigens. Disease progression and treatment may involve exosomes. Finally, exosomes modulate autoimmune responses, improve transplant graft acceptance, and boost pregnant immunological tolerance. They aid immunological tolerance and immune-related illness treatments by transporting bioactive chemicals across cells [95, 106]. The

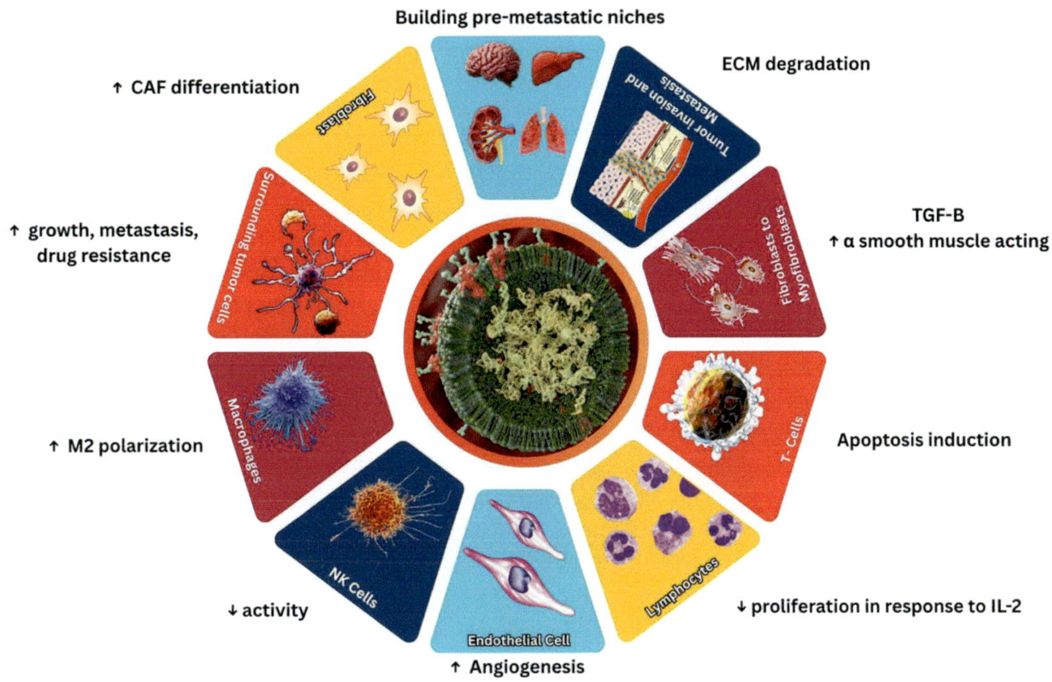

FIGURE 13.1 Immunoregulatory roles of exosomes in health and disease.

pro-inflammatory role of autoimmune-associated exosomes regulates autoimmune processes. Autoreactive immune cell exosomes convey self-antigens and inflammatory mediators, exacerbating illness. dsDNA and autoantigens in exosomes cause immune responses against self-components in systemic lupus erythematosus (SLE) [107].

MSC exosomes deliver immunosuppressive chemicals to maintain immunological homeostasis. Exosomal miR-21 and miR-146a decrease pro-inflammatory signaling in RA and MS [108]. In experimental models of autoimmune disorders, engineered exosomes encoding anti-inflammatory drugs, notably curcumin or disease-targeted miRNAs, delivered benefits in reducing inflammation [109]. Clinical implications show that exosome-based therapies may be personalized for autoimmune illnesses. Clinical trials are testing exosome-originated biomarkers for early diagnosis and disease surveillance.

13.4 THERAPEUTIC OPPORTUNITIES

13.4.1 Exosomes as Biomarkers: Diagnostic Applications in Cancer, Autoimmune Disorders, and Neurodegenerative Disease

Exosomes originating from diverse cellular subtypes carry distinct signaling molecules due to their composition. Knowing exosome contents and their connection with illness manifestation makes them potential therapeutic targets and biomarkers [111]. It is present in many bodily fluids and has high sensitivity, low invasiveness, and easy material access compared to other tests [112], which can help diagnose disorders, detect diseases, and evaluate therapy efficacy [113, 114] as depicted in Figure 13.2.

13.4.1.1 Cancer

Five exosomes store cancer cell data [115] that modulate cell proliferation, angiogenesis, metastasis, and immune responses. Early tumor identification affects therapy success and prognoses. They may indicate tumor activity in vivo by liquid biopsy. Research suggests exosomes can detect cancer early. Jafari et al. [116] assert that miRNA is the most extensively investigated element of EVs, especially in terms of its usage as an early biomarker for cancer identification [117]. Elevated and downregulated circulating exomiRs may predict poor gastrointestinal cancer survival, according to a meta-analysis [118]. In several malignancies, serum-specific exomiRs alter prognosis. High miR-122 expression in breast and non-small cell lung cancer (NSCLC) is associated with distant metastases and poor overall and progression-free survival [119]. Exosomal miR-5100 in hypoxic head and neck squamous cell carcinoma (HNSCC) activates CAF (cancer-associated fibroblast) via the quaking (RNA-binding protein)/protein kinase B/signal transducer and activator of transcription 3 (QKI/AKT/STAT3) axis, promoting metastasis [120]. miR-5100 contributes to cancer expansion and could be a useful target for diagnosing and treating HNSCC. Exosome proteins and nucleic acids may be cancer biomarkers [121]. Plasma exosome-derived Cx43 may predict five-year recurrence-free and survival rates in melanoma patients [122]. Cancer cell exosomes can measure chemotherapy and targeted therapy resistance [123].

13.4.1.2 Diagnostic Applications in Cancer

When tissue samples cannot be surgically removed, a liquid biopsy is preferred by scientists. Researchers can use it for noninvasive diagnosis. Liquid biopsy of EVs can detect cancer and related illnesses early. EVs are stabilized better in plasma and bodily fluids than ctDNA (circulating tumor DNA) or RNA, offering potential advantages for diagnostic use. Researchers use ultracentrifugation, ExoQuick, and Invitrogen Total Exosome Isolation kits to precipitate exosomes. The kits are not diagnostic or therapeutic. Multiple approaches have been devised to segregate and examine EVs and their markers. Several ISEV position papers standardize exosome extraction from small biofluid quantities, including blood plasma [124, 125]. Tumor-derived (TD)-EVs contain tumor DNA, proteins, and abnormal microRNAs [126]. Pancreatic and melanoma biomarkers in EVs are noninvasive glypican-1 (GPC1) and programmed death-ligand 1 (PD-L1) [127]. EVs from patients harbor melanoma-specific proteins like tyrosinase-related protein-2, late-onset antigen, Hsp70, caveolin-1, and mesenchymal-epithelial transition factor (MET), implying their application as biomarkers [128]. TD-EV miRNA profiles can detect ovarian, colorectal, glioblastoma, prostate, and colon cancers [129]. Circular RNAs (circRNAs) and miRNAs are prevalent in EVs sourced from cancer cells and patient blood, possibly assisting as distinctive biomarkers for cancer detection [130].

13.4.1.3 Liquid Biopsy Based on Exosomes

As a noninvasive diagnostic method, liquid biopsy detects CTCs (circulating tumor cells), ctDNA, and tumor-derived RNA in fluids like blood and saliva, outperforming traditional biopsies. Exosomes hold a key part owing to their stability, high abundance (~10^9/mL), and ability to protect nucleic acids, accurately reflecting parental cell status [131]. Unlike rapidly degrading ctDNA and cfRNA (cell-free RNA), exosomes remain stable in plasma [125]. Standardized protocols for exosome extraction via ultracentrifugation have been established [3]. Exosome-based diagnostics monitor molecular markers, aiding in precision therapy and disease tracking [132]. Exosomal miRNAs, mRNA, long noncoding RNA (lncRNA), circRNA, and proteins like mutant Kirsten rat sarcoma (kras), gpc1, cluster of differentiation (CD82), carbohydrate antigen (CA19–9), and cytoskeleton-associated protein (CKAP4) serve as promising cancer biomarkers for pancreatic, lung, breast, and prostate cancers [130, 133]. NK cell–derived EVs (NKEVs) display diagnostic and therapeutic potential in cancers like hepatocellular carcinoma (HCC) and NSCLC, with microfluidic approaches and tools like the GBM ImmunoProfiler linking NKEV profiles to tumor progression and immune checkpoint markers such as PD-L1 and CTLA4 [134–136].

13.4.1.4 Autoimmune Disorder

Immune-related exosomes are forthcoming as noninvasive biomarkers for autoimmune diseases, offering diagnostic and prognostic insights. In SLE, serum and urinary exosomes carry miR-21, miR-155, and miR-146a, correlating with lupus nephritis severity and renal injury risk [137]. Urinary miR-29c and miR-26a signal early fibrosis and podocyte damage, while signal intensities of miR-31, miR-135b-5p, and miR-107 rise post-treatment, indicating recovery potential [138]. In RA, exosomal miR-335-5p, miR-486-5p, lncRNAs like HOX transcript antisense intergenic RNA (HOTAIR) and nuclear enriched abundant transcript (NEAT1), and proteins such as serine/arginine-rich splicing factor 4 (SRSF4) and dihydropyrimidinase-like 3 (DPYSL3) aid early diagnosis [139, 140]. miR-548a-3p and have-miR-155-5p levels reflect treatment response and disease activity [141]. In MS, exosomal proteins (alpha-1-microglobulin/bikunin precursor [AMBP], fibrinogen beta chain (FIBB), gelsolin (GELS), myelin basic protein (MBP)] and miRNAs help monitor disease progression and recurrence [142, 143]. Exosomes also assist in diagnosing promyelocytic leukemia, Sjögren's syndrome, Inflammatory Bowel Disease, myasthenia gravis, autoimmune skin disorders, preeclampsia, and transplant rejection [18, 144], supporting their role in personalized immune-related disease management.

13.4.1.5 Neurodegenerative Diseases

The ability of exosomes to traverse the BBB renders them valuable for early, noninvasive detection and monitoring of CNS diseases [145]. Their protein, miRNA, and circRNA profiles are promising biomarkers for neurodegenerative disorders, which begin decades before clinical symptoms. Exosomal markers (P-S96-tau, P-T181-tau, Aβ1–42) may forecast Alzheimer's disease up to ten years early, while synaptic proteins (growth-associated protein [GAP43], synaptosomal-associated protein [SNAP25], neurogranin) signal cognitive decline five to seven years before onset [146]. Neuronal EVs isolated from plasma, containing soluble α-synuclein, can identify Parkinson's disease, with predictive features, including olfactory or cognitive impairment and rapid eye movement sleep behavior issues [147, 148]. Neuron-derived exosomes expressing SNAP25, synaptophysin-I binding protein, and complement proteins (C1q, C3b, C3d, CD59, CD46) can assess Alzheimer's progression. Additionally, reduced excitatory synaptic proteins in exosomes correlate with cognitive impairment, and the α-*syn* oligomer/total α-*syn* ratio reflects Parkinson's severity, while mitochondrial respiration markers help gauge functional recovery [149].

13.4.2 Exosome-Based Therapies: Engineered Exosomes for Drug Delivery, Gene Therapy, and Vaccine Development

Exosomes are nanoscale vesicular entities with diverse specialized contents that can target specific tissues, have low immunogenicity, transport hydrophilic and hydrophobic biomolecules, and cross biological barriers to invade tissue structures. Exosomes can perform several disease-treatment roles due to this benefit.

13.4.2.1 Exosomes in Drug Delivery

Drug delivery using liposomes, micellar nanoparticles, and superparamagnetic iron oxide nanoparticles faces limitations like premature clearance, biocompatibility issues, and long-term toxicity [150]. Exosomes provide a superior, biocompatible alternative capable of traversing the BBB, improving drug efficacy and safety [151]. For instance, paclitaxel- and adriamycin-encapsulated exosomes from brain endothelial cells reduced tumor markers in a zebrafish brain cancer model, while dopamine-loaded exosomes treated Parkinson's in mice with less toxicity than free dopamine [152]. Exosomes have delivered chemotherapeutics like adriamycin, paclitaxel, cisplatin, and gemcitabine, enhancing cytotoxicity and tumor regression [153, 154]. They also improve anti-inflammatory effects when loaded with curcumin or rosmarinic acid, and BDNF (brain-derived neurotrophic factor)-loaded exosomes aid stroke recovery. Methotrexate in grapefruit-derived exosomes reduces intestinal inflammation, while HES and anthocyanidins in milk exosomes boost oral bioavailability and antitumor efficacy in melanoma, providing a cost-effective strategy [155]. Additionally, HER2-positive cancer cell-derived exosomes loaded with trastuzumab enhance targeted therapy, and engineered exosomes show promise in gene delivery.

13.4.2.2 Targeted Therapy for EV

Exosomes are evolving as ideal targeted drug delivery systems owing to their inherent cell-specific homing ability, influenced by their cellular origin [156]. Their targeting capabilities can be further enhanced via direct (drug/gene/ligand loading) and indirect (physical/genetic) modifications [157, 158]. Strategies such as 1,2-distearoyl-sn-glycero-3-phosphoethanolamine–polyethylene glycol (DSPE–PEG) lipid insertion allow ligands like amino acid–polyethylene glycol (AA-PEG) and angiopep-2 to be incorporated into exosomal membranes for lung and brain tumor targeting, respectively [159, 160]. In triple-negative breast cancer (TNBC), cyclic Arg–Gly–Asp–D-Tyr–Lys peptide (c(RGDyK)) – modified exosomes have rendered miR-588 to remodel the tumor microenvironment, while similar systems loaded with curcumin reduced inflammation in ischemic brain models [161, 162]. Cholesterol-modified exosomes improved brain targeting of miR-210 in stroke therapy [163]. Receptor-ligand fusion techniques using lysosome-associated membrane glycoprotein (LAMP-2B) with rabies virus glycoprotein (RVG), arginine–glycine–aspartic acid (RGD), anti-HER2 affibody (zHER), or IL-3 enabled targeting of Alzheimer's disease, chronic myeloid leukemia (CML), and breast cancer [164, 165]. Tetraspanins (CD63, CD9, CD81) and ApoA-1-CD63 fusions enhanced liver-specific delivery [166]. To mitigate the limited output and heterogeneity of natural exosomes, exosome-mimetic nanosystems (EMNs) produced via extrusion and liposome fusion have been developed [167], with integrins like α6β4 promoting organ tropism [168]. Hybrid exosomes (e.g., RBC-leukocyte) and nanoparticle–exosome conjugates carrying miRNAs or gold-doxorubicin further improve tumor-specific delivery [169]. Their tetraspanin-mediated targeting, immune evasion, and modifiability make exosomes superior to traditional delivery platforms such as liposomes [170].

13.4.2.3 Exosomes in Gene Therapy

Direct and indirect engineering strategies have revolutionized exosome-based targeted drug delivery. Direct methods involve loading exosomes with drugs, genetic materials, or ligands to enhance specificity and therapeutic impact while minimizing off-target effects [158, 162]. For example, DSPE–PEG lipid insertion enables hydrophobic ligand integration – DSPE–PEG-AA-PEG targets lung cancer via sigma receptors [171], while DSPE–DSPE-DSPE-DSPE-DSPE-PEG-angiopep-2 facilitates blood–brain barrier penetration in glioblastoma [160]. In TNBC and ischemic models, c(RGDyK)- or cholesterol-modified exosomes deliver miR-588, curcumin, or miR-210 to remodel the tumor microenvironment or support neuroprotection [161, 163]. Receptor–ligand strategies and peptide fusion, such as cardiac homing peptides or magnetic nanocrystal-based systems, further improve targeting [172, 173]. Genetic modifications of donor cells involve fusing targeting ligands with exosomal membrane proteins (Lamp2b, C1C2 domain of lactadherin, tetraspanins) to direct exosome delivery – for example, RVG-Lamp2b delivers siRNA to neurons, RGD-Lamp2b enhances tumor vasculature targeting, and IL-3-decorated exosomes carry imatinib to CML cells [164, 165, 174]. Apolipoprotein A-I (ApoA-1)/CD63 and human hyaluronidase (PH20)-loaded exosomes further aid liver targeting and tumor penetration [166]. To overcome low yields of natural exosomes, EMNs generated via extrusion or liposome fusion offer scalable, customizable platforms enriched with organ-tropic molecules like integrin α6β4 [168, 175]. Hybrid EMNs from red blood cells or leukocytes mimic natural targeting and reduce immune response, while nanoparticle–exosome hybrids (e.g., gold–doxorubicin systems) enhance precision drug delivery and reduce toxicity in cancer therapies [169, 176].

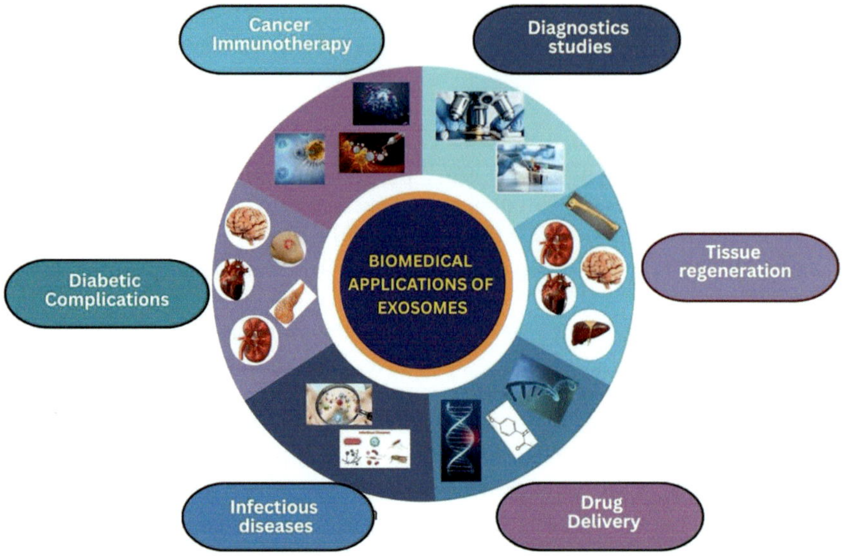

FIGURE 13.2 Biomedical and therapeutic applications of exosomes.

13.4.2.4 Exosomes in Cancer Vaccines

Exosomes exhibit potent immunomodulatory traits, rendering them valuable tools for cancer immunotherapy and vaccination. Immune cell–derived exosomes, encompassing those from DCs, NK cells, CD4+ T cells, M1 macrophages, chimeric antigen receptor T cells (CAR-T), and tumors (TEXs), can activate innate and adaptive immunity by facilitating immune synapse formation [177]. Genetically engineered exosomes expressing IL-2, IL-18, or superantigens enhance tumor-specific immunity [178, 179], while HSP-enriched TEXs and adjuvant-loaded exosomes further boost T and NK cell responses [82, 180]. Advanced platforms like gemcitabine-Exos and 0xaliplatin/siRNA-loaded vesicles target EGFR + TNBC and immunosuppressive microenvironments [177, 181]. Dendritic cell–derived exosomes (DEXs) activate CD4+, CD8+, and NK cells, especially when pulsed with tumor antigens or linked with anti-CTLA-4 antibodies [182, 183], and those expressing TNF-related apoptosis-inducing ligand (TRAIL), TNF, or FasL promote tumor apoptosis [77, 184]. Exosomes also serve as stable, low-immunogenic drug and RNAi carriers [62], with some (e.g., NK and M1 macrophage exosomes) acting as immunoadjuvants [185, 186]. Clinically, DEXs have shown efficacy in phase I/II trials for NSCLC and melanoma by boosting NK activity and post-chemotherapy immune surveillance [187], though PD-L1 upregulation during production can reduce efficacy [188, 189]. Future strategies involve engineering low PD-L1 DCs and combining DEXs with checkpoint inhibitors or viral protein-antigen fusions (e.g., vesicular stomatitis virus glycoprotein) to enhance cancer vaccine potency.

13.4.3 Challenges in Translation

13.4.3.1 Scalability, Standardization, Safety, and Immunogenicity Concerns

A 2024 Business Communications Company (BCC) research report predicts the global exosome diagnostics (ExoDx), therapies, and research tools market will grow from $227.5 million in 2023 to $1.3 billion by 2028, reflecting the growing significance of exosome research in biomedicine. However, clinical translation faces regulatory and technical constraints. This limitation is primarily due to the absence of standardized approaches for exosome preparation, purification, measurement, and storage across companies, with no universal industry standards [190]. Large-scale production remains limited by cell culture constraints, despite emerging techniques like ultrafiltration, size-exclusion chromatography, and aqueous two-phase systems, which lack comparative or scale-up guidelines. Additionally, exosome heterogeneity in nucleic acid and protein content from different parental cells complicates quality control. Regulatory frameworks for extracellular vesicle supervision are underdeveloped, further delaying progress. Technical issues such as low yields, high costs, and difficulty in modifying exosomes owing to their diminutive size and complex composition hinder therapeutic advancement. Attaining efficient and reproducible cargo loading remains a challenge, and heterogeneity complicates quantification and compositional control. Interdisciplinary collaboration among scientists, engineers, and clinicians is crucial to overcoming these barriers and advancing exosomes toward clinical application [191].

13.5 CASE STUDIES AND CLINICAL APPLICATIONS

A multitude of researchers have investigated the genesis of exosomes, their biocompatibility, their significance in life events that involve cell-to-cell interaction and immune response, as well as their capacity to transport therapeutic medications. These traits make exosomes important in translational nanomedicine, especially in theranostics for cancer, cardiovascular, viral, and neurological diseases. They provide tailored medical care by providing real-time, noninvasive illness diagnostic and progression information [192]. Das et al. demonstrated that exosomes can bypass phagocytosis and carry therapeutic substances across biological barriers, including the BBB to brain or neural cells, reducing systemic side effects and toxicity [193]. Cully supported exosome-based cancer, autoimmune, and neurological medicines in clinical trials [194]. The following sections describe exosome clinical uses in several medical fields and are presented in Table 13.2.

13.5.1 Cancer Immunotherapy

By focusing on exosomes' ability to more precisely target cancer cells while reducing side effects in comparison to traditional chemotherapy, Wang et al. explored their capacity in cancer treatment through gene therapy and immunotherapy. [195]. The oncogenic components and particular antigens carried by TEXs can increase the immune response against tumors [196], and some studies suggest they can help identify and track cancer [197]. Exosomes show great promise for individualized cancer treatment due to their ability to transport chemotherapy drugs straight to the tumor cells and their compatibility with a range of imaging modalities [198].

Research indicates that exosomes can combat cancer effectively. Munagala et al. discovered that drug-loaded exosomes considerably reduced the growth of breast (MDAMB-231 and T47D) and lung (A549 and H1299) cancer cell lines, with IC_{50} values that were lower in the exosomal preparations than in the free medicines. [199]. It was reported by Saari et al. and Kim et al. that exosomes could enhance cytotoxicity in drug-resistant cells [200]. Aqil et al. delivered celastrol via exosomes to NSCLC cell lines, demonstrating reduced IC_{50} values and minimal toxicity to normal functions, suggesting exosomes as a viable delivery method. They also found that exosomal curcumin formulations significantly increased bioavailability and reduced cancer cell growth while minimizing systemic toxicity [201, 202].

Munagala et al. also studied anthocyanidin-containing exosomes, which showed enhanced anticancer effects with significantly lower IC_{50} values [203]. Morse et al. developed dendritic cell exosomes with cancer antigens that improved NK cell efficacy in advanced NSCLC patients, leading to a 50% increase in survival [204]. Additionally, zebrafish studies indicated that exosomes could enhance drug transport to brain metastases [205]. Dorayappan et al. (2016) suggested that ovarian cancer exosomes might assist in diagnosis and treatment. From in vitro investigations to in vivo models and preclinical trials, modified exosomes are making great strides, along with innovations like chimeric exosomes and surface modifications enhancing their therapeutic potential. While exosomes show promise in treating various cancers, including lung, endometrial, and cervical cancers, their mechanisms and efficacy are still under investigation [206].

Exosomes are produced by humans, plants, animals, and microorganisms. Jo et al. tested Lactobacillus plantarum exosomes on aging skin, finding improvements in skin hydration and reduction of wrinkles [207]. Another study on beetroot extract exosome-like nanoparticles showed beneficial properties for skin regeneration [208]. Unfortunately, there is a lack of extensive clinical trials for cosmetic exosomes. Additional study is necessary to establish production standards and identify the most effective dosages.

TABLE 13.2
Emerging Clinical Applications of Exosome-Based Technologies

Targeted Clinical Field	Therapeutic Role
Cancer immunotherapy	• They are utilized for targeted cancer gene therapy and immunotherapy [195] • TEXs enhance antitumor responses by carrying oncogenic materials and antigens [196] • Exosomes can be used to identify and track cancer [197] because they are compatible with imaging techniques [198] • Delivering anticancer drugs directly to targeted cancer cells improves anticancer effects, making the drugs more effective and less harmful; they exhibit a lower IC_{50} compared to free drugs and demonstrate better efficacy in drug-resistant cells [203, 205]
Autoimmune and inflammatory diseases	• Exosomes facilitate intercellular communication through proteins, lipids, and nucleic acids [209] • They present self-antigens and modulate immune responses in conditions such as RA, SLE, and MS [209, 210]. • They transport citrullinated peptides, nuclear antigens, microRNAs, and immunomodulatory proteins for immune regulation and anti-inflammatory properties [212]
Infectious diseases	• Exosome-based vaccines to transport antigens, modify immune responses, and provide long-term protection [237].
Neurological disorders	• It has the capacity to encourage the growth and repair of neurons, boost neurotrophic factors [213, 214], and penetrate the blood–brain barrier for precise distribution [215]
Cardiovascular diseases	• Exosomal microRNAs promote angiogenesis and vascular repair in post-myocardial infarction [216, 217] • Exosomal markers are used to detect atherosclerosis [218].
Regenerative medicine	• MSC exosomes are used for tissue regeneration and immune modulation [219, 220] in lung diseases, eye diseases, OA, fertility, hair regeneration, and skin rejuvenation [221–223, 226, 228] • Promote fibroblast activity and collagen production [232, 234, 235] with improvements in skin hydration and wrinkles [207, 236]

13.5.2 Autoimmune and Inflammatory Diseases

Exosomes, nanosized EVs released by numerous cells, have become significant in the development and regulation of inflammatory and autoimmune illnesses. These vesicles communicate with other cells by transferring bioactive substances like genetic material, lipids, and proteins. In autoimmune diseases like RA, SLE, and MS, exosomes derived from immune cells (DCs, B cells, and T cells) can present self-antigens, modulate immune responses, and influence inflammatory cascades. Exosomes generated from synovial fibroblasts in RA, for example, are known to transport citrullinated peptides and inflammatory cytokines that promote T-cell activation and joint inflammation, thereby contributing to disease progression and tissue damage [209]. Similarly, in SLE, exosomes enriched with nuclear antigens and RNA components can trigger plasmacytoid DCs and type I interferon production, leading to systemic inflammation and autoimmunity [210].

Exosomes have been studied for their function in antigen presentation and myelin-specific responses of the immune system in MS. Findings suggest that exosomes can aggravate or alleviate sickness depending on their biological source and payload [211]. Anti-inflammatory and immunosuppressive effects are found in MSC-derived exosomes, offering a potential cell-free therapeutic approach for modulating immune responses in RA and SLE [212]. Due to their ability to transport and release immunomodulatory proteins or regulatory microRNAs, exosomes present a new avenue for drug administration and treatment. Therefore, comprehending the dual function of exosomes in illness development and treatment is essential for their incorporation into therapies and diagnostic procedures for autoimmune and inflammatory diseases.

13.5.3 Neurological Disorders

Although there are no cures for neurological disorders like Parkinson's and Alzheimer's, exosomes' bioactive substances may help neurons grow and repair [213]. Researchers observed that exosome-derived neurotrophic chemicals can increase neuron development and repair and improve mice's cognitive function following injection [214]. The capacity of exosomes to penetrate the barrier that separates blood from brain and to transport medicinal compounds to brain regions to treat brain diseases was investigated by Salarpour et al. who concluded by highlighting the potential of exosomes as customized delivery systems for brain disorders such as Alzheimer's, Parkinson's, and stroke [215]. Qu et al. developed biocompatible blood exosomes to facilitate the transportation of medications through the blood–brain barrier for the treatment of Parkinson's disease. Exosomes accumulated in brain tissue between 1 hour and 10 hours after injection, peaking 4–8 hours later. In mice, dopamine-containing blood exosomes increased brain concentration to 1.02 ± 0.15 nmol/g and boosted brain dopamine dispersion 15-fold. In Parkinson's disease mice, exosomes increased dopaminergic neurons and reduced disease severity [152]. Precision gene editing was also utilized to change exosomes, so they could be transplanted into cells to treat Parkinson's disease.

13.5.4 Cardiovascular Diseases

Cardiovascular studies show that myocardial infarction and coronary heart disease kill most people worldwide. Exosomes from mesenchymal progenitor cells contain microRNA-132, which induces vascular formation in myocardial infarction [216]; whereas MiR-126 exosomes diminish vascular permeability and increase angiogenesis and endothelial cell proliferation. Mice injected with miR-126 exosomes following myocardial infarction may grow and repair cardiomyocytes [217]. Lu et al. studied exosome indicators for atherosclerosis detection. Atherosclerosis patients have increased exosomal miR-126-3p and PD-L1, but decreased miR-145-5p. Exosome-derived indicators were employed in a diagnostic model to identify atherosclerosis patients from healthy controls, suggesting they may be clinically relevant in coronary artery disease patients [218].

13.5.5 Regenerative Medicine

The potential of stem cell exosomes to facilitate cellular renewal and healing has been revealed. Rezvani et al. recommended using MSC-derived exosomes to regenerate tissue and modify the immune system's response [219]. For the purpose of regenerative medicine, Codispoti et al. employed nanosized bioarchived exosomes derived from MSC, demonstrating promise in preclinical studies [220]. Popowski explored exosomes' biogenesis and their role in lung regenerative medicine for conditions like COPD (chronic obstructive pulmonary disease) and lung cancer [221]. Sanghani et al. investigated exosome treatments for eye diseases, finding efficacy in diabetic retinopathy and glaucoma [222]. Osteoarthritis (OA), a leading cause of impairment, has been targeted with exosome-based therapies. Vitha et al. found exosomes beneficial for osteoporosis and tendon injuries [223]. Early clinical trials indicated that exosomes could enhance chondrocyte proliferation and joint function in OA patients [224]. Dimik et al. studied exosomes in fertility and reproductive health, noting improvements in in vitro fertilization [225]. Exosomes are also being researched for hair regeneration. Kost et al. examined their effects on hair growth [226], while Ariyoshi et al. found that mouse buccal mucosa exosomes accelerated DNA repair and hair development post-radiation [227]. In skin rejuvenation, platelet-derived exosomes reduced signs of aging [228], and bio-pulsation with MSC exosomes enhanced skin healing and collagen synthesis [229]. Exosomes have shown potential in improving skin health by affecting fibroblast activity and collagen production [230, 231].

The function of senescent fibroblasts could be restored by 3D spheroid-originating exosomes, as reported by Hu et al. [232], while Wu et al. showed that human umbilical cord (hucMSC) – exosomes are capable of reducing inflammation and photodamage in the epidermis [233]. Gao et al. discovered that exosomes from adipose-derived stem cells improved cell survival and reduced DNA damage in UVB (ultraviolet B)-irradiated cells [234]. Clinical trials using microneedling with ADSC (adipose-derived stem cells) – exosomes showed enhanced skin hydration and reduced wrinkles [235]. A study on human platelet-derived exosome serum indicated improvements in skin health after topical application [228]. Chernoff reported enhanced skin texture and evenness after a combination treatment involving placental-derived MSC exosomes. Han et al. found that bovine colostrum exosomes improved collagen and photodamage recovery [236]. Exosomes are produced by various organisms, including plants and microorganisms. Jo et al. tested *Lactobacillus plantarum* exosomes on aging skin, finding improvements in hydration and wrinkle reduction [207]. Beetroot extract exosome-like nanoparticles showed promising properties for skin regeneration [208]. However, clinical trials in cosmetic dermatology are limited; additional research is required to standardize exosome production and ascertain the most effective dosages.

13.5.6 Status of Clinical Studies

First identified in sheep reticulocytes, exosomes are present in a variety of physiological fluids, such as blood, urine, and breast milk [238]. The biomolecules they contain are diverse, encompassing proteins, RNAs, and lipids, with databases like Vesiclepedia and Exocarta providing detailed information. According to Global Data (www.globaldata.com), exosomes are being researched for various conditions, including cancer (54%), transmissible illnesses (13%), immunological disorders (8%), and neurological disorders (13%). Most of the 420 exosomal drugs in clinical trials are still in preclinical or discovery phases, with only a small percentage progressing through phases 1, 2, and 3 [239]. Clinical trials have explored exosomes for various applications, such as a phase I trial using exosomes derived from plasma for the purpose of wound healing (NCT02565264) and exosome-delivered curcumin for colon cancer (NCT01294072). Several recent investigations have demonstrated that exosomes derived from MSC can be used to treat conditions like type I diabetes and COVID (coronavirus disease)-19, with early evidence indicating their effectiveness in reducing COVID-19 inflammation [240]. Both plant and human exosomes have been investigated, with trials involving ginger, aloe, and grape-derived exosomes aimed at improving health outcomes in various conditions (NCT03493984, NCT01668849). However, the understanding of plant-derived exosomes is limited compared to human tissue exosomes. Currently, only two exosomes are clinically available: patisiran for treating adult polyneuropathy and ibudilast for asthma. While the results of ongoing clinical trials are encouraging, further investigation is required to elucidate the ways they function, their safety, along their applications [240]. The ISEV has made efforts to standardize exosome extraction and characterization, as highlighted in the 2017 EV-TRACK database [3].

13.5.7 Drug Delivery System

Degradation and poor cellular absorption hinder systemic delivery of functional proteins. In certain cases, exosome-mediated transport is possible. Due to their low immunogenicity, physiological endurance, and blood–brain barrier crossing, exosomes may be excellent drug delivery carriers. Stable exosome lipid bilayers protect cargo from the immune system and hydrolases. Exosomes can transport water-soluble and non-water-soluble pharmaceuticals, such as chemotherapeutic medications, immunomodulators, antisense oligonucleotides, siRNA, and others. They are organotrophic and tumor-targeting, and delivering drugs through exosomes is feasible intravenously, intraperitoneally, cerebrally, or intranasally [241]. Multiple findings indicated that extracellular vesicle-driven transport stabilizes drug circulation, increasing drug accumulation in recipient cells and reducing prescription needs [242]. Using exosomes to deliver drugs could revolutionize cancer treatment. Selective exosome surface modification to express targeting ligands (proteins or peptides) and/or therapeutic medicines reduces off-target effects by targeting cells or tissues [243]. Many industries use exosome biotechnology. To engineer exosomes for personalized drug delivery, Codiak BioSciences established the engEx™ Platform in Cambridge, Massachusetts [244].

13.5.8 Diagnostic Applications

CTCs and tumor antigens lack sensitivity and specificity, limiting their therapeutic potential. Solid biopsy limits precision and personalized therapy, but liquid biopsy can diagnose and forecast noninvasively. All body fluids contain exosomes, which have prolonged biological activity and higher sensitivity and specificity, thereby rendering them suitable for use in liquid biopsy in the screening, prognosis, and management of illness. Recent studies suggest that exosomes are promising cancer biomarkers. Imaging diagnosis is needed for personalized treatment. Real-time radiography and cancer cell imaging use tissue-bound exosomes because they deliver payloads and resist degradation. Besides accuracy, early screening and diagnosis improve cancer survival and recovery. Cancer biomarkers are exosome molecular payloads – nucleic acids, mRNAs, miRNAs, long noncoding RNAs (lncRNAs), circRNAs, proteins, and lipids in bodily fluids.

Kirsten rat sarcoma (KRAS) mutations in circulating exoDNA were higher in early pancreatic cancer patients than in healthy controls, suggesting tumor detection. Exosome-based EGFR T790M detection may reduce NSCLC biopsies [245]. Recent research shows that tetraspanins, CD171, and CD151 are the most important molecules for histopathological lung cancer diagnosis. Bijnsdorp et al. found elevated integrin beta-1 (ITGB1) and integrin alpha-3 (ITGA3) exosomal biomarkers in prostate cancer urine samples utilizing mass spectrometry proteome analysis. Exosomal proteins are biomarkers for cancer, pulmonary, cardiovascular, and neurological disorders like Alzheimer's and Parkinson's [246]. In 2016, ExoDx Lung (anaplastic lymphoma kinase) was the first cancer diagnostic device to detect *Echinoderm microtubule*–associated protein-like 4-anaplastic lymphoma kinase (EML4ALK) mutations in real time using exosome miRNAs from body fluid biopsy [247]. The noninvasive ExoDx Prostate® (IntelliScore) (EPI) test detects prostate cancer [248]. Avalon GloboCare and Genexosome Technologies jointly developed the salivary-based exosome miRNA biomarker miR-185 for oral cancer diagnosis and treatment. Roche and PureTech Health recommended oral antisense. PureTech discovered that milk exosome-based intravenous injections are safer and more effective [249]. Many exosome-based diagnostic applications have shown potential; however, their applicability is debatable due to in vitro studies. Standardizing and identifying clinical exosomes is crucial. Most exosomes transport chemicals, but few biomarkers. Wide-ranging biomarker validation may aid disease panel development.

13.6 FUTURE DIRECTIONS

13.6.1 Emerging Technologies

Emerging evidence from EV research, with a focus on exosomes, supports their application in noninvasive cancer diagnostics via liquid biopsy and their versatility as carriers for therapeutic agents. Liquid biopsies offer noninvasive sampling and, with robust, automated genomic and proteomic protocols, can enable rapid diagnostics [250]. However, challenges remain in standardizing exosome isolation and characterization. Discriminating TEXs from those of normal cells is difficult due to overlapping markers and inconsistent antigen expression. Although antibody capture methods targeting TEX-specific antigens exist, variability in surface antigen presentation affects reproducibility. Quantification of exosomes is another concern; most studies report protein concentration per ISEV guidelines, but clinical translation, particularly in drug delivery, necessitates particle-based measurements (particles/mL), akin to viral particle quantification in gene therapy. Additionally, there are concerns about using autologous versus allogeneic exosomes – autologous sources support personalized therapy and specific tumor antigen targeting but face scalability issues, while allogeneic exosomes offer ease of production and enhanced immune responses, though they may induce unwanted immune reactions. Further comparative studies are needed to assess their clinical suitability. Large-scale production remains a technical hurdle, and the influence of exosome size (50–150 nm) on pharmacokinetics, biodistribution, and therapeutic efficacy must also be examined. Overcoming these challenges through continued research and technological advancements will be essential for the widespread clinical application of exosomes [3, 251].

13.6.2 Ethical and Regulatory Considerations

The clinical translation of exosome-based diagnostics and therapeutics necessitates strict adherence to ethical and regulatory standards, especially concerning manufacturing processes, quality control, and product consistency. Due to the natural heterogeneity and biological origin of exosomes, establishing standardized manufacturing protocols – covering isolation, purification, storage, and scalability – is a critical challenge. Regulatory agencies like the FDA and European Medicines Agency (EMA) currently lack specific, harmonized guidelines for exosome-based products, often categorizing them under biologics or advanced therapy medicinal products (ATMPs) [252, 253]. The ISEV has recommended minimal criteria for characterization and quantification, but further global consensus is necessary to ensure product reproducibility, safety, and efficacy in clinical applications [3]. GMP compliance is also essential, particularly when exosomes are used in therapeutic contexts such as regenerative medicine or immune modulation. Beyond technical considerations, ethical issues such as informed patient consent and the implications of commercialization are central to the responsible deployment of exosome technologies. When exosomes are derived from human tissues or cells – especially autologous or donor-derived sources – patients must be thoroughly made aware of the origin, purpose, storage, and promising secondary uses of their biological material [254]. Additionally, the commercialization of exosome-based therapies introduces questions around ownership, accessibility, and equity. High development and production costs could limit access to novel therapies, particularly in low-resource settings, raising concerns about healthcare disparities. Clear regulations concerning intellectual property, benefit-sharing, and equitable access must be developed to balance innovation with social responsibility [255]. Addressing these ethical and regulatory constraints is fundamental to ensuring the safe, fair, and effective inclusion of exosome technologies into modern medicine.

13.6.3 Exosomes in Preclinical and Clinical Trials

As drug delivery platforms, exosomes have exhibited considerable promise across both in vitro and in vivo experimental settings. Clinical evaluation of plant-derived exosomes for targeted curcumin delivery in colon cancer is ongoing in a phase I trial (NCT01294072). Similarly, dendritic cell–secreted exosomes, containing MHC class I/II molecules, tetraspanins, and CD86, have demonstrated the ability to stimulate innate and adaptive immune pathways. Phase I clinical trials and preclinical studies confirmed their role in enhancing antitumor immunity [256]. Furthermore, a phase II trial is underway using tumor antigen-loaded dexosomes as a cancer vaccine (NCT01159288), and another phase II trial showed that IFN-γ–stimulated dexosomes can promote antitumor activity via NK cells [68]. The accumulated evidence underscores the viability of exosome-mediated therapeutics and sets the stage for their clinical integration in cancer management.

This chapter highlights the impact of exosomes as natural, nonimmunogenic drug delivery vehicles and theranostic agents for cancer, offering advantages over synthetic liposomes such as organotropism, efficient cellular uptake, and the ability to carry multiple payloads simultaneously. Exosomes are gaining attention due to their biocompatibility and scalability when derived from biological materials. However, challenges remain in achieving efficient, large-scale isolation of homogeneous exosome populations and identifying the most therapeutically effective subsets. Additionally, improved understanding of exosome biogenesis and physiology is essential for optimizing therapeutic loading. While the concept of "organic exosomes" is still emerging, ongoing research and recent advancements suggest their clinical application in cancer diagnosis and therapy is approaching feasibility [3, 39, 257].

EVs offer promising avenues in clinical applications attributed to their biocompatibility, ability to traverse the BBB, and suitability for infiltrating solid tumors [258, 259]. However, their clinical translation faces several challenges. The biological distribution of exosomes in treated individuals remains poorly understood, complicating standardization in their isolation, quantification, and analysis [260–262]. Storage issues also arise, as exosomes tend to aggregate and their cargo can degrade during freeze-thaw cycles; such issues can be mitigated with the inclusion of stabilizers like trehalose, glucose, and sucrose. Selecting appropriate sources for exosome isolation is crucial, as it significantly impacts therapeutic outcomes. Moreover, naive exosomes often fail to elicit strong in vivo responses, prompting interest in engineering exosomes for enhanced therapeutic efficacy [39]. Strategies include loading drugs into exosomes, exemplified by doxorubicin-encapsulated dendritic cell–derived exosomes (DEXs) for tumor treatment. While exosomes can modulate immune responses beneficially, they may also contribute to tissue damage in diseases like asthma [263]. Nonetheless, their unique properties – such as low clearance rates and efficient drug transfer capabilities – underscore their potential as drug delivery vehicles [264]. Elucidating the diverse functions of exosomes is essential for their effective therapeutic application. Despite current limitations, ongoing research aims to harness exosomes' full potential in treating various diseases.

13.7 CONCLUSION

Exosome therapeutic applications are growing and could revolutionize disease treatment as theragnostic, that is, for both diagnosis and treatment. It was also found that exosome-based theranostics might improve tumor, heart, and nervous system diagnosis and treatment [264], making them a popular pharmacological research topic. Currently, over 400 clinical research studies use exosomes for direct delivery of therapeutic agents or as an indicator for assessing treatment efficacy. Clinically, they showed promise for biomarker development in early illness diagnosis as well as a novel natural medication delivery system due to the following benefits, like better bioavailability, reduced toxicity profiles, and good pharmacokinetics [265]. Similarly, stems, leaves, seeds, and fruit juices act as sources for plant-derived, biocompatible, and nontoxic exosomes, which can be scaled up for transporting drugs [266]. Precision modification technology, such as precision gene editing, could be used to accurately modify exosomes for more accurate regulation in their clinical applications [213]. Despite the progress in exosome research, their heterogeneity, immune rejection potential, need for improved manufacturing and characterization, and cargo stability and bioavailability remain challenges and uncertainties for clinical use. Additionally, exosome medical usage regulations are still evolving [267]. Exosomes made by different donors have varying functional features; hence, donor selection must be considered when generating them therapeutically. Recent research shows that the exosome source affects quality, safety, and efficacy. Age affects donors' ability to regulate, multiply, and differentiate. The donor's health, medical history, and cell collection and selection will also be examined. Exosomes are small particles that can be taken up by various cells and tissues, making it difficult to track their fate and determine their efficacy [3].

Good manufacturing practice also applies to exosomes regardless of clinical purpose. GMP-grade exosome production requires high-quality components, cells, growth conditions, manufacturing procedures, and qualified staff for strict control and monitoring. Purification and quality control release assays are also needed after manufacture to ensure the product meets the highest standards before being given to patients. All this slows exosome-based medicines from bench to bedside. For exosome-based therapies to succeed, these obstacles must be overcome. The investigation of exosomes constitutes a dynamic field of research. Still, many other potential uses of exosomes warrant further study, such as gene therapy for genetic abnormalities, vaccine development for infectious and malignant diseases, and their effectiveness in combination with other drugs. Continued technological and experimental advancements are expected to provide significant insights into their heterogeneity and biological functions while also improving our capacity to exploit their therapeutic and diagnostic potential. Finally, exosomes undoubtedly constitute a potential instrument in medicine and may offer solutions to several contemporary medical difficulties.

ABBREVIATIONS

AA-PEG	amino acid–polyethylene glycol
ADSC	adipose-derived stem cells
AIM2	absent in melanoma
ALIX	apoptosis-linked gene 2-interacting protein X
AMBP	alpha-1-microglobulin/bikunin precursor
ApoA-1	apolipoprotein A-I
ATMPs	advanced therapy medicinal products
BCC	business communications company
BDNF	brain-derived neurotrophic factor
c(RGDyK)	cyclic Arg–Gly–Asp–D-Tyr–Lys peptide
CA19–9	carbohydrate antigen
CAF	cancer-associated fibroblast
CAR-T	chimeric antigen receptor T cells
CD24	cluster of differentiation 24
CD82	cluster of Differentiation
cfRNA	cell-free RNA
cGAS	cyclic GMP–AMP synthase
circRNAs	circular RNAs
CKAP4	cytoskeleton-associated protein
CLL	chronic lymphocytic leukemia
CML	chronic myeloid leukemia
COPD	chronic obstructive pulmonary disease
COVID	coronavirus disease
CTCs	circulating tumor cells
ctDNA	circulating tumor DNA
CTL	cytotoxic T lymphocyte
CTLA-4	cytotoxic T-lymphocyte-associated protein 4
DAMPs	damage-associated molecular patterns
DCs	dendritic cells
DEXs	dendritic cell–derived exosomes
DPYSL3	dihydropyrimidinase-like 3
dsDNA	double-stranded DNA
DSPE–PEG	1,2-distearoyl-sn-glycero-3-phosphoethanolamine–polyethylene glycol
EBV	Epstein–Barr virus
EGFR	epidermal growth factor receptor
EMA	European Medicines Agency
EML4ALK	echinoderm microtubule-associated protein-like 4-anaplastic lymphoma kinase
EMNs	exosome-mimetic nanosystems
ESCRT	endosomal sorting complexes required for transport
EVs	extracellular vesicles
ExoDx	lung exosome diagnostics lung (anaplastic lymphoma kinase)
FasL	Fas ligand
FDA	Food and Drug Administration
FIBB	fibrinogen beta chain
FLT3L	FMS-like tyrosine kinase 3 ligand
FoxP3	Forkhead box P3
GADPH	glyceraldehyde-3-phosphate dehydrogenase
GAP43	growth-associated protein
GELS	gelsolin
GM-CSF	granulocyte-macrophage colony-stimulating factor
ICAM-1	intercellular adhesion molecule 1
GPC1	glypican-1
GVHD	graft-versus-host disease
HCC	hepatocellular carcinoma
HCV	hepatitis C virus
HER2	human epidermal growth factor receptor 2
HGF	hepatocyte growth factor
HLA-G	human leukocyte antigen G
HNSCC	head and neck squamous cell carcinoma
HOTAIR	HOX transcript antisense intergenic RNA
HSC	heat shock cognate
hucMSC	human umbilical cord
IFN-γ	interferon gamma
IgG	immunoglobulin G
IL-4	interleukin-4
ILVs	intraluminal vesicles
IRF3	interferon regulatory factor 3
ISEV	International Society for Extracellular Vesicles
ITGA3	integrin alpha-3
ITGB1	integrin beta-1
KRAS	Kirsten rat sarcoma
LAMP-2B	lysosome-associated membrane glycoprotein
lncRNAs	long noncoding RNAs
MBP	myelin basic protein
MDA5	melanoma differentiation-associated protein 5
mekk2	mitogen-activated protein kinase kinase kinase 2
MET	mesenchymal-epithelial transition factor
MHC II	major histocompatibility complex class I
MS	multiple sclerosis
MSC	mesenchymal stem cells,
MVBs	multivesicular bodies
NEAT1	nuclear enriched abundant transcript
NF-κB	nuclear factor kappa
NK	natural killer
NKEVs	NK cell-derived EVs
NSCLC	non-small cell lung cancer
OA	osteoarthritis
PAMPs	pathogen-associated molecular patterns
PD-L1	programmed death-ligand 1
pEXOs	placental exosomes
PH20	human hyaluronidase
PI3K/AKT	phosphatidylinositol 3-kinase/protein kinase B pathway
QKI/AKT/STAT3	quaking (RNA-binding protein)/protein kinase B/signal transducer and activator of transcription 3
RGD	arginine-glycine-aspartic acid
RIG-I	retinoic acid–inducible gene I
RVG	rabies virus glycoprotein
SLE	systemic lupus erythematous
SNAP25	synaptosomal-associated protein
SRSF4	serine/arginine-rich splicing factor 4
TAMs	tumor-associated macrophages

TBK1-IKKε	TANK-binding kinase 1-IκB kinase epsilon
TD	tumor derived
TEXs	tumor-derived exosomes
TGF-β	transforming growth factor beta
TLR9	Toll-like receptor 9
TNBC	triple-negative breast cancer
TNF-α	tumor necrosis factor alpha
Tol-DCs	tolerogenic dendritic cells
TRAIL	TNF-related apoptosis-inducing ligand
trNK	tissue-resident NK
TSG101	tumor susceptibility gene
UVB	ultraviolet B
VPS4	vacuolar protein sorting–associated protein 4
VTA1	vesicle trafficking 1
zHER	Anti-HER2 affibody

ACKNOWLEDGMENT

All the authors want to credit their affiliated institutes.

AI DISCLOSURE STATEMENT

During the preparation of this work, the authors used Grammarly, QuillBot, and ChatGPT in order to enhance grammatical structure and narrative coherence. After using this tool/service, the authors reviewed and edited the content as needed and take full responsibility for the content of the publication.

CONFLICT OF INTEREST

None.

REFERENCES

[1] M. Naseri, M. Bozorgmehr, M. Zöller, E. Ranaei Pirmardan, Z. Madjd, Tumor-derived exosomes: The next generation of promising cell-free vaccines in cancer immunotherapy, *Oncoimmunology* 9 (2020) 1–18. https://doi.org/10.1080/2162402X.2020.1779991.

[2] I.A. Batista, S.A. Melo, Exosomes and the future of immunotherapy in pancreatic cancer, *Int. J. Mol. Sci.* 20 (2019). https://doi.org/10.3390/ijms20030567.

[3] C. Théry, K.W. Witwer, E. Aikawa, M.J. Alcaraz, J.D. Anderson, R. Andriantsitohaina, A. Antoniou, T. Arab, F. Archer, G.K. Atkin-Smith, D.C. Ayre, J.M. Bach, D. Bachurski, H. Baharvand, L. Balaj, S. Baldacchino, N.N. Bauer, A.A. Baxter, M. Bebawy, C. Beckham, A. Bedina Zavec, A. Benmoussa, A.C. Berardi, P. Bergese, E. Bielska, C. Blenkiron, S. Bobis-Wozowicz, E. Boilard, W. Boireau, A. Bongiovanni, F.E. Borràs, S. Bosch, C.M. Boulanger, X. Breakefield, A.M. Breglio, M. Brennan, D.R. Brigstock, A. Brisson, M.L.D. Broekman, J.F. Bromberg, P. Bryl-Górecka, S. Buch, A.H. Buck, D. Burger, S. Busatto, D. Buschmann, B. Bussolati, E.I. Buzás, J.B. Byrd, G. Camussi, D.R.F. Carter, S. Caruso, L.W. Chamley, Y.T. Chang, A.D. Chaudhuri, C. Chen, S. Chen, L. Cheng, A.R. Chin, A. Clayton, S.P. Clerici, A. Cocks, E. Cocucci, R.J. Coffey, A. Cordeiro-da-Silva, Y. Couch, F.A.W. Coumans, B. Coyle, R. Crescitelli, M.F. Criado, C. D'Souza-Schorey, S. Das, P. de Candia, E.F. De Santana, O. De Wever, H.A. del Portillo, T. Demaret, S. Deville, A. Devitt, B. Dhondt, D. Di Vizio, L.C. Dieterich, V. Dolo, A.P. Dominguez Rubio, M. Dominici, M.R. Dourado, T.A.P. Driedonks, F.V. Duarte, H.M. Duncan, R.M. Eichenberger, K. Ekström, S.E.L. Andaloussi, C. Elie-Caille, U. Erdbrügger, J.M. Falcón-Pérez, F. Fatima, J.E. Fish, M. Flores-Bellver, A. Försönits, A. Frelet-Barrand, F. Fricke, G. Fuhrmann, S. Gabrielsson, A. Gámez-Valero, C. Gardiner, K. Gärtner, R. Gaudin, Y.S. Gho, B. Giebel, C. Gilbert, M. Gimona, I. Giusti, D.C.I. Goberdhan, A. Görgens, S.M. Gorski, D.W. Greening, J.C. Gross, A. Gualerzi, G.N. Gupta, D. Gustafson, A. Handberg, R.A. Haraszti, P. Harrison, H. Hegyesi, A. Hendrix, A.F. Hill, F.H. Hochberg, K.F. Hoffmann, B. Holder, H. Holthofer, B. Hosseinkhani, G. Hu, Y. Huang, V. Huber, S. Hunt, A.G.E. Ibrahim, T. Ikezu, J.M. Inal, M. Isin, A. Ivanova, H.K. Jackson, S. Jacobsen, S.M. Jay, M. Jayachandran, G. Jenster, L. Jiang, S.M. Johnson, J.C. Jones, A. Jong, T. Jovanovic-Talisman, S. Jung, R. Kalluri, S. ichi Kano, S. Kaur, Y. Kawamura, E.T. Keller, D. Khamari, E. Khomyakova, A. Khvorova, P. Kierulf, K.P. Kim, T. Kislinger, M. Klingeborn, D.J. Klinke, M. Kornek, M.M. Kosanović, Á.F. Kovács, E.M. Krämer-Albers, S. Krasemann, M. Krause, I.V. Kurochkin, G.D. Kusuma, S. Kuypers, S. Laitinen, S.M. Langevin, L.R. Languino, J. Lannigan, C. Lässer, L.C. Laurent, G. Lavieu, E. Lázaro-Ibáñez, S. Le Lay, M.S. Lee, Y.X.F. Lee, D.S. Lemos, M. Lenassi, A. Leszczynska, I.T.S. Li, K. Liao, S.F. Libregts, E. Ligeti, R. Lim, S.K. Lim, A. Linē, K. Linnemannstöns, A. Llorente, C.A. Lombard, M.J. Lorenowicz, Á.M. Lörincz, J. Lötvall, J. Lovett, M.C. Lowry, X. Loyer, Q. Lu, B. Lukomska, T.R. Lunavat, S.L.N. Maas, H. Malhi, A. Marcilla, J. Mariani, J. Mariscal, E.S. Martens-Uzunova, L. Martin-Jaular, M.C. Martinez, V.R. Martins, M. Mathieu, S. Mathivanan, M. Maugeri, L.K. McGinnis, M.J. McVey, D.G. Meckes, K.L. Meehan, I. Mertens, V.R. Minciacchi, A. Möller, M. Møller Jørgensen, A. Morales-Kastresana, J. Morhayim, F. Mullier, M. Muraca, L. Musante, V. Mussack, D.C. Muth, K.H. Myburgh, T. Najrana, M. Nawaz, I. Nazarenko, P. Nejsum, C. Neri, T. Neri, R. Nieuwland, L. Nimrichter, J.P. Nolan, E.N.M. Nolte-'t Hoen, N. Noren Hooten, L. O'Driscoll, T. O'Grady, A. O'Loghlen, T. Ochiya, M. Olivier, A. Ortiz, L.A. Ortiz, X. Osteikoetxea, O. Ostegaard, M. Ostrowski, J. Park, D.M. Pegtel, H. Peinado, F. Perut, M.W. Pfaffl, D.G. Phinney, B.C.H. Pieters, R.C. Pink, D.S. Pisetsky, E. Pogge von Strandmann, I. Polakovicova, I.K.H. Poon, B.H. Powell, I. Prada, L. Pulliam, P. Quesenberry, A. Radeghieri, R.L. Raffai, S. Raimondo, J. Rak, M.I. Ramirez, G. Raposo, M.S. Rayyan, N. Regev-Rudzki, F.L. Ricklefs, P.D. Robbins, D.D. Roberts, S.C. Rodrigues, E. Rohde, S. Rome, K.M.A. Rouschop, A. Rughetti, A.E. Russell, P. Saá, S. Sahoo, E. Salas-Huenuleo, C. Sánchez, J.A. Saugstad, M.J. Saul, R.M. Schiffelers, R. Schneider, T.H. Schøyen, A. Scott, E. Shahaj, S. Sharma, O. Shatnyeva, F. Shekari, G.V. Shelke, A.K. Shetty, K. Shiba, P.R.M. Siljander, A.M. Silva, A. Skowronek, O.L. Snyder, R.P. Soares, B.W. Sódar, C. Soekmadji, J. Sotillo, P.D. Stahl, W. Stoorvogel, S.L. Stott, E.F. Strasser, S. Swift, H. Tahara, M. Tewari, K. Timms, S. Tiwari, R. Tixeira, M. Tkach, W.S. Toh, R. Tomasini, A.C. Torrecilhas, J.P. Tosar, V. Toxavidis, L. Urbanelli, P. Vader, B.W.M. van Balkom, S.G. van der Grein, J. Van Deun, M.J.C. van Herwijnen, K. Van Keuren-Jensen, G. van Niel, M.E. van Royen, A.J. van Wijnen, M.H. Vasconcelos, I.J. Vechetti, T.D. Veit, L.J. Vella, É.

Velot, F.J. Verweij, B. Vestad, J.L. Viñas, T. Visnovitz, K.V. Vukman, J. Wahlgren, D.C. Watson, M.H.M. Wauben, A. Weaver, J.P. Webber, V. Weber, A.M. Wehman, D.J. Weiss, J.A. Welsh, S. Wendt, A.M. Wheelock, Z. Wiener, L. Witte, J. Wolfram, A. Xagorari, P. Xander, J. Xu, X. Yan, M. Yáñez-Mó, H. Yin, Y. Yuana, V. Zappulli, J. Zarubova, V. Žėkas, J. ye Zhang, Z. Zhao, L. Zheng, A.R. Zheutlin, A.M. Zickler, P. Zimmermann, A.M. Zivkovic, D. Zocco, E.K. Zuba-Surma, Minimal information for studies of extracellular vesicles 2018 (MISEV2018): A position statement of the International Society for Extracellular Vesicles and update of the MISEV2014 guidelines, *J. Extracell. Vesicles* 7 (2018). https://doi.org/10.1080/20013078.2018.1535750.

[4] M. Tkach, C. Théry, Communication by extracellular vesicles: Where we are and where we need to go, *Cell* 164 (2016) 1226–1232. https://doi.org/10.1016/j.cell.2016.01.043.

[5] R. Kalluri, V.S. Lebleu, Discovery of double-stranded genomic DNA in circulating exosomes, *Cold Spring Harb. Symp. Quant. Biol.* 81 (2016) 275–280. https://doi.org/10.1101/sqb.2016.81.030932.

[6] P. Nilsson, M. Sekiguchi, T. Akagi, S. Izumi, T. Komori, K. Hui, K. Sörgjerd, M. Tanaka, T. Saito, N. Iwata, T.C. Saido, Autophagy-related protein 7 deficiency in amyloid β (Aβ) precursor protein transgenic mice decreases Aβ in the multivesicular bodies and induces Aβ accumulation in the golgi, *Am. J. Pathol.* 185 (2015) 305–313. https://doi.org/10.1016/j.ajpath.2014.10.011.

[7] C. He, S. Zheng, Y. Luo, B. Wang, Exosome theranostics: Biology and translational medicine, *Theranostics* 8 (2018) 237–255. https://doi.org/10.7150/thno.21945.

[8] M. Cordonnier, G. Chanteloup, N. Isambert, R. Seigneuric, P. Fumoleau, C. Garrido, J. Gobbo, Exosomes in cancer theranostic: Diamonds in the rough, *Cell Adhes. Migr.* 11 (2017) 151–163. https://doi.org/10.1080/19336918.2016.1250999.

[9] R.L. Williams, S. Urbé, The emerging shape of the ESCRT machinery, *Nat. Rev. Mol. Cell Biol.* 8 (2007) 355–368. https://doi.org/10.1038/nrm2162.

[10] J.H. Hurley, ESCRT complexes and the biogenesis of multivesicular bodies, *Curr. Opin. Cell Biol.* 20 (2008) 4–11. https://doi.org/10.1016/j.ceb.2007.12.002.

[11] M. Colombo, G. Raposo, C. Théry, Biogenesis, secretion, and intercellular interactions of exosomes and other extracellular vesicles, *Annu. Rev. Cell Dev. Biol.* 30 (2014) 255–289. https://doi.org/10.1146/annurev-cellbio-101512-122326.

[12] H. Bu, D. He, X. He, K. Wang, Exosomes: Isolation, analysis, and applications in cancer detection and therapy, *Chembiochem* 20 (2019) 451–461. https://doi.org/10.1002/cbic.201800470.

[13] M.W.A. Hussain, S. Jahangir, B. Ghosh, F. Yesmin, A. Anis, S.N. Satil, F. Anwar, M.H. Rashid, Exosomes for regulation of immune responses and immunotherapy, *J. Nanotheranostics* 3 (2022) 55–85. https://doi.org/10.3390/jnt3010005.

[14] Q.Y. Ye, Y. Cui, H.Y. Wang, L.Y. Li, J.B. Chen, X.F. Zhu, Z.J. Xue, R.H. Zhang, Exosomal communication: A pivotal regulator of bone homeostasis and a potential therapeutic target, *Front. Pharmacol.* 15 (2024) 1–15. https://doi.org/10.3389/fphar.2024.1516125.

[15] Z. Wang, Z. Zhao, B. Gao, L. Zhang, Exosome mediated biological functions within skeletal microenvironment, *Front. Bioeng. Biotechnol.* 10 (2022) 1–11. https://doi.org/10.3389/fbioe.2022.953916.

[16] W. Qin, S.L. Dallas, Exosomes and extracellular RNA in muscle and bone aging and crosstalk, *Curr. Osteoporos. Rep.* 17 (2019) 548–559. https://doi.org/10.1007/s11914-019-00537-7.

[17] S. Mathivanan, H. Ji, R.J. Simpson, Exosomes: Extracellular organelles important in intercellular communication, *J. Proteomics* 73 (2010) 1907–1920. https://doi.org/10.1016/j.jprot.2010.06.006.

[18] R. Zhang, Y. Wei, T. Wang, X. Nie, Z. Shi, Y. Deng, D. Li, Exosomal miRNAs in autoimmune skin diseases, *Front. Immunol.* 14 (2023) 1–8. https://doi.org/10.3389/fimmu.2023.1307455.

[19] B.T. Pan, K. Teng, C. Wu, M. Adam, R.M. Johnstone, Electron microscopic evidence for externalization of the transferrin receptor in vesicular form in sheep reticulocytes, *J. Cell Biol.* 101 (1985) 942–948. https://doi.org/10.1083/jcb.101.3.942.

[20] G. Raposo, H.W. Nijman, W. Stoorvogel, R. Leijendekker, C. V. Harding, C.J.M. Melief, H.J. Geuze, B lymphocytes secrete antigen-presenting vesicles, *J. Exp. Med.* 183 (1996) 1161–1172. https://doi.org/10.1084/jem.183.3.1161.

[21] H. Valadi, K. Ekström, A. Bossios, M. Sjöstrand, J.J. Lee, J.O. Lötvall, Exosome-mediated transfer of mRNAs and microRNAs is a novel mechanism of genetic exchange between cells, *Nat. Cell Biol.* 9 (2007) 654–659. https://doi.org/10.1038/ncb1596.

[22] H. Kalra, R.J. Simpson, H. Ji, E. Aikawa, P. Altevogt, P. Askenase, V.C. Bond, F.E. Borràs, X. Breakefield, V. Budnik, E. Buzas, G. Camussi, A. Clayton, E. Cocucci, J.M. Falcon-Perez, S. Gabrielsson, Y.S. Gho, D. Gupta, H.C. Harsha, A. Hendrix, A.F. Hill, J.M. Inal, G. Jenster, E.M. Krämer-Albers, S.K. Lim, A. Llorente, J. Lötvall, A. Marcilla, L. Mincheva-Nilsson, I. Nazarenko, R. Nieuwland, E.N.M. Nolte-'t Hoen, A. Pandey, T. Patel, M.G. Piper, S. Pluchino, T.S.K. Prasad, L. Rajendran, G. Raposo, M. Record, G.E. Reid, F. Sánchez-Madrid, R.M. Schiffelers, P. Siljander, A. Stensballe, W. Stoorvogel, D. Taylor, C. Thery, H. Valadi, B.W.M. van Balkom, J. Vázquez, M. Vidal, M.H.M. Wauben, M. Yáñez-Mó, M. Zoeller, S. Mathivanan, Vesiclepedia: A compendium for extracellular vesicles with continuous community annotation, *PLoS Biol.* 10 (2012) 8–12. https://doi.org/10.1371/journal.pbio.1001450.

[23] J. Lötvall, A.F. Hill, F. Hochberg, E.I. Buzás, D. Di Vizio, C. Gardiner, Y.S. Gho, I. V. Kurochkin, S. Mathivanan, P. Quesenberry, S. Sahoo, H. Tahara, M.H. Wauben, K.W. Witwer, C. Théry, Minimal experimental requirements for definition of extracellular vesicles and their functions: A position statement from the International Society for Extracellular Vesicles, *J. Extracell. Vesicles* 3 (2014) 1–6. https://doi.org/10.3402/jev.v3.26913.

[24] R. Xu, A. Rai, M. Chen, W. Suwakulsiri, D.W. Greening, R.J. Simpson, Extracellular vesicles in cancer – implications for future improvements in cancer care, *Nat. Rev. Clin. Oncol.* 15 (2018) 617–638. https://doi.org/10.1038/s41571-018-0036-9.

[25] E. Cocucci, J. Meldolesi, Ectosomes and exosomes: Shedding the confusion between extracellular vesicles, *Trends Cell Biol.* 25 (2015) 364–372. https://doi.org/10.1016/j.tcb.2015.01.004.

[26] J. Huotari, A. Helenius, Endosome maturation, *EMBO J.* 30 (2011) 3481–3500. https://doi.org/10.1038/emboj.2011.286.

[27] B.A. Edgar, N. Zielke, C. Gutierrez, Endocycles: A recurrent evolutionary innovation for post-mitotic cell growth, *Nat. Rev. Mol. Cell Biol.* 15 (2014) 197–210. https://doi.org/10.1038/nrm3756.

[28] P.I. Hanson, A. Cashikar, Multivesicular body morphogenesis, *Annu. Rev. Cell Dev. Biol.* 28 (2012) 337–362. https://doi.org/10.1146/annurev-cellbio-092910-154152.
[29] S. Stuffers, C. Sem Wegner, H. Stenmark, A. Brech, Multivesicular endosome biogenesis in the absence of ESCRTs, *Traffic* 10 (2009) 925–937. https://doi.org/10.1111/j.1600-0854.2009.00920.x.
[30] T.S. Chen, R.C. Lai, M.M. Lee, A.B.H. Choo, C.N. Lee, S.K. Lim, Mesenchymal stem cell secretes microparticles enriched in pre-microRNAs, *Nucleic Acids Res.* 38 (2009) 215–224. https://doi.org/10.1093/nar/gkp857.
[31] L. Urbanelli, A. Magini, S. Buratta, A. Brozzi, K. Sagini, A. Polchi, B. Tancini, C. Emiliani, Signaling pathways in exosomes biogenesis, secretion and fate, *Genes (Basel).* 4 (2013) 152–170. https://doi.org/10.3390/genes4020152.
[32] R.J. Simpson, J.W. Lim, R.L. Moritz, S. Mathivanan, Exosomes: proteomic insights and diagnostic potential, *Expert Rev. Proteomics* 6 (2009) 267–283. https://doi.org/10.1586/epr.09.17.
[33] H. Zhao, A. Achreja, E. Iessi, M. Logozzi, D. Mizzoni, R. Di Raimo, D. Nagrath, S. Fais, The key role of extracellular vesicles in the metastatic process, *Biochim. Biophys. Acta Rev. Cancer* 1869 (2018) 64–77. https://doi.org/10.1016/j.bbcan.2017.11.005.
[34] Q.F. Han, W.J. Li, K.S. Hu, J. Gao, W.L. Zhai, J.H. Yang, S.J. Zhang, Exosome biogenesis: Machinery, regulation, and therapeutic implications in cancer, *Mol. Cancer* 21 (2022) 1–26. https://doi.org/10.1186/s12943-022-01671-0.
[35] J. Meldolesi, Exosomes and ectosomes in intercellular communication, *Curr. Biol.* 28 (2018) R435–R444. https://doi.org/10.1016/j.cub.2018.01.059.
[36] R.M. Johnstone, M. Adam, J.R. Hammond, L. Orr, C. Turbide, Vesicle formation during reticulocyte maturation. Association of plasma membrane activities with released vesicles (exosomes), *J. Biol. Chem.* 262 (1987) 9412–9420. https://doi.org/10.1016/s0021-9258(18)48095-7.
[37] C. Théry, L. Zitvogel, S. Amigorena, Exosomes: Composition, biogenesis and function, *Nat. Rev. Immunol.* 2 (2002) 569–579. https://doi.org/10.1038/nri855.
[38] J. Qin, Q. Xu, Functions and application of exosomes, *Acta Pol. Pharm.* 71 (2014) 537–543. PMID: 25272880.
[39] R. Kalluri, V.S. LeBleu, 外泌体的生物学、功能和生物医学应用, *Science (80-.).* 367 (2020). https://doi.org/10.1126/science.aau6977.The.
[40] M. Hazawa, K. Tomiyama, A. Saotome-Nakamura, C. Obara, T. Yasuda, T. Gotoh, I. Tanaka, H. Yakumaru, H. Ishihara, K. Tajima, Radiation increases the cellular uptake of exosomes through CD29/CD81 complex formation, *Biochem. Biophys. Res. Commun.* 446 (2014) 1165–1171. https://doi.org/10.1016/j.bbrc.2014.03.067.
[41] W. Zhang, R. Liu, Y. Chen, M. Wang, J. Du, Crosstalk between oxidative stress and exosomes, *Oxid. Med. Cell. Longev.* 2022 (2022). https://doi.org/10.1155/2022/3553617.
[42] L. Wang, L. Hu, X. Zhou, Z. Xiong, C. Zhang, H.M.A. Shehada, B. Hu, J. Song, L. Chen, Exosomes secreted by human adipose mesenchymal stem cells promote scarless cutaneous repair by regulating extracellular matrix remodelling, *Sci. Rep.* 7 (2017) 1–12. https://doi.org/10.1038/s41598-017-12919-x.
[43] J. De Toro, L. Herschlik, C. Waldner, C. Mongini, Emerging roles of exosomes in normal and pathological conditions: New insights for diagnosis and therapeutic applications, *Front. Immunol.* 6 (2015) 1–12. https://doi.org/10.3389/fimmu.2015.00203.
[44] M. Chivet, C. Javalet, F. Hemming, K. Pernet-Gallay, K. Laulagnier, S. Fraboulet, R. Sadoul, Exosomes as a novel way of interneuronal communication, *Biochem. Soc. Trans.* 41 (2013) 241–244. https://doi.org/10.1042/BST20120266.
[45] H. Sheldon, E. Heikamp, H. Turley, R. Dragovic, P. Thomas, C.E. Oon, R. Leek, M. Edelmann, B. Kessler, R.C.A. Sainson, I. Sargent, J.L. Li, A.L. Harris, New mechanism for Notch signaling to endothelium at a distance by delta-like 4 incorporation into exosomes, *Blood* 116 (2010) 2385–2394. https://doi.org/10.1182/blood-2009-08-239228.
[46] F.J. Barrat, K.B. Elkon, K.A. Fitzgerald, Importance of nucleic acid recognition in inflammation and autoimmunity, *Annu. Rev. Med.* 67 (2016) 323–336. https://doi.org/10.1146/annurev-med-052814-023338.
[47] V. Hornung, R. Hartmann, A. Ablasser, K.P. Hopfner, OAS proteins and cGAS: Unifying concepts in sensing and responding to cytosolic nucleic acids, *Nat. Rev. Immunol.* 14 (2014) 521–528. https://doi.org/10.1038/nri3719.
[48] S.R. Paludan, A.G. Bowie, K.A. Horan, K.A. Fitzgerald, Recognition of herpesviruses by the innate immune system, *Nat. Rev. Immunol.* 11 (2011) 143–154. https://doi.org/10.1038/nri2937.
[49] D. Torralba, F. Baixauli, C. Villarroya-Beltri, I. Fernández-Delgado, A. Latorre-Pellicer, R. Acín-Pérez, N.B. Martín-Cófreces, Á.L. Jaso-Tamame, S. Iborra, I. Jorge, G. González-Aseguinolaza, J. Garaude, M. Vicente-Manzanares, J.A. Enríquez, M. Mittelbrunn, F. Sánchez-Madrid, Priming of dendritic cells by DNA-containing extracellular vesicles from activated T cells through antigen-driven contacts, *Nat. Commun.* 9 (2018) 1–17. https://doi.org/10.1038/s41467-018-05077-9.
[50] A. Takahashi, R. Okada, K. Nagao, Y. Kawamata, A. Hanyu, S. Yoshimoto, M. Takasugi, S. Watanabe, M.T. Kanemaki, C. Obuse, E. Hara, Exosomes maintain cellular homeostasis by excreting harmful DNA from cells, *Nat. Commun.* 8 (2017) 1–14. https://doi.org/10.1038/ncomms15287.
[51] Y. Kitai, T. Kawasaki, T. Sueyoshi, K. Kobiyama, K.J. Ishii, J. Zou, S. Akira, T. Matsuda, T. Kawai, DNA-containing exosomes derived from cancer cells treated with topotecan activate a STING-dependent pathway and reinforce antitumor immunity, *J. Immunol.* 198 (2017) 1649–1659. https://doi.org/10.4049/jimmunol.1601694.
[52] P. Kurywchak, J. Tavormina, R. Kalluri, The emerging roles of exosomes in the modulation of immune responses in cancer, *Genome Med.* 10 (2018) 1–4. https://doi.org/10.1186/s13073-018-0535-4.
[53] S.R. Baglio, M.A.J. Van Eijndhoven, D. Koppers-Lalic, J. Berenguer, S.M. Lougheed, S. Gibbs, N. Léveillé, R.N.P.M. Rinkel, E.S. Hopmans, S. Swaminathan, S.A.W.M. Verkuijlen, G.L. Scheffer, F.J.M. Van Kuppeveld, T.D. De Gruijl, I.E.M. Bultink, E.S. Jordanova, M. Hackenberg, S.R. Piersma, J.C. Knol, A.E. Voskuyl, T. Wurdinger, C.R. Jiménez, J.M. Middeldorp, D.M. Pegtel, Sensing of latent EBV infection through exosomal transfer of 5'pppRNA, *Proc. Natl. Acad. Sci. USA.* 113 (2016) E587–E596. https://doi.org/10.1073/pnas.1518130113.
[54] M. Lundholm, M. Schröder, O. Nagaeva, V. Baranov, A. Widmark, L. Mincheva-Nilsson, P. Wikström, Prostate tumor-derived exosomes down-regulate NKG2D expression on natural killer cells and CD8+ T cells: Mechanism of immune evasion, *PLoS One* 9 (2014). https://doi.org/10.1371/journal.pone.0108925.
[55] F. Haderk, R. Schulz, M. Iskar, L.L. Cid, T. Worst, K. V Willmund, A. Schulz, U. Warnken, J. Seiler, A. Benner,

M. Nessling, T. Zenz, M. Göbel, J. Dürig, S. Diederichs, J. Paggetti, E. Moussay, S. Stilgenbauer, M. Zapatka, P. Lichter, M. Seiffert, Tumor-derived exosomes modulate PD-L1 expression in monocytes, *Sci. Immunol.* 2 (2017). https://doi.org/10.1126/sciimmunol.aah5509.

[56] H. Zhang, T. Deng, R. Liu, M. Bai, L. Zhou, X. Wang, S. Li, X. Wang, H. Yang, J. Li, T. Ning, Di. Huang, H. Li, L. Zhang, G. Ying, Y. Ba, Exosome-delivered EGFR regulates liver microenvironment to promote gastric cancer liver metastasis, *Nat. Commun.* 8 (2017). https://doi.org/10.1038/ncomms15016.

[57] D. Daassi, K.M. Mahoney, G.J. Freeman, The importance of exosomal PDL1 in tumour immune evasion, *Nat. Rev. Immunol.* 20 (2020) 209–215. https://doi.org/10.1038/s41577-019-0264-y.

[58] Y. Wang, Q. Lin, H. Zhang, S. Wang, J. Cui, Y. Hu, J. Liu, M. Li, K. Zhang, F. Zhou, Y. Jing, Z. Geng, J. Su, M2 macrophage-derived exosomes promote diabetic fracture healing by acting as an immunomodulator, *Bioact. Mater.* 28 (2023) 273–283. https://doi.org/10.1016/j.bioactmat.2023.05.018.

[59] M.S. Baig, A. Roy, S. Rajpoot, D. Liu, R. Savai, S. Banerjee, M. Kawada, S.M. Faisal, R. Saluja, U. Saqib, T. Ohishi, K.K. Wary, Tumor-derived exosomes in the regulation of macrophage polarization, *Inflamm. Res. Off. J. Eur. Histamine Res. Soc. . . . [et al.]* 69 (2020) 435–451. https://doi.org/10.1007/s00011-020-01318-0.

[60] Q. Yan, J. Liu, Y. Liu, Z. Wen, D. Jin, F. Wang, L. Gao, Tumor-associated macrophage-derived exosomal miR21-5p promotes tumor angiogenesis by regulating YAP1/HIF-1α axis in head and neck squamous cell carcinoma, *Cell. Mol. Life Sci.* 81 (2024) 179. https://doi.org/10.1007/s00018-024-05210-6.

[61] T. Yang, P. Martin, B. Fogarty, A. Brown, K. Schurman, R. Phipps, V.P. Yin, P. Lockman, S. Bai, Exosome delivered anticancer drugs across the blood-brain barrier for brain cancer therapy in *Danio rerio*, *Pharm. Res.* 32 (2015) 2003–2014. https://doi.org/10.1007/s11095-014-1593-y.

[62] S. Kamerkar, V.S. LeBleu, H. Sugimoto, S. Yang, C.F. Ruivo, S.A. Melo, J.J. Lee, R. Kalluri, Exosomes facilitate therapeutic targeting of oncogenic KRAS in pancreatic cancer, *Nature* 546 (2017) 498–503. https://doi.org/10.1038/nature22341.

[63] P. Guermonprez, J. Valladeau, L. Zitvogel, C. Théry, S. Amigorena, Antigen presentation and T cell stimulation by dendritic cells, *Annu. Rev. Immunol.* 20 (2002) 621–667. https://doi.org/10.1146/annurev.immunol.20.100301.064828.

[64] K. Liu, G.D. Victora, T.A. Schwickert, P. Guermonprez, M. Meredith, K. Yao, F. Chu, G.J. Randolph, Y. Alexander, M. Nussenzweig, *NIH Public Access* 324 (2010) 392–397. https://doi.org/10.1126/science.1170540.

[65] S.H. Naik, P. Sathe, H.-Y. Park, D. Metcalf, A.I. Proietto, A. Dakic, S. Carotta, M. O'Keeffe, M. Bahlo, A. Papenfuss, J.-Y. Kwak, L. Wu, K. Shortman, Development of plasmacytoid and conventional dendritic cell subtypes from single precursor cells derived in vitro and in vivo, *Nat. Immunol.* 8 (2007) 1217–1226. https://doi.org/10.1038/ni1522.

[66] C. Théry, M. Boussac, P. Véron, P. Ricciardi-Castagnoli, G. Raposo, J. Garin, S. Amigorena, Proteomic analysis of dendritic cell-derived exosomes: A secreted subcellular compartment distinct from apoptotic vesicles, *J. Immunol.* 166 (2001) 7309–7318. https://doi.org/10.4049/jimmunol.166.12.7309.

[67] N. Chaput, J. Taïeb, N. Schartz, C. Flament, S. Novault, F. André, L. Zitvogel, The potential of exosomes in immunotherapy of cancer, *Blood Cells Mol. Dis.* 35 (2005) 111–115. https://doi.org/10.1016/j.bcmd.2005.05.009.

[68] S. Viaud, C. Théry, S. Ploix, T. Tursz, V. Lapierre, O. Lantz, L. Zitvogel, N. Chaput, Dendritic cell-derived exosomes for cancer immunotherapy: What's next? *Cancer Res.* 70 (2010) 1281–1285. https://doi.org/10.1158/0008-5472.CAN-09-3276.

[69] S.H. Kim, N.R. Bianco, W.J. Shufesky, A.E. Morelli, P.D. Robbins, MHC class II+ exosomes in plasma suppress inflammation in an antigen-specific and Fas ligand/Fas-dependent manner, *J. Immunol.* 179 (2007) 2235–2241. https://doi.org/10.4049/jimmunol.179.4.2235.

[70] D.G. Meckes, N. Raab-Traub, Microvesicles and viral infection, *J. Virol.* 85 (2011) 12844–12854. https://doi.org/10.1128/jvi.05853-11.

[71] R.D. Wiley, S. Gummuluru, Immature dendritic cell-derived exosomes can mediate HIV-1 trans infection, *Proc. Natl. Acad. Sci. USA.* 103 (2006) 738–743. https://doi.org/10.1073/pnas.0507995103.

[72] C. Di Vito, J. Mikulak, D. Mavilio, On the way to become a natural killer cell, *Front. Immunol.* 10 (2019) 1–15. https://doi.org/10.3389/fimmu.2019.01812.

[73] G. Habif, A. Crinier, P. André, E. Vivier, E. Narni-Mancinelli, Targeting natural killer cells in solid tumors, *Cell. Mol. Immunol.* 16 (2019) 415–422. https://doi.org/10.1038/s41423-019-0224-2.

[74] Q. Li, Q. Huang, T. Huyan, Y. Wang, Q. Huang, J. Shi, Bifacial effects of engineering tumour cell-derived exosomes on human natural killer cells, *Exp. Cell Res.* 363 (2018) 141–150. https://doi.org/10.1016/j.yexcr.2017.12.005.

[75] Z. Wang, J.Q. Chen, J. lu Liu, L. Tian, Exosomes in tumor microenvironment: Novel transporters and biomarkers, *J. Transl. Med.* 14 (2016) 1–9. https://doi.org/10.1186/s12967-016-1056-9.

[76] U. Gehrmann, T.I. Näslund, S. Hiltbrunner, P. Larssen, S. Gabrielsson, Harnessing the exosome-induced immune response for cancer immunotherapy, *Semin. Cancer Biol.* 28 (2014) 58–67. https://doi.org/10.1016/j.semcancer.2014.05.003.

[77] S. Munich, A. Sobo-Vujanovic, W.J. Buchser, D. Beer-Stolz, N.L. Vujanovic, Dendritic cell exosomes directly kill tumor cells and activate natural killer cells via TNF superfamily ligands, *Oncoimmunology* 1 (2012) 1074–1083. https://doi.org/10.4161/onci.20897.

[78] Y. Jiang, H. Jiang, K. Wang, C. Liu, X. Man, Q. Fu, Hypoxia enhances the production and antitumor effect of exosomes derived from natural killer cells, *Ann. Transl. Med.* 9 (2021) 473–473. https://doi.org/10.21037/atm-21-347.

[79] A. Shoae-Hassani, A.A. Hamidieh, M. Behfar, R. Mohseni, S.A. Mortazavi-Tabatabaei, S. Asgharzadeh, NK cell-derived exosomes from NK cells previously exposed to neuroblastoma cells augment the antitumor activity of cytokine-activated NK cells, *J. Immunother.* 40 (2017) 265–276. https://doi.org/10.1097/CJI.0000000000000179.

[80] E. McMullan, D. Joladarashi, R. Kishore, Unpacking exosomes: A therapeutic frontier for cardiac repair, *Curr. Cardiol. Rep.* 27 (2025). https://doi.org/10.1007/s11886-025-02225-8.

[81] A. Anel, A. Gallego-Lleyda, D. de Miguel, J. Naval, L. Martínez-Lostao, Role of exosomes in the regulation of t-cell mediated immune responses and in autoimmune disease, *Cells* 8 (2019). https://doi.org/10.3390/cells8020154.

[82] J.M. Pitt, F. André, S. Amigorena, J.C. Soria, A. Eggermont, G. Kroemer, L. Zitvogel, Dendritic cell-derived exosomes for cancer therapy, *J. Clin. Invest.* 126 (2016) 1224–1232. https://doi.org/10.1172/JCI81137.

[83] M. Nakayama, Antigen presentation by MHC-dressed cells, *Front. Immunol.* 5 (2014) 1–8. https://doi.org/10.3389/fimmu.2014.00672.
[84] J. Ruiz-Navarro, V. Calvo, M. Izquierdo, Extracellular vesicles and microvilli in the immune synapse, *Front. Immunol.* 14 (2023) 1–17. https://doi.org/10.3389/fimmu.2023.1324557.
[85] S.L. Schell, Z.S.M. Rahman, miRNA-Mediated control of B cell responses in immunity and SLE, *Front. Immunol.* 12 (2021) 1–20. https://doi.org/10.3389/fimmu.2021.683710.
[86] J. Pérez-Boza, D.M. Pegtel, Exosomes take (germinal) center stage, *EMBO Rep.* 21 (2020) 1–3. https://doi.org/10.15252/embr.202050190.
[87] W.J. Chen, TGF-β regulation of T cells, *Annu. Rev. Immunol.* 41 (2023) 483–512. https://doi.org/10.1146/annurev-immunol-101921-045939.
[88] A. Hazrati, S. Soudi, K. Malekpour, M. Mahmoudi, A. Rahimi, S.M. Hashemi, R.S. Varma, Immune cells-derived exosomes function as a double-edged sword: Role in disease progression and their therapeutic applications, *Biomark. Res.* 10 (2022) 1–25. https://doi.org/10.1186/s40364-022-00374-4.
[89] M. Capello, J. V. Vykoukal, H. Katayama, L.E. Bantis, H. Wang, D.L. Kundnani, C. Aguilar-Bonavides, M. Aguilar, S.C. Tripathi, D.S. Dhillon, A.A. Momin, H. Peters, M.H. Katz, H. Alvarez, V. Bernard, S. Ferri-Borgogno, R. Brand, D.G. Adler, M.A. Firpo, S.J. Mulvihill, J.J. Molldrem, Z. Feng, A. Taguchi, A. Maitra, S.M. Hanash, Exosomes harbor B cell targets in pancreatic adenocarcinoma and exert decoy function against complement-mediated cytotoxicity, *Nat. Commun.* 10 (2019). https://doi.org/10.1038/s41467-018-08109-6.
[90] C. Ni, Q.Q. Fang, W.Z. Chen, J.X. Jiang, Z. Jiang, J. Ye, T. Zhang, L. Yang, F.B. Meng, W.J. Xia, M. Zhong, J. Huang, Breast cancer-derived exosomes transmit lncRNA SNHG16 to induce CD73+γδ1 treg cells, *Signal Transduct. Target. Ther.* 5 (2020). https://doi.org/10.1038/s41392-020-0129-7.
[91] A.M. Zonouz, M.G. Rahbardar, M. Alibolandi, Exosome-based platforms for treatment of multiple sclerosis, *Brain Res. Bull.* 222 (2025) 111256. https://doi.org/10.1016/j.brainresbull.2025.111256.
[92] C. Macedo, L.M. Tran, A.F. Zahorchak, H. Dai, X. Gu, R. Ravichandran, T. Mohanakumar, B. Elinoff, M.A. Styn, A. Humar, F.G. Lakkis, D.M. Metes, W. Angus, *HHS Public Access*, 21 (2022) 2372–2386. https://doi.org/10.1111/ajt.16393.Donor-derived.
[93] Y. Hong Zhang, J. Chen, W. Fei Mao, L. Xi Yan, F. Long, S. Hao Li, R. Xian Zhang, Harnessing miR-16-5p-loaded exosomes from adipose-derived stem cells to restore immune homeostasis in SLE patients, *Immunobiology* 230 (2025) 152885. https://doi.org/10.1016/j.imbio.2025.152885.
[94] S. Zhou, J. Huang, Y. Zhang, H. Yu, X. Wang, Exosomes in action: Unraveling their role in autoimmune diseases and exploring potential therapeutic applications, *Immune Netw.* 24 (2024) 1–17. https://doi.org/10.4110/in.2024.24.e12.
[95] K. Bai, C.L. Lee, X. Liu, J. Li, D. Cao, L. Zhang, D. Hu, H. Li, Y. Hou, Y. Xu, A.S.Y. Kan, K.W. Cheung, E.H.Y. Ng, W.S.B. Yeung, P.C.N. Chiu, Human placental exosomes induce maternal systemic immune tolerance by reprogramming circulating monocytes, *J. Nanobiotechnol.* 20 (2022) 1–16. https://doi.org/10.1186/s12951-022-01283-2.
[96] Q. Sun, H. Chang, H. Wang, L. Zheng, Y. Weng, D. Zheng, D. Zheng, Regulatory roles of extracellular vesicles in pregnancy complications, *J. Adv. Res.* (2025). https://doi.org/10.1016/j.jare.2025.02.010.
[97] S.Q. Cao, T.X. Jiang, Y.Y. Guo, R. Lin, L. Lin, MiR-519d-3p from placenta-derived exosomes induce immune intolerance regulating immune cells, contributing to the pathogenesis of preeclampsia, *Immunol. Invest.* 00 (2025) 1–22. https://doi.org/10.1080/08820139.2025.2450234.
[98] K. Bai, X. Li, J. Zhong, E.H.Y. Ng, W.S.B. Yeung, C.L. Lee, P.C.N. Chiu, Placenta-derived exosomes as a modulator in maternal immune tolerance during pregnancy, *Front. Immunol.* 12 (2021) 671093. https://doi.org/10.3389/fimmu.2021.671093.
[99] E. Delorme-Axford, R.B. Donker, J.F. Mouillet, T. Chu, A. Bayer, Y. Ouyang, T. Wang, D.B. Stolz, S.N. Sarkar, A.E. Morelli, Y. Sadovsky, C.B. Coyne, Human placental trophoblasts confer viral resistance to recipient cells, *Proc. Natl. Acad. Sci. USA.* 110 (2013) 12048–12053. https://doi.org/10.1073/pnas.1304718110.
[100] N.G. Than, A. Balogh, R. Romero, E. Kárpáti, O. Erez, A. Szilágyi, I. Kovalszky, M. Sammar, S. Gizurarson, J. Matkó, P. Závodszky, Z. Papp, H. Meiri, Placental protein 13 (PP13) – a placental immunoregulatory galectin protecting pregnancy, *Front. Immunol.* 5 (2014) 348. https://doi.org/10.3389/fimmu.2014.00348.
[101] F. Tan, X. Li, Z. Wang, J. Li, K. Shahzad, J. Zheng, Clinical applications of stem cell-derived exosomes, *Signal Transduct. Target. Ther.* 9 (2024). https://doi.org/10.1038/s41392-023-01704-0.
[102] G. Benichou, H.H. Lancia, Intercellular transfer of MHC molecules in T cell alloimmunity and allotransplantation, *Biomed. J.* 47 (2024) 100749. https://doi.org/10.1016/j.bj.2024.100749.
[103] Z. Shafeghat, M. Dorfaki, S. Dehrouyeh, F.L. Arab, M. Roozbehani, R. Falak, F. Faraji, R. Jafari, Mesenchymal stem cell-derived exosomes for managing graft-versus-host disease: An updated view, *Transpl. Immunol.* 81 (2023) 101957. https://doi.org/10.1016/j.trim.2023.101957.
[104] N. Sailliet, M. Ullah, A. Dupuy, A.K.A. Silva, F. Gazeau, H. Le Mai, S. Brouard, Extracellular vesicles in transplantation, *Front. Immunol.* 13 (2022). https://doi.org/10.3389/fimmu.2022.800018.
[105] M. Budoni, A. Fierabracci, R. Luciano, S. Petrini, V. Di Ciommo, M. Muraca, The immunosuppressive effect of mesenchymal stromal cells on B lymphocytes is mediated by membrane vesicles, *Cell Transplant.* 22 (2013) 369–379. https://doi.org/10.3727/096368911x582769b.
[106] Y. Ji, L. Mi, M. Zhao, X. He, Y. Hu, Y. Gao, C. Yin, K. Xu, Innovative diagnosis and therapeutic modalities: Engineered exosomes in autoimmune disease, *Int. J. Nanomed.* 19 (2024) 3943–3956. https://doi.org/10.2147/IJN.S452184.
[107] J. Dieker, J. Tel, E. Pieterse, A. Thielen, N. Rother, M. Bakker, J. Fransen, H.B.P.M. Dijkman, J.H. Berden, J.M. de Vries, L.B. Hilbrands, J. van der Vlag, Circulating apoptotic microparticles in systemic lupus erythematosus patients drive the activation of dendritic cell subsets and prime neutrophils for NETosis, *Arthritis Rheumatol. (Hoboken, N.J.)* 68 (2016) 462–472. https://doi.org/10.1002/art.39417.
[108] X. Kou, X. Xu, C. Chen, M.L. Sanmillan, T. Cai, Y. Zhou, C. Giraudo, A. Le, S. Shi, The Fas/Fap-1/Cav-1 complex regulates IL-1RA secretion in mesenchymal stem cells to accelerate wound healing, *Sci. Transl. Med.* 10 (2018) eaai8524. https://doi.org/10.1126/scitranslmed.aai8524.
[109] D. Li, C. Zhang, Z. Gao, N. Xia, C. Wu, C. Liu, H. Tian, X. Mei, Curcumin-loaded macrophage-derived exosomes effectively improve wound healing, *Mol. Pharm.* 20 (2023) 4453–4467. https://doi.org/10.1021/acs.molpharmaceut.3c00062.

[110] M. Yáñez-Mó, P.R.M. Siljander, Z. Andreu, A.B. Zavec, F.E. Borràs, E.I. Buzas, K. Buzas, E. Casal, F. Cappello, J. Carvalho, E. Colás, A. Cordeiro-Da Silva, S. Fais, J.M. Falcon-Perez, I.M. Ghobrial, B. Giebel, M. Gimona, M. Graner, I. Gursel, M. Gursel, N.H.H. Heegaard, A. Hendrix, P. Kierulf, K. Kokubun, M. Kosanovic, V. Kralj-Iglic, E.M. Krämer-Albers, S. Laitinen, C. Lässer, T. Lener, E. Ligeti, A. Line, G. Lipps, A. Llorente, J. Lötvall, M. Mančček-Keber, A. Marcilla, M. Mittelbrunn, I. Nazarenko, E.N.M. Nolte-'t Hoen, T.A. Nyman, L. O'Driscoll, M. Olivan, C. Oliveira, É. Pállinger, H.A. Del Portillo, J. Reventós, M. Rigau, E. Rohde, M. Sammar, F. Sánchez-Madrid, N. Santarém, K. Schallmoser, M.S. Ostenfeld, W. Stoorvogel, R. Stukelj, S.G. Van Der Grein, M. Helena Vasconcelos, M.H.M. Wauben, O. De Wever, Biological properties of extracellular vesicles and their physiological functions, *J. Extracell. Vesicles* 4 (2015) 1–60. https://doi.org/10.3402/jev.v4.27066.

[111] S. Miaomiao, W. Xiaoqian, S. Yuwei, C. Chao, Y. Chenbo, L. Yinghao, H. Yichen, S. Jiao, C. Kuisheng, Cancer-associated fibroblast-derived exosome microRNA-21 promotes angiogenesis in multiple myeloma, *Sci. Rep.* 13 (2023) 1–13. https://doi.org/10.1038/s41598-023-36092-6.

[112] D. Yu, Y. Li, M. Wang, J. Gu, W. Xu, H. Cai, X. Fang, X. Zhang, Exosomes as a new frontier of cancer liquid biopsy, *Mol. Cancer* 21 (2022) 56. https://doi.org/10.1186/s12943-022-01509-9.

[113] W. Olejarz, G. Kubiak-Tomaszewska, A. Chrzanowska, T. Lorenc, Exosomes in angiogenesis and anti-angiogenic therapy in cancers, *Int. J. Mol. Sci.* 21 (2020) 1–25. https://doi.org/10.3390/ijms21165840.

[114] H. Luo, H. Zhang, J. Mao, H. Cao, Y. Tao, G. Zhao, Z. Zhang, N. Zhang, Z. Liu, J. Zhang, P. Luo, Y. Xia, Y. Cheng, Z. Xie, Q. Cheng, G. Liu, Exosome-based nano-immunotherapy targeting TAMs, a promising strategy for glioma, *Cell Death Dis.* 14 (2023). https://doi.org/10.1038/s41419-023-05753-9.

[115] C. Jiang, N. Zhang, X. Hu, H. Wang, Tumor-associated exosomes promote lung cancer metastasis through multiple mechanisms, *Mol. Cancer* 20 (2021) 1–17. https://doi.org/10.1186/s12943-021-01411-w.

[116] A. Jafari, K. Karimabadi, A. Rahimi, G. Rostaminasab, M. Khazaei, L. Rezakhani, T. Ahmadi jouybari, The emerging role of exosomal miRNAs as biomarkers for early cancer detection: A comprehensive literature review, *Technol. Cancer Res. Treat.* 22 (2023). https://doi.org/10.1177/15330338231205999.

[117] H. Li, Screening and identification of serum exosomal protein ZNF587B in liquid biopsy for ovarian cancer diagnosis, *Am. J. Cancer Res.* 14 (2024) 1904–1913. https://doi.org/10.62347/rbtm1834.

[118] E. Gheytanchi, F. Tajik, M. Razmi, S. Babashah, W.C.S. Cho, K. Tanha, M. Sahlolbei, R. Ghods, Z. Madjd, Circulating exosomal microRNAs as potential prognostic biomarkers in gastrointestinal cancers: A systematic review and meta-analysis, *Cancer Cell Int.* 23 (2023). https://doi.org/10.1186/s12935-023-02851-8.

[119] A.A. Saleh, S.E. Soliman, M.S.E.-D. Habib, S.F. Gohar, G.S. Abo-Zeid, Potential value of circulatory microRNA122 gene expression as a prognostic and metastatic prediction marker for breast cancer, *Mol. Biol. Rep.* 46 (2019) 2809–2818. https://doi.org/10.1007/s11033-019-04727-5.

[120] F. Lan, Q. Qing, Q. Pan, M. Hu, H. Yu, X. Yue, Serum exosomal miR-301a as a potential diagnostic and prognostic biomarker for human glioma, *Cell. Oncol. (Dordrecht, Netherlands)* 41 (2018) 25–33. https://doi.org/10.1007/s13402-017-0355-3.

[121] Y. Duan, M. Zhou, B. Ye, K. Yue, F. Qiao, Y. Wang, Q. Lai, Y. Wu, J. Cao, Y. Wu, X. Wang, C. Jing, Hypoxia-induced miR-5100 promotes exosome-mediated activation of cancer-associated fibroblasts and metastasis of head and neck squamous cell carcinoma, *Cell Death Dis.* 15 (2024). https://doi.org/10.1038/s41419-024-06587-9.

[122] Y. Shen, M. Li, L. Liao, S. Gao, Y. Wang, Plasma exosome-derived connexin43 as a promising biomarker for melanoma patients, *BMC Cancer* 23 (2023) 1–8. https://doi.org/10.1186/s12885-023-10705-9.

[123] Y. Bao, S. Xu, J. Zhou, C. Zhao, S. Dai, Y. Zhang, M. Rao, Exosomal miR-93 derived from hepatocellular carcinoma cell promotes the sorafenib resistance of hepatocellular carcinoma through PTEN/PI3K/Akt pathway, *J. Biochem. Mol. Toxicol.* 38 (2024) e23666. https://doi.org/10.1002/jbt.23666.

[124] B. Mateescu, E.J.K. Kowal, B.W.M. van Balkom, S. Bartel, S.N. Bhattacharyya, E.I. Buzás, A.H. Buck, P. de Candia, F.W.N. Chow, S. Das, T.A.P. Driedonks, L. Fernández-Messina, F. Haderk, A.F. Hill, J.C. Jones, K.R. Van Keuren-Jensen, C.P. Lai, C. Lässer, I. di Liegro, T.R. Lunavat, M.J. Lorenowicz, S.L.N. Maas, I. Mäger, M. Mittelbrunn, S. Momma, K. Mukherjee, M. Nawaz, D.M. Pegtel, M.W. Pfaffl, R.M. Schiffelers, H. Tahara, C. Théry, J.P. Tosar, M.H.M. Wauben, K.W. Witwer, E.N.M. Nolte-'t Hoen, Obstacles and opportunities in the functional analysis of extracellular vesicle RNA – an ISEV position paper, *J. Extracell. Vesicles* 6 (2017). https://doi.org/10.1080/20013078.2017.1286095.

[125] K.W. Witwer, E.I. Buzás, L.T. Bemis, A. Bora, C. Lässer, J. Lötvall, E.N. Nolte-'t Hoen, M.G. Piper, S. Sivaraman, J. Skog, C. Théry, M.H. Wauben, F. Hochberg, Standardization of sample collection, isolation and analysis methods in extracellular vesicle research, *J. Extracell. Vesicles* 2 (2013) 1–25. https://doi.org/10.3402/jev.v2i0.20360.

[126] GBD 2013 Risk Factors Collaborators, M.H. Forouzanfar, L. Alexander, H.R. Anderson, V.F. Bachman, S. Biryukov, M. Brauer, R. Burnett, D. Casey, M.M. Coates, A. Cohen, K. Delwiche, K. Estep, J.J. Frostad, K.C. Astha, H.H. Kyu, M. Moradi-Lakeh, M. Ng, E.L. Slepak, B.A. Thomas, J. Wagner, G.M. Aasvang, C. Abbafati, A. Abbasoglu Ozgoren, F. Abd-Allah, S.F. Abera, V. Aboyans, B. Abraham, J.P. Abraham, I. Abubakar, N.M. Abu-Rmeileh, T.C. Aburto, T. Achoki, A. Adelekan, K. Adofo, A.K. Adou, J.C. Adsuar, A. Afshin, E.E. Agardh, M.J. Al Khabouri, F.H. Al Lami, S.S. Alam, D. Alasfoor, M.I. Albittar, M.A. Alegretti, A.V. Aleman, Z.A. Alemu, R. Alfonso-Cristancho, S. Alhabib, R. Ali, M.K. Ali, F. Alla, P. Allebeck, P.J. Allen, U. Alsharif, E. Alvarez, N. Alvis-Guzman, A.A. Amankwaa, A.T. Amare, E.A. Ameh, O. Ameli, H. Amini, W. Ammar, B.O. Anderson, C.A. Antonio, P. Anwari, S. Argeseanu Cunningham, J. Arnlöv, V.S. Arsenijevic, A. Artaman, R.J. Asghar, R. Assadi, L.S. Atkins, C. Atkinson, M.A. Avila, B. Awuah, A. Badawi, M.C. Bahit, T. Bakfalouni, K. Balakrishnan, S. Balalla, R.K. Balu, A. Banerjee, R.M. Barber, S.L. Barker-Collo, S. Barquera, L. Barregard, L.H. Barrero, T. Barrientos-Gutierrez, A.C. Basto-Abreu, A. Basu, S. Basu, M.O. Basulaiman, C. Batis Ruvalcaba, J. Beardsley, N. Bedi, T. Bekele, M.L. Bell, C. Benjet, D.A. Bennett, H. Benzian, E. Bernabé, T.J. Beyene, N. Bhala, A. Bhalla, Z.A. Bhutta, B. Bikbov, A.A. Bin Abdulhak, J.D. Blore, F.M.

Blyth, M.A. Bohensky, B. Bora Başara, G. Borges, N.M. Bornstein, D. Bose, S. Boufous, R.R. Bourne, M. Brainin, A. Brazinova, N.J. Breitborde, H. Brenner, A.D. Briggs, D.M. Broday, P.M. Brooks, N.G. Bruce, T.S. Brugha, B. Brunekreef, R. Buchbinder, L.N. Bui, G. Bukhman, A.G. Bulloch, M. Burch, P.G. Burney, I.R. Campos-Nonato, J.C. Campuzano, A.J. Cantoral, J. Caravanos, R. Cárdenas, E. Cardis, D.O. Carpenter, V. Caso, C.A. Castañeda-Orjuela, R.E. Castro, F. Catalá-López, F. Cavalleri, A. Çavlin, V.K. Chadha, J.C. Chang, F.J. Charlson, H. Chen, W. Chen, Z. Chen, P.P. Chiang, O. Chimed-Ochir, R. Chowdhury, C.A. Christophi, T.W. Chuang, S.S. Chugh, M. Cirillo, T.K. Claßen, V. Colistro, M. Colomar, S.M. Colquhoun, A.G. Contreras, C. Cooper, K. Cooperrider, L.T. Cooper, J. Coresh, K.J. Courville, M.H. Criqui, L. Cuevas-Nasu, J. Damsere-Derry, H. Danawi, L. Dandona, R. Dandona, P.I. Dargan, A. Davis, D.V. Davitoiu, A. Dayama, E.F. de Castro, V. De la Cruz-Góngora, D. De Leo, G. de Lima, L. Degenhardt, B. del Pozo-Cruz, R.P. Dellavalle, K. Deribe, S. Derrett, D.C. Des Jarlais, M. Dessalegn, G.A. deVeber, K.M. Devries, S.D. Dharmaratne, M.K. Dherani, D. Dicker, E.L. Ding, K. Dokova, E.R. Dorsey, T.R. Driscoll, L. Duan, A.M. Durrani, B.E. Ebel, R.G. Ellenbogen, Y.M. Elshrek, M. Endres, S.P. Ermakov, H.E. Erskine, B. Eshrati, A. Esteghamati, S. Fahimi, E.J. Faraon, F. Farzadfar, D.F. Fay, V.L. Feigin, A.B. Feigl, S.M. Fereshtehnejad, A.J. Ferrari, C.P. Ferri, A.D. Flaxman, T.D. Fleming, N. Foigt, K.J. Foreman, U.F. Paleo, R.C. Franklin, B. Gabbe, L. Gaffikin, E. Gakidou, A. Gamkrelidze, F.G. Gankpé, R.T. Gansevoort, F.A. García-Guerra, E. Gasana, J.M. Geleijnse, B.D. Gessner, P. Gething, K.B. Gibney, R.F. Gillum, I.A. Ginawi, M. Giroud, G. Giussani, S. Goenka, K. Goginashvili, H. Gomez Dantes, P. Gona, T. Gonzalez de Cosio, D. González-Castell, C.C. Gotay, A. Goto, H.N. Gouda, R.L. Guerrant, H.C. Gugnani, F. Guillemin, D. Gunnell, R. Gupta, R. Gupta, R.A. Gutiérrez, N. Hafezi-Nejad, H. Hagan, M. Hagstromer, Y.A. Halasa, R.R. Hamadeh, M. Hammami, G.J. Hankey, Y. Hao, H.L. Harb, T.N. Haregu, J.M. Haro, R. Havmoeller, S.I. Hay, M.T. Hedayati, I.B. Heredia-Pi, L. Hernandez, K.R. Heuton, P. Heydarpour, M. Hijar, H.W. Hoek, H.J. Hoffman, J.C. Hornberger, H.D. Hosgood, D.G. Hoy, M. Hsairi, G. Hu, H. Hu, C. Huang, J.J. Huang, B.J. Hubbell, L. Huiart, A. Husseini, M.L. Iannarone, K.M. Iburg, B.T. Idrisov, N. Ikeda, K. Innos, M. Inoue, F. Islami, S. Ismayilova, K.H. Jacobsen, H.A. Jansen, D.L. Jarvis, S.K. Jassal, A. Jauregui, S. Jayaraman, P. Jeemon, P.N. Jensen, V. Jha, F. Jiang, G. Jiang, Y. Jiang, J.B. Jonas, K. Juel, H. Kan, S.S. Kany Roseline, N.E. Karam, A. Karch, C.K. Karema, G. Karthikeyan, A. Kaul, N. Kawakami, D.S. Kazi, A.H. Kemp, A.P. Kengne, A. Keren, Y.S. Khader, S.E. Khalifa, E.A. Khan, Y.H. Khang, S. Khatibzadeh, I. Khonelidze, C. Kieling, D. Kim, S. Kim, Y. Kim, R.W. Kimokoti, Y. Kinfu, J.M. Kinge, B.M. Kissela, M. Kivipelto, L.D. Knibbs, A.K. Knudsen, Y. Kokubo, M.R. Kose, S. Kosen, A. Kraemer, M. Kravchenko, S. Krishnaswami, H. Kromhout, T. Ku, B. Kuate Defo, B. Kucuk Bicer, E.J. Kuipers, C. Kulkarni, V.S. Kulkarni, G.A. Kumar, G.F. Kwan, T. Lai, A. Lakshmana Balaji, R. Lalloo, T. Lallukka, H. Lam, Q. Lan, V.C. Lansingh, H.J. Larson, A. Larsson, D.O. Laryea, P.M. Lavados, A.E. Lawrynowicz, J.L. Leasher, J.T. Lee, J. Leigh, R. Leung, M. Levi, Y. Li, Y. Li, J. Liang, X. Liang, S.S. Lim, M.P. Lindsay, S.E. Lipshultz, S. Liu, Y. Liu, B.K. Lloyd, G. Logroscino, S.J. London, N. Lopez, J. Lortet-Tieulent, P.A. Lotufo, R. Lozano, R. Lunevicius, J. Ma, S. Ma, V.M. Machado, M.F. MacIntyre, C. Magis-Rodriguez, A.A. Mahdi, M. Majdan, R. Malekzadeh, S. Mangalam, C.C. Mapoma, M. Marape, W. Marcenes, D.J. Margolis, C. Margono, G.B. Marks, R.V. Martin, M.B. Marzan, M.T. Mashal, F. Masiye, A.J. Mason-Jones, K. Matsushita, R. Matzopoulos, B.M. Mayosi, T.T. Mazorodze, A.C. McKay, M. McKee, A. McLain, P.A. Meaney, C. Medina, M.M. Mehndiratta, F. Mejia-Rodriguez, W. Mekonnen, Y.A. Melaku, M. Meltzer, Z.A. Memish, W. Mendoza, G.A. Mensah, A. Meretoja, F.A. Mhimbira, R. Micha, T.R. Miller, E.J. Mills, A. Misganaw, S. Mishra, N. Mohamed Ibrahim, K.A. Mohammad, A.H. Mokdad, G.L. Mola, L. Monasta, J.C. Montañez Hernandez, M. Montico, A.R. Moore, L. Morawska, R. Mori, J. Moschandreas, W.N. Moturi, D. Mozaffarian, U.O. Mueller, M. Mukaigawara, E.C. Mullany, K.S. Murthy, M. Naghavi, Z. Nahas, A. Naheed, K.S. Naidoo, L. Naldi, D. Nand, V. Nangia, K.M. Narayan, D. Nash, B. Neal, C. Nejjari, S.P. Neupane, C.R. Newton, F.N. Ngalesoni, Jde D. Ngirabega, G. Nguyen, N.T. Nguyen, M.J. Nieuwenhuijsen, M.I. Nisar, J.R. Nogueira, J.M. Nolla, S. Nolte, O.F. Norheim, R.E. Norman, B. Norrving, L. Nyakarahuka, I.H. Oh, T. Ohkubo, B.O. Olusanya, S.B. Omer, J.N. Opio, R. Orozco, R.S. Pagcatipunan Jr., A.W. Pain, J.D. Pandian, C.I. Panelo, C. Papachristou, E.K. Park, C.D. Parry, A.J. Paternina Caicedo, S.B. Patten, V.K. Paul, B.I. Pavlin, N. Pearce, L.S. Pedraza, A. Pedroza, L. Pejin Stokic, A. Pekericli, D.M. Pereira, R. Perez-Padilla, F. Perez-Ruiz, N. Perico, S.A. Perry, A. Pervaiz, K. Pesudovs, C.B. Peterson, M. Petzold, M.R. Phillips, H.P. Phua, D. Plass, D. Poenaru, G.V. Polanczyk, S. Polinder, C.D. Pond, C.A. Pope, D. Pope, S. Popova, F. Pourmalek, J. Powles, D. Prabhakaran, N.M. Prasad, D.M. Qato, A.D. Quezada, D.A. Quistberg, L. Racapé, A. Rafay, K. Rahimi, V. Rahimi-Movaghar, S.U. Rahman, M. Raju, I. Rakovac, S.M. Rana, M. Rao, H. Razavi, K.S. Reddy, A.H. Refaat, J. Rehm, G. Remuzzi, A.L. Ribeiro, P.M. Riccio, L. Richardson, A. Riederer, M. Robinson, A. Roca, A. Rodriguez, D. Rojas-Rueda, I. Romieu, L. Ronfani, R. Room, N. Roy, G.M. Ruhago, L. Rushton, N. Sabin, R.L. Sacco, S. Saha, R. Sahathevan, M.A. Sahraian, J.A. Salomon, D. Salvo, U.K. Sampson, J.R. Sanabria, L.M. Sanchez, T.G. Sánchez-Pimienta, L. Sanchez-Riera, L. Sandar, I.S. Santos, A. Sapkota, M. Satpathy, J.E. Saunders, M. Sawhney, M.I. Saylan, P. Scarborough, J.C. Schmidt, I.J. Schneider, B. Schöttker, D.C. Schwebel, J.G. Scott, S. Seedat, S.G. Sepanlou, B. Serdar, E.E. Servan-Mori, G. Shaddick, S. Shahraz, T.S. Levy, S. Shangguan, J. She, S. Sheikhbahaei, K. Shibuya, H.H. Shin, Y. Shinohara, R. Shiri, K. Shishani, I. Shiue, I.D. Sigfusdottir, D.H. Silberberg, E.P. Simard, S. Sindi, A. Singh, G.M. Singh, J.A. Singh, V. Skirbekk, K. Sliwa, M. Soljak, S. Soneji, K. Søreide, S. Soshnikov, L.A. Sposato, C.T. Sreeramareddy, N.J. Stapelberg, V. Stathopoulou, N. Steckling, D.J. Stein, M.B. Stein, N. Stephens, H. Stöckl, K. Straif, K. Stroumpoulis, L. Sturua, B.F. Sunguya, S. Swaminathan, M. Swaroop, B.L. Sykes, K.M. Tabb, K. Takahashi, R.T. Talongwa, N. Tandon, D. Tanne, M. Tanner, M. Tavakkoli, B.J. Te Ao, C.M. Teixeira, M.M. Téllez Rojo, A.S. Terkawi, J.L. Texcalac-Sangrador, S.V. Thackway, B. Thomson, A.L. Thorne-Lyman, A.G. Thrift, G.D. Thurston, T. Tillmann, M. Tobollik, M. Tonelli, F. Topouzis, J.A. Towbin, H. Toyoshima, J. Traebert, B.X. Tran, L. Trasande, M. Trillini, U. Trujillo, Z.T. Dimbuene, M. Tsilimbaris, E.M. Tuzcu, U.S. Uchendu, K.N. Ukwaja, S.B. Uzun, S. van de Vijver, R.

Van Dingenen, C.H. van Gool, J. van Os, Y.Y. Varakin, T.J. Vasankari, A.M. Vasconcelos, M.S. Vavilala, L.J. Veerman, G. Velasquez-Melendez, N. Venketasubramanian, L. Vijayakumar, S. Villalpando, F.S. Violante, V.V. Vlassov, S.E. Vollset, G.R. Wagner, S.G. Waller, M.T. Wallin, X. Wan, H. Wang, J. Wang, L. Wang, W. Wang, Y. Wang, T.S. Warouw, C.H. Watts, S. Weichenthal, E. Weiderpass, R.G. Weintraub, A. Werdecker, K.R. Wessells, R. Westerman, H.A. Whiteford, J.D. Wilkinson, H.C. Williams, T.N. Williams, S.M. Woldeyohannes, C.D. Wolfe, J.Q. Wong, A.D. Woolf, J.L. Wright, B. Wurtz, G. Xu, L.L. Yan, G. Yang, Y. Yano, P. Ye, M. Yenesew, G.K. Yentür, P. Yip, N. Yonemoto, S.J. Yoon, M.Z. Younis, Z. Younoussi, C. Yu, M.E. Zaki, Y. Zhao, Y. Zheng, M. Zhou, J. Zhu, S. Zhu, X. Zou, J.R. Zunt, A.D. Lopez, T. Vos, C.J.L. Murray, Global, regional, and national comparative risk assessment of 79 behavioural, environmental and occupational, and metabolic risks or clusters of risks in 188 countries, 1990–2013: a systematic analysis for the Global Burden of Disease Study 2013, *Lancet* 386 (2015) 2287–2323. https://doi.org/10.1016/S0140-6736(15)00128-2.

[127] B.L. Chen, J.B. Guo, M.S. Liu, X. Li, J. Zou, X. Chen, L.L. Zhang, Y.S. Yue, X.Q. Wang, Effect of traditional Chinese exercise on gait and balance for stroke: A systematic review and meta-analysis, *PLoS One* 10 (2015) 1–14. https://doi.org/10.1371/journal.pone.0135932.

[128] H. Peinado, M. Alečković, S. Lavotshkin, I. Matei, B. Costa-Silva, G. Moreno-Bueno, M. Hergueta-Redondo, C. Williams, G. García-Santos, C. Ghajar, A. Nitadori-Hoshino, C. Hoffman, K. Badal, B.A. Garcia, M.K. Callahan, J. Yuan, V.R. Martins, J. Skog, R.N. Kaplan, M.S. Brady, J.D. Wolchok, P.B. Chapman, Y. Kang, J. Bromberg, D. Lyden, Melanoma exosomes educate bone marrow progenitor cells toward a pro-metastatic phenotype through MET, *Nat. Med.* 18 (2012) 883–891. https://doi.org/10.1038/nm.2753.

[129] L. Manterola, E. Guruceaga, J.G. Pérez-Larraya, M. González-Huarriz, P. Jauregui, S. Tejada, R. Diez-Valle, V. Segura, N. Samprón, C. Barrena, I. Ruiz, A. Agirre, Á. Ayuso, J. Rodríguez, Á. González, E. Xipell, A. Matheu, A. López De Munain, T. Tuñón, I. Zazpe, J. García-Foncillas, S. Paris, J.Y. Delattre, M.M. Alonso, A small noncoding RNA signature found in exosomes of GBM patient serum as a diagnostic tool, *Neuro. Oncol.* 16 (2014) 520–527. https://doi.org/10.1093/neuonc/not218.

[130] Y. Li, Q. Zheng, C. Bao, S. Li, W. Guo, J. Zhao, D. Chen, J. Gu, X. He, S. Huang, Circular RNA is enriched and stable in exosomes: A promising biomarker for cancer diagnosis, *Cell Res.* 25 (2015) 981–984. https://doi.org/10.1038/cr.2015.82.

[131] H. Im, K. Lee, R. Weissleder, H. Lee, C.M. Castro, Novel nanosensing technologies for exosome detection and profiling, *Lab Chip* 17 (2017) 2892–2898. https://doi.org/10.1039/c7lc00247e.

[132] G. Siravegna, S. Marsoni, S. Siena, A. Bardelli, Integrating liquid biopsies into the management of cancer, *Nat. Rev. Clin. Oncol.* 14 (2017) 531–548. https://doi.org/10.1038/nrclinonc.2017.14.

[133] Y. Dou, D.J. Cha, J.L. Franklin, J.N. Higginbotham, D.K. Jeppesen, A.M. Weaver, N. Prasad, S. Levy, R.J. Coffey, J.G. Patton, B. Zhang, Circular RNAs are down-regulated in KRAS mutant colon cancer cells and can be transferred to exosomes, *Sci. Rep.* 6 (2016) 1–11. https://doi.org/10.1038/srep37982.

[134] D. Ishwar, R. Haldavnekar, S. Das, B. Tan, K. Venkatakrishnan, Glioblastoma associated natural killer cell EVs generating tumour-specific signatures: Noninvasive GBM liquid biopsy with self-functionalized quantum probes, *ACS Nano* 16 (2022) 10859–10877. https://doi.org/10.1021/acsnano.2c03055.

[135] H. Chi, X. Xie, Y. Yan, G. Peng, D.F. Strohmer, G. Lai, S. Zhao, Z. Xia, G. Tian, Natural killer cell-related prognosis signature characterizes immune landscape and predicts prognosis of HNSCC, *Front. Immunol.* 13 (2022) 1–18. https://doi.org/10.3389/fimmu.2022.1018685.

[136] Y.T. Kang, Z. Niu, T. Hadlock, E. Purcell, T.W. Lo, M. Zeinali, S. Owen, V.G. Keshamouni, R. Reddy, N. Ramnath, S. Nagrath, On-chip biogenesis of circulating NK cell-derived exosomes in non-small cell lung cancer exhibits antitumoral activity, *Adv. Sci.* 8 (2021) 1–13. https://doi.org/10.1002/advs.202003747.

[137] L. Tan, M. Zhao, H. Wu, Y. Zhang, X. Tong, L. Gao, L. Zhou, Q. Lu, J. Zeng, Downregulated serum exosomal miR-451a expression correlates with renal damage and its intercellular communication role in systemic lupus erythematosus, *Front. Immunol.* 12 (2021) 1–14. https://doi.org/10.3389/fimmu.2021.630112.

[138] E. Garcia-vives, C. Solé, T. Moliné, M. Vidal, I. Agraz, J. Ordi-ros, J. Cortés-hernández, The urinary exosomal miRNA expression profile is predictive of clinical response in lupus nephritis, *Int. J. Mol. Sci.* 21 (2020). https://doi.org/10.3390/ijms21041372.

[139] L. Huang, L. Liang, Z. Ji, S. Chen, M. Liu, Q. Huang, Z. Huang, S. Sun, J. Ding, J. Chen, X. Huang, S. Zheng, W. Deng, Y. Huang, T. Li, Proteomics profiling of CD4 + T-cell-derived exosomes from patients with rheumatoid arthritis, *Int. Immunopharmacol.* 122 (2023) 110560. https://doi.org/10.1016/j.intimp.2023.110560.

[140] Z.-Q. Shuai, Z.-X. Wang, J.-L. Ren, X.-K. Yang, B. Xu, Differential expressions and potential clinical values of lncRNAs in the plasma exosomes of rheumatoid arthritis, *Int. Immunopharmacol.* 128 (2024) 111511. https://doi.org/10.1016/j.intimp.2024.111511.

[141] T.L. Liao, I.C. Chen, H.W. Chen, K.T. Tang, W.N. Huang, Y.H. Chen, Y.M. Chen, Exosomal microRNAs as biomarkers for viral replication in tofacitinib-treated rheumatoid arthritis patients with hepatitis C, *Sci. Rep.* 14 (2024) 1–10. https://doi.org/10.1038/s41598-023-50963-y.

[142] G. Galazka, M.P. Mycko, I. Selmaj, C.S. Raine, K.W. Selmaj, Multiple sclerosis: Serum-derived exosomes express myelin proteins, *Mult. Scler.* 24 (2018) 449–458. https://doi.org/10.1177/1352458517696597.

[143] G. Torres Iglesias, M. Fernández-Fournier, L. Botella, D. Piniella, F. Laso-García, M. Carmen Gómez-de Frutos, B. Chamorro, I. Puertas, A. Tallón Barranco, B. Fuentes, M. Alonso de Leciñana, E. Alonso-López, S.B. Bravo, M. Eugenia Miranda-Carús, A. Montero-Calle, R. Barderas, E. Díez-Tejedor, M. Gutiérrez-Fernández, L. Otero-Ortega, Brain and immune system-derived extracellular vesicles mediate regulation of complement system, extracellular matrix remodeling, brain repair and antigen tolerance in multiple sclerosis, *Brain. Behav. Immun.* 113 (2023) 44–55. https://doi.org/10.1016/j.bbi.2023.06.025.

[144] S. Arora, N. Verma, Exosomal microRNAs as potential biomarkers and therapeutic targets in corneal diseases, *Mol. Vis.* 30 (2024) 92–106. https://pubmed.ncbi.nlm.nih.gov/38601014.

[145] L. Jia, Q. Qiu, H. Zhang, L. Chu, Y. Du, J. Zhang, C. Zhou, F. Liang, S. Shi, S. Wang, W. Qin, Q. Wang, F. Li, Q. Wang, Y. Li, L. Shen, Y. Wei, J. Jia, Concordance between the assessment of Aβ42, T-tau, and P-T181-tau in peripheral blood neuronal-derived exosomes and cerebrospinal fluid, *Alzheimer's Dement.* 15 (2019) 1071–1080. https://doi.org/10.1016/j.jalz.2019.05.002.

[146] M.S. Fiandaca, D. Kapogiannis, M. Mapstone, A. Boxer, E. Eitan, J.B. Schwartz, E.L. Abner, R.C. Petersen, H.J. Federoff, B.L. Miller, E.J. Goetzl, Identification of preclinical Alzheimer's disease by a profile of pathogenic proteins in neurally derived blood exosomes: a case-control study, *Alzheimers Dement.* 11 (2015) 600–607.e1. https://doi.org/10.1016/j.jalz.2014.06.008.

[147] A. Kluge, J. Bunk, E. Schaeffer, A. Drobny, W. Xiang, H. Knacke, S. Bub, W. Lückstädt, P. Arnold, R. Lucius, D. Berg, F. Zunke, Detection of neuron-derived pathological α-synuclein in blood, *Brain* 145 (2022) 3058–3071. https://doi.org/10.1093/brain/awac115.

[148] S. Yan, C. Jiang, A. Janzen, T.R. Barber, A. Seger, M. Sommerauer, J.J. Davis, K. Marek, M.T. Hu, W.H. Oertel, G.K. Tofaris, Neuronally derived extracellular vesicle α-synuclein as a serum biomarker for individuals at risk of developing Parkinson disease, *JAMA Neurol.* 81 (2024) 59–68. https://doi.org/10.1001/jamaneurol.2023.4398.

[149] Y. Wang, Y. Liu, Z. Jin, C. Liu, X. Yu, K. Chen, D. Meng, A. Liu, B. Fang, Association between mitochondrial function and rehabilitation of Parkinson's disease: Revealed by exosomal mRNA and lncRNA expression profiles, *Front. Aging Neurosci.* 14 (2022). https://doi.org/10.3389/fnagi.2022.909622.

[150] M.J. Mitchell, M.M. Billingsley, R.M. Haley, M.E. Wechsler, N.A. Peppas, R. Langer, Engineering precision nanoparticles for drug delivery, *Nat. Rev. Drug Discov.* 20 (2021) 101–124. https://doi.org/10.1038/s41573-020-0090-8.

[151] S. Sen, J. Xavier, N. Kumar, M.Z. Ahmad, O.P. Ranjan, Exosomes as natural nanocarrier-based drug delivery system: Recent insights and future perspectives, *3 Biotech.* 13 (2023) 1–25. https://doi.org/10.1007/s13205-023-03521-2.

[152] M. Qu, Q. Lin, L. Huang, Y. Fu, L. Wang, S. He, Y. Fu, S. Yang, Z. Zhang, L. Zhang, X. Sun, Dopamine-loaded blood exosomes targeted to brain for better treatment of Parkinson's disease, *J. Control. Release Off. J. Control. Release Soc.* 287 (2018) 156–166. https://doi.org/10.1016/j.jconrel.2018.08.035.

[153] S. Kalimuthu, P. Gangadaran, R.L. Rajendran, L. Zhu, J.M. Oh, H.W. Lee, A. Gopal, S.H. Baek, S.Y. Jeong, S.W. Lee, J. Lee, B.C. Ahn, A new approach for loading anticancer drugs into mesenchymal stem cell-derived exosome mimetics for cancer therapy, *Front. Pharmacol.* 9 (2018) 1–10. https://doi.org/10.3389/fphar.2018.01116.

[154] X. Zhang, L. Liu, M. Tang, H. Li, X. Guo, X. Yang, The effects of umbilical cord-derived macrophage exosomes loaded with cisplatin on the growth and drug resistance of ovarian cancer cells, *Drug Dev. Ind. Pharm.* 46 (2020) 1150–1162. https://doi.org/10.1080/03639045.2020.1776320.

[155] D.N. Kumar, A. Chaudhuri, D. Dehari, A.M. Gamper, D. Kumar, A.K. Agrawal, Enhanced therapeutic efficacy against melanoma through exosomal delivery of hesperidin, *Mol. Pharm.* 21 (2024) 3061–3076. https://doi.org/10.1021/acs.molpharmaceut.4c00490.

[156] X. Luan, K. Sansanaphongpricha, I. Myers, H. Chen, H. Yuan, D. Sun, Engineering exosomes as refined biological nanoplatforms for drug delivery, *Acta Pharmacol. Sin.* 38 (2017) 754–763. https://doi.org/10.1038/aps.2017.12.

[157] S. Sadeghi, F.R. Tehrani, S. Tahmasebi, A. Shafiee, S.M. Hashemi, Exosome engineering in cell therapy and drug delivery, *Inflammopharmacology* 31 (2023) 145–169. https://doi.org/10.1007/s10787-022-01115-7.

[158] W. Nan, C. Zhang, H. Wang, H. Chen, S. Ji, Direct modification of extracellular vesicles and its applications for cancer therapy: A mini-review, *Front. Chem.* 10 (2022) 1–7. https://doi.org/10.3389/fchem.2022.910341.

[159] Q. Liu, D. Li, X. Pan, Y. Liang, Targeted therapy using engineered extracellular vesicles: Principles and strategies for membrane modification, *J. Nanobiotechnol.* 21 (2023) 1–28. https://doi.org/10.1186/s12951-023-02081-0.

[160] S.-F. Liang, F.-F. Zuo, B.-C. Yin, B.-C. Ye, Delivery of siRNA based on engineered exosomes for glioblastoma therapy by targeting STAT3, *Biomater. Sci.* 10 (2022) 1582–1590. https://doi.org/10.1039/d1bm01723c.

[161] Z. Zhang, X. Luo, X. Xue, M. Pang, X. Wang, L. Yu, J. Qian, X. Li, M. Tian, A. Lu, C. Lu, Y. Liu, Engineered exosomes carrying miR-588 for treatment of triple negative breast cancer through remodeling the immunosuppressive microenvironment, *Int. J. Nanomedicine* 19 (2024) 743–758. https://doi.org/10.2147/IJN.S440619.

[162] T. Tian, H.-X. Zhang, C.-P. He, S. Fan, Y.-L. Zhu, C. Qi, N.-P. Huang, Z.-D. Xiao, Z.-H. Lu, B.A. Tannous, J. Gao, Surface functionalized exosomes as targeted drug delivery vehicles for cerebral ischemia therapy, *Biomaterials* 150 (2018) 137–149. https://doi.org/10.1016/j.biomaterials.2017.10.012.

[163] H. Zhang, J. Wu, J. Wu, Q. Fan, J. Zhou, J. Wu, S. Liu, J. Zang, J. Ye, M. Xiao, T. Tian, J. Gao, Exosome-mediated targeted delivery of miR-210 for angiogenic therapy after cerebral ischemia in mice, *J. Nanobiotechnol.* 17 (2019) 1–13. https://doi.org/10.1186/s12951-019-0461-7.

[164] Z. Zhao, T. Shuang, Y. Gao, F. Lu, J. Zhang, W. He, L. Qu, B. Chen, Q. Hao, Targeted delivery of exosomal miR-484 reprograms tumor vasculature for chemotherapy sensitization, *Cancer Lett.* 530 (2022) 45–58. https://doi.org/10.1016/j.canlet.2022.01.011.

[165] Y. Liang, L. Duan, J. Lu, J. Xia, Engineering exosomes for targeted drug delivery, *Theranostics* 11 (2021) 3183–3195. https://doi.org/10.7150/THNO.52570.

[166] Y. Hong, G.H. Nam, E. Koh, S. Jeon, G.B. Kim, C. Jeong, D.H. Kim, Y. Yang, I.S. Kim, Correction to: Exosome as a vehicle for delivery of membrane protein therapeutics, PH20, for enhanced tumor penetration and antitumor efficacy (*Advanced Functional Materials*, (2018), 28, 5, (1703074), 10.1002/adfm.201703074), *Adv. Funct. Mater.* 28 (2018) 2010292. https://doi.org/10.1002/adfm.201801301.

[167] A.J. Vázquez-Ríos, Á. Molina-Crespo, B.L. Bouzo, R. López-López, G. Moreno-Bueno, M. de la Fuente, Exosome-mimetic nanoplatforms for targeted cancer drug delivery, *J. Nanobiotechnol.* 17 (2019) 85. https://doi.org/10.1186/s12951-019-0517-8.

[168] Y.-Y. Jhan, D. Prasca-Chamorro, G. Palou Zuniga, D.M. Moore, S. Arun Kumar, A.K. Gaharwar, C.J. Bishop, Engineered extracellular vesicles with synthetic lipids via membrane fusion to establish efficient gene delivery, *Int. J. Pharm.* 573 (2020) 118802. https://doi.org/10.1016/j.ijpharm.2019.118802.

[169] A. Srivastava, N. Amreddy, V. Pareek, M. Chinnappan, R. Ahmed, M. Mehta, M. Razaq, A. Munshi, R. Ramesh, Progress in extracellular vesicle biology and their application in cancer medicine, *Wiley Interdiscip. Rev. Nanomed.*

Nanobiotechnol. 12 (2020) 1–20. https://doi.org/10.1002/wnan.1621.

[170] D. Jafari, S. Shajari, R. Jafari, N. Mardi, H. Gomari, F. Ganji, M. Forouzandeh Moghadam, A. Samadikuchaksaraei, Designer exosomes: A new platform for biotechnology therapeutics, *BioDrugs* 34 (2020) 567–586. https://doi.org/10.1007/s40259-020-00434-x.

[171] M.S. Kim, M.J. Haney, Y. Zhao, D. Yuan, I. Deygen, N.L. Klyachko, A.V. Kabanov, E.V. Batrakova, Engineering macrophage-derived exosomes for targeted paclitaxel delivery to pulmonary metastases: In vitro and in vivo evaluations, *Nanomedicine* 14 (2018) 195–204. https://doi.org/10.1016/j.nano.2017.09.011.

[172] H. Qi, C. Liu, L. Long, Y. Ren, S. Zhang, X. Chang, X. Qian, H. Jia, J. Zhao, J. Sun, X. Hou, X. Yuan, C. Kang, Blood exosomes endowed with magnetic and targeting properties for cancer therapy, *ACS Nano* 10 (2016) 3323–3333. https://doi.org/10.1021/acsnano.5b06939.

[173] A. Vandergriff, K. Huang, D. Shen, S. Hu, M.T. Hensley, T.G. Caranasos, L. Qian, K. Cheng, Targeting regenerative exosomes to myocardial infarction using cardiac homing peptide, *Theranostics* 8 (2018) 1869–1878. https://doi.org/10.7150/thno.20524.

[174] J. Bai, J. Duan, R. Liu, Y. Du, Q. Luo, Y. Cui, Z. Su, J. Xu, Y. Xie, W. Lu, Engineered targeting tLyp-1 exosomes as gene therapy vectors for efficient delivery of siRNA into lung cancer cells, *Asian J. Pharm. Sci.* 15 (2020) 461–471. https://doi.org/10.1016/j.ajps.2019.04.002.

[175] K.L. Zhang, Y.J. Wang, J. Sun, J. Zhou, C. Xing, G. Huang, J. Li, H. Yang, Artificial chimeric exosomes for anti-phagocytosis and targeted cancer therapy, *Chem. Sci.* 10 (2019) 1555–1561. https://doi.org/10.1039/c8sc03224f.

[176] G. Wang, W. Hu, H. Chen, X. Shou, T. Ye, Y. Xu, Cocktail strategy based on NK cell-derived exosomes and their biomimetic nanoparticles for dual tumor therapy, *Cancers (Basel).* 11 (2019). https://doi.org/10.3390/cancers11101560.

[177] Q. Cheng, Z. Dai, G. Smbatyan, A.L. Epstein, H.J. Lenz, Y. Zhang, Eliciting anti-cancer immunity by genetically engineered multifunctional exosomes, *Mol. Ther.* 30 (2022) 3066–3077. https://doi.org/10.1016/j.ymthe.2022.06.013.

[178] X. Wang, J. Xia, L. Yang, J. Dai, L. He, Recent progress in exosome research: Isolation, characterization and clinical applications, *Cancer Gene Ther.* 30 (2023) 1051–1065. https://doi.org/10.1038/s41417-023-00617-y.

[179] F. Yang, M. Wang, X. Guan, Exosomes and mimics as novel delivery platform for cancer therapy, *Front. Pharmacol.* 13 (2022) 1–9. https://doi.org/10.3389/fphar.2022.1001417.

[180] W. Xiao, W. Dong, C. Zhang, G. Saren, P. Geng, H. Zhao, Q. Li, J. Zhu, G. Li, S. Zhang, M. Ye, Effects of the epigenetic drug MS-275 on the release and function of exosome-related immune molecules in hepatocellular carcinoma cells, *Eur. J. Med. Res.* 18 (2013) 1–7. https://doi.org/10.1186/2047-783X-18-61.

[181] W. Zhou, Y. Zhou, X. Chen, T. Ning, H. Chen, Q. Guo, Y. Zhang, P. Liu, Y. Zhang, C. Li, Y. Chu, T. Sun, C. Jiang, Pancreatic cancer-targeting exosomes for enhancing immunotherapy and reprogramming tumor microenvironment, *Biomaterials* 268 (2021) 120546. https://doi.org/10.1016/j.biomaterials.2020.120546.

[182] C. Liu, X. Liu, X. Xiang, X. Pang, S. Chen, Y. Zhang, E. Ren, L. Zhang, X. Liu, P. Lv, X. Wang, W. Luo, N. Xia, X. Chen, G. Liu, A nanovaccine for antigen self-presentation and immunosuppression reversal as a personalized cancer immunotherapy strategy, *Nat. Nanotechnol.* 17 (2022) 531–540. https://doi.org/10.1038/s41565-022-01098-0.

[183] C.D. Phung, T.T. Pham, H.T. Nguyen, T.T. Nguyen, W. Ou, J.-H. Jeong, H.-G. Choi, S.K. Ku, C.S. Yong, J.O. Kim, Anti-CTLA-4 antibody-functionalized dendritic cell-derived exosomes targeting tumor-draining lymph nodes for effective induction of antitumor T-cell responses, *Acta Biomater.* 115 (2020) 371–382. https://doi.org/10.1016/j.actbio.2020.08.008.

[184] F. Pi, D.W. Binzel, T.J. Lee, Z. Li, M. Sun, P. Rychahou, H. Li, F. Haque, S. Wang, C.M. Croce, B. Guo, B.M. Evers, P. Guo, Nanoparticle orientation to control RNA loading and ligand display on extracellular vesicles for cancer regression, *Nat. Nanotechnol.* 13 (2018) 82–89. https://doi.org/10.1038/s41565-017-0012-z.

[185] N. Shimasaki, E. Coustan-Smith, T. Kamiya, D. Campana, Expanded and armed natural killer cells for cancer treatment, *Cytotherapy* 18 (2016) 1422–1434. https://doi.org/10.1016/j.jcyt.2016.06.013.

[186] L. Zhu, S. Kalimuthu, P. Gangadaran, J.M. Oh, H.W. Lee, S.H. Baek, S.Y. Jeong, S.W. Lee, J. Lee, B.C. Ahn, Exosomes derived from natural killer cells exert therapeutic effect in melanoma, *Theranostics* 7 (2017) 2732–2745. https://doi.org/10.7150/thno.18752.

[187] L. Cheng, Y. Wang, L. Huang, Exosomes from M1-polarized macrophages potentiate the cancer vaccine by creating a pro-inflammatory microenvironment in the lymph node, *Mol. Ther.* 25 (2017) 1665–1675. https://doi.org/10.1016/j.ymthe.2017.02.007.

[188] G. Chen, A.C. Huang, W. Zhang, G. Zhang, M. Wu, W. Xu, Z. Yu, J. Yang, B. Wang, H. Sun, H. Xia, Q. Man, W. Zhong, L.F. Antelo, B. Wu, X. Xiong, X. Liu, L. Guan, T. Li, S. Liu, R. Yang, Y. Lu, L. Dong, S. McGettigan, R. Somasundaram, R. Radhakrishnan, G. Mills, Y. Lu, J. Kim, Y.H. Chen, H. Dong, Y. Zhao, G.C. Karakousis, T.C. Mitchell, L.M. Schuchter, M. Herlyn, E.J. Wherry, X. Xu, W. Guo, Exosomal PD-L1 contributes to immunosuppression and is associated with anti-PD-1 response, *Nature* 560 (2018) 382–386. https://doi.org/10.1038/s41586-018-0392-8.

[189] M. Poggio, T. Hu, C.C. Pai, B. Chu, C.D. Belair, A. Chang, E. Montabana, U.E. Lang, Q. Fu, L. Fong, R. Blelloch, Suppression of exosomal PD-L1 induces systemic anti-tumor immunity and memory, *Cell* 177 (2019) 414–427.e13. https://doi.org/10.1016/j.cell.2019.02.016.

[190] I.L. Colao, R. Corteling, D. Bracewell, I. Wall, Manufacturing exosomes: A promising therapeutic platform, *Trends Mol. Med.* 24 (2018) 242–256. https://doi.org/10.1016/j.molmed.2018.01.006.

[191] D. Yang, W. Zhang, H. Zhang, F. Zhang, L. Chen, L. Ma, L.M. Larcher, S. Chen, N. Liu, Q. Zhao, P.H.L. Tran, C. Chen, R.N. Veedu, T. Wang, Progress, opportunity, and perspective on exosome isolation – efforts for efficient exosome-based theranostics, *Theranostics* 10 (2020) 3684–3707. https://doi.org/10.7150/thno.41580.

[192] Y. Lin, Y. Lu, X. Li, Biological characteristics of exosomes and genetically engineered exosomes for the targeted delivery of therapeutic agents, *J. Drug Target.* 28 (2020) 129–141. https://doi.org/10.1080/1061186X.2019.1641508.

[193] C.K. Das, B.C. Jena, I. Banerjee, S. Das, A. Parekh, S.K. Bhutia, M. Mandal, Exosome as a novel shuttle for delivery of therapeutics across biological barriers, *Mol. Pharm.* 16 (2019) 24–40. https://doi.org/10.1021/acs.molpharmaceut.8b00901.

[194] M. Cully, Exosome-based candidates move into the clinic, *Nat. Rev. Drug Discov.* 20 (2021) 6–7. https://doi.org/10.1038/d41573-020-00220-y.

[195] H. Chen, L. Wang, X. Zeng, H. Schwarz, H.S. Nanda, X. Peng, Y. Zhou, Exosomes, a new star for targeted delivery, *Front. Cell Dev. Biol.* 9 (2021). https://doi.org/10.3389/fcell.2021.751079.

[196] T.L. Whiteside, Tumor-derived exosomes and their role in cancer progression, *Adv. Clin. Chem.* 74 (2016) 103–141. https://doi.org/10.1016/bs.acc.2015.12.005.

[197] S. Araujo-Abad, M. Saceda, C. de Juan Romero, Biomedical application of small extracellular vesicles in cancer treatment, *Adv. Drug Deliv. Rev.* 182 (2022) 114117. https://doi.org/10.1016/j.addr.2022.114117.

[198] G.E. Melling, E. Carollo, R. Conlon, J.C. Simpson, D.R.F. Carter, The challenges and possibilities of extracellular vesicles as therapeutic vehicles, *Eur. J. Pharm. Biopharm. Off. J. Arbeitsgemeinschaft Fur Pharm. Verfahrenstechnik e.V* 144 (2019) 50–56. https://doi.org/10.1016/j.ejpb.2019.08.009.

[199] R. Munagala, F. Aqil, J. Jeyabalan, R.C. Gupta, Bovine milk-derived exosomes for drug delivery, *Cancer Lett.* 371 (2016) 48–61. https://doi.org/10.1016/j.canlet.2015.10.020.

[200] H. Saari, E. Lázaro-Ibáñez, T. Viitala, E. Vuorimaa-Laukkanen, P. Siljander, M. Yliperttula, Microvesicle- and exosome-mediated drug delivery enhances the cytotoxicity of Paclitaxel in autologous prostate cancer cells, *J. Control. Release Off. J. Control. Release Soc.* 220 (2015) 727–737. https://doi.org/10.1016/j.jconrel.2015.09.031.

[201] F. Aqil, H. Kausar, A.K. Agrawal, J. Jeyabalan, A.-H. Kyakulaga, R. Munagala, R. Gupta, Exosomal formulation enhances therapeutic response of celastrol against lung cancer, *Exp. Mol. Pathol.* 101 (2016) 12–21. https://doi.org/10.1016/j.yexmp.2016.05.013.

[202] F. Aqil, R. Munagala, J. Jeyabalan, A.K. Agrawal, R. Gupta, Exosomes for the enhanced tissue bioavailability and efficacy of curcumin, *AAPS J.* 19 (2017) 1691–1702. https://doi.org/10.1208/s12248-017-0154-9.

[203] R. Munagala, F. Aqil, J. Jeyabalan, A.K. Agrawal, A.M. Mudd, A.H. Kyakulaga, I.P. Singh, M.V. Vadhanam, R.C. Gupta, Exosomal formulation of anthocyanidins against multiple cancer types, *Cancer Lett.* 393 (2017) 94–102. https://doi.org/10.1016/j.canlet.2017.02.00.

[204] M.A. Morse, J. Garst, T. Osada, S. Khan, A. Hobeika, T.M. Clay, N. Valente, R. Shreeniwas, M.A. Sutton, A. Delcayre, D.H. Hsu, J.B. Le Pecq, H.K. Lyerly, A phase I study of dexosome immunotherapy in patients with advanced non-small cell lung cancer, *J. Transl. Med.* 3 (2005) 1–8. https://doi.org/10.1186/1479-5876-3-9.

[205] M. Wu, G. Wang, W. Hu, Y. Yao, X.F. Yu, Emerging roles and therapeutic value of exosomes in cancer metastasis, *Mol. Cancer* 18 (2019) 1–11. https://doi.org/10.1186/s12943-019-0964-8.

[206] K.D.P. Dorayappan, J.J. Wallbillich, D.E. Cohn, K. Selvendiran, The biological significance and clinical applications of exosomes in ovarian cancer, *Gynecol. Oncol.* 142 (2016) 199–205. https://doi.org/10.1016/j.ygyno.2016.03.036.

[207] C.S. Jo, C.H. Myung, Y.C. Yoon, B.H. Ahn, J.W. Min, W.S. Seo, D.H. Lee, H.C. Kang, Y.H. Heo, H. Choi, I.K. Hong, J.S. Hwang, The effect of lactobacillus plantarum extracellular vesicles from Korean women in their 20s on skin aging, *Curr. Issues Mol. Biol.* 44 (2022) 526–540. https://doi.org/10.3390/cimb44020036.

[208] E. Mahdipour, Beta vulgaris juice contains biologically active exosome-like nanoparticles, *Tissue Cell* 76 (2022) 101800. https://doi.org/10.1016/j.tice.2022.101800.

[209] S.E. Headland, H.R. Jones, L. V Norling, A. Kim, P.R. Souza, E. Corsiero, C.D. Gil, A. Nerviani, F.D. Accio, C. Pitzalis, S.M. Oliani, L.Y. Jan, M. Perretti, Joint in inflammatory arthritis, *Sci. Transl. Med.* 7 (2015). https://doi.org/10.1126/scitranslmed.aac5608.Neutrophil-derived.

[210] C. Zheng, L. Xie, H. Qin, X. Liu, X. Chen, F. Lv, L. Wang, X. Zhu, J. Xu, The role of extracellular vesicles in systemic lupus erythematosus, *Front. Cell Dev. Biol.* 10 (2022). https://doi.org/10.3389/fcell.2022.835566.

[211] M. Sáenz-Cuesta, I. Osorio-Querejeta, D. Otaegui, Extracellular vesicles in multiple sclerosis: What are they telling us? *Front. Cell. Neurosci.* 8 (2014) 1–9. https://doi.org/10.3389/fncel.2014.00100.

[212] B. Zhang, M. Zhao, Q. Lu, Extracellular vesicles in rheumatoid arthritis and systemic lupus erythematosus: functions and applications, *Front. Immunol.* 11 (2020) 575712. https://doi.org/10.3389/fimmu.2020.575712.

[213] S.D. Robinson, M. Samuels, W. Jones, N. Stewart, M. Eravci, N.K. Mazarakis, D. Gilbert, G. Critchley, G. Giamas, Confirming size-exclusion chromatography as a clinically relevant extracellular vesicles separation method from 1mL plasma through a comprehensive comparison of methods, *BMC Methods* 1 (2024) 1–17. https://doi.org/10.1186/s44330-024-00007-2.

[214] J. Noboa-Velástegui, J.C. León, J. Castro, A. Fletes, P. Madrigal, I. Álvarez, R. Navarro, Comparison of methods for isolating exosomes from plasma subjects with normal and high fat percentages †, *Life* 15 (2025) 0–14. https://doi.org/10.3390/life15030410.

[215] S. Salarpour, M. Barani, A. Pardakhty, M. Khatami, N.P.S. Chauhan, The application of exosomes and exosome-nanoparticle in treating brain disorders, *J. Mol. Liq.* 350 (2022) 118549. https://doi.org/10.1016/j.molliq.2022.118549.

[216] M. Dehghani, S.M. Gulvin, J. Flax, T.R. Gaborski, Systematic evaluation of PKH labelling on extracellular vesicle size by nanoparticle tracking analysis, *Sci. Rep.* 10 (2020) 1–10. https://doi.org/10.1038/s41598-020-66434-7.

[217] J. Li, F. Yu, G. Chen, J. Liu, X.-L. Li, B. Cheng, X.-M. Mo, C. Chen, J.-F. Pan, Moist-retaining, self-recoverable, bioadhesive, and transparent in situ forming hydrogels to accelerate wound healing, *ACS Appl. Mater. Interfaces* 12 (2020) 2023–2038. https://doi.org/10.1021/acsami.9b17180.

[218] M. Lu, S. Yuan, S. Li, L. Li, M. Liu, S. Wan, The exosome-derived biomarker in atherosclerosis and its clinical application, *J. Cardiovasc. Transl. Res.* 12 (2019) 68–74. https://doi.org/10.1007/s12265-018-9796-y.

[219] M. Mendt, K. Rezvani, E. Shpall, Mesenchymal stem cell-derived exosomes for clinical use, *Bone Marrow Transplant.* 54 (2019) 789–792. https://doi.org/10.1038/s41409-019-0616-z.

[220] B. Codispoti, M. Marrelli, F. Paduano, M. Tatullo, NANOmetric BIO-banked MSC-derived exosome (NANOBIOME) as a novel approach to regenerative medicine, *J. Clin. Med.* 7 (2018). https://doi.org/10.3390/jcm7100357.

[221] K. Popowski, H. Lutz, S. Hu, A. George, P.U. Dinh, K. Cheng, Exosome therapeutics for lung regenerative medicine, *J. Extracell. Vesicles* 9 (2020). https://doi.org/10.1080/20013078.2020.1785161.

[222] A. Sanghani, P. Andriesei, K.N. Kafetzis, A.D. Tagalakis, C. Yu-Wai-Man, Advances in exosome therapies in ophthalmology–From bench to clinical trial, *Acta Ophthalmol.* 100 (2022) 243–252. https://doi.org/10.1111/aos.14932.

[223] A.E. Vitha, A.W. Kollefrath, C.-Y.C. Huang, F. Garcia-Godoy, Characterization and therapeutic uses of exosomes: A new potential tool in orthopedics, *Stem Cells Dev.* 28 (2019) 141–150. https://doi.org/10.1089/scd.2018.0205.

[224] Y.S. Kim, C.H. Sung, S.H. Chung, S.J. Kwak, Y.G. Koh, Does an injection of adipose-derived mesenchymal stem cells loaded in fibrin glue influence rotator cuff repair outcomes? A clinical and magnetic resonance imaging study, *Am. J. Sports Med.* 45 (2017) 2010–2018. https://doi.org/10.1177/0363546517702863.

[225] M. Dimik, P. Abeysinghe, J. Logan, M. Mitchell, The exosome: A review of current therapeutic roles and capabilities in human reproduction, *Drug Deliv. Transl. Res.* 13 (2023) 473–502. https://doi.org/10.1007/s13346-022-01225-3.

[226] Y. Kost, A. Muskat, N. Mhaimeed, R.S. Nazarian, K. Kobets, Exosome therapy in hair regeneration: A literature review of the evidence, challenges, and future opportunities, *J. Cosmet. Dermatol.* 21 (2022) 3226–3231. https://doi.org/10.1111/jocd.15008.

[227] K. Ariyoshi, Y. Hiroyama, N. Fujiwara, T. Miura, K. Kasai, A. Nakata, Y. Fujishima, V.S. Ting Goh, M.A. Yoshida, Extracellular vesicles released from irradiated neonatal mouse cheek tissue increased cell survival after radiation, *J. Radiat. Res.* 62 (2021) 73–78. https://doi.org/10.1093/jrr/rraa106.

[228] S.L. Proffer, C.R. Paradise, E. Degrazia, Y. Halaas, K.K. Durairaj, M. Somenek, A. Sivly, A.J. Boon, A. Behfar, S.P. Wyles, Efficacy and tolerability of topical platelet exosomes for skin rejuvenation: Six-week results, *Aesthetic Surg. J.* 42 (2022) 1185–1193. https://doi.org/10.1093/asj/sjac149.

[229] J.S. Shieh, Y.T. Chin, H.C. Chiu, Y.Y. Hsieh, H.R. Cheng, H. Gu, F.W. Chang, Bio-pulsed stimulation effectively improves the production of avian mesenchymal stem cell-derived extracellular vesicles that enhance the bioactivity of skin fibroblasts and hair follicle cells, *Int. J. Mol. Sci.* 23 (2022). https://doi.org/10.3390/ijms232315010.

[230] M. Xiong, Q. Zhang, W. Hu, C. Zhao, W. Lv, Y. Yi, Y. Wang, H. Tang, M. Wu, Y. Wu, The novel mechanisms and applications of exosomes in dermatology and cutaneous medical aesthetics, *Pharmacol. Res.* 166 (2021). https://doi.org/10.1016/j.phrs.2021.105490.

[231] G.H. Yang, Y.B. Lee, D. Kang, E. Choi, Y. Nam, K.H. Lee, H.J. You, H.J. Kang, S.H. An, H. Jeon, Overcome the barriers of the skin: Exosome therapy, *Biomater. Res.* 25 (2021) 1–13. https://doi.org/10.1186/s40824-021-00224-8.

[232] S. Hu, Z. Li, J. Cores, K. Huang, T. Su, P.U. Dinh, K. Cheng, Needle-free injection of exosomes derived from human dermal fibroblast spheroids ameliorates skin photoaging, *ACS Nano* 13 (2019) 11273–11282. https://doi.org/10.1021/acsnano.9b04384.

[233] P. Wu, B. Zhang, X. Han, Y. Sun, Z. Sun, L. Li, X. Zhou, Q. Jin, P. Fu, W. Xu, H. Qian, HucMSC exosome-delivered 14-3-3(Formula presented) alleviates ultraviolet radiation-induced photodamage via SIRT1 pathway modulation, *Aging (Albany. NY).* 13 (2021) 11542–11563. https://doi.org/10.18632/aging.202851.

[234] J.X. Liang, X. Liao, S.H. Li, X. Jiang, Z.H. Li, Y.D. Wu, L.L. Xiao, G.H. Xie, J.X. Song, H.W. Liu, Antiaging properties of exosomes from adipose-derived mesenchymal stem cells in photoaged rat skin, *Biomed. Res. Int.* 2020 (2020) 6406395. https://doi.org/10.1155/2020/6406395.

[235] G.H. Park, H.H. Kwon, J. Seok, S.H. Yang, J. Lee, B.C. Park, E. Shin, K.Y. Park, Efficacy of combined treatment with human adipose tissue stem cell-derived exosome-containing solution and microneedling for facial skin aging: A 12-week prospective, randomized, split-face study, *J. Cosmet. Dermatol.* 22 (2023) 3418–3426. https://doi.org/10.1111/jocd.15872.

[236] G. Han, H. Kim, D.E. Kim, Y. Ahn, J. Kim, Y.J. Jang, K. Kim, Y. Yang, S.H. Kim, The potential of bovine colostrum-derived exosomes to repair aged and damaged skin cells, *Pharmaceutics* 14 (2022). https://doi.org/10.3390/pharmaceutics14020307.

[237] P. Santos, F. Almeida, Exosome-based vaccines: History, current state, and clinical trials, *Front. Immunol.* 12 (2021) 1–15. https://doi.org/10.3389/fimmu.2021.711565.

[238] A. Kulshreshtha, T. Ahmad, A. Agrawal, B. Ghosh, Proinflammatory role of epithelial cell-derived exosomes in allergic airway inflammation, *J. Allergy Clin. Immunol.* 131 (2013) 1194–1203.e14. https://doi.org/10.1016/j.jaci.2012.12.1565.

[239] S.A. Ibrahim, Y.S. Khan, Histology, extracellular vesicles. In: *StatPearls* [Internet]. Treasure Island (FL): StatPearls Publishing (2023). Available from: https://www.ncbi.nlm.nih.gov/books/NBK562256.

[240] A. Gupta, S. Kashte, M. Gupta, H.C. Rodriguez, S.S. Gautam, S. Kadam, Mesenchymal stem cells and exosome therapy for COVID-19: Current status and future perspective, *Hum. Cell* 33 (2020) 907–918. https://doi.org/10.1007/s13577-020-00407-w.

[241] M.J. Haney, N.L. Klyachko, Y. Zhao, R. Gupta, E.G. Plotnikova, Z. He, T. Patel, A. Piroyan, M. Sokolsky, A.V. Kabanov, E.V. Batrakova, Exosomes as drug delivery vehicles for Parkinson's disease therapy, *J. Control Release* 207 (2015) 18–30. https://doi.org/10.1016/j.jconrel.2015.03.033.

[242] F. Susa, T. Limongi, B. Dumontel, V. Vighetto, V. Cauda, Engineered extracellular vesicles as a reliable tool in cancer nanomedicine, *Cancers (Basel).* 11 (2019). https://doi.org/10.3390/cancers11121979.

[243] J.L. Hood, S.A. Wickline, A systematic approach to exosome-based translational nanomedicine, *Wiley Interdiscip. Rev. Nanomed. Nanobiotechnol.* 4 (2012) 458–467. https://doi.org/10.1002/wnan.1174.

[244] X. Yang, X. Xie, S. Liu, W. Ma, Z. Zheng, H. Wei, C.-Y. Yu, Engineered exosomes as theranostic platforms for cancer treatment, *ACS Biomater. Sci. Eng.* 9 (2023) 5479–5503. https://doi.org/10.1021/acsbiomaterials.3c00745.

[245] K. Allenson, J. Castillo, F.A. San Lucas, G. Scelo, D.U. Kim, V. Bernard, G. Davis, T. Kumar, M. Katz, M.J. Overman, L. Foretova, E. Fabianova, I. Holcatova, V. Janout, F. Meric-Bernstam, P. Gascoyne, I. Wistuba, G. Varadhachary, P. Brennan, S. Hanash, D. Li, A. Maitra, H. Alvarez, High prevalence of mutant KRAS in circulating exosome-derived DNA from early-stage pancreatic cancer patients, *Ann. Oncol.* 28 (2017) 741–747. https://doi.org/10.1093/annonc/mdx004.

[246] I. V. Bijnsdorp, A.A. Geldof, M. Lavaei, S.R. Piersma, R.J.A. van Moorselaar, C.R. Jimenez, Exosomal ITGA3 interferes with non-cancerous prostate cell functions and is increased in urine exosomes of metastatic prostate cancer patients, *J. Extracell. Vesicles* 2 (2013) 1–10. https://doi.org/10.3402/jev.v2i0.22097.

[247] Y. Zhang, J. Bi, J. Huang, Y. Tang, S. Du, P. Li, Exosome: A review of its classification, isolation techniques, storage, diagnostic and targeted therapy applications, *Int. J. Nanomed.* 15 (2020) 6917–6934. https://doi.org/10.2147/IJN.S264498.

[248] J. McKiernan, M.J. Donovan, E. Margolis, A. Partin, B. Carter, G. Brown, P. Torkler, M. Noerholm, J. Skog, N. Shore, G. Andriole, I. Thompson, P. Carroll, A prospective adaptive utility trial to validate performance of a novel urine exosome gene expression assay to predict high-grade prostate cancer in patients with prostate-specific antigen 2–10 ng/ml at initial biopsy, *Eur. Urol.* 74 (2018) 731–738. https://doi.org/10.1016/j.eururo.2018.08.019.

[249] A.K. Agrawal, F. Aqil, J. Jeyabalan, W.A. Spencer, J. Beck, B.W. Gachuki, S.S. Alhakeem, K. Oben, R. Munagala, S. Bondada, R.C. Gupta, Milk-derived exosomes for oral delivery of paclitaxel, *Nanomedicine* 13 (2017) 1627–1636. https://doi.org/10.1016/j.nano.2017.03.001.

[250] B. Zhou, K. Xu, X. Zheng, T. Chen, J. Wang, Y. Song, Y. Shao, S. Zheng, Application of exosomes as liquid biopsy in clinical diagnosis, *Signal Transduct. Target. Ther.* 5 (2020). https://doi.org/10.1038/s41392-020-00258-9.

[251] J.A. Welsh, E. Van Der Pol, G.J.A. Arkesteijn, M. Bremer, A. Brisson, F. Coumans, F. Dignat-George, E. Duggan, I. Ghiran, B. Giebel, A. Görgens, A. Hendrix, R. Lacroix, J. Lannigan, S.F.W.M. Libregts, E. Lozano-Andrés, A. Morales-Kastresana, S. Robert, L. De Rond, T. Tertel, J. Tigges, O. De Wever, X. Yan, R. Nieuwland, M.H.M. Wauben, J.P. Nolan, J.C. Jones, MIFlowCyt-EV: A framework for standardized reporting of extracellular vesicle flow cytometry experiments, *J. Extracell. Vesicles* 9 (2020). https://doi.org/10.1080/20013078.2020.1713526.

[252] M. Riazifar, E.J. Pone, J. Lötval, W. Zhao, Stem cell extracellular vesicles: Extended messages of regeneration, *Annu. Rev. Pharmacol. Toxicol.* 57 (2017) 125–154. https://doi.org/10.1146/annurev-pharmtox-061616-030146.Stem.

[253] T. Lener, M. Gimona, L. Aigner, V. Bo¨rger, E. Buzas, G. Camussi, N. Chaput, E. Al, Applying extracellular vesicles based therapeutics in clinical trials an ISEV position paper, *J. Extracell. Vesicles* 4 (201AD) 30087. https://doi.org/10.1002/9780470612217.ch1.

[254] S. Terai, M. Asonuma, A. Hoshino, M. Kino-oka, T. Ochiya, K. Okada, Y. Sato, Y. Takahashi, M. Tobita, A. Tsuchiya, Guidance on the clinical application of extracellular vesicles, *Regen. Ther.* 29 (2025) 43–50. https://doi.org/10.1016/j.reth.2025.02.008.

[255] K. Cheng, R. Kalluri, Guidelines for clinical translation and commercialization of extracellular vesicles and exosomes based therapeutics, *Extracell. Vesicle* 2 (2023) 100029. https://doi.org/10.1016/j.vesic.2023.100029.

[256] B. Escudier, T. Dorval, N. Chaput, F. André, M.P. Caby, S. Novault, C. Flament, C. Leboulaire, C. Borg, S. Amigorena, C. Boccaccio, C. Bonnerot, O. Dhellin, M. Movassagh, S. Piperno, C. Robert, V. Serra, N. Valente, J.B. Le Pecq, A. Spatz, O. Lantz, T. Tursz, E. Angevin, L. Zitvogel, Vaccination of metastatic melanoma patients with autologous dendritic cell (DC) derived-exosomes: Results of the first phase 1 clinical trial, *J. Transl. Med.* 3 (2005) 1–13. https://doi.org/10.1186/1479-5876-3-10.

[257] A. Esmaeili, M. Alini, M. Baghaban Eslaminejad, S. Hosseini, Engineering strategies for customizing extracellular vesicle uptake in a therapeutic context, *Stem Cell Res. Ther.* 13 (2022) 1–16. https://doi.org/10.1186/s13287-022-02806-2.

[258] M.F.S. Lindenbergh, W. Stoorvogel, Antigen presentation by extracellular vesicles from professional antigen-presenting cells, *Annu. Rev. Immunol.* 36 (2018) 435–459. https://doi.org/10.1146/annurev-immunol-041015-055700.

[259] L. Czernek, M. Düchler, Functions of cancer-derived extracellular vesicles in immunosuppression, *Arch. Immunol. Ther. Exp. (Warsz).* 65 (2017) 311–323. https://doi.org/10.1007/s00005-016-0453-3.

[260] G. van Niel, G. D'Angelo, G. Raposo, Shedding light on the cell biology of extracellular vesicles, *Nat. Rev. Mol. Cell Biol.* 19 (2018) 213–228. https://doi.org/10.1038/nrm.2017.125.

[261] F. Fleissner, M. Bonn, S.H. Parekh, Microscale spatial heterogeneity of protein structural transitions in fibrin matrices, *Sci. Adv.* 2 (2016) 1–9. https://doi.org/10.1126/sciadv.1501778.

[262] M.A. Antonyak, M.J. Lukey, R.A. Cerione, Lipid-filled vesicles modulate macrophages, *Science* 363 (2019) 931–932. https://doi.org/10.1126/science.aaw6765.

[263] R.E. Veerman, G. Güçlüler Akpinar, M. Eldh, S. Gabrielsson, Immune cell-derived extracellular vesicles – functions and therapeutic applications, *Trends Mol. Med.* 25 (2019) 382–394. https://doi.org/10.1016/j.molmed.2019.02.003.

[264] A. Tan, H. de la Peña, A.M. Seifalian, The application of exosomes as a nanoscale cancer vaccine, *Int. J. Nanomed.* 5 (2010) 889–900. https://doi.org/10.2147/IJN.S13402.

[265] F.X. Li, J.J. Liu, F. Xu, X. Lin, J.Y. Zhong, F. Wu, L.Q. Yuan, Role of tumor-derived exosomes in bone metastasis, *Oncol. Lett.* 18 (2019) 3935–3945. https://doi.org/10.3892/ol.2019.10776.

[266] Z.L. Tan, J.F. Li, H.M. Luo, Y.Y. Liu, Y. Jin, Plant extracellular vesicles: A novel bioactive nanoparticle for tumor therapy, *Front. Pharmacol.* 13 (2022) 1–10. https://doi.org/10.3389/fphar.2022.1006299.

[267] H. Pinto, E. Sánchez-Vizcaíno Mengual, Exosomes in the real world of medical aesthetics: A review, *Aesthetic Plast. Surg.* 48 (2024) 2513–2527. https://doi.org/10.1007/s00266-023-03844-8.

14 Clinical Translation of Exosome Therapies

Current Trials and Outcomes

Monika Bhairam[1], *Shilpa Sahu*[1], *Shiv Shankar Shukla*[1], *Ravindra Kumar Pandey*[1], *and Bina Gidwani*[1]

[1]Columbia Institute of Pharmacy, Raipur, Chhattisgarh, India

14.1 INTRODUCTION

Exosomes, diminutive membrane-bound vesicles, are naturally synthesized by many cells, including diverse type and released into body fluid [1]. Previously deemed cellular waste, exosomes are now recognized as significant contribute to intercellular communication. They transport a varied array of bioactive substance, allowing them to affect numerous physiological and pathological processes in recipient cells. They enable intracellular communication by transmitting biological signals, encompassing diverse noncoding RNAs, coding RNAs, DNA strands, proteins, to neighboring cells [2]. Among the various types of extracellular vesicle (EVs), exosome are secreted directly from plasma membranes and released by deceased cells [3]. Exosomes were formerly characterized as undesirable proteins produced by cells, appearing as membrane vesicles [4]. Every type of cell secretes exosomes, which are extracellular vesicles that are nanoscale. They are typically 30–150 nm in diameter and are engaged in intercellular communication by use of lipids, proteins, RNA (including microRNAs and mRNA), and other molecules between cells [5]. These vesicles are essential in numerous biological activities. Exosomes facilitate communication between cells by transferring these bioactive molecules, allowing them to influence the behavior of target cells [6]. This can have effects on processes such as immune responses, cancer metastasis, and tissue regeneration [7].

14.2 CLINICAL TRIALS ON EXOSOME THERAPIES

Numerous ongoing clinical trials reflect growing interest in exosome therapies and their effectiveness in humans [8]. Findings from ongoing trials are crucial for understanding the safety and efficacy and for shaping future regulatory frameworks [9]. Over 100 exosome therapies are in various stages, with more than 15 human trials currently underway [10]. Several companies are at the forefront, including Direct Biologics, Capricor Therapeutics, BrainStorm Cell Therapeutics and Rion, with expectations of entering the market in the coming years. Many other companies like Aegle Therapeutics, ZEO ScientifiX, StemXO Therapeutics, CodiakBioSciences (though it has faced challenges), Aruna Bio, and Evox Therapeutics are also developing novel exosome therapies [11]. Some investigations and clinical trials in different areas are described in Table 14.1.

14.2.1 Oncological Treatment

Exosomes are under examination as diagnostic indicators for the early diagnosis and surveillance of cancer. Multiple clinical trials are investigating exosomes for various solid tumors, including lung, breast, pancreatic, colorectal, ovarian, and brain malignancies. This research examines exosome-mediated medication delivery (such as paclitaxel-loaded exosomes), immunotherapy (utilizing dendritic cell–derived exosomes), and RNA-based therapies (distribution of siRNA or mRNA). Some examples are as follows: (1) A phase I trial (NCT06245746) is evaluating umbilical cord mesenchymal stem cell–derived exosomes for acute myeloid leukemia. (2) Several trials (e.g., NCT04427475, NCT05427227) are investigating exosomes in combination with standard cancer treatments for non-small cell lung carcinoma and gastrointestinal cancers. (3) A phase I/II trial (NCT03608631) is using mesenchymal stromal cell-derived exosomes to deliver $KRAS^{G12D}$ siRNA in pancreatic cancer [12].

14.2.2 Inflammatory and Autoimmune Diseases

Exosomes are now under investigation in clinical research for the treatment of several disorders [17–21]. Examples include Direct Biologics is undertaking phase I clinical trials of ExoFlo for ulcerative colitis and Crohn's disease. A phase I clinical research assessed MSC-derived exosomes for refractory perianal fistulas in patients with Crohn's disease, demonstrating encouraging outcomes without adverse effects. Early-phase clinical trials have investigated the safety and efficacy of autologous adipose-derived exosomes in patients with osteoarthritis, indicating pain alleviation and enhanced joint function. Aegle Therapeutics is conducting a phase I/II clinical trial with AGLE-102 for

 DOI: 10.1201/9781003623632-14

TABLE 14.1
Ongoing and Completed Clinical Trials Involving Exosome-Based Therapeutics

Experimental Work	Objective	Study Design	Findings	References
Allogeneic extracellular vesicles produced from platelets are being tested in humans for the first time as a possible treatment for delayed wound healing	Evaluate the safety and efficacy of platelet-derived extracellular vesicles EVs for treating delayed wound healing	First-in-human clinical trial	Improved wound healing	[13]
Assessing mesenchymal stem cell–derived exosomes' safety and effectiveness in treating refractory perianal fistulas in patients with IBD: phase I clinical trial	To assess the safety and effectiveness of exosomes in in IBD patients	Phase I clinical trial with 5 patients	3 patients achieved complete healing; no adverse effects reported	[14]
The Exonerate research used an exosome-based liquid biopsy to forecast response depth and survival outcomes in metastatic colorectal cancer patients treated with cetuximab and panitumumab	Develop an exosome-based liquid biopsy to predict treatment response in metastatic colorectal cancer	Clinical study analyzing exosomal biomarkers	Exosome analysis predicted treatment response and survival outcomes	[15]
Preclinical investigation and initial human intra-articular validation of clinical-grade extracellular vesicles derived from umbilical cord mesenchymal stromal cells as therapeutic interventions for knee osteoarthritis	Evaluate the safety and efficacy of UC-MSC-derived EVs in treating knee osteoarthritis	Preclinical development followed by first-in-human intra-articular injection	Demonstrated safety and potential therapeutic effects in knee osteoarthritis	[16]
Assessing the viability of using exosome technology to treat COVID-19: balancing the views of fact and fiction	Assess the potential of exosome-based therapies in treating COVID-19	Clinical trial using inhaled exosomes overexpressing CD24	Significant improvements in respiratory parameters and inflammatory markers	[17]
Exosomes produced from mesenchymal stromal/stem cells (MSCs) in clinical trials	Review ongoing clinical trials involving MSC-derived exosomes across various conditions	Compilation of multiple clinical trials	Highlighted diverse applications and therapeutic potential of MSC-derived exosomes	[18]
Extracellular vesicles derived from umbilical cord mesenchymal stem cells can effectively mitigate the progression of chronic kidney diseases	Source of MSC: umbilical cord blood in 40 patients	Single-center, randomized, placebo-controlled, phase II/III clinical pilot investigation	Marked enhancement in eGFR, serum creatinine concentration, blood urea levels, and UACR during the 12-month trial duration	[19]
Exosomes made from human adipose tissue-derived stem/stromal cells and their ability to brighten skin: a prospective, split-face, randomized placebo-controlled investigation	Source of MSC: Adipose tissue for disease, Skin hyperpigmentation	A prospective, split-face, randomized placebo-controlled study	Significantly reduced the amount of melanin for two months	[20]
An extracellular vesicle isolate product's RB-Approved Pilot Safety Study Assessing the Management of Osteoarthritis in Combat-Related Injuries	Source of MSC bone marrow (ExoFlo) for osteoarthritis	Pilot study	After a six-month follow-up, the average patient improved 77% in BPI, 80% in ODI, 76% in LEFS, 51% in UEFS, and 77% in QD	[21]
MSC exosomes and GvHD	Source of MSC: bone marrow for GvHD	Case report	The clinical symptoms of GvHD markedly improved quickly following MSC-Exos therapy. The cutaneous and mucosal graft-versus-host disease exhibited significant control within two weeks, remaining stable even after four months	[22]

TABLE 14.2
From Historical Milestones to Current Modifications and Applications across Various Fields

History	Year	Work Done by	Description
First time found in maturing reticulocytes	1983	Heuser, Stahl, Pan, and Johnstone	Initially, they thought that the way for cells to dispose of unwanted products [26]
Reported as small vesicles	1987	Appeared in *Journal of Cell Biology* and *Cell*	Release of transferrin receptors on 50 nm vesicles from reticulocytes [17]
Name given Exosome	1987	Rose Johnstone	Exosome vesicles are released by the fusion of MVBs [27].
Early era	1980s to early 2000s	Various researchers involved	Research focused on exosome biogenesis, such as cellular waste disposal [28]
Exosome biology	2001	Done by various researchers	The field of exosome biology has *grown rapidly*, giving rise to societies, specialized journals, and more than a thousand publications [29]
First paper	1983	C. Harding, J. Heuser, and P. Stahl,	Three-way collaboration outlined transferrin receptor recycling in reticulocytes and its role in iron uptake [30].
Dendritic cells produce exosomes	1998	Zitvogel	Explains the exosomes' capacity to stimulate the T cell response [31]
Mid-era	2001–2020	Done by various researchers	Clinical trials involving exosomes were conducted to understand their therapeutic potential during that time [32].
Intercellular communication and exosomes in CNS	2006–2018	Done by various researchers	Diagnostic and therapeutic potentials of exosomes in CNS diseases [33]
The application of pharmaceutical and clinical research for microRNA	2008–2019	Done by various researchers, *Preclinical research models are being expanded*	Preclinical research models are being expanded, and bioinformatics tools for locating miRNA-binding sites and related mechanisms [34]
Future	2021 to present	Done by various researchers	In the current scenario, the focus is on the unique composition and targeting capabilities of exosomes. Have to discover the possibilities for their use as disease biomarkers and therapeutic agents, particularly in areas like cancer treatment and regenerative medicine [35]
Exosome-based vaccines	2021	Patrick Santos and Fausto Almeida	Ability to deliver molecules and transfer signals over long distances to modulate various physiological functions [36]
Application in cancer and extracellular vesicle biology	2020	Akhil Srivastava, Narsireddy Amreddy et al.	Clinical trials of exosomes and cancer are ongoing; some are completed, some are recruiting, and others have unknown status [37]

patients with recessive dystrophic epidermolysis bullosa, utilizing the immune-modulatory and anti-inflammatory characteristics of MSC-derived exosomes [23].

14.2.3 Neurological Disorders

Exosomes are being evaluated for their potential to treat neurological disorders. The ability of some exosomes to cross the blood–brain barrier makes them attractive candidates for treating neurological and neurodegenerative diseases. Some examples areas follows: While specific details of numerous active clinical trials are still emerging, preclinical studies and early-phase research suggest effectiveness in treating neurological disorders by promoting neuroprotection, reducing inflammation, and enhancing drug delivery to the central nervous system [24].

14.2.4 Other Conditions

Exosomes are also being investigated for regenerative medicine, like in skin disease and cardiovascular disorders. Direct Biologics is conducting a global phase III clinical research, EXTINGUISH ARDS, to assess the efficacy of ExoFlo in treating moderate to severe acute respiratory distress syndrome (ARDS) in hospitalized people. They also offer expanded access trials for severe ARDs and solid abdominal organ transplants. Capricor Therapeutics announced encouraging results from the 36-month follow-up of the HOPE-2 OLE study for deramiocel (cardiosphere-derived cell exosomes) for Duchenne Muscular Dystrophy; the HOPE-3 trial is now fully enrolled in the phase III HOPE-3 trial in Table 14.2 [25].

14.3 EFFICACY AND SAFETY OUTCOMES IN CLINICAL STUDIES

Exosomes have shown promise as treatment agents for a variety of illnesses, from regenerative medicine to cancer, with extensive preclinical evidence supporting their potential. Recent years have seen a rise in clinical trials exploring exosomes derived from both human and plant sources. Exosomes are primarily investigated as biomarkers, therapeutic agents, DDS, and cancer vaccines. Out of 116 registered trials, 50% focus on biomarkers, 28.4% on therapies, 5.2% on drug delivery, 14.7% on basic analysis, and 1.7% on vaccine development. Clinical use of exosomes requires compliance with GMP, involving standardized production, purification, and characterization methods [38].

Understanding the potential and the limitations of these delivery systems to enhance the therapeutic efficacy. Exosome-based treatments demonstrate the potential for targeting a variety of diseases in clinical trials [39].

14.3.1 Cancer Immunotherapy

In phase I clinical investigation, the safety and effectiveness of autologous dendritic cell–derived extracellular vesicles loaded with MAGE tumors in patients with non-small cell lung cancer were evaluated. The effective immunotherapeutic impact was revealed by research studies that showed the EVs were well-loaded and that more than half of the patients exhibited an increase in natural killer cell activity [40].

14.3.2 Adverse Events (AEs)

The same study produced a 4.4% (95% CI: 0.7–22.2%) total AE rate [41]. Fascinatingly, neither engineered nor nonengineered EV products, whether autologous or allogeneic, clearly influenced SAE occurrence [42].

14.3.3 Inflammatory Diseases

Exosomes generated from mesenchymal stem cells (MSCs) exhibit potential in the management of inflammatory disorders. A study on refractory perianal fistulas revealed that exosome therapy is an effective approach, with patients showing considerable improvement and no adverse effects reported [43].

14.3.4 Minimal Occurrence of Significant Adverse Events (SAEs)

A systematic review and meta-analysis of multiple clinical studies revealed a low incidence of serious adverse events (SAEs) at 0.7% (95% CI: 0.1–5.2%) [44].

14.3.5 Neurological Disorders

Preclinical and clinical trials suggest that exosome-based therapies may improve the treatment of neurological disorders by promoting neurogenesis and reducing inflammation [45].

14.3.6 Immunogenicity

Exosome therapies exhibit low immunogenicity, thereby minimum risk of immune rejection and facilitating their suitability for repeated administrations [46].

The current body of clinical evidence indicates that exosome-based therapies are both effective and safe across a range of medical conditions. Standardization in exosome isolation, characterization, and administration protocols are also important for treatments [47].

14.4 CHALLENGES IN EXOSOME ISOLATION, CHARACTERIZATION, AND PRODUCTION

Considering the increasing interest in exosome-based medicines, several challenges hinder their clinical translation, particularly in the areas of isolation, characterization, and large-scale production [48].

14.4.1 Isolation Challenges

Exosomes are generally extracted from complex bodily fluids, which also contain various other extracellular vesicles (EVs) and proteins [44]. Common isolation methods, including ultracentrifugation, size-exclusion chromatography, and polymer-based precipitation, frequently yield heterogeneous mixtures, low yields, or the co-isolation of non-exosomal particles [49]. Achieving high purity and scalability continues to be a major challenge (Figure 14.1) [50].

14.4.2 Characterization Challenges

Absence of precise methodologies to ascertain the functionality of exosomes. Contemporary methods, such as nanoparticle tracking analysis (NTA), electron microscopy, flow cytometry, and Western blotting, provide valuable insights into size, morphology, and surface markers; however, they frequently fall short of delivering comprehensive profiling. The variability in characterization protocols among studies hinders accurate comparison and reproducibility [51].

14.4.3 Production Barriers

The production (clinical-grade) of exosomes in accordance with GMP standards presents significant complexities. It necessitates selective source cells, optimization of culture conditions,

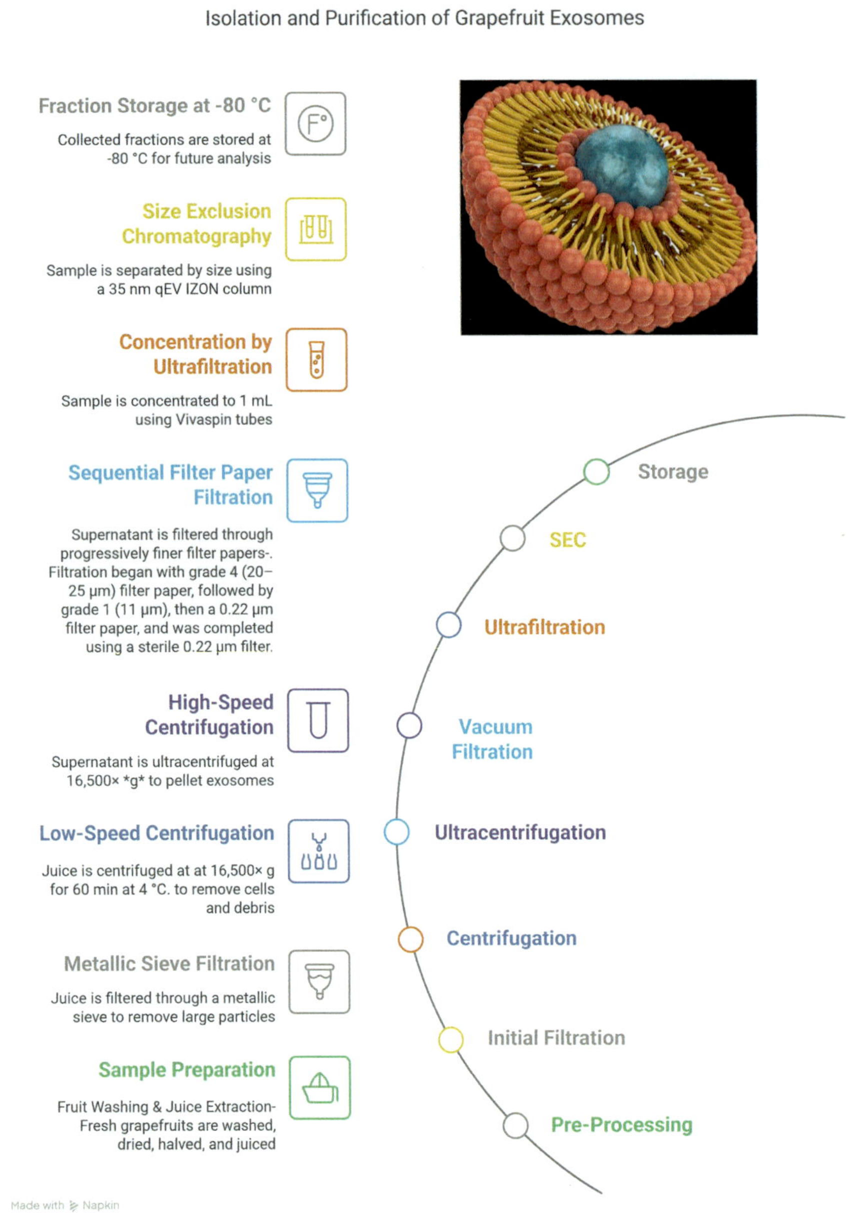

FIGURE 14.1 Isolation and purification of exosomes.

and utilization of scalable bioreactors [52]. Post-production requires stringent purification and quality control measures to guarantee safety and efficacy. Furthermore, ensuring batch-to-batch consistency remains a considerable challenge [53].

14.4.4 Functional Heterogeneity

Exosomes exhibit significant variability based on the type and physical condition of the generating cells and outside factors, which leads to functional heterogeneity. The standardization of effective therapeutic exosomes and their mode of action is hampered by this diversity.

14.4.5 Regulatory Uncertainty

Regulatory frameworks are constantly evolving due to the difficulty of exosomes and the lack of standardized production and testing procedures. To guarantee reliable and secure clinical use, precise rules pertaining to classification, quality assurance, and safety testing are urgently required. The successful clinical translation of exosome medicines depends on resolving these technological and regulatory issues. Standardization, technological innovation in isolation, and regulatory alignment are all necessary to advance this exciting subject [54].

14.5 REGULATORY FRAMEWORK FOR EXOSOME-BASED THERAPEUTICS

Exosome-based treatments represent a novel and complex class of biologics; the regulatory environment for them is still evolving [40]. Exosomes provide regulatory obstacles that are essential for clinical translation, origin, shape, and function, which make them incompatible with conventional classifications of medicines, biologics, or gene and cell treatments.

14.5.1 Classification as Advanced Therapy Medicinal Products (ATMPs)

The EMA has initiated the inclusion of specific exosome-based therapies within the category of Advanced Therapy Medicinal Products (ATMPs) in the European Union. These include gene therapies, somatic cell therapies, and tissue-engineered products.

14.5.2 Guidelines from the US FDA

In 2020, the FDA cautioned against unapproved exosome products and underscored the necessity for Investigational New Drug (IND) applications and adherence to Good Manufacturing Practices (GMP) for clinical use [55].

14.5.3 Manufacturing Standards–GMP Compliance

For clinical applications, exosome production must adhere to GMP standards:

- Controlled source cell selection
- Standardized culture conditions
- Validated isolation and purification methods
- Batch-to-batch consistency
- Robust quality control testing (e.g., sterility, potency, purity)

14.5.4 Quality and Safety Testing

Given the complexity of exosomes, regulatory agencies emphasize the importance of the following:

- Characterization of exosome composition (proteins, lipids, RNA cargo)
- Determination of biological activity
- Evaluation of potential toxicity, immunogenicity, and tumorigenicity
- Stability testing for storage and shelf life

14.5.5 International Harmonization

Global agreement on regulatory requirements for exosome therapies is currently deficient. To expedite the approval process and encourage worldwide clinical use, it is essential to harmonize regulations like the FDA, EMA, and other international organizations. Because of the complexity and diversity of these biologics, the regulatory framework for exosome-based therapies is changing. To determine the safety, quality, and efficacy while encouraging innovation, comprehensive and standardized regulations are necessary [56]. Exosome medicines follow clearly defined regulatory pathways to progress from research to clinical use.

14.6 ETHICAL CONSIDERATIONS AND LONG-TERM SAFETY CONCERNS

The clinical application of exosome-based therapies raises several ethical and long-term safety considerations that must be addressed to ensure responsible development and research application.

Global agreements on regulatory requirements for exosome therapies are currently lacking. To ensure their responsible development, the clinical usages of these therapies present a number of ethical and long-term safety concerns that require careful evaluation [43]. Patients should be completely informed about the origin of exosomes, whether from synthetic, human, or stem cell sources, as well as the risks, advantages, and uncertainties involved, in order to give their informed consent [57].

In addition to the ethical challenges raised by the use of embryonic stem cells, ethical sourcing of donor cells, particularly from human tissue or stem cells, requires consent from patients, donor anonymity, and strict limitations on commercial exploitation. Misuse is also another serious issue since uncontrolled promotion of exosome therapies, especially in antiaging or cosmetic formulations, can put patients in danger and undermine public confidence. Strong regulatory control is therefore essential for the use of all these [58].

The ethical dilemma of ensuring fair access across various socioeconomic groups is further complicated by the high cost and complexity of exosome manufacture, which may hamper access to wealthy populations. This issue needs to be addressed by a considerate policy and healthcare planning.

14.6.1 Long-Term Safety Concerns

Extended safety issues: There is currently a lack of long-term safety data, despite early clinical trials suggesting that exosome treatments are generally well tolerated. Possible issues include the following:

- *Unintended immune responses or chronic inflammation.*
- *Uncontrolled proliferation or tumorigenicity*, especially those derived from stem cells.

- *Off-target effects* due to the broad bioactivity of exosomal cargo. Ongoing post-treatment surveillance and long-term follow-up studies are essential to detect and manage delayed adverse effects [59].

14.6.2 Ethical Use of Engineered Exosomes

The utilization of genetically edited exosomes raises more ethical concerns around biological manipulation, unintended consequences, and biosafety. More ethical questions about biological modification, undesirable effects, and biosafety raise additional concerns about the use of genetically modified exosomes. To monitor the use and containment of such technology, clear restrictions must be put in place. The development of exosome-based therapies must prioritize patient safety and ethical responsibility. The responsible, moral, and socially beneficial production and use of these innovative medications will be made easier by addressing these issues through strict informed consent procedures, open research practices, fair access programs, and ongoing monitoring [49].

14.7 ADVANCEMENTS IN EXOSOME ENGINEERING FOR ENHANCED THERAPEUTICS

The potential of exosome engineering as an effective and tailored therapeutic platform has been greatly increased by current developments. These developments seek to modify exosomes' inherent qualities, improve their therapeutic effectiveness, and expand their use in a variety of illnesses. Exosome engineering is the technique of modifying the exosome surface or adding certain materials to them in order to enhance their functionality [60]. Numerous novel techniques, like chemical, physical, and biological ones, are used to develop and modify exosomes. Cells releasing these exosomes are genetically modified by biological processes. By genetically altering the cells, this technique produces exosomes with specific characteristics. Various chemical moieties conjugating to the exosome surface are essential for chemical methods, as depicted in Figure 14.2.

This surface modification enables targeted interactions with specific cell types or molecular targets. Physical techniques like cell membrane permeabilization, extrusion, electroporation, sonication, freeze-thaw, and hypotonic dialysis are employed to load cargo. Researchers are experimenting with various combinations of all these engineering technologies to achieve maximum incredible results. Combining surface modification or alteration techniques with the delivery of therapeutic materials, including proteins, nucleic acids, and other therapeutic agents, can result in increased therapeutic potential [61].

14.8 FUTURE DIRECTIONS IN CLINICAL TRANSLATION OF EXOSOME THERAPIES

Exosome-based therapies have a bright future ahead of them, with ongoing advancements expected to overcome current limitations and expand clinical applications. The next phase of clinical translation is being influenced by several important trajectories [49].

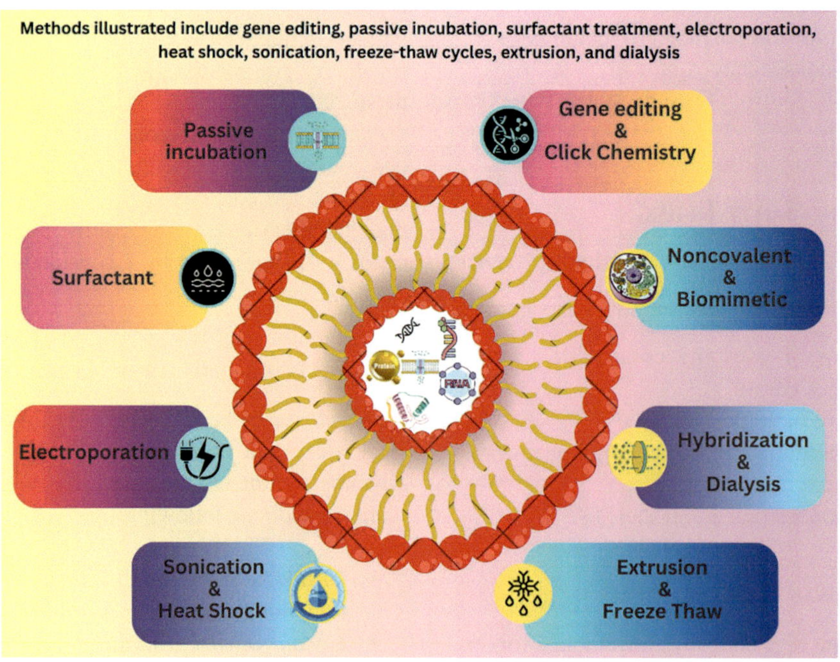

FIGURE 14.2 Diagrammatic illustration of exosome engineering surface modification.

14.8.1 Ethical and Social Considerations

Future development will also require frameworks that address ethical issues, including donor consent, equitable access, and transparency in clinical use. Public engagement and education will role in fostering acceptance, responsible application, and next generation of exosome research is focused on translating laboratory innovation into safe, effective, and accessible therapies. With continued interdisciplinary collaboration, regulatory clarity, and technological integration, exosome-based therapeutics are set to become a cornerstone of next-generation precision medicine [50].

14.8.2 Exosome-Based Diagnostics and Theragnostic

In addition to therapies, exosomes are being explored as diagnostic and theragnostic tools. This dual functionality positions exosomes at the center of the future of precision and personalized medicine [51].

14.8.3 Clinical Trials Expansion

We can expect a significant increase in the number and diversity of clinical trials focusing on exosomes, with a shift toward large-scale, multi-center, and randomized controlled trials. These will yield more reliable statistics on long-term safety, efficacy, and therapeutic potential in specific disease contexts [52].

14.8.4 Integration with Emerging Technologies

The use of exosome treatments in conjunction with other cutting-edge technologies, including CRISPR-based gene editing, artificial intelligence (for biomarker discovery and patient matching), and nanotechnology. These synergies may improve targeting, payload delivery, and disease monitoring.

14.8.5 Personalized Exosome Therapies

The use of autologous (patient-derived) exosomes could pave the way for highly personalized therapies, minimizing immunogenicity and maximizing therapeutic compatibility. Exosomes derived from patient-specific stem cells or immune cells could be customized for targeted treatments in cancer and neurodegenerative diseases.

14.8.6 Standardization and Regulatory Harmonization

A primary objective for the future is the implementation of standardized methods for exosome separation, characterization, and quality assurance. Aligning regulatory frameworks among international bodies like the FDA and EMA will facilitate approval processes and guarantee safety and efficacy across borders [34].

14.8.7 Large-Scale GMP Manufacturing

Creating scalable, cost-efficient, and good manufacturing practice (GMP)-compliant methodologies is crucial for facilitating extensive clinical use. Advancements in bioreactor technology, cell culture systems, and purification methods will be essential for the commercial-scale production of clinical-grade exosomes [24].

14.9 CONCLUSION

Exosome therapies are rapidly emerging in the field of pharmaceuticals as a transformative approach in modern medicine, offering novel avenues for regeneration, targeted treatment, and immunomodulation. The fundamental concept and mechanism of intercellular communication of exosomes underscores their unique role as natural carriers of bioactive molecules, capable of influencing diverse physiological and pathological processes. Present use of regenerative medicine and drug delivery underscores their therapeutic potential and adaptability, especially in cases where traditional treatments are inadequate. Their inherent immunomodulatory characteristics increase their potential as treatments for autoimmune disorders, cancer immunotherapy, and transplant tolerance. Clinical trials conducted thus far have yielded promising information concerning the efficacy and safety of exosome therapies; however, the sample size and indications are diverse.

Initial results indicate their promise; nonetheless, comprehensive long-term data and extensive multicenter investigations are important to develop their therapeutic significance. Standardized protocols for exosome isolation, characterization, and large-scale production are essential to ensure uniformity and repeatability in therapeutic applications. Ethical and regulatory frameworks are continuing to develop to tackle the intricacies of these biologics, especially with quality control, long-term safety, and classification.

Advances in exosome engineering are helping the research community to overcome some of the barriers, enabling improvements in targeting, enhancing loading efficacy, and increasing therapeutic specificity. As these technologies mature, they will likely accelerate the translation of exosome-based products from bench to bedside. Clinical exosomes therapy's future depends on a multidisciplinary strategy that incorporates bioengineering, clinical science, regulatory policy, and ethical supervision. To fully realize the therapeutic potential of exosomes and make them a multidimensional modality in precision medicine, more research findings and thorough clinical evaluation are required. This chapter offers a comprehensive overview of exosomes and their clinical translation, serving as a valuable guide for researchers, clinicians, and industry stakeholders navigating this evolving field.

ABBREVIATIONS

ARDS	acute respiratory distress syndrome
ATMPs	advanced therapy medicinal products
BPI	Brief Pain Inventory
CD24	cluster of differentiation 24
CNS	central nervous system
COVID-19	coronavirus disease 2019
eGFR	estimated glomerular filtration rate
EMA	European Medicines Agency
EVs	extracellular vesicles
EXONERATE	study name (no expansion unless defined by study)
Extinguish ARDS	study name (phase III clinical trial of ExoFlo)
GMP	good manufacturing practice
GVHD	graft-versus-host disease
G12D	a specific point mutation at codon 12, where glycine (G) is replaced by aspartic acid (D)
HOPE-2 OLE	Halt cardiomyOPathy progression, Phase II HOPE trial with an Open Label Extension
IBD	inflammatory bowel disease
IND	investigational new drug
IRB	Institutional Review Board
KRAS G12D	Kirsten rat sarcoma viral oncogene homolog
LEFS	lower extremity functional scale
MAGE	melanoma antigen gene
miRNA	microRNA
mRNA	messenger RNA
MSCs	mesenchymal stromal/stem cells
MVBs	multivesicular bodies
NTA	nanoparticle tracking analysis
ODI	Oswestry Disability Index
OLE	open-label extension
QD	Quebec Disability Questionnaire
siRNA	small interfering RNA
UACR	urine albumin-to-creatinine ratio
UC	umbilical cord
UC-MSC	umbilical cord–derived mesenchymal stem/stromal cells
UEFS	upper extremity functional scale

ACKNOWLEDGMENT

The authors acknowledge the support of their affiliated institutions.

AI DISCLOSURE STATEMENT

During the preparation of this work the authors used Grammarly, QuillBot, and ChatGPT in order to enhance grammatical structure and narrative coherence. After using this tool/service, the authors reviewed and edited the content as needed and take full responsibility for the content of the publication.

CONFLICT OF INTEREST

The authors declare no conflict of interest.

REFERENCES

[1] Boukouris S, Mathivanan S. Exosomes in bodily fluids are a highly stable resource of disease biomarkers. *Proteomics Clin Appl*. 2015 Apr 19;9(3–4):358–67.

[2] Urbanelli L, Magini A, Buratta S, Brozzi A, Sagini K, Polchi A, et al. Signaling pathways in exosomes biogenesis, secretion and fate. *Genes (Basel)*. 2013 Mar 28;4(2):152–70. doi: 10.3390/genes4020152.

[3] Han C, Sun X, Liu L, Jiang H, Shen Y, Xu X, et al. Exosomes and their therapeutic potentials of stem cells. *Stem Cells Int*. 2016 Jan 6;2016:7653489. doi: 10.1155/2016/7653489.

[4] Akers JC, Gonda D, Kim R, Carter BS, Chen CC. Biogenesis of extracellular vesicles (EV): Exosomes, microvesicles, retrovirus-like vesicles, and apoptotic bodies. *J Neurooncol*. 2013 May;113(1):1–11.

[5] Sato-Kuwabara Y, Sonia A. Melo, Fernando A. Soares, George A. Calin. The fusion of two worlds: Non-coding RNAs and extracellular vesicles – diagnostic and therapeutic implications (review). *Int J Oncol*. 2015 Jan;46(1):17–27.

[6] Dai J, Su Y, Zhong S, Cong L, Liu B, Yang J, et al. Exosomes: Key players in cancer and potential therapeutic strategy. *Signal Transduct Target Ther*. 2020 Aug 5;5(1):145.

[7] Kalluri R, LeBleu VS. The biology, function, and biomedical applications of exosomes. *Science (1979)*. 2020 Feb 7;367(6478).

[8] Tan F, Li X, Wang Z, Li J, Shahzad K, Zheng J. Clinical applications of stem cell-derived exosomes. *Signal Transduct Target Ther*. 2024 Jan 12;9(1):17.

[9] Steinbach OC. An industry update: The latest developments in therapeutic delivery covering July 2019. *Ther Deliv*. 2019 Dec 18;10(12):753–6.

[10] Rahnama M, Heidari M, Poursalehi Z, Golchin A. Global trends of exosomes application in clinical trials: A scoping review. *Stem Cell Rev Rep*. 2024 Nov 28;20(8):2165–93.

[11] Gilligan K, Dwyer R. Engineering exosomes for cancer therapy. *Int J Mol Sci*. 2017 May 24;18(6):1122.

[12] Shen Z, Huang W, Liu J, Tian J, Wang S, Rui K. Effects of mesenchymal stem cell-derived exosomes on autoimmune diseases. *Front Immunol*. 2021 Sep 27;12.

[13] Johnson J, Law SQK, Shojaee M, Hall AS, Bhuiyan S, Lim MBL, et al. First-in-human clinical trial of allogeneic, platelet-derived extracellular vesicles as a potential therapeutic for delayed wound healing. *J Extracell Vesicles*. 2023 Jul 23;12(7).

[14] Nazari H, Alborzi F, Heirani-Tabasi A, Hadizadeh A, Asbagh RA, Behboudi B, et al. Evaluating the safety and efficacy of mesenchymal stem cell-derived exosomes for treatment of refractory perianal fistula in IBD patients: Clinical trial phase I. *Gastroenterol Rep (Oxf)*. 2022 Dec 9;10:goac075. doi: 10.1093/gastro/goac075.

[15] Xu C, Mannucci A, Esposito F, Oliveres H, Alonso-Orduña V, Yubero A, et al. An exosome-based liquid biopsy predicts depth of response and survival outcomes to cetuximab and panitumumab in metastatic colorectal cancer: The EXONERATE study. *Clin Cancer Res*. 2025 Mar 17;31(6):1002–15.

[16] Figueroa-Valdés AI, Luz-Crawford P, Herrera-Luna Y, Georges-Calderón N, García C, Tobar HE, et al. Clinical-grade extracellular vesicles derived from umbilical cord mesenchymal stromal cells: Preclinical development and first-in-human intra-articular validation as therapeutics for knee osteoarthritis. *J Nanobiotechnol*. 2025 Jan 13;23(1):13.

[17] Memon MA, Shahidan WNS. Evaluating the feasibility of exosome technologies in COVID-19 treatment: Navigating the intersection of reality and fantasy perspectives. *AME Med J*. 2024 Sep;9:25.

[18] Lotfy A, AboQuella NM, Wang H. Mesenchymal stromal/stem cell (MSC)-derived exosomes in clinical trials. *Stem Cell Res Ther*. 2023 Apr 7;14(1):66.

[19] Grootjans H, Verschuuren EAM, van Gemert JP, Kerstjens HAM, Bakker SJL, Berger SP, et al. Chronic kidney disease after lung transplantation in a changing era. *Transplant Rev*. 2022 Dec;36(4):100727.

[20] Nautiyal A, Wairkar S. Management of hyperpigmentation: Current treatments and emerging therapies. *Pigment Cell Melanoma Res*. 2021 Nov 3;34(6):1000–14.

[21] Mohseni R, Mahdavi Sharif P, Khosravi A, Taheri AR, Behfar M, Zarrabi M, et al. The application of umbilical cord blood-derived platelet gel for skin ulcers associated with chronic graft-versus-host disease in pediatrics: A randomized trial. *Transplant Cell Ther*. 2024 Jul;30(7):694.e1–694.e10.

[22] Bradley JA, Bolton EM, Pedersen RA. Stem cell medicine encounters the immune system. *Nat Rev Immunol*. 2002 Nov;2(11):859–71.

[23] Zhu X, Ma D, Yang B, An Q, Zhao J, Gao X, et al. Research progress of engineered mesenchymal stem cells and their derived exosomes and their application in autoimmune/inflammatory diseases. *Stem Cell Res Ther*. 2023 Apr 11;14(1):71.

[24] Su Y, Guo H, Liu Q. Effects of mesenchymal stromal cell-derived extracellular vesicles in acute respiratory distress syndrome (ARDS): Current understanding and future perspectives. *J Leukoc Biol*. 2021 Jun 29;110(1):27–38.

[25] Uddin MJ, Mohite P, Munde S, Ade N, Oladosu TA, Chidrawar VR, et al. Extracellular vesicles: The future of therapeutics and drug delivery systems. *Intelligent Pharm*. 2024 Jun;2(3):312–28.

[26] Johnstone R, Bianchini A, Teng K. Reticulocyte maturation and exosome release: Transferrin receptor containing exosomes shows multiple plasma membrane functions. *Blood*. 1989 Oct 1;74(5):1844–51.

[27] Johnstone RM, Adam M, Hammond JR, Orr L, Turbide C. Vesicle formation during reticulocyte maturation. Association of plasma membrane activities with released vesicles (exosomes). *J Biol Chem*. 1987 Jul;262(19):9412–20.

[28] Harding CV, Heuser JE, Stahl PD. Exosomes: Looking back three decades and into the future. *J Cell Biol*. 2013 Feb 18;200(4):367–71.

[29] Harding C, Heuser J, Stahl P. Receptor-mediated endocytosis of transferrin and recycling of the transferrin receptor in rat reticulocytes. *J Cell Biol*. 1983 Aug 1;97(2):329–39.

[30] Zitvogel L, Fernandez N, Lozier A, Wolfers J, Regnault A, Raposo G, et al. Dendritic cells or their exosomes are effective biotherapies of cancer. *Eur J Cancer*. 1999 Jul;35:S36–8.

[31] Rezaie J, Feghhi M, Etemadi T. A review on exosomes application in clinical trials: Perspective, questions, and challenges. *Cell Commun Signal*. 2022 Sep 19;20(1):145.

[32] Kawikova I, Askenase PW. Diagnostic and therapeutic potentials of exosomes in CNS diseases. *Brain Res*. 2015 Aug;1617:63–71.

[33] Hanna J, Hossain GS, Kocerha J. The potential for microRNA therapeutics and clinical research. *Front Genet*. 2019 May 16;10.

[34] Mosquera-Heredia MI, Morales LC, Vidal OM, Barceló E, Silvera-Redondo C, Vélez JI, et al. Exosomes: Potential disease biomarkers and new therapeutic targets. *Biomedicines*. 2021 Aug 20;9(8):1061.

[35] Santos P, Almeida F. Exosome-based vaccines: History, current state, and clinical trials. *Front Immunol*. 2021 Jul 14;12:711565. doi: 10.3389/fimmu.2021.711565.

[36] Verma AK, Dewangan K, Daunday L, Naurange K, Verma K, Bhiaram M. Spirulina as functional food: Insights into cultivation, production, and health benefits. *J Appl Pharm Res*. 2024 Oct 31;12(5):28–50.

[37] Queen D, Avram MR. Exosomes for treating hair loss: A review of clinical studies. *Dermatol Surg*. 2025 Apr;51(4):409–15.

[38] Xiao Q, Tan M, Yan G, Peng L. Revolutionizing lung cancer treatment: Harnessing exosomes as early diagnostic biomarkers, therapeutics and nano-delivery platforms. *J Nanobiotechnol*. 2025 Mar 21;23(1):232.

[39] Rahman E, Webb WR, Rao P, Abu-Farsakh HN, Upton AE, Yu N, et al. Exosomes exposed: Overview systematic review on evidence versus expectation in aesthetic and regenerative medicine. *Aesthetic Plast Surg*. 2025 Jan 29;49(2):557–68.

[40] Choudhury A, Deka H, Dey BK, Bhairam M, Sengupta K. Nanomaterials for skin repair and regeneration. In: Kumar JK, Singh SK, editors. Nanostructured Materials for Tissue Engineering. Amsterdam: Elsevier; 2023. p. 497–510.

[41] Villarreal-Gómez LJ, Origel-Lucio S, Hernández-Hernández DA, Pérez-González GL. Use of exosomes for cosmetics applications. *Cosmetics*. 2025 Jan 13;12(1):9.

[42] Bai L, Yu L, Ran M, Zhong X, Sun M, Xu M, et al. Harnessing the potential of exosomes in therapeutic interventions for brain disorders. *Int J Mol Sci*. 2025 Mar 11;26(6):2491.

[43] Bhairam M, Gidwani B, Pandey RK, Shukla SS. 3D Printing for in vitro and in vivo skin disease models. *In: 3D Printing and Microfluidics in Dermatology*. 1st ed. Boca Raton: CRC Press; 2024. p. 100–24.

[44] Majewska L, Dorosz K, Kijowski J. Efficacy of rose stem cell-derived exosomes (RSCEs) in skin treatment: From healing to hyperpigmentation management: Case series and review. *J Cosmet Dermatol*. 2025 Jan 15;24(1).

[45] Karami MH, Abdouss M. Advancements in utilizing exosomes for cancer therapy through drug delivery systems: Recent developments. *Nanomed J*. 2025 Apr 1;12(2):181.

[46] Geng JX, Lu YF, Zhou JN, Huang B, Qin Y. Exosome technology: A novel and effective drug delivery system in the field of cancer therapy. *World J Gastrointest Oncol*. 2025 Mar 15;17(3).

[47] Youssef E, Palmer D, Fletcher B, Vaughn R. Exosomes in precision oncology and beyond: From bench to bedside in diagnostics and therapeutics. *Cancers (Basel)*. 2025 Mar 10;17(6):940.

[48] Mehdizadeh S, Mamaghani M, Hassanikia S, Pilehvar Y, Ertas YN. Exosome-powered neuropharmaceutics: Unlocking the blood-brain barrier for next-gen therapies. *J Nanobiotechnol*. 2025 May 3;23(1):329.

[49] Jiang M, Zhang K, Meng J, Xu L, Liu Y, Wei R. Engineered exosomes in service of tumor immunotherapy: From optimizing tumor-derived exosomes to delivering CRISPR/Cas9 system. *Int J Cancer*. 2025 Mar 30;156(5): 898–913.

[50] Bhairam M, Pandey RK, Shukla SS, Gidwani B. Enhanced solubility and bioavailability of ranolazine by solid dispersion method in vitro and in vivo evaluation. *Int J Health Sci (Qassim)*. 2022 Jul 7;3729–48.
[51] Jangam TC, Desai SA, Patel VP, Pagare NB, Raut ND. Exosomes as therapeutic and diagnostic tools: Advances, challenges, and future directions. *Cell Biochem Biophys*. 2025 Mar 24;
[52] Ma C, Wei Tang, Jiaye Wang, Shiyu Yang, Jin Hou, Meng Guo, et al. Application of engineered exosomes in tumor therapy. *Am J Transl Res*. 2025;17(2):736–47.
[53] Aldali F, Deng C, Nie M, Chen H. Advances in therapies using mesenchymal stem cells and their exosomes for treatment of peripheral nerve injury: State of the art and future perspectives. *Neural Regen Res*. 2025 Nov;20(11):3151–71.
[54] Lou Y, Wang Y, Lu J, Chen X. MicroRNA-targeted nanoparticle delivery systems for cancer therapy: Current status and future prospects. *Nanomedicine*. 2025 May 19;20(10):1181–94.
[55] Rawojć K, Ahmed MM, Mukhtiar A, Łukowiak M, Kisielewicz K. Nanomedicine-enhanced radiotherapy for glioblastoma: Advances in targeted therapy and adaptive treatment strategies. *Pharmaceutics*. 2025 Apr 11;17(4):508.
[56] Chao T, Zhao J, Gao R, Wang H, Guo J, Gao Z, et al. Exosome surface modification and functionalization: A narrative review of emerging technologies and their application potential in precision medicine. *Adv Technol Neurosci*. 2025 Mar;2(1):27–33.
[57] Chen Y, Chen QW, Fu FS, Yan JH, Su J, Zhang XZ, et al. Antibacterial peptides-produced cell hydrogel eliminates intracellular pathogens and remodels immune microenvironment for osteomyelitis therapy. *ACS Nano*. 2025 May 20;19(19):18433–49.
[58] Wang B, Chen P, Li W, Chen Z. Exosomes in stroke management: A promising paradigm shift in stroke therapy. *Neural Regen Res*. 2026 Jan;21(1):6–22.
[59] Chen B da, Zhao Y, Wu J long, Zhu Z guan, Yang X dong, Fang R peng, et al. Exosomes in skin flap survival: Unlocking their role in angiogenesis and tissue regeneration. *Biomedicines*. 2025 Feb 4;13(2):353.
[60] Zhou Y, Wei Y, Tian X, Wei X. Cancer vaccines: Current status and future directions. *J Hematol Oncol*. 2025 Feb 17;18(1):18.
[61] Ziaran S, Danisovic L, Hammer N. Editorial: Tissue engineering and regenerative medicine: Advances, controversies, and future directions. *Front Bioeng Biotechnol*. 2025 Mar 14;13:1568490. doi: 10.3389/fbioe.2025.1568490.

Index

Note: Page numbers in *italics* indicate a figure and page numbers in **bold** indicate a table on the corresponding page.

D

E

F

G

H

I

J

K

L

M